SMART FOOD INDUSTRY

The Blockchain for Sustainable Engineering

Volume I - Fundamentals, Technologies, and Management

Editors

Eduardo Jacob-Lopes
Food Science and Technology Department
Federal University of Santa Maria
Santa Maria, RS, Brazil

Leila Queiroz Zepka
Food Science and Technology Department
Federal University of Santa Maria
Santa Maria, RS, Brazil

Mariany Costa Deprá
Food Science and Technology Department
Federal University of Santa Maria
Santa Maria, RS, Brazil

CRC Press
Taylor & Francis Group
Boca Raton London New York

CRC Press is an imprint of the
Taylor & Francis Group, an **informa** business
A SCIENCE PUBLISHERS BOOK

Cover image taken from Shutterstock.

First edition published 2024
by CRC Press
2385 NW Executive Center Drive, Suite 320, Boca Raton FL 33431

and by CRC Press
4 Park Square, Milton Park, Abingdon, Oxon, OX14 4RN

CRC Press is an imprint of Taylor & Francis Group, LLC

Library of Congress Cataloging-in-Publication Data (applied for)

ISBN: 978-1-032-13840-4 (hbk)
ISBN: 978-1-032-13841-1 (pbk)
ISBN: 978-1-003-23105-9 (ebk)

DOI: 10.1201/9781003231059

Typeset in Times New Roman
by Radiant Productions

Preface

In the last 20 years, the food industry has gone from being an admired and thriving industry to a target under attack. Among the potential responsible for driving this change in scenario, the triple challenge is intertwined: how to increase food production and healthiness, while economic strength remains, in parallel with minimizing adverse impacts on the environment?

To shed light on these answers, a close look at the entire complex food production system becomes crucial to finding the biggest levers toward sustainable change. However, achieving the goal of sustainability involves so many factors—from economics to ecology—that investigating one or even a handful of variables at a time often ignores large parts of the problem. However, it is better to grow in new shoes than to shrink into them.

Furthermore, while simple definitions of sustainability are time-independent, in practice how quickly should we seek to move from the status quo to a sustainable food industry? History suggests that change is slow; of course, no transition is a walk in the park. However, to accelerate this transition, it is imperative to build links between industry and research. Therefore, the big leap will be in facing the challenges and opportunities through sustainable engineering based on knowledge and know-how.

In light of this, the present book, *Smart Food Industry: The Blockchain for Sustainable Engineering: Volume I: Fundamentals, Technologies, and Management*, provides a complete study of the food industry from the perspective of sustainable engineering. Divided into four parts, this book discusses the (i) fundamentals of sustainable food, (ii) conventional technologies in the food industry, (iii) sustainability emerging technologies in food industries, and (iv) sustainable management in food industries. Once the reader begins to explore each section more fully, they will find that the chapters will elucidate the fundamental basics in the area of food and sustainability—i.e., the transition between conventional technologies and the recent technological changes applied to the food industry. Associated with this transition, new visions projected for the food industry will be discussed, aiming at the consolidation of sustainable industrial systems. For a complete overview of future trends in the smart food industry and blockchain sustainable engineering, this book project also addresses waste management strategies, aiming at achieving sustainable development. It is hoped that the information plurality shared in this book will allow new knowledge gaps to be filled and that the reader will find here a foundation for their research and interests.

Eduardo Jacob-Lopes
Leila Queiroz Zepka
Mariany Costa Deprá

Contents

Part III - Sustainability Emerging Technologies in Food Industries

Part IV - Sustainable Management in Food Industries

Part I

Fundamentals, Technologies, and Management

1

An Overview of the Trajectory of the Food Industry

Addressing Expanding Societal Needs and Demands

Delia B. Rodriguez-Amaya and *Jaime Amaya-Farfan**

1. Introduction

Food processing is a vital operation to provide society with a sufficient, safe, and nutritious food supply. A robust agricultural production sector is essential, but it has to be complemented by food processing to provide the world with food security.

Today, the world depends on a dynamic and highly complex food supply chain, which is characterized by the globalization of the market, the flow of raw materials, ingredients, and products among countries, and the immense diversification of food products. Food processors are confronted with more demanding customer expectations, a competitive market, and the drive for more efficient and environmentally friendly processes.

This chapter discusses the evolution of food processing and the trajectory of the food industry, which is guided mainly by society's needs and demands for food.

2. Evolution of Food Processing

Processing technologies are discussed in detail in the other chapters of this book. These processes are briefly mentioned in this chapter to demonstrate how they have evolved with time in answer to societal needs and demands.

Food processing began in prehistoric times (Fig. 1) when men discovered fire. The first use of fire was estimated to occur between 2 million and 200,000 years ago (Knorr and Watzke 2019). As agriculture and animal husbandry progressed, the preservation of foods became essential to avoid losses and provide for times of scarcity. Primitive forms of processing were then introduced, such as fermenting, sun drying, and preserving with salt. Initially, the preservation of food was done at home. Only within the past century has large-scale food processing become an industrial process.

University of Campinas, São Paulo, Brazil.
* Corresponding author: jaf@unicamp.br

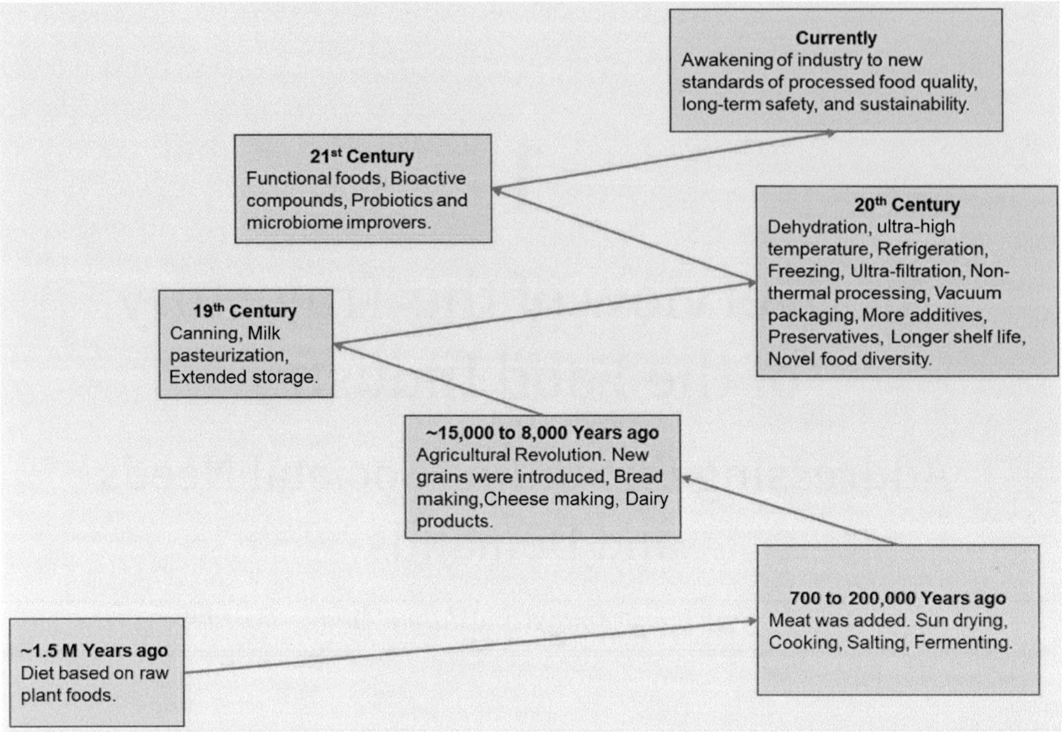

Figure 1. Evolution of Food Processing

Archaeological evidence suggests that brewing beer started around 13,000 years ago (Liu et al. 2018) and bread-making around 14,000 years ago (Arranz-Otaegui et al. 2018). Cheesemaking began in Europe approximately 8,000 years ago (Salque et al. 2013).

In the early 1800s, Nicholas Appert, responding to Napoleon Bonaparte's command, invented a canning process to preserve foodstuffs of animal and vegetable origin to feed the French army. In the 1860s, Louis Pasteur proved that food spoilage could be attributed to microorganisms and developed a preservation method using mild heating (pasteurization).

Welch and Mitchell (2000) characterized the evolution of food processing in the 1900s. The first half of the century was marred by world wars, economic depression, and post-war austerity. However, by 1950, many common infectious diseases were under control, and the diet was generally nutrient-adequate. The second half of the century witnessed increasing economic prosperity and scientific advances. Technologies such as chilling/freezing and freeze drying were increasingly used, and the food industry became more sophisticated and employed automation and computerization. New developments occurred in drying, heat processing, controlled and modified atmosphere packaging, ingredients, and quality assurance. By 1999, the food industry provided foods that were safe, nutritious, palatable, and also increasingly convenient and healthy.

Huebbe and Rimbach (2020) distinguished three major socioeconomic transitions that might have driven the evolution of food processing most:

- the change from hunting and gathering to settled societies with agriculture and livestock farming approximately 15,000 to 10,000 years ago;
- the Industrial Revolution during the 18th and 19th centuries with improvements in food processing, including the introduction of steam and rolling mills for refined flour production;
- the exploitation of sustainable and efficient protein and food sources that will ensure high-quality food production for the growing world population.

Cereals and cereal products have a long history of use. They have to be processed in some way in order to be consumed by humans. Most scientists believed that the domestication of grains began about 10,000 years ago, but more recent findings put this time point much earlier (Thielecke et al. 2021). An ancient site in Israel contained a collection of cereals, dating back to about 23,000 years ago (Weiss et al. 2004). A large assemblage of starch granules was retrieved from the surfaces of Middle Stone Age tools from Mozambique, which indicated that men relied on grass seeds, including sorghum, from at least 105,000 years ago (Mercader 2009). Starch grains from various wild plants were found on the surfaces of grinding tools at three sites (in Italy, Russia, and the Czech Republic), which suggested that vegetable food processing, and possibly the production of flour, was a common and widespread practice across Europe from, at least, approximately 30,000 years ago (Revedin et al. 2010).

Cow's milk utilization has undergone many changes over the past century. The introduction of pasteurized milk has provided consumers with a consistently safe product. In Denmark and Sweden, commercial pasteurization of milk was common as early as the mid-1880s (Rankin et al. 2017). Homogenization and advances in the packaging and transport of milk improved the milk supply (Tunick 2009). Other developments included the concentration of milk and whey, availability and quality of ice cream products, lactose-reduced milk, popularization of yogurt, and advances in butter packaging, low-fat ice cream, and cheese manufacture.

Food packaging has evolved from simply a container to hold food to something that can play an active role in food quality (Risch 2009). Packages have been developed to protect the food, including barriers to oxygen, moisture, and off-flavors. Active packaging includes some microwave packaging as well as the packaging with built-in oxygen absorbers to remove oxygen from the atmosphere surrounding the product or provide antimicrobials to the surface of the food.

Thermal processing (e.g., pasteurization and sterilization) has been the most widely employed and investigated food preservation technique. However, due to the unwanted changes provoked by conventional thermal processing, technologies involving rapid and more uniform heating have been introduced, such as microwave, radio frequency and ohmic heating, to replace or complement conventional heating. In all cases, precision processing has been a growing concern. Moreover, non-thermal processing methods have been developed in the last 30 to 40 years, such as high-pressure processing, pulsed electric field, and pulsed light treatment. Initially aimed at preserving the pleasing natural flavors of food, these technologies were quickly found to also maintain health-promoting bioactive compounds.

The adoption of nonthermal processing alternatives to conventional heating techniques in the food industry is based mainly on their potential to achieve preservation while maintaining the fresh-like characteristics and health benefits of final products. However, keeping the balance between the efficiency of microbial/enzymatic inactivation and the maintenance of sensory and nutritional characteristics has been a great challenge (Xia et al. 2020). Thus, thermal processing continues to be extensively used. It is almost the only method capable of single-handedly achieving the necessary microbiological inactivation to guarantee food safety and preservation, especially in relation to bacterial spores (Kubo et al. 2021).

A survey was conducted among food experts to investigate the extent of non-thermal food processing technology used in the United States (Khouryieh 2021). High-pressure processing (35.6%) was the most commonly used technology, followed by a pulsed electric field (20%). Rapidly increasing novel technologies included cold atmospheric plasma (14.1%) and oscillating magnetic fields (14.1%). More than 70% of the respondents indicated that the main factor for choosing non-thermal food processing technology was better nutrient and sensory properties. High-investment (41%) was the major limitation.

3. Societal Role of Food Processing

Table 1 summarizes the diverse benefits society derives from food processing, particularly thermal processing (Rodriguez-Amaya et al. 2021; van Boekel et al. 2010).

Table 1. Summary of the Beneficial Effects of Food Processing.

- Reduction of food losses throughout the food supply chain.
- Availability of seasonal foods throughout the year.
- Availability of foods in places far from agricultural production.
- Destruction of food-borne microbes and toxins.
- Retention of nutrients and bioactive compounds in the final products.
- Inactivation of anti-nutritional enzymes and substances.
- Improved digestibility and bioavailability of nutrients and bioactive compounds.
- Improved texture and flavor.
- Greater convenience.
- Offering a variety of foods at accessible prices.
- Sustainable production of food.

3.1 Food Security

Food security exists when all people, at all times, have physical, social, and economic access to sufficient, safe, and nutritious food that meets their dietary needs and food preferences for an active and healthy life (FAO, IFAD, WFP 2012). Therefore, food security encompasses availability and access to food, nutritional security, and food safety.

3.1.1 Availability and Access to Food

Along with adequate packaging, food processing has facilitated transport to places far from the site of agricultural production. Processing seasonal crops during harvest reduces post-harvest losses and makes seasonal food available during the year. Extending the shelf-life of foods provides ample time for distribution and storage.

Without processing, the population would have limited food availability and would be restricted to what is produced locally. Food supply would be irregular and lack diversity, and it would be plentiful during harvest seasons and scarce between seasons. The uneven distribution of food is directly related to hunger in some areas of the world and uncontrollable waste in others. As part of its societal role, food processing has come to mitigate these seasonal and regional disparities in distribution and make foods more accessible to all social classes by regulating supply and prices.

The number of hungry people in the world continues to be unacceptably high. Yet, about one-third of food produced for human consumption is lost or wasted globally, amounting to 1.3 billion tons per year with an estimated value of US$ 1 trillion annually (Gustavsson et al. 2011).

Food is lost or wasted throughout the supply chain, from agricultural production to household consumption. Termed as ‘food losses’, in low-income countries, foods are mostly lost during the early and middle stages of the food supply chain. Termed as ‘food waste’, in medium- and high-income countries, food is wasted significantly at the distribution and consumption stages. Food waste refers to food that is fit for consumption but is left to spoil or discarded by processors, retailers, and consumers.

Food losses and wastes have negative economic effects, impede development, hinder social progress, contribute to unnecessary emissions, and not only undermine food security but also waste valuable nutrients, energy, and natural resources (Gustavsson et al. 2011). Reducing food losses and wastes may, thus, present a great opportunity in enhancing the sustainability of the food system and simultaneously improve food security and nutrition.

Suggestions for the reduction of food loss in developing countries include investment in infrastructure, transportation and storage facilities, farmer education, diversification of production, increased use and efficiency of processing and packaging, and improved market facilities. In developed countries, consumers, processors, and restaurants are urged to take the following actions: enhanced consciousness, planning of purchasing, better consumption habits, and reduced restaurant portions.

Expert interviews with representatives from 13 German food processing companies revealed that causes of food waste in the processing sector can be categorized as losses resulting from processing operations and quality assurance and products not fulfilling quality demands from trade (Raak et al. 2017).

Food processing has reduced the overall cost of food production. Mass production of food tends to make meals less expensive than home preparations from raw ingredients. Due to the more efficient utilization of raw materials, the systematic direction of by-products for recycling, and the nutritionally planned apportioning of calories, industrial processing has consequently facilitated life in big urban centers throughout the world. The availability of processed foods made it possible to diminish home food losses, in addition to allowing urban dwellers to have smaller areas for food preparation and storage in better sanitized homes.

3.1.2 Nutritional Security

Food processing can enhance nutritional quality in many ways, such as:

- Unstable nutrients can be preserved.
- Foods can be enriched or fortified with important vitamins and minerals.
- Antinutrients can be removed or transformed into inactive forms.
- Nutrient bioavailability can be improved.

Numerous papers have reported losses of labile nutrients and bioactive compounds during the thermal processing of foods; the extent of which depends mainly on the processing technique and conditions (especially time and temperature). There have been continued efforts to optimize processing to achieve maximum retention of these valuable food components (Lund 1982; Ling et al. 2015; Preedy 2014; Rodriguez-Amaya 2015; Rodriguez-Amaya and Amaya-Farfan 2018).

Processing methods with the ability to preserve the more labile nutrients, such as freezing, have been perfected to bring out the characteristics naturally present in foods. Freezing is usually done with freshly harvested crops when the nutritional quality is at its best. Thus, in many cases, frozen fruits and vegetables sold at markets have higher nutrient content than raw produce (Godoy et al. 2021).

Alternative food preservation technologies have been developed during the past several decades to meet the demand of consumers for fresh-like foods. These processes have also been shown to maintain or result in only slight or insignificant losses of nutrients and bioactive compounds (e.g., Al-juhaimi et al. 2018). As discussed above, these technologies include thermal processes, such as microwave and ohmic heating, which are much faster than the traditional canning method and thus result in reduced losses of heat-labile health-promoting compounds. Methods that do not use heat as a primary mode of inactivating microorganisms in foods have also been introduced, such as high-pressure and high-intensity pulsed electric field processing (Bermúdez-Aguirre and Barbosa-Cánovas 2011; Chauhan 2019; Jan et al. 2017)

Enrichment (replacing nutrients lost in processing) and fortification (adding nutrients in higher amounts than naturally found in the food) have been used globally as public health strategies to mitigate population-level nutrient deficiencies. The early targets were deficiencies, such as goiter, rickets, beriberi, and pellagra. More recently, folate and neural tube defects, zinc and child growth, and selenium and cancer have been addressed (Samaniego-Vaesken et al. 2012). Examples of enriched foods are grain products, especially bread. Examples of fortified foods include ready-to-eat cereals (fortified with B vitamins, folate, iron, and other nutrients) and milk (fortified with vitamins A and D).

Using dietary data from 1989 to 1991, Berner et al. (2001) confirmed that in the U.S., fortification made a major contribution to intakes of all nutrients examined (nine vitamins and minerals), except calcium, in all age/gender groups, especially in children. The breakfast cereal category was responsible for nearly all the intake of nutrients from fortified foods, except vitamin C, for which juice-type beverages made a greater contribution.

Weaver et al. (2014) concluded that processed foods are nutritionally important to American diets. Analyses of the National Health and Nutrition Examination Survey from 2003 to 2008 showed that processed foods provided nutrients specified in the 2010 *Dietary Guidelines for Americans,* contributing 55% of dietary fiber, 48% of calcium, 43% of potassium, 34% of vitamin D, 64% of iron, 65% of folate, and 46% of vitamin B-12.

It is now widely recognized that a healthy diet means eating a variety of nutritious foods from different food groups. Foods differ in their nutrient composition, and no single food can provide all the nutrients needed. Worded differently, a more varied diet is more likely to provide the nutrients required for good health. The impressively varied modern diet has been made possible through food processing.

Another benefit of thermal processing is the deactivation of thermolabile anti-nutritional factors. For example, heating inactivates protease inhibitors found in peas, beans, or potatoes (Damodaran 2008). These inhibitors are globular proteins that tightly bind the human digestive enzymes trypsin and chymotrypsin, thus annulling their action. Prolonged heating also inactivates lectins, which bind and damage intestinal mucosa cells and interfere with the absorption of peptides and amino acids.

It is well known that heat treatment enhances the digestibility of food. For example, denatured proteins are generally more digestible than proteins that are not denatured. As shown with carotenoids (Rock et al. 1998; Stahl and Sies 1992), mild processing may boost the bioavailability of nutrients and bioactive compounds. This is attributed to the softening or breaking of cell walls/membranes and denaturing proteins complexed with carotenoids, thereby facilitating their release from the food matrices. Processing conditions should, therefore, be optimized to increase bioavailability while minimizing the degradation of the carotenoids.

3.1.3 Food Safety

Processed foods are subject to norms set by regulatory organs in both developed and developing countries. Moreover, the food industry's attention to food safety is not only in terms of compliance with legislation but also motivated by financial liabilities. The consequences of a food safety failure can be commercially devastating to the manufacturer, including product recalls, damage to reputation, and punitive lawsuits. Consumer confidence in the safety of food products is one of the key elements in brand loyalty and determines its success and competitiveness.

Processing ensures the safety of foods by reducing the microbial load, particularly harmful microorganisms. Drying, pickling, and smoking reduce the water activity and alter the pH of foods, thereby restricting the growth of pathogenic and spoilage microorganisms and retarding enzymatic reactions. Other techniques such as canning and pasteurization can destroy microorganisms through heat treatment. Processing also deals with chemical and physical hazards.

The greatest food safety threats come from pathogenic microorganisms. Foodborne illnesses are a burden to public health, contributing significantly to the cost of health care. For example, the American food supply is considered among the safest in the world, but the Food and Drug Administration estimates that there are about 48 million cases of food-borne illnesses annually—the equivalent of infecting 1 in 6 Americans each year. Moreover, each year these illnesses result in an estimated 128,000 hospitalizations and 3,000 deaths (FDA 2019).

Processing reduces the incidence of foodborne diseases. Major examples of these diseases together with the food sources (FDA 2019) are shown in Table 2. Unprocessed food, such as fresh

Table 2. Major Foodborne Diseases and Foods Most Commonly Affected.

Disease	Target Food
Salmonellosis	eggs, poultry, meat, unpasteurized milk or juice, contaminated raw fruits, and vegetables
Campylobacteriosis	unpasteurized milk, raw or undercooked poultry, and contaminated water
Hemorrhagic colitis or *E. coli* O157:H7 infection	undercooked beef (especially hamburger), unpasteurized milk and juice, raw fruits and vegetables (e.g. sprouts), contaminated water
Listeriosis	raw and undercooked meats, unpasteurized milk, soft cheeses made with unpasteurized milk, ready-to-eat deli meats, and undercooked hot dogs
Botulism	improperly canned foods, especially home-canned vegetables, fermented fish, baked potatoes in aluminum foil

Reference: FDA (2019)

produce and raw meat, are more likely to harbor pathogenic microorganisms capable of causing these illnesses. Processing reduces and even eliminates microbial contamination responsible for food-borne diseases, while packaging and post-processing storage control recontamination.

3.1.4 Major Foodborne Diseases and Foods Most Commonly Affected

In the last 40 years, food safety research has resulted in an increased understanding of a range of health effects from foodborne chemicals, and technological developments have improved the US food safety from farm to fork by offering new ways to manage risks (Wu and Rodricks 2020).

Various food processing operations, including sorting, trimming, cleaning, cooking, baking, frying, roasting, flaking, and extrusion, have variable effects on mycotoxins (Kaushik 2015). In general, the processes are known to reduce mycotoxin concentrations significantly but do not completely eliminate them. Usually, the processes that utilize the higher temperatures have greater effects. Roasting and extrusion processing result in the lowest mycotoxin concentrations, especially if temperatures reach 150°C or higher (Sipos et al. 2021).

The use of additives in food processing represents another safety concern. Food additives are added for a specific purpose, such as to extend shelf life, ensure food safety, add nutritional value, or improve food sensorial quality. They are important in preserving the freshness, taste, appearance, texture, and wholesomeness of foods. For example, antioxidants prevent fats and oils from becoming rancid and emulsifiers stop peanut butter from separating into solid and liquid phases. The use of additives is subject to laws and regulatory practices; approved additives are permitted for use in food products at specific levels.

Attempts to deal with food allergenicity through food processing have yielded mixed results (Amaya-Farfan 2021; Sathe et al. 2005). The allergenic activity may be unchanged, decreased, or even increased by food processing (Besler et al. 2001). The identification of specific variables that can be used to reliably determine how processing can influence protein allergenicity has been difficult so far (Thomas et al. 2007).

3.2 Consumers' Demands

Figure 2 illustrates society's constantly increasing needs and expectations of foods, reflecting the widening responsibilities of food processors (Rodriguez-Amaya et al. 2021). Aside from being available, affordable, nutritious, diverse, and safe, as discussed above, today's food should be attractive, tasty, convenient, health-promoting, and environmentally sustainable. The factors that influence consumers' food choices include quality, price, appearance, taste, health, family preferences, habits, safety, production methods, country of origin, brand name, availability, and food allergen avoidance.

Figure 2. Evolution of Societal Needs and Demands.

As consumers become increasingly conscious about the safety of their food, its origin, and the sustainability of the processes that have produced and delivered the food (Wognum et al. 2011), the scope of such terms as health, safety, and quality could be expanded to include their long-term reach. Precision food processing, a term first used to ensure high-retention/bioavailability of nutrients and bioactive compounds, in coordination with the destruction of pathogens and antinutritional factors, is a strategy that could be adapted to meet the present and future emerging consumer demands related to good health. The scope of precision processing can be further developed to include not only the heating rate, intensity, and duration but also a more profound selection of process(es) and the order of addition of ingredients, which is based on recent knowledge of human metabolism and technological advancements. Examples can be drawn from the areas of agriculture, nutrition, and pharmaceutical design (Hartel et al. 2021; Singh et al. 2020). Although investing in industrial innovation at a time when energy, components for automation, and raw materials are in short supply and may presently seem as unthinkable, the food industry is still in a position to make food more health-promoting.

3.2.1 Palatability and Other Sensory Attributes

Flavor, appearance, and texture remain as overriding considerations for consumers' food acceptance. Sensory features are still at the forefront of most consumers' preferences; sometimes it is being considered more important than the potential health benefits of food (Falguera et al. 2012).

Impressively, food processors have been very creative in changing basic raw materials into a vast range of attractive and tasty foods, providing interesting and varied diets. Natural drives need to be invoked to understand why food preferences compel the industry to offer more sensory-enticing foods (Ostan et al. 2010). Food processors, however, should be open to the changing times and become aware of the new corporate responsibilities by extending the characteristics of quality and safety beyond their classical meanings to include health quality and the long-term innocuousness that food products should have.

3.2.2 Convenience

With more women joining the workforce and dealing with the fast pace and pressures of the modern world, consumers are now looking for ways to ease the burden of food preparation. Processing and packaging technologies have provided a range of convenience foods, allowing consumers to enjoy varied and nutritious meals that take little time to prepare and also practically eliminating the need for after-meal clean-ups. Moreover, consumers are saving time by shopping less frequently by stocking a wide range of processed foods.

Convenience food products include complete meals for almost instant serving from freezer to microwave or conventional heating to table, frozen pizzas ready for the oven, special mixes for pastries and bread, bagged salads, and sliced and canned fruits and vegetables. Time constraints from work, childcare, and commuting are cited as the main reasons why the US consumers, for instance, turn to convenience foods (Rahkovsky et al. 2021).

Few authors have studied the total energy and carbon emissions as potential costs to the convenience delivered by industrialized foods. Sonesson et al. (2005) estimated that the difference in environmental impact among homemade, semi-prepared, and ready-to-eat meals in Sweden was small. While the ready-to-eat meal used the most energy, the homemade meal produced higher emissions causing more eutrophication and global warming.

3.3 Functional Foods

As consumers have become more health conscious and interested in maintaining or improving their health through their diets, research and development of functional foods have emerged worldwide. Defined as conventional or modified foods that deliver potentially positive effects on health beyond basic nutrition, these foods may help reduce the risk of chronic, non-communicable diseases (e.g., cancer and cardiovascular diseases). A processor, however, should beware of the possibility that a modification introduced with the intention of adding benefits could result in the introduction of a negative consequence detectable in the consumer only years later.

In developed countries, research and the market have focused mostly on the following aspects (Rodriguez-Amaya et al. 2021):

- Producing bioactive ingredients (e.g., fiber, omega-3 fatty acids, isolated soy and milk proteins, probiotics, tomato concentrate, tea extracts, and fruit extracts).
- Manufacturing bioactive-enriched foods (e.g., prebiotic-enriched foods and dairy products with probiotics)
- Processing naturally functional foods (e.g., flax, nuts, cranberry, and whole grains).

Developing countries concentrate more on:

- Optimization of naturally functional traditional foods (e.g., quinoa, amaranth seeds, and yerba mate).
- Processing native, often unexploited, plant species with high levels of bioactive components (e.g., tropical fruits).

This area has grown in the past two decades under the consensus that an appropriate and effective regulatory framework should be in operation in order to guarantee high-quality, safe, and stable functional foods with proven efficacy.

Modern food manufacturing has also provided foods for individuals with specific health conditions, offering foods that have been formulated or modified to meet their needs. Examples are sugar-free foods sweetened with natural sweeteners, such as stevia and thaumatin for diabetic and celiac patients, and gluten-free and lactose-free foods for those who are sensitive to these food components.

3.4 Sustainable Food Systems

Sustainability has gained importance in the food industry (e.g., Mattson and Sonesson 2003). Given the very large environmental and social footprint that the food industry has globally, managing sustainability in the food supply chain is critical (Krystallis et al. 2012). While ensuring that there is enough food to meet the needs of the world's population, the significant contribution of the global food system to climate-changing greenhouse gas emissions must be addressed (Garnett

2013). Smith and Gregory (2013) concluded that the status quo is not an option, and tinkering with the current production systems is unlikely to deliver the food and ecosystem services needed in the future; radical changes in production and consumption are likely to be required over the coming decades.

Lindgren et al. (2018) emphasized the urgent need to develop and implement policies and practices that provide universal access to healthy food choices for a growing world population while reducing the environmental footprint of the global food system. Two challenges to achieving healthy sustainable diets for a global population are cited. The first challenge is the reduction in the yield and nutritional quality of crops (in particular vegetables and fruits) due to climate change; the second is the trade-offs between food production and industrial crops.

Sustainability requires maximum utilization of all raw materials including their by-products and integration of activities throughout the entire production-to-consumption stages (Floros et al. 2010). To maximize the conversion of raw materials into consumer products, postharvest losses should be reduced and the utilization of processing by-products and wastes should be increased. Food processors are striving to minimize the environmental impact of processing, including efforts to reduce air, water, and solid waste emissions and reduce the environmental impact of packaging by using recycled and recyclable materials and reducing the weight of the packaging.

The importance of this social responsibility and appropriate corporate governance can be translated into the utilization of agro-industrial by-products and wastes, thus mitigating the increasing environmental demand and instead gaining economic benefits by generating high-value bioactive compounds (Kaur et al. 2022). For example, the seeds and peels of many fruits and vegetables, discarded during processing, may contain large amounts of valuable compounds, often in higher concentrations than in the parts retained for industrial processing. Tomato peel is five times richer in lycopene than the pulp (Jurić et al. 2019; Machmudah et al. 2012). Components in apple pomace, such as dietary fiber and phenolic compounds, may be extracted and subsequently utilized in the food chain (Rabetafika et al. 2014). Wheat bran is currently used as animal feed, but wheat bran proteins have also been explored as a source of amino acids and bioactive peptides or as inhibitors of enzymes of industrial interest (Balandrán-Quintana et al. 2015).

4. Present and Future Challenges

4.1 Unhealthy Diets

While food processing has provided a basis to develop healthier diets, according to the standard principles of the 20th century as discussed above, it has been blamed in recent times for being conducive to a new type of unhealthy diet. The past decades have witnessed alarming increases in obesity and chronic diseases, such as diabetes, systemic inflammation, cardiovascular diseases, and cancer. An unhealthy diet—high in fat, added sugar and salt, and low in fiber—may increase the risk for these diseases. The same assessment, which showed processed foods provided nutrients to the American population (Weaver et al. 2014), also concluded that processed foods contributed constituents that need to be limited, as stated by the 2010 *Dietary Guidelines for Americans:* 52% of saturated fat, 75% of added sugars, and 57% of sodium. Eating only refined grains may increase the risk of developing type-2 diabetes, cardiovascular diseases, and weight gain (Liu et al. 2003. Ye et al. 2012). Substituting part of the overly processed ingredients will lead to eating relatively more whole-grain foods. Most consumers will need to reduce their current consumption of refined grain products to no more than one-third to one-half of all grains in order to meet the targets for whole-grain foods (Williams 2012).

Food companies have paid attention to this situation. Bread and cereal products are now available that are made from whole conventional grains but also multi-grain mixes, including non-conventional health grains such as amaranth, quinoa, and chia. Food processing techniques have been developed to offer low-fat or fat-free, gluten-free, low-salt, low-sugar, and high-fiber versions of foods.

It must be recalled that, besides extending the storage life of cereals and cereal products, one major objective of processing whole grains was to improve the nutritive value by removing growth-stunting phytates and excess fiber (Sreenivasan 1946). Considering that different sectors of free societies tend to develop independently and in an uncoordinated fashion before public policies are established, the use of refined flours of wheat, rice, maize, and other grains gradually conquered the popular diet market in more developed areas of the world.

Several countries, working with industry voluntarily or mandatorily, have introduced salt (Dötsch et al. 2009; He and MacGregor 2009; He et al. 2014; Webster et al. 2014, 2015) and sugar (MacGregor and Hashem 2014; Moore et al. 2020) reduction programs. In Australia, salt levels in bread were estimated to be reduced by 9%, cereals by 25%, and processed meat by 8% during the period from 2010 to 2013 (Trevena et al. 2014). The United Kingdom has successfully implemented a salt reduction program through gradual reformulation on a voluntary basis (He et al. 2014).

High consumption of added sugars through soft drinks has been a health concern globally. The sugar content of these drinks has been reduced mostly by their replacement with non-nutritive sweeteners (Silva et al. 2021). Natural and artificial sweeteners have also been used in dairy products (e.g., ice cream, yogurt, and flavored milk) (McCain et al. 2018)

Partially hydrogenated fat has been used to obtain desirable texture and palatability and increase the resistance of oils to oxidation during deep frying. It is produced by hydrogenating vegetable oils, thereby increasing the degree of saturation of the fatty acids. This product, however, introduces unnatural trans-fatty acids in such foods as snacks and deep-fried foods, baked goods, margarine products, crackers, cookies, pie crusts, doughnuts, and frozen pizza.

Trans-fat increases the risk of developing heart disease, stroke, and diabetes (Bhardwaj et al. 2011; Brownell and Pomeranz 2014; Dietz and Scanlon 2012; Mozaffarian et al. 2006; Stender and Dyerberg 2004). Thus, regulations for limiting and removing trans-fatty acids from the food supply have been implemented across the world. In 2003, Denmark was the first country to introduce a law that limited trans-fatty acid content in food.

Weaver et al. (2014) enumerated the following challenges for food processing, especially in relation to nutrition and health; these included reducing calories, enhancing gut health, reducing salt intake, enhancing health benefits of foods, improving food safety, reducing food waste, reducing allergens, promoting fresh but stable foods, and producing age-specific products

4.2 Constraints in Food Production

Agro-food production is faced with enormous obstacles: less land available for agricultural production; limited access to water; higher costs of production, transport, and storage; climate change; soil degradation and desertification; more resistant pests; and overexploitation of fisheries. With continuing population growth, the global demand for food will increase considerably amid stiff competition for land, water, and energy, along with the urgent need to reduce the impact of the food system on the environment. Climate change will have far-reaching impacts on crops, livestock, and fisheries production and will modify the prevalence of crop pests (Campbell et al. 2016).

In addition to the unsolved global constraints of the 21st century, both food production and the food industry are now confronted with the challenges of urgently remediating world food scarcity that followed the COVID-19 pandemic. This mega calamity has upset all human activities, from restricting the movement of workers, redefining consumers' demands, closing food production facilities, and limiting or canceling food trade policies to applying financial pressures on the food supply chain (Aday and Aday 2020).

The World Economic Forum (WEF 2022) has stated that some regions of the world are more affected than others by food constraints. About one in five people in Africa (21%) faced hunger in 2020; the proportion increased by 3% in one year. This was more than double the rate of any other region. In Latin America and the Caribbean, one in 10 people faced hunger in 2020.

The challenge of supplying healthy diets to 9 billion people in 2050 will in part be met through an increase in agricultural production. However, reducing food losses throughout the supply chain from production to consumption and ensuring sustainable enhancements in the preservation, nutrient content, safety, and shelf life of foods through food processing will also be essential (Augustin et al. 2016).

Augustin et al. (2016) have concluded that environmental sustainability is critical, and both the agro-food production and the food processing sectors are challenged to use fewer resources to produce greater quantities of existing foods and develop innovative new foods that are nutritionally appropriate for the promotion of health and well-being, have long shelf lives, and are conveniently transportable. Another challenge is the huge and widening food security gap between industrialized and developing countries. Science-based improvements in agricultural production, food science and technology, and food distribution systems are critically important in decreasing this gap (Floros et al. 2010). Lack of technology, unskilled labor, and underdeveloped infrastructure remain formidable challenges in developing countries.

4.3 Unintentional Consequences of Thermal Processing

The list of unintentional chemical reactions occurring during thermal processing has grown with the advancement of chemical-analytical technology and biomolecular science. An unintentional, undesired consequence of thermal processing is the formation of toxic compounds, such as acrylamide, furan, heterocyclic aromatic amines, and polycyclic aromatic hydrocarbons (van Boekel et al. 2010). The occurrence, mechanism of formation, influencing factors, health effects, and mitigation strategies for each of these compounds have been the subjects of intense investigations (Gloria et al. 2021). Applying the concept of precision processing, thermal processing should be controlled so that the desirable effects are enhanced while those that are harmful are prevented or at least minimized.

5. Future Perspectives

With the world population approaching 9 billion inhabitants by 2050, food production, which relies on large amounts of water and energy, must become more efficient (Finley 2020). Food production and delivery must also find innovative ways to reduce food waste, environmental pollutants, and greenhouse gas production.

According to an IFT Scientific Review (Floros et al. 2010), the solution to the challenge of meeting the food demands of our future world population lies clearly in the following principal thrusts:

- Increased agricultural productivity everywhere but particularly among poor farmers of whom there are hundreds of millions.
- Increased economic development and education, both for their own merits, because these promote infrastructure gains in transportation and water management.
- Much-increased efforts in environmental and water conservation and other improvements.
- Continued improvements in food and beverage processing and packaging to deliver safe, nutritious, and affordable food.
- Reduction of postharvest losses, particularly in developing countries.

All these goals must be achieved if we are to deliver a sustainable diet. The recent constraints imposed by the COVID-19 pandemic and the Russian invasion of Ukraine are so great that they may impose a decline on the quantity and quality of food in general in many parts of the world. Improvement of processed foods, however, is an endeavor that must continue, but with the use of new technologies to make the methods of assessing food innocuousness faster and less expensive. New approaches can predict how a process or an ingredient, which is deemed safe within one or five years, can impact health along a 10 or 20-year span. Animal or human molecular predictors, such as inflammatory pathway biomarkers and intestinal microbiome shifts (Carvalho et al. 2018; Machado et al. 2019), are today plausible means to implement a large-scale reverse engineering endeavor to reevaluate most current, as well as new processes and ingredients to make them healthier to humans in the long term.

The food systems approach is considered the sustainable solution for a sufficient supply of healthy food (van Berkum et al. 2018). Food systems consist of all the processes associated with food production and food utilization: growing, harvesting, packing, processing, transporting, marketing, consuming, and disposing of food remains.

A systems-based approach that includes improving natural resource use, reducing environmental impact, examining new food resources, enhancing consumer trust and understanding, and developing profitable market opportunity-led solutions for food and nutrition security is required (Knorr and Augustin 2021).

In 2010, FAO defined sustainable diets as diets with low environmental impacts, which contribute to food and nutrition security and healthy life for present and future generations. Sustainable diets are protective and respectful of biodiversity and ecosystems, culturally acceptable, accessible, economically fair, and affordable; it is also nutritionally adequate, safe, and healthy while optimizing natural and human resources. The major determinants of sustainable diets fall into five categories: 1) agricultural, 2) health, 3) sociocultural, 4) environmental, and 5) socioeconomic (Johnston et al. 2014). Promoting sustainable diets will require an inclusive approach that reflects the multidisciplinary determinants.

Gustafson et al. (2016) noted that sustainability considerations had been absent from most food security assessments conducted. In addition, previous food security work had generally focused only on achieving adequate calories, rather than addressing dietary diversity and micronutrient adequacy. In response to the limitations of previous assessments, seven metrics were proposed, and each is based on a combination of multiple indicators for use in characterizing sustainable nutrition outcomes of food systems: (1) food nutrient adequacy; (2) ecosystem stability; (3) food affordability and availability; (4) sociocultural wellbeing; (5) food safety; (6) resilience; and (7) loss and waste reduction.

There has been a constant call for integrated multi-sectorial (academia, government, industry, and consumers), and multidisciplinary (agriculture, food science and technology, nutritional sciences, medical sciences, environmental sciences, social sciences, and economics) approach to address the complex and multifaceted challenge to feed the world and minimize global food insecurity (Dwyer et al. 2012; Godfray et al. 2010; Johnston et al. 2014; Knorr et al. 2020; Lowe et al. 2008; van Mil. et al. 2014, Wu et al. 2014). The challenges to be overcome before achieving objectives of multi-sectorial interest are formidable in free-enterprise societies, and they may constitute the main reason for sensible states of integration not to be adopted with greater celerity.

Nevertheless, the increasing demands of a growing world population, the more frequent occurrence of global upheavals, along with the mounting need for safeguarding nature, have led to the development of the second generation of technologies that use bio-waste as a resource for diverse industrial sectors (Brandão et al. 2021). For instance, circular bioeconomy, a production and consumption model designed to promote more sustainable growth over time, has attracted much attention. It is based on the optimization of resources, reduction of the consumption of raw materials, and recovery of waste by recycling or giving it a second life as a new product. However, this model has not yet been significantly manifested in the market; its full potential is still far from being realized.

For decades, food processing has confronted and resolved significant challenges, providing great effective benefits for humankind. Processed foods are not perfect and probably will never be for everyone but with the aid of the most advanced science, responsible food processors have been constantly striving to improve them to meet society's ever-expanding needs and demands. Processed foods will continue to be an essential part of our future.

References

Aday, S. and Aday, M.S. 2020. Impact of COVID-19 on the food supply chain, Food Qual Saf. 4: 167–180.

Al-juhaimi, F., Ghafoor, K., Özcan, M.M., Jahurul, M.H.A., Babiker, E.E., Jinap, S. et al. 2018. Effect of various food processing and handling methods on preservation of natural antioxidants in fruits and vegetables. J. Food Sci. Technol. 55: 3872–3880.

Amaya-Farfan, J. 2021. Denaturation of proteins, generation of bioactive peptides, and alterations of amino acids. pp. 21–84. *In*: Rodriguez-Amaya, D.B. and Amaya-Farfan, J. (eds.). Chemical Changes during Processing and Storage of Foods. Implications for Food Quality and Human Health. Elsevier—Academic Press, London, UK.

Arranz-Otaegui, A., Carretero, L.G., Ramsey, M.N., Fuller, D.G. and Richter, T. 2018. Archaeobotanical evidence reveals the origins of bread 14,400 years ago in northeastern Jordan. Proc. Natl. Acad. Sci. USA 115: 7925–7930.

Augustin, M.A., Riley, M., Stockmann, R., Bennett, L., Kahl, A., Lockett, T. et al. 2016. Role of food processing in food and nutrition security. Trends Food Sci. Technol. 56: 115–125.

Balandrán-Qunitana, R.R., Mercado-Ruiz, J.N. and Mendoza-Wilson, A.M. 2015. Wheat bran proteins: a review of their uses and potential. Food Rev. Int. 31: 279–293.

Brandão, A.S., Gonçalves, A. and J.M.R.C.A. Santos. 2021. Circular bioeconomy strategies: From scientific research to commercially viable products J. Clean. Prod. 295: 126407.

Bermúdez-Aguirre, D. and Barbosa-Cánovas, G.V. 2011. An update on high hydrostatic pressure, from the laboratory to industrial applications. Food Eng. Rev. 3: 44–61.

Berner, L.A., Clydesdale, F.M. and Douglass, J.S. 2001. Fortification contributed greatly to vitamin and mineral intakes in the United States 1889–1991. J. Nutr. 131: 2177–2183.

Besler, M., Steinhart, H. and Paschke, A. 2001. Stability of food allergens and allergenicity of processed foods. J. Chromatogr. B 756: 207–228.

Bhardwaj, S., Passi, S.J. and Misra, A. 2011. Overview of trans fatty acids: biochemistry and health effects. Diabetes Metab Syndr. 5: 161–164.

Brownell, K.D. and Pomeranz, J.L. 2014. The trans-fat ban—food regulation and long-term health. N. Engl. J. Med. 370: 1773–1775.

Campbell, B.M., Vermeulen, S.J., Aggarwal, P.K., Ramirez-Villegas, J., Rosenstock, T., Sebastian, L. et al. 2016. Reducing risks to food security from climate change. Glob Food Sec. 11: 34–43.

Carvalho, G.C.B.C., Moura, C.S., Roquetto, A.R., Barrera-Arellano, D., Yamada, A.T., Santos, A.D. et al. 2018. Impact of trans-fats on heat-shock protein expression and the gut microbiota profile of mice. J. Food Sci. 83: 489–498.

Chauhan, O.P. [ed.]. 2019. Non-thermal Processing of Foods, first ed. CRC Press, Boca Raton, USA.

Damodaran, S. 2008. Amino acids, peptides, and proteins. pp. 217–329. *In*: Damodaran, S., Parkin, K.L. and Fennema, O.R. (eds.). Fennema's Food Chemistry, Fourth ed., CRC Press, Boca Raton, USA.

Dietz, W.H. and Scanlon, K.S. 2012. Eliminating the use of partially hydrogenated oil in food production and preparation. JAMA 308: 143–144.

Dötsch M., Busch, J., Batenberg, M., Liem, G., Tareilus, E., Mueller, R. et al. 2009. Strategies to reduce sodium consumption: a food industry perspective. Crit. Rev. Food. Sci. Nutr. 49: 841–851.

Dwyer, J.T., Fulgoni III, V.L., Clemens, R.A., Schmidt, D.B. and Freedman, M.R. 2012. Is "processed" a four-letter word? The role of processed foods in achieving dietary guidelines and nutrient recommendations. Adv. Nutr. 3: 536–548.

Falguera, V., Aliguer, N. and Falguera, M. 2012. An integrated approach to current trends in food consumption: Moving toward functional and organic products? Food Control. 26: 274–281.

FAO. 2010. Sustainable diets and biodiversity. Directions and solutions for policy, research and action. Proceedings of the International Scientific Symposium: Biodiversity and Sustainable Diets United Against Hunger, Rome, Italy, November 3–5, 2010. http://www.fao.org/docrep/016/i3004e/i3004e.pdf (accessed on March 30, 2022). International Food Information Council Washington, DC Search for other works by this author on: Oxford Academic Google Scholar

FAO, WFP, Oxford Academic IFAD. 2012. The State of Food Insecurity in the World. Economic growth is necessary but not sufficient to accelerate reduction of hunger and malnutrition. FAO, Rome, Italy.

FDA. 2019. Foodborne Illness-causing Organisms in the US. What you need to know. https://www.fda.gov/downloads/Food/Food borne Illness Contaminants/UCM187482.pdf (accessed on March 21, 2022).

Finley, J.W. 2020. Evolution and future needs of food chemistry in a changing world. J. Agric. Food Chem. 68: 12956–12971.

Floros, J.D., R. Newsome, W. Fisher, G.V. Barbosa-Canovas, H. Chen, C.P. Dunne et al. 2010. Feeding the world today and tomorrow: the importance of food science and technology: an IFT scientific review. Crit. Rev. Food Sci. F. 9: 572–599.

Garnett, T. 2013. Food sustainability: problems, perspectives and solutions. Proc. Nutr. Soc. 72: 29–39.

Gloria, M.B.A., Masson, L., Amaya-Farfan , J. and Rodriguez-Amaya, D.B. 2021. Generation of process-induced toxicants. pp. 453–535. *In*: Rodriguez-Amaya, D.B. and Amaya-Farfan, J. (eds.). Chemical Changes during Processing and Storage of Foods. Implications for Food Quality and Human Health. Elsevier–Academic Press, London, UK.

Godfray, H.C.J., Beddington, J.R., Crute, I.R., Haddad, L., Lawrence, D., Muir, J.F. et al. 2010. Food Security: The challenge of feeding 9 billion people. Science 327(5967): 812–818.

Godoy, H.T., Amaya-Farfan, J. and Rodriguez-Amaya, D.B. 2021. Degradation of vitamins. pp. 329–383. *In*: Rodriguez-Amaya, D.B. and Amaya-Farfan, J. (eds.). Chemical Changes during Processing and Storage of Foods. Implications for Food Quality and Human Health. Elsevier–Academic Press, London, UK.

Gustafson, D., Gutman, A., Leet, W., Drewnowski, A., Franzo, J. and Ingram, J. 2016. Seven food system metrics of sustainable nutrition security. Sustainability 8: 196.
Gustavsson, I., Cederberg, C., Sonesson, U., van Otterdijk, R. and Meybeck, A. 2011. Global Food Losses and Food Waste: Extent, Causes and Prevention. FAO, Rome.
Hartl, D., de Luca, V., Kostikova, A., Laramie, J., Kennedy, S., Ferrero, E. et al. 2021. Translational precision medicine: an industry perspective. J. Transl. Med. 19: 245.
He, F.J., Brinsden, H.C. and MacGregor, G.A. 2014. Salt reduction in the United Kingdom: a successful experiment in public health. J. Hum. Hypertens. 28: 345–352
He, F.J. and MacGregor, G.A. 2009. A comprehensive review on salt and health and current experience of worldwide salt reduction programmes. J. Hum. Hypertens 23: 363–84.
Huebbe, P. and Rimbach, G. 2020. Historical reflection of food processing and the role of legumes as part of a healthy balanced diet. Foods 9(8): 1056.
Jan, A., Sood, M., Sofi, S.A. and Norzom, T. 2017. Non-thermal processing in food applications: a review. Int. J. Food Sci. Nutr. 2: 171–180.
Johnston, J.L., Fanzo, J.C. and Cogill, B. 2014. Understanding sustainable diets: descriptive analysis of the determinants and processes that influence diets. and their impacts on health, food security, and environmental sustainability. Adv. Nutr. 5: 418–429.
Jurić, S., Ferrari, G., Velikov, K.P. and Donsì, F. 2019. High-pressure homogenization treatment to recover bioactive compounds from tomato peels. J. Food Eng. 262: 170–180.
Kaur, B., Panesar, P.S., Anal, A.K. and Chu-Ky, S. 2022. Recent trends in the management of mango by-products. Food Rev. Inter. DOI:10.1080/87559129.2021.2021935
Kaushik, G. 2015. Effect of processing on mycotoxin content in grains. Crit. Rev. Food Sci. Nutr. 55: 1672–1683.
Khouryieh, H.A. 2021. Novel and emerging technologies used by the U.S. food processing industry. Innov. Food Sci. Emerg. Technol. 67: 102559.
Knorr, D. and Augustin, M.A. 2021. From value chains to food webs: The quest for lasting food systems. Trends. Food Sci. Technol. 110: 812–821.
Knorr, D., Augustin, M.A. and Tiwari, B. 2020. Advancing the role of food processing for improved integration in sustainable food chains. Front. Nutr. 7: 34.
Knorr, D. and Watze, H. 2019. Food processing at a crossroad. Front. Nutr. https://doi.org/10.3389/fnut.2019.00085
Krystallis, A., Grunert, K.G., de Barcellos, M.D., Perrea, T. and Verbeke, W. 2012. Consumer attitudes towards sustainability aspects of food production: insights from three continents. J. Mark. Manag 28: 334–372.
Kubo, M.T.K., Baicu, A., Erdogdu, F., Poças, M.F., Silva, C.L.M., Simpson, R. et al. 2021. Thermal processing of food: challenges, innovations and opportunities. A position paper. Food Rev. Inter. DOI: 10.1080/87559129.2021.2012789.
Lindgren, E., Harris, F., Dangour, A.D., Gasparatos, A., Hiramatsu, M. and Javadi, F. 2018. Sustainable food systems—a health perspective. Sustain. Sci. 13: 1505–1517.
Ling, B., Tang, J., Kong, F., Mitcham, E.J. and Wang, S. 2015. Kinetics of food quality changes during thermal processing: a review. Food Bioprocess. Tech. 8: 343–358.
Liu, L., Wang, J., Rosenberg, D., Zhao, H., Lengyel, G. and Nadel, D. 2018. Fermented beverage and food storage in 13,000 y-old stone mortars at Raqefet Cave, Israel: investigating Natufian ritual feasting. J. Archaeol. Sci. Rep. 21: 783–793.
Liu, S., Willet, W.C., Manson, J.E., Hu, F.B., Rosner, B. and Colditz, G. 2003. Relation between changes in intakes of dietary fiber and grain products and changes in weight and development of obesity among middle-aged women. Am. J. Clin. Nutr. 78: 920–927.
Lowe, P., Phillipson, J. and Lee, R.P. 2008. Socio-technical innovation for sustainable food chains: roles for social science. Trends Food Sci. Technol. 19: 226–233.
Lund, D.B. 1982. Influence of processing on nutrients in foods. J. Food Protect. 45: 367–373.
MacGregor, G.A. and Hashem, K.M. 2014. Action on sugar—lessons from UK salt reduction programme. Lancet 383(9921): 929–931.
Machado, K.I.A., Roquetto, A.R., Moura, C.S., Lopes, A.S., Cristianini, M. and Amaya-Farfan, J. 2019. Comparative impact of thermal and high isostatic pressure inactivation of gram-negative microorganisms on the endotoxic potential of reconstituted powder milk. LWT - Food Sci. Technol. 106:78–82.
Machmudah, S., Zakaria, S. Winardi, M. Sasaki, M. Goto, N. Kusumoto et al. 2012. Lycopene extraction from tomato peel by-product containing tomato seed using supercritical carbon dioxide. J. Food Eng. 108: 290–296.
Mattsson, B. and Sonesson, U. (eds.). 2003. Environmentally-friendly Food Processing. Woodhead Publishing, Salt Lake City, USA.
McCain, H.R., Kaliappan, S. and Drake, M.A. 2018. Sugar reduction in dairy products. J. Dairy Sci. 101: 8619–8640.
Mercader, J. 2009. Mozambican grass seed consumption during the Middle Stone Age. Science 326: 1680–1683.
Moore, J.B., Sutton, E.H. and Hancock, N. 2020. Sugar reduction in yogurt products sold in the UK between 2016 and 2019. Nutrients 12: 17.
Mozaffarian, D., Katan, M.B., Ascherio, A., Stampfer, M.J. and Willett, W.C. 2006. Trans fatty acids and cardiovascular disease. N. Engl. J. Med. 354: 1601–1613.
Nature editorial. 2022. War in Ukraine and the challenge to global food security. Nature 604: 217–218.

Ostan, I., Poljsak, B., Simcic, M. and Tijskens, L.M. 2010. Appetite for the selfish gene. Appetite. 54:442–449.

Preedy, V. 2014. Processing and Impact on Active Components in Food, first ed. Elsevier, Amsterdam, The Netherlands.

Raak, N., Symmank, C., Zahn, S., Aschemann-Witzel, J. and Rohm, H. 2017. Processing- and product-related causes for food waste and implications for the food supply chain. Waste Management 61: 461–472.

Rabetafika, H.N., Bchir, B., Blecker, C. and Richel, A. 2104. Fractionation of apple by products as source of new ingredients: Current situation and perspectives. Trends Food Sci. Technol. 40: 99–114.

Rahkovsky, I., Young, J. and Carlson, A. 2021. What drives consumers to purchase convenience foods? Food Economics Division, Economic Research Service, US Department of Agriculture. August 2021.

Rankin, S.A., Bradley, R.L., Miller, G. and Mildenhall, K.B. 2017. A 100-year review: a century of dairy processing advancement—pasteurization, cleaning and sanitation, and sanitary equipment design. J. Dairy Sci. 100: 9903-9915.

Revedin, A., Aranguren, B., Becattini, R., Longo, L., Merconi, E., Lippi, M.M. et al. 2010. Thirty thousand-year-old evidence of plant food processing. Proc. Natl. Acad. Sci. USA 107: 18815–18819.

Risch, S.J. 2009. Food packaging history and innovations. J. Agric. Food Chem. 57(18): 8089–92. doi: 10.1021/jf900040r. PMID: 19719135.

Rock, C.L., Lovalvo, J.L., Emenhiser, C., Ruffin, M.T., Flatt, S.W. and Schwartz, S.J. 1998. Bioavailability of β-carotene is lower in raw than in processed carrots and spinach in women. J. Nutr. 128: 913–916.

Rodriguez-Amaya, D.B. 2015. Food Carotenoids: Chemistry, Biology and Technology. IFT Press-Wiley, Oxford, UK.

Rodriguez-Amaya, D.B. and Amaya-Farfan, J. 2018. Nutritional and functional attributes of fruit products. pp. 45–66. *In*: Rosenthal, A., Deliza, R., Welti-Chanes, J. and Barbosa-Cánovas, G.V. (eds.). Fruit Preservation. Novel and Conventional Technologies. Springer Science + Business Media, New York, USA.

Rodriguez-Amaya, D.B., Amaya-Farfan, J. and Lund, D.B. 2021. Societal role of food processing: envisaging the future. pp. 1–20. *In*: Rodriguez-Amaya, D.B. and Amaya-Farfan, J. (eds.). Chemical Changes during Processing and Storage of Foods. Implications for Food Quality and Human Health. Elsevier–Academic Press, London, UK.

Salque, M., Bogucki, P.I., Pyzel, J., Sobkowiak-Tabaka, I., Grygiel, R. and Szmyt, M. et al. 2013. Earliest evidence for cheese making in the sixth millennium BC in northern Europe. Nature 493: 522–525.

Samaniego-Vaesken, M.L., Alonso-Aperte, E. and Varela-Moreiras, G. 2012. Vitamin food fortification today. Food Nutr. Res. 56: 5459.

Sathe, S.K., Teuber, S.S. and Roux, K.H. 2005. Effects of food processing on the stability of food allergens. Biotech. Adv. 23: 423–429.

Silva, P.D., Cruz, R. and Casal, S. 2021. Sugars and artificial sweeteners in soft drinks: A decade of evolution in Portugal. Food Control. 120: 107481.

Singh, P., Pandey, P.C., Petropoulos, G.P., Pavlides, A., Srivastava, P.K., Koutsias, N. et al. 2020. Hyperspectral remote sensing in precision agriculture: present status, challenges, and future trends. pp. 121–146. *In*: Pandy, P.C., Srivastava, P.K., Balzter, H., Bhattacharya, B. and Petropoulos, G. (eds.). Hyperspectral Remote Sensing: Theory and Applications. Elsevier.

Sipos, P., Peles, F., Brassó, D.L., Béri, B., Pusztahelyi, T., Pócsi, I. et al. 2021. Physical and chemical methods for reduction in aflatoxin content of feed and food. Toxins 13: 204–221.

Smith, P. and Gregory, P.J. 2013. Climate change and sustainable food production. Proc. Nutr. Soc. 72: 21–28.

Sonesson, U., Mattsson, B., Nybrant, T. and Ohlsson, T. 2005. Industrial processing versus home cooking: an environmental comparison between three ways to prepare a meal. AMBIO: J. Human Environ. 34: 414–421.

Sreenivasan, A. 1946. Nutritional improvement of rice. Curr. Sci. 15: 180–184.

Stahl, W. and Sies, H. 1992. Uptake of lycopene and its geometrical isomers is greater from heat-processed than from unprocessed tomato juice in humans. J. Nutr. 122: 2161–2166.

Stender, S. and Dyerberg, J. 2004. Influence of trans fatty acids on health. Ann. Nutr. Metab. 48: 61–66.

Thielecke, F., Lecedrf, J. and Nugent, A. 2021. Processing in the food chain: Do cereals have to be processed to add value to the human diet? Nutr. Res. Rev. 34: 159–173.

Thomas, K., Herouet-Guicheney, C., Ladics, G., Bannon, G., Cockburn, A., Crevel, R. et al. 2007. Evaluating the effect of food processing on the potential human allergenicity of novel proteins: international workshop report. Food Chem. Toxicol. 45: 1116–1122.

Trevena, H., Neal, B., Dunford, E. and Wu, J.H.Y. 2014. An evaluation of the effects of the Australian Food and Health Dialogue targets on the sodium content of bread, breakfast cereals and processed meats. Nutrients 6: 3802–3817.

Tunick, M.H. 2009. Dairy innovations over the past 100 years. J. Agric. Food Chem. 57: 8093–8097.

Van Berkum, S., Dengerink, J. and Ruben, R. 2018. The Food Systems Approach: Sustainable Solutions for a Sufficient Supply of Healthy Food. Wageningen Economic Research, Wageningen, The Netherlands.

Van Boekel. M., Fogliano, V., Pellegrini, N., Stanton, C., Scholz, G., Lalljie, S. et al. 2010. A review on the beneficial aspects of food processing. Mol. Nutr. Food Res. 54: 1215–1247.

Van Mil, H.G.J., Foegeding, E.A., Windhab, E.J., Perrot, N., van der Linden, E. 2014. A complex system approach to address world challenges in food and agriculture. Trends Food Sci. Technol. 40: 20–32.

Weaver, C.M., Dwyer, J., Fulgani III, V.L., King, J.C., Leveille, G.A. and MacDonald, R.S. 2014. Processed foods: contribution to nutrition. Am. J. Clin. Nutr. 99: 1525–1542.

Webster, J., Trieu, K., Dunford, E. and Hawkes, C. 2014. Target salt 2025: a global overview of national programs to encourage the food industry to reduce salt in foods. Nutrients 6: 3274–3287.
Webster, J., Treiu, K., Dunford, E., Nowson, C., Jolly, A.-A., Greenlands, R. et al. 2015. Salt reduction in Australia: from advocacy to action. Cardiovasc. Diagn. Ther. 5: 207–218.
Weiss, E, Kislev, M.E., Simchoni, O. and Nadel, D. 2004. Small-grained wild grasses as staple food at the 23 000-year-old site of Ohalo II, Israel. Econ. Bot. 58: S125–S134.
Welch, R.W. and Mitchell, P.C. 2000. Food processing: a century of change. Br. Med. Bull. 56: 1–17.
Williams, P.G. 2012. Evaluation of the evidence between consumption of refined grains and health outcomes. Nutr. Rev. 70: 80–99.
Wognum, P.M., Bremmers, H., Trienekens, J.H., van der Vorst, J.G.A.J. and Bloemhof, J.M. 2011. Systems for sustainability and transparency of food supply chains—current status and challenges. Adv. Eng. Inform. 25: 65–76.
World Economic Forum. 2022. https://www.weforum.org/agenda/2022/03/global-food-security-challenges-solutions. Accessed on May 5: 2022.
Wu, F. and Rodricks, J.V. 2020. Forty years of food safety risk assessment: a history and analysis. Risk Anal. 40: 2218–2230.
Wu, S.-H., Ho, C.-T., Nah, S.-L. and Chau, C.-F. 2014. Global hunger: A challenge to agricultural, food, and nutritional sciences. Crit. Rev. Food Sci. Nutr. 54: 151–162.
Xia, Q., Zheng, Y., Liu, Z., Cao, J., Chen, X. and Liu, L. 2020. Nonthermally driven volatilome evolution of food matrices: The case of high pressure processing. Trends Food Sci. Technol. 106: 365–381.
Ye, E.Q., Chacko, S.A., Chou, E.L., Kugizaki, M. and Liu, S. 2012. Greater whole-grain intake is associated with lower risk of type 2 diabetes, cardiovascular disease, and weight gain. J. Nutr. 142: 1304–1313.

2

Concepts for Food Sustainable Production

Mariany Costa Deprá, Leila Queiroz Zepka and *Eduardo Jacob-Lopes**

1. Background

A crucial ecosystem service for modern society is the ability to use natural resources for food production and thus aim to ensure food security. After all, food is essential to life. However, the main paradox that haunts food production is the need to meet the population's demand for food; at the same time, production systems have been accused of pushing the safe limits of planetary borders and, consequently, putting ecosystem risk (Springmann et al. 2018).

To meet and reinforce this premise, the data suggest that around 70% of the demands for freshwater, 50% of arable land, and 26% of greenhouse gas emissions are attributed to agriculture, industry, and food processing activities. In parallel, the food industry needs to adapt to the demand, production, and distribution of food in the next 20 to 40 years. This is because it is assumed that the population growth trend will increase from nearly 7 billion today to 8 billion in 2030 and probably more than 9 billion in 2050 (Rupawalla et al. 2021). Furthermore, globalization will continue to expose the food system to new economic and political pressures. Although the diagnosis is crystal clear and there is no intention to mask it, the estimates presented are not favorable and, unfortunately, there is no easy or immediate solution to overcome these impasses.

However, despite the past neglect, which gave us unpromising prospects, the convergence between the supporting roles in relation to environmental protection and food production has been an emerging target in discussions that promote sustainability as a collective objective. Hence, the call for sustainable food production has been promoted in a number of influential publications (Dawson et al. 2019).

Indeed, the ability to sustainably produce food can be a powerful force for change, resulting in a wide range of benefits, going beyond the obvious ones, such as the nexus among the environment, society, and economy. However, the path to changing the food production system is immensely complex and fraught with many challenges, which we are just beginning to tackle. From this

Food Science and Technology Department, Federal University of Santa Maria, UFSM, Roraima Avenue 1000, 97105-900, Santa Maria, RS, Brazil.
* Corresponding author: ejacoblopes@gmail.com

perspective, different sets of strategies and priorities to address potential paths and point out levels of problems have been suggested by decision-makers in the food and sustainable production sector. Among them, the concepts of a sustainable future through the establishment of sustainable development goals (SDGs) have become guidelines for new production models. Through them, there is the expectation of achieving greater international commitment to intensify food production in a sustainable way; it establishes global and consistent networks between governments in order to consolidate the implementation of sustainable guidelines and cover people and communities as a whole. This is because the lack of close links and no integration can drastically put control policies at risk, and all these actions can be increasingly inefficient and ineffective and can frustrate the non-success of the proposed objectives.

But after all, what is sustainable food production? How fast should the shift be from the status quo to sustainable food production?

Conceptually, sustainable food production can be understood as a production method that uses processes and systems that do not pollute, conserve non-renewable energy and natural resources, and are economically efficient in addition to being safe for workers, communities, and consumers; the systems do not compromise the needs of future generations as well. However, the speed at which the transition to a new sustainable food production chain will take place requires a stronger link between the connection of science and industrial application in the field of sustainability since process transfer is often characterized by the success of a long and uncertain form, requiring immediate investments to anticipate future needs (Kaiser et al. 2021).

In view of the above, this chapter discusses how food consumption and production developed in the past and what specific changes, in the short and medium term, can be evolved to help achieve a sustainable production system. Here, through a timeline approach, the main drivers of the past and future evolution of food production were reviewed. Subsequently, sustainability criteria and concepts, through the relationship between sustainable development goals and the production and consumption of food, were occasionally discussed. In the end, an integrative compilation of the criteria beyond sustainability was carried out in which we reported some of the main challenges, opportunities, and global implications that hindered the establishment of a sustainable food production system.

2. Sustainable Food Production: Where We Started From and Where We are Going

Since prehistory, when life on Earth began, human beings have fought for food. Over the centuries, humans began to benefit from agricultural practices and raised animals to make food more accessible and less labor-intensive. It is believed that, at that time, the "hunter-gatherer" mode was sufficient for small population groups in favorable environments. However, as the population grew and people moved to areas less endowed with food, they looked to more reliable sources of nutrition.

Thus, based on agriculture and livestock, the technology, as unpretentious as it was, began to gain space and at a slow pace played a fundamental role in increasing food production. At the level of exemplification, the development of a simple heavy plow, capable of breaking through dense and moist soils, became the stage to open up ways for the production of grain sources along the continents in the middle of the Middle Ages. At the end of this period, it was possible to see those small populations could easily make use of seasonal rotations. In addition, it is estimated that this activity led to an increase in productivity of at least one-third of total food production.

However, if local food management was a matter of easy maintenance, the discovery of the New World unleashed the largest and fastest spread of new crops the world had ever seen. Thus, the need to preserve food, whether for transport or later consumption, gained prominence. Consequently, science and technology played an increasingly important role in food production in the 18th and 19th centuries.

Since then, the trajectory of food processing began, mainly led by pasteurization and canning techniques. These two processes have become vital to the history of the food industry. In short, pasteurization was particularly important for beverages, such as juices and especially milk, due to their susceptibility to bacterial growth. Without this process, the story of food processing would not have advanced much. Long-term food storage and transportation around the world would have been extremely limited. Likewise, canning had become important with the onset of World War I and the high demand for cheap, durable, and transportable food for soldiers.

Later, World War II and the space race in the mid-19th century helped to accelerate the development of ready-to-eat packaged meals. During this period, the working middle class also began to expand around the world, bringing greater demand for fast meals with a long shelf life. Subsequently, the Green Revolution was one of the profound transformations the world underwent after World War II. This innovation was known for improving agricultural production and increasing food production from the 1960s and 1970s.

Since then, driven by revolutionary advances in food production, new processes, as well as new ingredients, have contributed to the history of food processing in the 20th century. Spray drying, evaporation, lyophilization, and the use of preservatives facilitated the packaging of different types of food and its conservation on the shelf. The addition of additives and artificial colors helped make these preserved foods even more palatable. In addition, the use of equipment such as the home oven, microwave, blender, and other appliances provided an easy way to prepare these meals quickly. Factories and mass-production techniques made it possible to produce and package food quickly. These theoretically simple developments paved the way for globally popular foods.

However, when we look at the evolutionary chain of food processing, it is possible to see the high demand for energy and arable land. Indeed, industrial food processes have traditionally been designed assuming the abundant and cheap availability of natural resources (van der Goot et al. 2016). Under this premise, concepts of sustainability and environmental preservation were introduced in the food industry. This is because, in practice, the exploitation of natural resources in the long term has brought profound destruction to the ecosystem. Among them, the increase in the food waste index linked to world hunger and the increase in global emissions of greenhouse gases are major causes of the disruption in the ecosystem. Therefore, the sustainable vision applied to the food industries came with the premise of producing healthy foods to nourish the population while maintaining the health of the ecosystem, thus avoiding the negative impact on the environment.

Given this scenario, the needs of consumers are currently far from being able to "preserve" the food for later consumption. Now, our main objective is to maintain the demand for food without entering the environmental collapse in which the ecosystem finds itself. In addition, the awakening of greater environmental and social awareness in consumers has created new habits and behaviors that even affect their consumption patterns. Among them, aspects such as animal health, labor used, and management of the supply chain are also on the radar of the more in-tuned public. Therefore, food industry advances must go hand in hand to meet the new consumers' demands.

3. Food Production and Consumption Role Under Sustainable Development Goals

To achieve universal sustainability, nations and public and private institutions, as well as researchers and critical stakeholders, have strategically met to propose plans for peace and prosperity in favor of society and the environment. However, it is important to emphasize that these initiatives are often designed for a volatile, uncertain, and complex ecosystem, such as the case of food production and consumption chains (Dwivedi et al. 2021).

In this context, sustainable practices implementation described in the United Nations 2030 Agenda, proposed the search for new ways to promote sustainability in the light of the 17 sustainable development goals (SDGs) (Fig. 1). This global initiative recognizes the interrelated nature of issues,

Figure 1. The 17 Sustainable Development Goals Under Pyramid Perspective.

such as poverty, inequality, gender equality, and conservation, as well as the need for intersectoral and intergovernmental cooperation for systemic change, through three pillars subdivided into the social, economic, and environmental spheres (Dantas et al. 2021).

However, when we take a closer look at these global targets, we can see that the food production chain can be seen as a key player in the quest for purpose compliance. As shown in Fig. 1, out of a total of 17 proposed universal goals, the food production industry, as well as the consumption and final disposal activity, has a direct relationship with eight of these goals. Thus, the following is a peculiar perspective on how these activities can compromise and assist in the fulfillment of the goals established for achieving universal sustainable development.

(i) Goal 2 – Zero Hunger: According to data reported by the United Nations Organization (ONU), it is estimated that almost 690 million people are hungry; this represents about 9% of the world's population. Furthermore, the prospects do not seem rosy. This is because if recent trends continue, it is likely that the size of the population that will be affected by hunger will exceed 840 million people in the next 10 years, representing a total of 10% of the global population. However, one of the measures designed to achieve this goal is to double agricultural productivity, which would imply an increase in the exploitation of natural resources. Thus, a potential alternative to avoid increasing environmental exploitation and consequently breaking the cycle of global destruction may be linked to better use of the food already produced. As is well known, more than a third of the food produced is wasted. In this way, planning and establishing food logistics strategies can allow a more correct destination for food that until then would not have been used.

(ii) Goal 6 – Clean Water and Sanitation: It is estimated that > 80% of wastewater resulting from human activities is discharged into rivers or the sea without any removal of pollution. Given this scenario, one of the main goals is to improve water quality by reducing pollution, eliminating waste, minimizing chemical and biological contamination and reducing by half the proportion of untreated wastewater, and substantially increasing recycling and safe reuse all around the world. In this regard, the food industry can play a key role, as the wastewater generated by food production and agricultural activities is an important source of environmental pollution; in addition, it is also among the most difficult and expensive wastes to manage due to its biological and biochemical characteristics that make an ideal treatment model difficult. Thus, assigning water recycling strategies to heat exchange equipment within the food industry and processing (e.g., pasteurizers and chillers) may be a promising alternative. Furthermore, applying efficient treatments to wastewater can effectively contribute to reducing the contamination of water bodies. Among them, the application of more effective treatments, such as the removal of organic matter through versatile microorganisms, such as microalgae (dos Santos et al. 2020) and anammox bacteria (Yang et al. 2021), has been recently studied and improved by research and development for large-scale application in industry.

(iii) Goal 7 – Affordable and Clean Energy: Currently, the main source of global energy comes from fossil fuels. These, in turn, have harmful profiles in relation to environmental and social indicators. Some research suggests that energy is the dominant contributor to climate change, accounting for around 60% of total global greenhouse gas emissions. From this perspective, large amounts of fossil fuel are needed to power heavy agricultural machinery, process food, refrigerate food during transport, produce packaging materials, and manufacture and transport chemical inputs. In addition, estimates show that the food industry is currently consuming 30% of the world's available energy with over 70% occurring beyond the farm gate and producing over 20% of the world's greenhouse gas emissions (approximately 31% of land-use change is included). Thus, ensuring the efficient use of energy is almost a matter of survival in the food sector since intelligent planning will seek the best way to use this input with energy awareness and productivity. Thus, expanding infrastructure and updating technology to provide modern and sustainable energy services through abundant and inexhaustible matrices, such as the use of solar panels and/or wind energy, will result in benefits not only reflected in the environment and society but also in exclusive gains from obtaining its own generation of renewable energy with the optimization of system costs and reduced carbon in relation to business-as-usual and solely renewable scenarios.

(iv) Goal 12 – Responsible Consumption and Production: Undeniably, the prospects for global collapse associated with the food-energy-water nexus are about to come true. Therefore, responsible consumption puts pressure on all production chains against the abusive exploitation of the workforce, child labor, deforestation, the waste of natural resources, damage to the environment, and the gas emissions that cause the greenhouse effect among so many other condemned practices. From this perspective, the food industry recognizes its leading role in global practices as it needs to circumvent the demand for food and its waste, which is combined with strategies that reduce the exploitation of natural resources in addition to being always alert to the wishes of consumers. For this, the food industries have been using a set of strategies to place products on the market with environmental responsibility, which meets the quality profile of consumers, encouraging them to join the sustainable global movement. Among some options, there is food traceability, through industry 4.0, which allows for greater operational intelligence through the automation of the production sector and enables a constant exchange of information between all stages of the production chain. Thus, through the Internet of Things (IoT), real-time access to the complete life cycle of the product will be able to help people in making daily purchasing decisions based on the transparency and reliability of food information.

(v) Goal 13 – Climate Action; Goal 14 – Life Below Water; Goal 15 – Life on Land: According to data from the Intergovernmental Panel on Climate Change (IPCC), approximately 26% of greenhouse gas emissions are attributed to agriculture, industry, and food processing activities (IPCC 2019; Mbow et al. 2019). As a result, carbon emissions from anthropogenic activities are causing ocean warming, acidification, and the loss of oxygen and marine biodiversity. However, the devastating repercussions of environmental resources also affect the soils. This is because it is currently estimated that the loss of arable land is 30 to 35 times the historical rate. To try to contain ecosystem ruin, food industries are taking steps to face extensive battles to become carbon and forest positive by providing 100% sustainable ingredients. Out of curiosity, in practice, great leaders of the global food market already have reports that include a reduction of ~-7.3% of GHG emissions, -16% of the water footprint, -8.4% of use demand and soil transformation in addition to a 21.7% increase in recycling and reuse of plastic materials (Mars 2021).

Despite all the challenges that encompass the consolidation of sustainable processes and products, the food industry has undeniably followed prosperous paths that go in accordance with the collective thinking for sustainability. However, it is evident that the trajectory has some gaps that need to be filled and polished. However, it is hoped that decision-makers will continue to engage in this race for the common good. So far, the political and economic implications are still difficult barriers to get around in a world led by industrial capitalism. However, it is important to be aware that through the union of academia and industry, associated with government policies, it will be possible to build and achieve an ideal model of sustainability in the food chain without incurring social, environmental, and economic damage.

4. Beyond Sustainability—Political Implications and Opportunities

Globalization, economic development, technological advances, and changes in consumer thinking have rapidly transformed food production systems (Turner et al. 2018). Therefore, an important common starting point among these existing changes is the collective search for sustainability.

In practice, the ultimate breadth of the call for sustainability is often questionable. Proof of this is that while many admit that truly sustainable societies are far beyond our reach, for the cultures most attracted to sustainability, balancing the three pillars is an urgent imperative that must be achieved (Blühdorn 2017). However, in a world full of relativity, sustainability is undeniably a unifying goal today.

Based on this, new perspectives are possible and necessary to change the unsustainable paradigm so far adopted by the production and consumption of food. This means acting in different dimensions of the production chain (Fig. 2), which go beyond sustainability criteria.

As can be seen in Fig. 2, the main starting point for building a sustainable structure in food production and consumption is the identification of the different sets of actors that are linked to

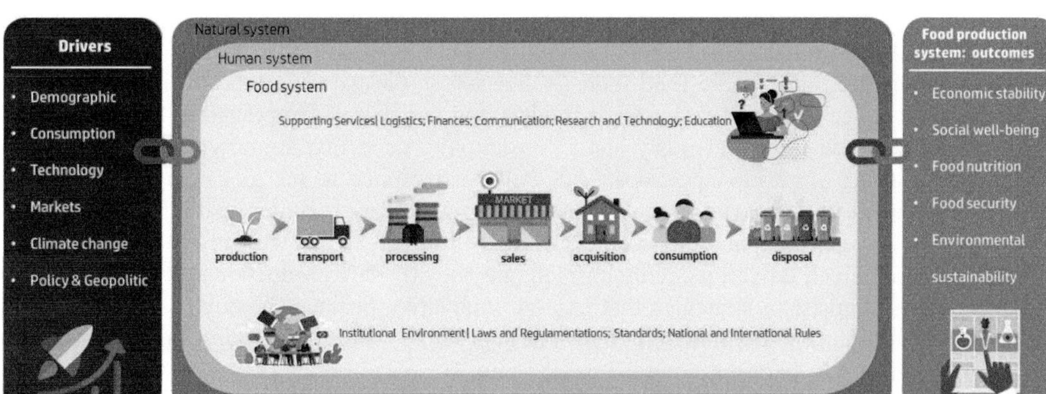

Figure 2. Interconnection Between Components and Actors of the Food Industry and its Action Drivers Towards Sustainability.

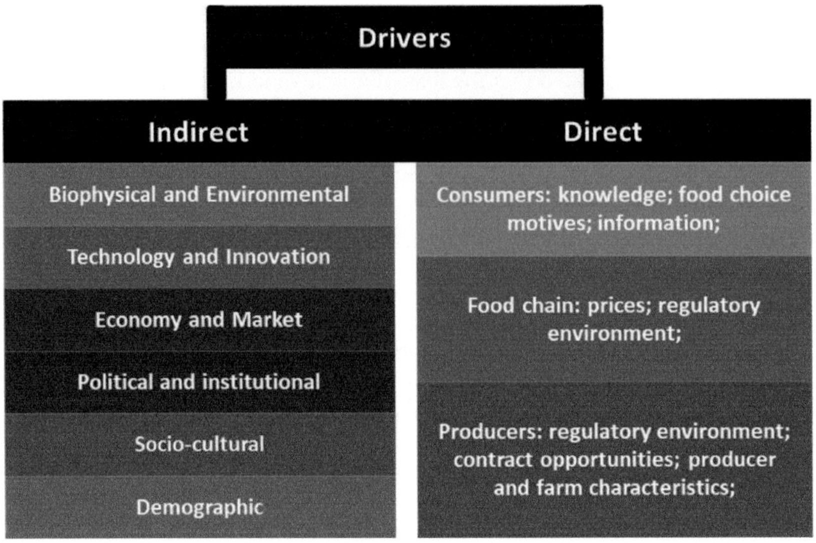

Figure 3. Indirect and Direct Drivers Applied to Food Sustainable Production.

the production system. Furthermore, it is important to characterize the direct and indirect factors (Fig. 3) that can drive the individual behavior of actors, which influence change within the production system itself (Lentz 2021). Consequently, the pre-results identified, through interactions and feedback loops that occur between the components of the process, can serve as a basis for shaping the drivers and local policies in order to reach a consensus with global policies and together move towards achieving sustainability.

Yet, one of the main challenges that need to be taken into account is the differences and diversities that comprise the globe. The extent to which joint actions are essential and everyone must be committed and work for the common interest, maximizing synergies and minimizing trade-offs, is initially led by government policies; it is beyond disseminating information and raising awareness about the benefits of sustainability. Indeed, it is necessary to understand that single-character policies are not adequate.In fact, current policy formulation is impacted by past policies and governance systems, which commonly advocated a productivity agenda, resulting in outdated projects and often presented to managements that are resistant to change (Kugelberg et al. 2021). Thus, it is of paramount importance to integrate traditional knowledge into the design of modern resource management systems (Table 1), which not only provide the basis for the sustainable concept of ecologically sound production systems but are also economically viable and highly productive (Marchetti et al. 2020).

Notoriously, not only in the food industry but across the world, a renewal in the values of products and services, as well as business models and commercial implications are being seen. Food products, which were previously just a consumption necessity, are now no longer simply a matter of satisfying a nutritional purpose, but rather correspond to the global logic. In fact, the catalytic principles that gave rise to the industrial redesign we are surfing require that a strong link be established between companies, consumers, and governments so that there is an effective formula for rebuilding efficient, environmentally correct, and socially safe production systems. Therefore, it is necessary to invest in strategies that explore innovative business models, as well as public policies and multi-stakeholder initiatives that seek to promote robust and sustainable practices in the tracking of food and its partners along the value chain.

Table 1. Tasks, Actions Towards Sustainability, and Major Research Directions From Novel Models of Sustainability Perspectives.

Proposed Tasks	**Actions Towards Sustainability**	**Major Research Directions**
Impact on food availability at the regional and local level	Providing nutritious food; encouraging local small-scale production	Performance analysis of family-based agroforestry and agro-silvo-pastoral systems; support to expand agroecology as a tool to optimize, restore and improve the productivity of small local farmers
Resource efficiency and conservation	Preserving and regenerating local resources; optimizing the performance of groecological practices and integrated agro-silvo-pastoral systems	Landscape functional ecology of biogeochemical and water cycles
Resilience to environmental variability and climate change	Strengthening farming systems' adaptability to environmental heterogeneity and constraints; introducing techniques controlling anthropogenic GHG emissions	Testing the performance and effectiveness of design and management processes in terms of production and residence times of GHGs, radiative forcing, and local warming; implementing long-term research and proxy analysis on the effects of FbtA on anthropogenic GHG emissions
Production of food crops vs. industrial and no-food crops	Controlling directly food quality; increasing responsibility and knowledge	Modeling the effectiveness of nature-based solutions for more nutritious food and better life quality; deepening agroecological methods for the progressive replacement of industrial agriculture, especially monocultures and intensive livestock systems
Shorter circuits between food production and consumption	Activating and supporting circular bioeconomy	Research on flows of components and dynamics of circular bioeconomy
Healthy and culturally appropriate food for people	Providing equal access to quality food	Perception and preferences analysis of food products; grounded research on food culture based on agroecology principles

5. Conclusions

The food industry has undergone profound changes in recent decades and is currently facing increasingly complex challenges. However, it is important to recognize that this chain, as important as it is, also plays a fundamental role in terms of sustainability principles. Thus, given the challenges mentioned throughout the chapter, and the many difficulties in homogenizing the discourse on the food industry and sustainability, the arguments described here are an attempt to delineate common flows and failures in the integration of innovation, sustainability, and responsibility by food companies, as well as a chance to illustrate how the reconfiguration of the current production model can contribute to the sustainable food industry. Therefore, to obtain a sustainable approach in the food industry, it is necessary that incentive policies make it possible to take the sustainability concept on a more integrative bias, which encourages industry and society to change their way of thinking and ensure a promising sustainable future.

References

Springmann, M., Clark, M., Mason-D'Croz, D., Wiebe, K., Bodirsky, B.L., Lassaletta, L. and Willett, W. 2018. Options for keeping the food system within environmental limits. Nature, 562(7728): 519–525.

Dawson, I.K., Park, S.E., Attwood, S.J., Jamnadass, R., Powell, W., Sunderland, T. and Carsan, S. 2019. Contributions of biodiversity to the sustainable intensification of food production. Global Food Security 21: 23–37.

Rupawalla, Z., Robinson, N., Schmidt, S., Li, S., Carruthers, S., Buisset, E. and Wolf, J. 2021. Algae biofertilisers promote sustainable food production and a circular nutrient economy—An integrated empirical-modelling study. Science of the Total Environment 796: 148913.

Kaiser, M., Goldson, S., Buklijas, T., Gluckman, P., Allen, K., Bardsley, A. and Lam, M.E. 2021. Towards Post-Pandemic Sustainable and Ethical Food Systems. Food Ethics 6(1): 1–19.

Marchetti, L., Cattivelli, V., Cocozza, C., Salbitano, F. and Marchetti, M. 2020. Beyond sustainability in food systems: perspectives from agroecology and social innovation. Sustainability 12(18): 7524.

Turner, C., Aggarwal, A., Walls, H., Herforth, A., Drewnowski, A., Coates, J. and Kadiyala, S. 2018. Concepts and critical perspectives for food environment research: A global framework with implications for action in low-and middle-income countries. Global Food Security 18: 93–101.

Blühdorn, I. 2017. Post-capitalism, post-growth, post-consumerism? Eco-political hopes beyond sustainability. Global Discourse 7(1): 42–61.

Kugelberg, S., Bartolini, F., Kanter, D.R., Milford, A.B., Pira, K., Sanz-Cobena, A. and Leip, A. 2021. Implications of a food system approach for policy agenda-setting design. Global Food Security 28: 100451.

Lentz, E.C. 2021. Food and agriculture systems foresight study: implications for gender, poverty, and nutrition. Q Open, 1(1): qoaa003.

van der Goot, A.J., Pelgrom, P.J., Berghout, J.A., Geerts, M.E., Jankowiak, L., Hardt, N.A. and Boom, R.M. 2016. Concepts for further sustainable production of foods. Journal of Food Engineering 168: 42–51.

Dwivedi, A., Agrawal, D., Jha, A., Gastaldi, M., Paul, S.K. and D'Adamo, I. 2021. Addressing the challenges to sustainable initiatives in value chain flexibility: implications for sustainable development goals. Global Journal of Flexible Systems Management 1–19. DOI: 10.1007/s40171-021-00288-4

Dantas, T.E.T., De-Souza, E.D., Destro, I.R., Hammes, G., Rodriguez, C.M.T. and Soares, S.R. 2021. How the combination of Circular Economy and Industry 4.0 can contribute towards achieving the Sustainable Development Goals. Sustainable Production and Consumption 26: 213–227. DOI: 10.1016/j.spc.2020.10.005

Yang, Y., Azari, M., Herbold, C.W., Li, M., Chen, H., Ding, X. and Gu, J.D. 2021. Activities and Metabolic Versatility of Distinct Anammox Bacteria in a Full-Scale Wastewater Treatment System. Water Research 117763. DOI: 10.1016/j.watres.2021.117763

dos Santos, A.M., Deprá, M.C., dos Santos, A.M., Cichoski, A.J., Zepka, L.Q. and Jacob-Lopes, E. 2020. Sustainability metrics on microalgae-based wastewater treatment system. Desalin. Water Treat. 185: 51–61. DOI: 10.5004/dwt.2020.25397

Mars 2021. Building a sustainable future. Accessed in August 8th, 2021 <https://www.mars.com/sustainability-plan>.

IPCC 2019. Summary for policymakers. *In*: Shukla, P.R., Skea, J., CalvoBuendia, E., Masson-Delmotte, V., Pörtner, H.-O., Roberts, D.C., Zhai, P., Slade, R., Connors, S., van Diemen, R., Ferrat, M., Haughey, E., Luz, S., Neogi, S., Pathak, M., Petzold, J., Portugal Pereira, J., Vyas, P., Huntley, E., Kissick, K., Belkacemi, M. and Malley, J. (eds.). Climate Change and Land: An IPCC special report on climate change, desertification, land degra-dation, sustainable land management, food security, and greenhousegas fluxes in terrestrial ecosystems. Paris: IPCC

Mbow, C., Rosenzweig, C, Barioni, L.G., Benton, T.G., Herrero, M., Krishnapillai, M., Liwenga, E., Pradhan, P., Rivera-Ferre, M.G., Sapkota, T., Tubiello, F.N. and Xu, Y. 2019. Food security. *In*: Shukla, P.R., Skea, J., Calvo Buendia, E., Masson-Delmotte, V., Pörtner, H.-O., Roberts, D.C., Zhai, P., Slade, R., Connors, S., van Diemen, R., Ferrat, M., Haughey, E., Luz, S., Neogi, S., Pathak, M., Petzold, J., Portugal Pereira, J., Vyas, P., Huntley, E., Kissick, K., Belkacemi, M., Malley, J. (eds.). Climate Change and Land: An IPCC special report on climate change, desertification, land degradation, sustainable land management, food security, and greenhouse gas fluxes in terrestrial ecosystem.

Mbow, C., Rosenzweig, C, Barioni, L.G., Benton, T.G., Herrero, M., Krishnapillai, M., Liwenga, E., Pradhan, P., Rivera-Ferre, M.G., Sapkota, T., Tubiello, F.N. and Xu, Y. 2019. Food security. *In*: Shukla, P.R., Skea, J., Calvo Buendia, E., Masson-Delmotte, V., Pörtner, H.-O., Roberts, D.C., Zhai, P., Slade, R., Connors, S., van Diemen, R., Ferrat, M., Haughey, E., Luz, S., Neogi, S., Pathak, M., Petzold, J., Portugal Pereira, J., Vyas, P., Huntley, E., Kissick, K., Belkacemi, M. and Malley, J. (eds.). Climate Change and Land: An IPCC special report on climate change, desertification, land degradation, sustainable land management, food security, and greenhouse gas fluxes in terrestrial ecosystem. https://www.ipcc.ch/site/assets/uploads/2019/11/08_Chapter-5.pdf

3

Recent Changes in the Food Supply Chain

Driving to Sustainability

Hêriş Golpîra

1. Introduction

Food—a basic human need and one of the pillars on which human society is built—has always been a product that has received worldwide attention. It also has a significant share in the regional and global economy. For instance, the share of the food and beverage industry in the EU's gross domestic product (GDP) is about 8%. This share was 10.7% in France in 2011 and 15.8% in China in 2010 (Zhu et al. 2018). The food industry in the United Kingdom (UK), with an 8% share of GDP, also accounts for 12.7% of the labor market (Fryer and Versteeg 2008). India's food industry, as the fifth-largest economy in terms of nominal GDP, is also the largest employer in the country (Bajpai, 2020; Memon et al. 2021). And the agriculture sector which accounts for over 25% of Iran's GDP involves about 33% of its employment (Rashidhalam 2019). In general, agriculture accounts for a larger share of GDP in developing countries, especially the poorer ones, which are also major exporters of agricultural products. However, the focus in developed countries in the food industry is more on the retail and processing stages (Sexton et al. 2007). Such a vertical food supply chain (FSC) should be even more in focus if we know that the food supply should increase by about 70% compared to its current level to meet the needs of the projected population of 9.6 billion in 2050 (Krishnan et al. 2020) with an increase of 49% to 70% of urbanization (DESA, 2018).

Rising food demand for the growing population has made the FSC a growing integrated business sector that dates back to the time when people first settled in Mesopotamia and the lower Egypt region (Gharehgozli et al. 2017). It, however, leads to higher prices and market fluctuations and imposes a greater burden on scarce natural resources, such as clean water, land, and energy (Dobbs et al. 2011; Parfitt et al. 2010). More specifically, the rise of bio-energy markets, which are based on using food crops, further threatens food security and puts more pressure on scarce agricultural resources. This is while such limitations as temperature, terrain, and soil and water qualities have made only a limited portion of the land suitable for agricultural use. Even the productivity of existing agricultural land has been affected by economic and political factors, including population growth, climate, and land-use patterns. This is while the rapid process of urbanization is constantly pushing for access to

Department of Industrial Engineering, Sanandaj Branch, Islamic Azad University, Sanandaj, Iran.
Email: herishgolpira@gmail.com

arable land (Khan et al. 2021). These are major threats to the sustainable development of the FSC, which covers all the aspects from farm to fork, including production/manufacturing, packaging, distributing, storing, and further processing or cooking for consumption (Chen et al. 2020).

In addition to general supply chain (SC) management considerations, FSCs—whether fresh agricultural products or processed food products—often have to deal naturally with issues related to perishability, product degradation and waste (Amorin et al. 2012; Göbel et al. 2015), strict governmental regulations on the safety, consumers' high variation on tastes and processes, and consequential operational constraints on their storage, processing, and distribution (Zhu et al. 2018). National attention to recent cases of contamination in fresh produce, changes in consumer awareness of health and want of accurate information about the chain, and new market-based regulations and standards have made the FSC design and operations more complex. The complexity is exacerbated when the SC becomes a global SC compassing two or more countries (Ahumada and Villalobos, 2009). More specifically, the globalized food system aggravates public concern about its sustainability due to the rising negative operational impacts on the environment, society, and economy (Yakovleva 2007). Such characteristics, in addition to the special attention of the United Nations (UN) 2030 Agenda (United-Nations 2016), enhance the attractiveness of driving toward sustainability in the FSC design and management. This is the field that the research at hand tries to study.

Previous review studies have focused on aspects, such as traceability, especially in food processing industries, food safety and consistency, and performance measurement. Sustainability, however, has rarely been considered as a means of integrating these issues. Through a systematic review, Dania et al. (2018) have extracted 10 key behavioral factors to enable an effective collaboration system for sustainable agri-food SC management, such as joint efforts, sharing activities, collaboration value, adaptation, trust, commitment, power, continuous improvement, coordination, and stability. Palazzo and Vollero (2021) have also reviewed the relevant research from 1997 to early 2021 to uncover research perspectives, gaps, and trends in this regard. The research shows that most of the research works in the field of FSC were published between 2014 and 2021. This is probably due to the increasing awareness of environmental issues and the need to reduce global hunger during this period. Because zero hunger and good health and well-being are defined as the second and third goals of the sustainable development goals of the 2030 Agenda. Chiaraluce et al. (2021) have performed a literature review only on the effects of Circular Economy (CE) approaches on the agri-food sector. This study with a special focus on the reuse and evaluation of waste and by-products shows the special importance of the issue and its continued development in the scientific community led in Europe. Through another systematic review, Chan et al. (2020) have identified four drivers for FSCs as follows:

1) Food security governance, especially for agriculture, logistics, retailers, and restaurants.
2) Effective management of the input resources, e.g., water, biosafety materials, and energy via various innovative interventions.
3) Output management, especially on food waste reduction and Greenhouse Gas (GHG) reduction.
4) Labor income gained through fair trade, which directly impacts farmworkers.

Quality management was also identified as a hypothetic generic intervention that could mediate the effect of these familiar drivers toward food security in the FSC. Some aspects such as the development of new approaches as well as the use of advanced Information Technology (IT) are not considered in these research works. Along with Golpîra et al. (2021), Rana et al. (2021) have provided a review of the applications of the blockchain in agri-food SC sustainability, providing a systematic perspective and research directions to address unresolved issues in this area. They highlight the important and influential role of governments, along with intergovernmental organizations and researchers, in providing policy guidance for sustainable FSCs. In doing so, they suggest introducing governmental and intergovernmental initiatives, strengthening public and private partnerships, creating platforms that integrate new technologies, including blockchain, the

Internet-of-Things (IoT), and artificial intelligence, and ultimately developing digital infrastructure and skills among the chain participants.

As can be seen, in none of the research works, the drivers of the sustainable FSC have been studied comprehensively and categorically. This research has been done to fill this research gap. In summary, this research seeks to answer these research questions: 1) what are the drivers of the sustainable FSC?; 2) how can these drivers be effectively classified?; 3) based on the study of the existing literature, what orientations in the future of the FSC, focusing on the studied drivers, may be considered by researchers and practitioners?

To answer these questions, an extensive review of 144 research papers that are conducted based on research published in Web of Science, by considering the keyword "Sustainable Food Supply Chain" in their title, is carried out in the research. The papers are studied and the drivers that have been highlighted are extracted and classified. In this way, a condensed summary of the broad field of FSC drivers is provided, especially for those readers who are not fully familiar with the diversity of the field. Beyond giving an overview, some current trends in project scheduling research are discussed as well. The rest of the paper is as follows: existing trends in sustainable FSC studies are graphically provided and further analyzed in Section 2; it is followed by a detailed literature review reported in Section 3, and the conclusion and some future potentials are introduced in Section 4.

2. Trends in Sustainable Food Supply Chain

As one can see in Fig. 1, the trend of paying attention to the concept of the sustainable FSC has been increasing from 2008 to 2021 with a growing trend of more than two papers per year. Another noteworthy point that can be seen in this figure is the significant growth of researchers' attention to the FSC in 2021 compared to previous years. This rapid growth can be justified by the identification of the first coronavirus in November 2019 in Wuhan, China; however, the outbreak of a pandemic is not new. Throughout history, humans have faced many epidemics. The common denominator of all epidemics has been their serious negative effects on the global economy, in general, and, more specifically, the FSC as one of the most important sectors of the economy. The COVID-19 epidemic virus is no exception to this rule and affects the entire process of food supply from the

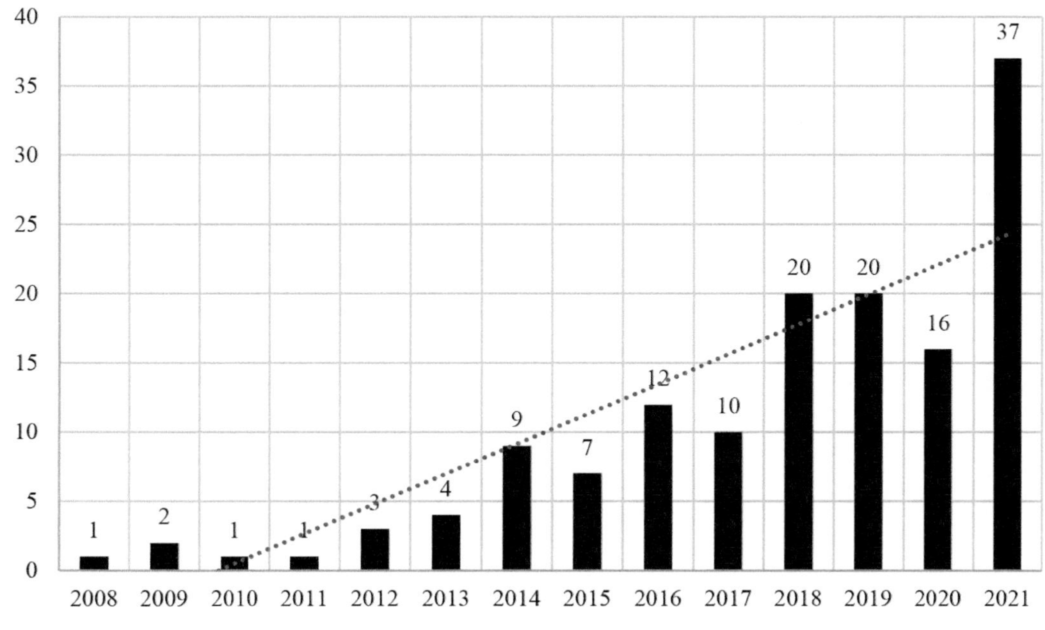

Figure 1. Distribution of Publications Per Year Through the Study Period (2008–2021).

field to the consumer. COVID-19 has led to restrictions on the movement of workers, changes in consumer demand, closures of food production facilities, restrictions on food trade policies, and financial pressures on the FSC. This epidemic has affected governments' approaches and policies in supporting all components of the chain, starting from the top to the bottom. Also, policy-making to prevent food price increases in the face of an epidemic has been on the agenda of governments (Aday and Aday, 2020). All of this has put more emphasis on the FSC, and more research has been done accordingly. These research works, as shown in Fig. 2, have been conducted mostly in the fields of environmental science and engineering. And given that integrating SCs with multiple and often contradictory goals in the presence of various constraints, and sometimes various uncertainties and risks without the use of Operations Research (OR) approaches is difficult and somewhat impossible (Golpîra et al. 2021; Golpîra et al. 2022), this branch of science has been of particular interest to researchers in different countries as shown in Fig. 3. As can be seen in the figure, researchers in a wide range of countries have researched the sustainable FSC. This range includes 1) developing countries, such as Iran, China, etc.; some are also underdeveloped in logistics and SC, are highly

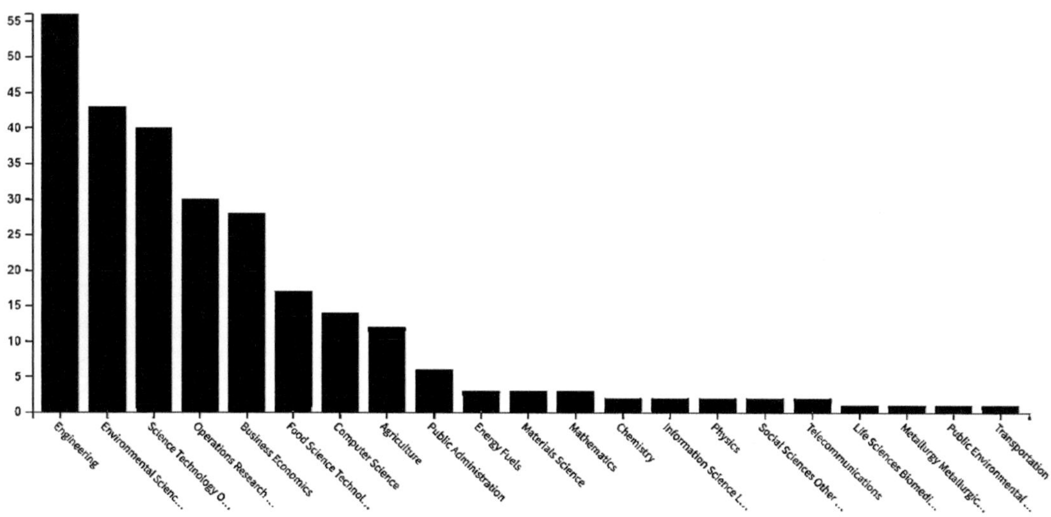

Figure 2. Distribution of Publications in Top Research Areas With the Highest Number of Papers on the Sustainable FSC (2008–2021).

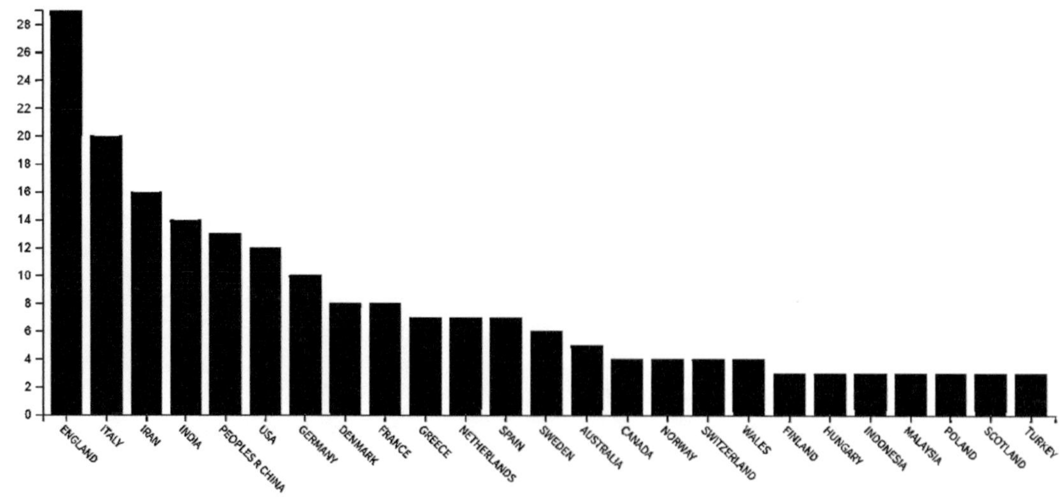

Figure 3. Distribution of Publications in 25 Countries With the Highest Number of Papers on the FSC (2008–2021).

populated and are even embodied with poor institutional governance, facing a higher unemployment rate, poverty, undernourishment, and crime, and 2) developed countries such as the United States (US) and Italy. In addition, this research area has been widely conducted not only in populous countries, such as China, India, and the US, but also in less populous countries, such as Greece. The subject becomes even more interesting if we know that since 1974, food waste in, for example, the US has continuously increased (Posmanik et al. 2017), while multiple sludges are also generated by around 13.84 million tons a year (Seiple et al. 2017). China annually generates 37–62% of municipal solid waste (Ltd FBIC 2017); average annual multiple sludges underwent a growth rate of 13% from 2007 to 2013 (Yang et al. 2015). The European Union (EU) generates around 88 million tons of food waste and around 11 million tons of multiple sludges a year (Scherhaufer et al. 2018; EUROSTAT 2017), and India generates around 6.24–35 million tons a year (Hafid et al. 2017). Tracking systems using blockchain technologies offer benefits to minimize waste during the production phase (Mangla et al. 2021). It is also capable of reducing food waste by improving SC efficiency and preventing food recalls (Li et al. 2021). By tracing food along with the SC, the level of safety of perishable food products is measured, and the path where they are grown, handled, and stored, and the conditions under which they are transported or processed is controlled. So, it is expected that using such tracking technology as blockchain lead to the development of a transparent and authentic chain of records of the food ecosystem (Kayikci et al. 2020). For example, Using blockchain technology has been able to help Walmart manage food waste and prevent food fraud (Kouhizadeh et al. 2020).

So, despite that every year more than 200 billion tons of food are shipped 35% by land, 60% by sea, and 5% by air, globally (Ackerley et al. 2010; Bendickson 2007), the main emphasis of researchers in this field, as shown in Fig. 4, has shifted from the field of energy and transportation to tracking systems based on blockchain technology in recent years. This issue is also emphasized and confirmed in Fig. 2, according to the number of papers that have been published in these areas. This is while the average kilometers traveled by food in the 1980s increased by 25% in the US and by 50% in the UK to an average of 2,500–4,000 kilometers at the beginning of the 21st century (Halweil 2002). Food transport negatively affects logistics costs and the environment. It increases the risk of food not being able to meet the quality, safety, and traceability requirements (Gharehgozli et al. 2017), thus leading to more waste as one of the important impacts of food long-time transport (Scholz et al. 2009). So the areas of food logistics and transportation should not be neglected. Accordingly, Parker et al. (2010) and Blackburn and Scudder (2009) have returned the physical and cost losses to the transportation and storage harvested crop quality deterioration. Widodo et al. (2006)have estimated that 20% to 60% of agricultural and food waste occurs during the harvest phase. Hence, mere attention to waste management in agriculture and food is not enough. And in recent years, beyond the concept of waste management, a more comprehensive concept called

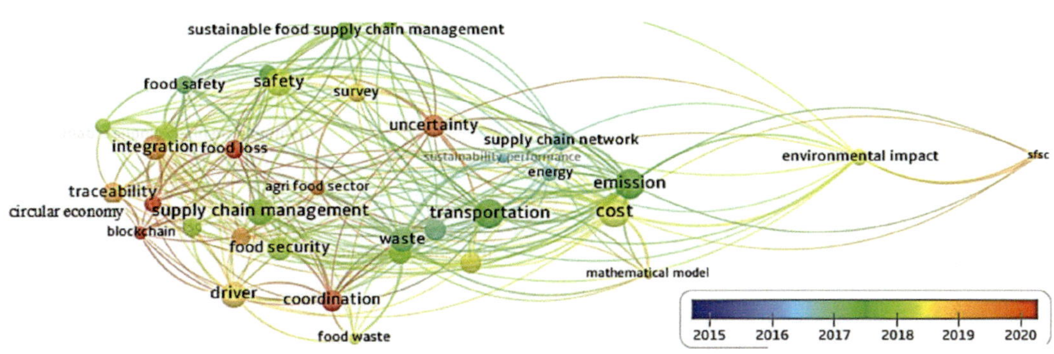

Figure 4. Keywords Co-Occurrence Network Created by VOSviewer (2008–2021).

CE in the field of FSC integration has been addressed. This further emphasizes the need to use mathematical optimization techniques and OR techniques for the sustainable FSC network design, which was mentioned earlier in Fig. 2. This is because the ability to integrate SC networks is generally created by using such approaches. They are also able to take into account the objective and subjective uncertainties that have become a trend in recent years, as shown in Fig. 4.

3. Detailed Literature Review and Concept Classification

According to what was examined in the previous section, through a comprehensive review, a novel comprehensive classification for the drivers of sustainable FSC is presented in this study. As one can see in Fig. 5, based on the literature, the drivers of sustainable FSC can be classified into three categories named the golden triangle of the sustainable FSC. The triangle includes the circularity point as the first vertex, digitalization as the second vertex, and socio-economics as the third vertex. These are considered the main drivers of sustainability in the FSC, which are completely discussed in the following subsections.

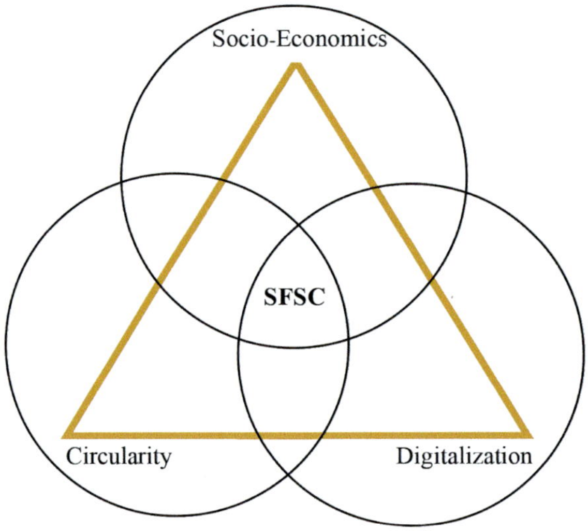

Figure 5. The Proposed Golden Triangle of the Sustainable FSC

3.1 Circularity

Responsible production and consumption, as well as availability and sustainable management of clean water and sanitation, are sustainable development goals called by the UN Foundation to protect the planet (SDG 2015). In this regard, reducing the environmental and climate footprint of current food systems has been recommended by the European Commission in the recently released Agriculture and Green Deal Report (2019). Circularity can help in doing so by restructuring the natural process, reducing losses, and recirculating and reusing materials and resources after their initial use, Verma and Jones (2021) aiming at reducing emissions, energy consumption, and waste (Atabaki et al. 2020). By this means, the CE is a strategy to deal with the unsustainable use of resources that empowers businesses to pay special attention to conserving them by better understanding natural inputs. Applying the principles of CE, such as waste evaluation, cost reduction, and greater resource efficiency, offers significant opportunities to reduce food waste and food loss (Garcia-Garcia et al. 2019; Kusumowardani et al. 2022; Prieto-Sandoval et al. 2019; Rizos et al. 2016), which accounts for 30% of the world's total food production (Sethi et al. 2020). The waste leads to a loss of resources, energy, and time invested in food production (Krishnan et al. 2020), therefore it can increase food prices due to reduced supply that reduces access to affordable food (Shafiee-Jood and Cai 2016). This wastage, on the other hand, is one of the main causes of GHG emissions (Schanes et al. 2018). These gases arise from the decomposition of food at landfills and the emissions produced in the growth, transportation, and retailing of food that is ultimately not consumed and wasted (Segre et al. 2014). Since such CE solutions, like closing the loop in the supply chain, do not always result by default in emission reductions (Gallego-Schmid et al. 2020), offering carbon credits can be considered as an efficient policy to directly reduce emissions (Katsikouli et al. 2021). So, environmental, policy and economy, and financial benefits are the three top drivers of circularity in the FSC. However, institutional, financial, and technological risks are the top three barriers in implementing CE practices in the FSC (Mehmood et al. 2021). Applying the circularity in the FSC may be even more challenging due to poor government policies, lack of farmers' knowledge and awareness, and lack of technology and techniques (Sharma et al. 2019). Accordingly, a blockchain-based database, in line with establishing digital agriculture, can be a step toward a reliable and efficient way to manage and verify farm-specific efforts in this regard.

3.2 Digital Agriculture

In line with moving toward sustainability, digitization (agriculture 4.0) and autonomous vehicles (agriculture 5.0) are by themselves capable of facilitating the journey from the current linear food and agricultural systems to the circular one (SmartAgrHub 2019; Precision Plants 2019; Agricultural Robotics 2018; Migliaccio et al. 2021; Basso et al. 2021). Such kinds of technological optimism of eco-modernity in line with the CE with a special focus on efficiency allow for simultaneous environmental protection, efficiency gain, and economic growth toward sustainability in the entire production and SC (Isenhour et al. 2022; Bhat 2021). Digitization of the SC and its management, from input to product transfer to markets and through cooperatives and digital companies, may reduce production costs and efficiency losses due to the intervention of brokers and traders. For example, digital innovation centers, such as FIWARE, OVH, Google Cloud, AgROBOfood, or Amazon AWS, act as facilitators for the introduction of smart agricultural products (SmartAgrHub 2019). Blockchain-based digital marketing technology used by Walmart seeks to shorten the SC and increase productivity and sustainability by connecting farmers to consumers to track the product supply and market locations in the e-grocery at the SC downstream (Trendov et al. 2019). At the SC upstream, since the quality and quantity of crops are highly dependent on climate, topography, and soil characteristics and are affected by available nutrients, soil types and health, insect resistance, irrigation quality and quantity, and site-specific analysis or technique is needed to obtain optimal yields. In addition to increasing crop production, farmers have to do multiple, mixed crops, year-round, or intensive crops. Hence, producers and farmers need the most advanced technologies and service-based techniques to be able to produce more food with limited land and resources. In doing so they must monitor their arable land on a near-real-time scale and move toward smart or digital agriculture (Sisinni et al. 2018). The IoT technology (Club of Bologna, 2018), cloud computing (Bosch 2019), networking (AEF ISOBUS 2020; Agritechnica 2019), smart mobile devices (AgGateway 2020; QuickConnect 2019), blockchain and QR technologies (Katsikouli et al. 2021), field robots and drones (hogan et al. 2017; Feldschwarm 2019; Koparam et al. 2020; Naio-Technologies 2019), satellites and GPS-computer mapping (USDA 2016), wireless sensor networks (National Academic Press 2019; Saiz-Rubio and Rovira-Más 2020), and tracked and automated steering systems (Fendt 2019; Deere 2019; Real Agriculture 2019) are all aiming at sustainability in the agri-food SC. Many such advanced technologies as the IoT play a significant role in reducing the usage of the chemicals, which are hazardous to human health (Stein et al. 2017; Wietzke et al. 2018). Relying on such an ICT-based technology as variable-rate technology (VRT) may increase profit and decrease environmental impact (Koparan et al. 2020; FAO 2019; EasyMix 2019; Rauch 2019; Kuhn 2019; Biooekonomie 2018). The adoption rate for fertilizers, soil amendments, seed, or plant protection chemicals in the US, UK, Australia (Lowenberg-DeBoer et al. 2019), China (Zhao and Huang, 2011), and Turkey (FAO 2019) for cereals from 2009 to 2012 was 14%, 31%, 15%,

0%, and 0%, respectively. VRT adoption rate for fertilizers in Brazil was 89% in 2012, Argentina 42% in 2018, and Canada 67% in 2015 (FAo 2019). VRT has been adopted by 19% of corn farms and 28% of cropland acres by 2010 (USDA 2016) and 11% of winter wheat in 2009 (Lowenverg-DeBoer et al. 2019). GPS-guided or auto-steered combine harvesters (Club of Bologna 2018; USDA 2016; Deere 2019), robotized vehicles (Harvest CROO 2020), and autonomous field modules (Feldschwarm, 2019) represent efforts to make farming resources more efficient. So, the adoption rate of GPS-guided and auto-steered tractors and combine harvesters in Nebraska and Canada in 2015 was 71% and 67%, respectively (Trendov et al. 2019). Such digitizations, in addition to having a direct impact on sustainability, also indirectly affect sustainability through the socio-economic benefits that come with improving the job profile of farmers and related occupations (Walter et al. 2017; Benyam et al. 2021).

3.3 Socioeconomic

Not only environmental issues but also socio-economic perspectives have been discussed in the context of the UN Millennium and Sustainable Development Goals (UN 2014). While the emphasis on environmental sustainability is on natural resource management, social sustainability is related to managing such resources as (1) internal human resources (health and safety), (2) external population (those affecting resource scarcity), and (3) stakeholder participation (those related to their expectations of companies' decision-making and macro-social performance in socio-economic and socio-environmental issues) (Ashby et al. 2012). Lower socio-economic status and halted economic activities may translate into poverty, hunger, crime, and undernourishment that negatively affect sustainability goals (Khan et al. 2021; Vogliano et al. 2021). Accordingly, any organization must explicitly link environmental, social, and economic goals, despite the conflicts they may have with each other, in a broader strategic perspective to ensure that it moves toward sustainability through the chain (Carter and Rogers 2008).

By 2050, the urban population is projected to be 2.4 billion more than the current urban population with higher mortality rates and shorter life expectancy in rural areas. Therefore, food, nutrition, and agriculture are expected to be directly affected by this increase in urbanization, which creates more logistical challenges (Gharehgozli et al. 2017). Population growth and urbanization mean that food production and consumption are no longer limited to small communities and rural areas, but demand increases and is transferred to urban areas. This change brings with it newly operational, tactical, and strategic challenges to food production and transportation that require academic research. Urbanists use their higher income to demand more processed foods, more fruits, vegetables, and better quality meat as part of their varied diets (Shariff et al. 2015; Drewnowski and Eichelsdoerfer 2010). Such a change in consumption patterns will change jobs along the FSC. More people work in transportation, wholesale, retail, food processing, and sales instead of agriculture. Therefore, traditional methods of food production and distribution are no longer responsive and local farms are no longer food producers. The strategic level of the global food transport fleet is changing with production in large GHGs and livestock centers, leading to changes in operational levels in routing and planning. Pollution, noise disturbances, traffic congestion, and safety problems in cities may increase to the point that local governments are forced to take measures such as limited-time access restrictions to reduce these problems. Creating these time windows creates conflicting interests and goals with stakeholders in the process and planning window policies that are consistent with municipalities' goals while enhancing environmental sustainability and distribution efficiency is challenging (Akyol and De Koster 2007). Despite few studies on urban freight transport (Quak and De Koster 2007; 2009), there is not much research on this topic, especially research focusing on the features of the digitalized FSC.

The increasing urbanization and economic development may also lead to a significant rise in municipal solid waste worldwide, which includes degradable organic waste such as food waste and multiple sludges (Kumar and Samadder 2020). The widely used disposal approaches have been incineration, landfill, and composting. However, none of these approaches are compatible with the CE concept (Baldi et al. 2019). The other approach is anaerobic digestion. It involves a series of microorganic actions that decompose the organic waste to produce biogas. It, despite its low efficiency (Ratanatamskul et al. 2014), can avoid the environmental pollution problems faced during the above treatment methods, such as intensive energy consumption, high GHG emissions, waste of land resources, and soil and water contamination (Kumar an samadder 2020; Singh et al. 2019; Yang et al. 2019).

As can be seen, the complexity of the interactions and interdependencies of the human-nature system makes it difficult to understand the feedback, tipping points, side effects, trade-offs, and benefits of ecosystem services from socio-economic factors (Ausseil et al. 2019). However, science and technology, trends in markets, policies, traditions, and customs relating to land tenure and land use, labor and food safety, and the values of individuals can be stated as the socioeconomic drivers of the FSC toward sustainability (Aramyan et al. 2021; Borman et al. 2022). Due to the changes in consumer lifestyles and preferences, income growth, diet changes, and increasing demand for branded, further processed foods focusing more on product quality has also become extremely important and is signaled by a firm's brand name toward sustainability (Hobbs and Young 2001). And finally, factors such as people's willingness to participate in urban agriculture, whether as active producers and consumers or passively approving urban farming practices in their neighborhood, and the dynamics of both types of acceptance over time can be other drivers toward sustainable FSC (Rich et al. 2018).

4. Conclusion

The FSC involves the supply, production, and distribution of plant and animal products from farm to fork. As a dynamic and complex system, FSC experiences the process of globalization through specialization, intensification, global sourcing and marketing, integrated processes, and homogenization of food production and consumption. However, environmental and social concerns exacerbate the negative effects on FSC sustainability due to new changes in FSC technology, production and marketing, consumption trends and demand evolution, and demographic change.

In this chapter, key indicators of FSC sustainability including environmental indicators, social factors, and economic indicators have been studied. The main drivers have been categorized as social economy, CE, and digitization in a new feature called the FSC Golden Triangle. Recent changes in the FSC have been examined with a focus on sustainability drivers. New trends in FSC are discussed in detail and the most important factors related to sustainability have also been explained.

References

Ackerley, N., Sertkaya, A. and Lange, R. 2010. Food transportation safety: characterizing risks and controls by use of expert. Food Protection Trends 30(4): 212–222.

Aday, S. and Aday, M.S. 2020. Impact of COVID-19 on the food supply chain. Food Quality and Safety 4(4): 167–180.

AEF ISOBUS, 2020. Available online: https://www.aef-online.org/about-us/isobus.html. [Internet]. 2020.

AgGateway, 2020. Available online: https://aggateway.org/AboutUs/Mission.aspx [Internet]. 2020.

Agricultural robotics: the future of robotic agriculture. Available online: https://arxiv.org/pdf/1806.06762. [Internet]. 2018.

Agriculture and Green Deal Report. https://ec.europa.eu/info/strategy/priorities-2019-2024/european-green-deal/agriculture-and-green-deal_en. [Internet]. 2019.

Agritechnica, 2019. Improve your performance with IsoMatch FarmCentre. Available online: https://agritechnica.kvernelandgroup.com/Kverneland-News/Improve-your-performance-with-IsoMatch-FarmCentre [Internet]. 2019.

Ahumada, O. and Villalobos, J.R. 2009. Application of planning models in the agri-food supply chain: A review. European Journal of Operational Research 196(1): 1–20.

Akyol, D.E. and De Koster, R.B. 2013. Non-dominated Time-Window Policies in City Distribution. Production and Operations Management 22(3): 739–751.

Amorim, P., Günther, H.-O. and Almada-Lobo, B. 2012. Multi-objective integrated production and distribution planning of perishable products. International Journal of Production Economics 138(1): 89–101.

Aramyan, L.H., Beekman, G., Galama, J. et al. 2021. Moving from niche to Norm: Lessons from food waste initiatives. Sustainability 13(14): 7667.

Ashby, A., Leat, M. and Hudson-Smith, M. 2012. Making connections: a review of supply chain management and sustainability literature. Supply Chain Management: An International Journal.

Atabaki, M.S., Mohammadi, M. and Naderi, B. 2020. New robust optimization models for closed-loop supply chain of durable products: Towards a circular economy. Computers & Industrial Engineering 146: 106520.

Ausseil, A.-G.E., Daigneault, A.J., Frame, B. et al. 2019. Towards an integrated assessment of climate and socio-economic change impacts and implications in New Zealand. Environmental Modelling & Software 119: 1–20.

Bajpai, P. 2020. The 5 Largest Economies in the World and their Growth in 2020. Nasdaq, https://www nasdaq com/articles/ the-5-largest-economies-in-the-world-and-their-growth-in-2020-2020-01-22 tarihinden. 30: 2020.

Baldi, F., Pecorini, I. and Iannelli, R. 2019. Comparison of single-stage and two-stage anaerobic co-digestion of food waste and activated sludge for hydrogen and methane production. Renewable Energy 143: 1755–1765.

Basso, B., Jones, J., Richard, T. et al. 2021. Designing Circularity into Corn-Soybean Systems. Resource Magazine 28(2): 10–14.

Bendickson, N. 2007. Transportation and food distribution security. TransActions, NAOSH Week 15–17.

Benyam, A.A., Soma, T. and Fraser, E. 2021. Digital agricultural technologies for food loss and waste prevention and reduction: Global trends, adoption opportunities and barriers. Journal of Cleaner Production 323: 129099.

Bhat, R. 2021. Sustainability challenges in the valorization of agri-food wastes and by-products. Valorization of Agri-Food Wastes and By-Products: Elsevier, pp. 1–27.

Biooekonomie, 2018. Digitization in agriculture—from precision farming to farming 4.0. Available online: https://www. biooekonomie-bw.de/en/articles/dossiers/digitisation-in-agriculture-from-precision-farming-to-farming-40 [Internet]. 2018.

Blackburn, J. and Scudder, G. 2009. Supply chain strategies for perishable products: the case of fresh produce. Production and Operations Management 18(2): 129–137.

Borman, G.D., de Boef, W.S., Dirks, F. et al. 2022. Putting food systems thinking into practice: Integrating agricultural sectors into a multi-level analytical framework. Global Food Security 32: 100591.

Bosch, 2019. IoT-based smart irrigation. Available online: https://www.bosch.com/stories/iot-based-smart-irrigation-system/ [Internet]. 2019.

Carter, C.R. and Rogers, D.S. 2008. A framework of sustainable supply chain management: moving toward new theory. International Journal of Physical Distribution & Logistics Management. 2008.

Chan, H.Y., Abdul Halim-Lim, S., Tan, T.B. et al. 2020. Exploring the Drivers and the Interventions towards Sustainable Food Security in the Food Supply Chain. Sustainability 12(19): 7890.

Chen, S., Brahma, S., Mackay, J. et al. 2020. The role of smart packaging system in food supply chain. Journal of Food Science 85(3): 517–525.

Chiaraluce, G., Bentivoglio, D. and Finco, A. 2021. Circular economy for a sustainable agri-food supply chain: A review for current trends and future pathways. Sustainability 13(16): 9294.

Club of Bolgna, 2018. Combine harvester with driving assist function. Available online: https://www.clubofbologna.org/ew/ ew_proceedings/2019_PPTX_S1.3_Okamoto_P.pdf [Internet]. 2018.

Dania, W.A.P., Xing, K. and Amer, Y. 2018. Collaboration behavioural factors for sustainable agri-food supply chains: A systematic review. Journal of Cleaner Production 186: 851–864.

Deere, 2019. Proactive throughput controller. Available online: https://www.deere.com/international/en/campaigns/ag-turf/ agritechnica-2019/. [Internet]. 2019.

DESA, U. 2018. 68% of the world population projected to live in urban areas by 2050, says UN. United Nafions Department of Economic and Social Affairs.

Dobbs, R., Oppenheim, J., Thompson, F. et al. M. 2011. Resource revolution: Meeting the world's energy, materials, food, and water needs, McKinsey Global Institute.

Drewnowski, A. and Eichelsdoerfer, P. 2010. Can low-income Americans afford a healthy diet? Nutrition Today 44(6): 246.

EasyMix. Available online: https://go.amazone.de/go2020/agritechnica/2019/neuheiten-en-us/duengetechnik-en-us/ easymix-en-us/ [Internet]. Amazonen-Werke H. Dreyer GmbH & Co.: 2019.

EUROSTAT. Environment and energy, environment, water, water statistics on national level, sewage sludge production and disposal 2017. Available from: http://ec.europa.eu/eurostat/tgm/refreshTableAction. do?tab=table&plugin=1&pcode=ten00030&language=en

FAO, 2019. Digital technologies in agriculture and rural areas—Status report. Food Agricultural Organization of the United Nations, Rome. 157 pp. Licence: cc by-nc-sa 3.0 igo. [Internet]. 2019.

Feldschwarm-autonomous field modules for resource-efficient agriculture. Available online: http://www.feldschwarm.de/ [Internet]. 2019.

Fendt, 2019. Global launch of the new Fendt tracked tractors: Fendt 900 Vario MT and Fendt 1100 MT. Available online: https://www.fendt.com/int/fendt-900-vario-mt-and-fendt-1100-mt. [Internet]. 2019.

Fryer, P.J. and Versteeg, C. 2008. Processing technology innovation in the food industry. Innovation. 10(1): 74–90.

Gallego-Schmid, A., Chen, H.-M., Sharmina, M. et al. 2020. Links between circular economy and climate change mitigation in the built environment. Journal of Cleaner Production 260: 121115.

Garcia-Garcia, G., Stone, J. and Rahimifard, S. 2019. Opportunities for waste valorisation in the food industry–A case study with four UK food manufacturers. Journal of Cleaner Production 211: 1339–1356.

Gharehgozli, A., Iakovou, E., Chang, Y. et al. 2017. Trends in global E-food supply chain and implications for transport: Literature review and research directions. Research in transportation business & management. 25: 2–14.
Göbel, C., Langen, N., Blumenthal, A. et al. 2015. Cutting food waste through cooperation along the food supply chain. Sustainability 7(2): 1429–1445.
Golpîra, H. and Javanmardan, A. 2021. Decentralized Decision System for Closed-Loop Supply Chain: A Bi-Level Multi-Objective Risk-Based Robust Optimization Approach. Computers & Chemical Engineering 154: 107472.
Golpîra, H. and Javanmardan, A. 2022. Robust Optimization of Sustainable Closed-Loop Supply Chain Considering Carbon Emission Schemes. Sustainable Production and Consumption.
Golpîra, H., Khan, S.A.R. and Safaeipour, S. 2021. A review of logistics internet-of-things: Current trends and scope for future research. Journal of Industrial Information Integration 100194.
Hafid, H.S., Shah, U.K.M., Baharuddin, A.S. et al. 2017. Feasibility of using kitchen waste as future substrate for bioethanol production: a review. Renewable and Sustainable Energy Reviews 74: 671–686.
Halweil, B. 2002. Home grown: the case for local food in a global market. Vol. 163. Worldwatch Institute.
Harvest CROO robotics. 2020. Available online: https://harvestcroo.com/. [Internet].
Hobbs, J.E. and Young, L.M. 2001. Vertical linkages in agri-foods supply chains in Canada and the United States. Citeseer.
Hogan, S., Kelly, M., Stark, B. et al. 2017. Unmanned aerial systems for agriculture and natural resources. California Agriculture 71(1): 5–14.
Isenhour, C., Haedicke, M., Berry, B. et al. 2022. Toxicants, entanglement, and mitigation in New England's emerging circular economy for food waste. Journal of Environmental Studies and Sciences 1–13.
Katsikouli, P., Wilde, A.S., Dragoni, N. et al. 2021. On the benefits and challenges of blockchains for managing food supply chains. Journal of the Science of Food and Agriculture 101(6): 2175–2181.
Kayikci, Y., Subramanian, N., Dora, M. et al. 2020. Food supply chain in the era of Industry 4.0: Blockchain technology implementation opportunities and impediments from the perspective of people, process, performance, and technology. Production Planning & Control 1–21.
Kearney, J. 2010. Food consumption trends and drivers. Philosophical transactions of the royal society B: biological sciences. 365(1554): 2793–2807.
Khan, N., Ray, R.L., Sargani, G.R. et al. 2021. Current Progress and Future Prospects of Agriculture Technology: Gateway to Sustainable Agriculture. Sustainability 13(9): 4883.
Khan, S.A.R., Razzaq, A., Yu, Z. et al. 2021. Disruption in food supply chain and undernourishment challenges: An empirical study in the context of Asian countries. Socio-Economic Planning Sciences 101033.
Koparan, C., Koc, A.B., Privette, C.V. et al. 2020. Adaptive water sampling device for aerial robots. Drones 4(1): 5.
Kouhizadeh, M., Zhu, Q. and Sarkis, J. 2020. Blockchain and the circular economy: potential tensions and critical reflections from practice. Production Planning & Control 31(11–12): 950–966.
Krishnan, R., Agarwal, R., Bajada, C. et al. 2020. Redesigning a food supply chain for environmental sustainability–An analysis of resource use and recovery. Journal of Cleaner Production 242: 118374.
Kuhn, 2019. Spraying. Available online: https://www.kuhn.com/com_en/range/spraying.html [Internet]. 2019.
Kumar, A. and Samadder, S. 2020. Performance evaluation of anaerobic digestion technology for energy recovery from organic fraction of municipal solid waste: A review. Energy 197: 117253.
Kusumowardani, N., Tjahjono, B., Lazell, J. et al. 2022. A circular capability framework to address food waste and losses in the agri-food supply chain: The antecedents, principles and outcomes of circular economy. Journal of Business Research 142: 17–31.
Li, K., Lee, J.-Y., Gharehgozli, A. 2021. Blockchain in food supply chains: a literature review and synthesis analysis of platforms, benefits and challenges. International Journal of Production Research 1–20.
Lowenberg-DeBoer, J. and Erickson, B. 2019. Setting the record straight on precision agriculture adoption. Agronomy Journal. 111(4): 1552–1569.
Ltd FBIC. 2016. Report of prospects and investment strategy planning analysis on china food waste treatment industry (2017–2022). Tech. rep., Institute of Forward Industry Research.
Mangla, S.K., Kazancoglu, Y., Ekinci, E. et al. 2021. Using system dynamics to analyze the societal impacts of blockchain technology in milk supply chainsrefer. Transportation Research Part E: Logistics and Transportation Review 149: 102289.
Mehmood, A., Ahmed, S., Viza, E. et al. 2021. Drivers and barriers towards circular economy in agri-food supply chain: A review. Business Strategy & Development 4(4): 465–481.
Memon, S.U.R., Pawase, V.R., Pavase, T.R. et al. 2021. Investigation of COVID-19 Impact on the Food and Beverages Industry: China and India Perspective. Foods. 10(5): 1069.
Migliaccio, K., Stockle, C., Boz, Z. et al. 2012. Circular Open-Field Systems for Fresh-Market Tomatoes. Resource Magazine 28(2): 15–18.
Naio-Technologies, 2019. Autonomous vegetables weeding robot, Dino. Available online: https://www.naio-technologies.com/en/agricultural-equipment/large-scale-vegetable-weeding-robot/ [Internet]. 2019.
National Academic Press, 2019. Science Breakthrough to Advance Food and Agricultural Research by 2030. Washington, DC: The National Academic Press. http//:doi.org/10.17226/25059 [Internet]. 2019.

Palazzo, M. and Vollero, A. 2021. A systematic literature review of food sustainable supply chain management (FSSCM): building blocks and research trends. The TQM Journal.
Parfitt, J., Barthel, M. and Macnaughton, S. 2010. Food waste within food supply chains: quantification and potential for change to 2050. Philosophical Transactions of the Royal Society B: Biological Sciences 365(1554): 3065–3081.
Parker, N., Fan, Y. and Ogden, J. 2010. From waste to hydrogen: An optimal design of energy production and distribution network. Transportation Research Part E: Logistics and Transportation Review 46(4): 534–545.
Posmanik, R., Labatut, R.A., Kim, A.H. et al. 2017. Coupling hydrothermal liquefaction and anaerobic digestion for energy valorization from model biomass feedstocks. Bioresource Technology 233: 134–143.
Precision Plants, 2019. Precision Planting Unveils FurrowForce and SmartDepth. Available online: https://www.agriculture.com/machinery/planting/precision-planting-unveils-2019-product-lineup [Internet]. 2019.
Prieto-Sandoval, V., Jaca, C., Santos, J. et al. 2019. Key strategies, resources, and capabilities for implementing circular economy in industrial small and medium enterprises. Corporate Social Responsibility and Environmental Management. 26(6): 1473–1484.
Quak, H. and de Koster, M.B. 2009. Delivering goods in urban areas: how to deal with urban policy restrictions and the environment. Transportation Science 43(2): 211–227.
Quak, H.J. and De Koster, M. 2007. Exploring retailers' sensitivity to local sustainability policies. Journal of Operations Management 25(6): 1103–1122.
QuickConnect, 2019. The automatic top link coupling system for the Sirius sprayers. Available online: https://lemken.com/en/technology/quickconnect/ [Internet]. 2019.
Rana, R.L., Tricase, C. and De Cesare, L. 2021. Blockchain technology for a sustainable agri-food supply chain. British Food Journal.
Rashidghalam, M. 2019. Introduction to sustainable agriculture and agribusiness in Iran. Sustainable Agriculture and Agribusiness in Iran: Springer, pp. 1–7.
Ratanatamskul, C., Onnum, G. and Yamamoto, K. 2014. A prototype single-stage anaerobic digester for co-digestion of food waste and sewage sludge from high-rise building for on-site biogas production. International Biodeterioration & Biodegradation 95: 176–180.
Rauch, 2019. Disk spreaders with universal drive shaft. Available online: https://rauch.de/en/fertiliser-spreaders/axis-m-20-2-emc-w.html [Internet]. 2019.
Real Agriculture, 2019. eAutoPower gearbox e8WD for 8R large tractors. Available online: https://www.realagriculture.com/2019/11/john-deere-teams-up-with-joskin-to-develop-award-winning-eautopower-gearbox/. [Internet]. 2019.
Rich, K.M., Rich, M. and Dizyee, K. 2018. Participatory systems approaches for urban and peri-urban agriculture planning: The role of system dynamics and spatial group model building. Agricultural Systems 160: 110–123.
Rizos, V., Behrens, A., Van der Gaast, W. et al. 2016. Implementation of circular economy business models by small and medium-sized enterprises (SMEs): Barriers and enablers. Sustainability 8(11): 1212.
Saiz-Rubio, V. and Rovira-Más, F. 2020. From smart farming towards agriculture 5.0: A review on crop data management. Agronomy 10(207).
Schanes, K., Dobernig, K. and Gözet, B. 2018. Food waste matters—A systematic review of household food waste practices and their policy implications. Journal of Cleaner Production 182: 978–991.
Scherhaufer, S., Moates, G., Hartikainen, H. et al. 2018. Environmental impacts of food waste in Europe. Waste management. 77: 98–113.
Scholz, A., Sonesson, U. and Tyedmers, P. 2009. Catch of the freezer. The New York Times 8.
SDG, 2015. Sustainable development goals. https://unfoundation.org/what-we-do/issues/sustainable-development-goals/?gclid=Cj0KCQjwl_SHBhCQARIsAFIFRVXPt5T_PheO_vVymEkTUXBcadT_YXRoXVDwxk02FarU2vvPrFfc60saAikhEALw_wcB [Internet]. 2015.
Segre, A., Falasconi, L., Politano, A. et al. 2014. Background paper on the economics of food loss and waste. Rome, Italy: Università di Bologna
Seiple, T.E., Coleman, A.M. and Skaggs, R.L. 2017. Municipal wastewater sludge as a sustainable bioresource in the United States. Journal of Environmental Management. 197: 673–680.
Sethi, G., Bedregal, L., Cassou, E. et al. 2020. Addressing Food Loss and Waste: A Global Problem with Local Solutions.
Sexton, R.J., Sheldon, I., McCorriston, S. et al. 2007. Agricultural trade liberalization and economic development: the role of downstream market power. Agricultural Economics 36(2): 253–270.
Shafiee-Jood, M. and Cai, X. 2016. Reducing food loss and waste to enhance food security and environmental sustainability. Environmental Science & Technology 50(16): 8432–8443.
Shariff, Z.M., Lin, K.G., Sariman, S. et al. 2015. The relationship between household income and dietary intakes of 1–10 year old urban Malaysian. Nutrition Research and Practice 9(3): 278–287.
Sharma, Y.K., Mangla, S.K., Patil, P.P. et al. 2019. When challenges impede the process: For circular economy-driven sustainability practices in food supply chain. Management Decision.
Singh, P.K., Verma, S.K., Ojha, S.K. et al. 2019. Intrinsic molecular insights to enhancement of biogas production from kitchen refuse using alkaline-microwave pretreatment. Scientific Reports 9(1): 1–12.
Sisinni, E., Saifullah, A., Han, S. et al. 2018. Industrial internet of things: Challenges, opportunities, and directions. IEEE Transactions on Industrial Informatics 14(11): 4724–4734.

SmartAgrHub. 2019. Needs Assessment Report. WP4, Report to identify, analyze, and assess the needs of farmers and DIHs in relation to digital transformation, 2nd release. Available online: https://www.smartagrihubs.eu/Deliverables/pdfs/D4.1-Needs-Assessment-Report-2nd-release-final.pdf.

Stein, K., Coulibaly, D., Stenchly, K. et al. 2017. Bee pollination increases yield quantity and quality of cash crops in Burkina Faso, West Africa. Scientific Reports. 7(1): 1–10.

Trendov, N.M., Varas, S. and Zenf, M. 2019. Digital technologies in agriculture and rural areas—Status report. Food Agricultural Organization of the United Nations, Rome. License: cc by-nc-sa 3.0 igo.

United-Nations, 2016. Sustainable Development Goals 2016. Available from: https://sdgs.un.org/goals

United-Nations. A/69/700 Follow-up to the outcome of the Millennium Summit, The road to dignity by 2030: ending poverty, transforming all lives and protecting the planet. Synthesis report of the Secretary-General on the post-2015 sustainable development agenda. 2014.

USDA, 2016. Farm profits and adoption of precision agriculture, ERR-217, U.S. Department of Agriculture, Economic Research Service [Internet]. 2016.

Verma, B. and Jones, J. 2021. Last Word From Linear to Circular, An Ambitious but Necessary Goal. Resource Magazine. 28(2): 38–39.

Vogliano, C., Murray, L., Coad, J. et al. 2021. Progress towards SDG 2: Zero hunger in melanesia—A state of data scoping review. Global Food Security 29: 100519.

Walter, A., Finger, R., Huber, R. et al. 2017. Opinion: Smart farming is key to developing sustainable agriculture. Proceedings of the National Academy of Sciences 114(24): 6148–6150.

Widodo, K.H., Nagasawa, H., Morizawa, K. et al. 2006. A periodical flowering–harvesting model for delivering agricultural fresh products. European Journal of Operational Research 170(1): 24–43.

Wietzke, A., Westphal, C., Gras, P. et al. 2018. Insect pollination as a key factor for strawberry physiology and marketable fruit quality. Agriculture, Ecosystems & Environment 258: 197–204.

Yakovleva, N. 2007. Measuring the sustainability of the food supply chain: a case study of the UK. Journal of Environmental Policy & Planning 9(1): 75–100.

Yang, F., Li, Y., Han, Y. et al. 2019. Performance of mature compost to control gaseous emissions in kitchen waste composting. Science of the Total Environment 657: 262–269.

Yang, G., Zhang, G. and Wang, H. 2015. Current state of sludge production, management, treatment and disposal in China. Water Research 78: 60–73.

Zhao, Q. and Huang, J. 2011. Agricultural science & technology in China: a roadmap to 2050. Springer.

Zhu, Z., Chu, F., Dolgui, A. et al. 2018. Recent advances and opportunities in sustainable food supply chain: a model-oriented review. International Journal of Production Research. 56(17): 5700–5722.

4

The Sustainability Roadmap for the Food Industry 4.0

Sneha Kumari,[1] *Venkataswamy Gurusamy Venkatesh*[2,*] and *Yangyan Shi*[3]

1. Introduction

The world has been witnessing continuous changes for ages. These changes signify that humankind has developed and attained evolution at every stage of life. The Industrial Revolution started with the first revolution focusing on mechanization. This revolution was followed by the second Industrial Revolution, where the prime aim was mass production. The third Industrial Revolution focused on information, technology, and automation, while the fourth Industrial Revolution talked about big data, cloud computing, cognitive computing, and the Internet of things (IoT) (Mavani et al. 2021). The Industrial Revolution brought in Food Industry 4.0 as society is in dire need of it.

During the past two decades, the food industry has undergone enormous changes. The population has been driven toward ready-to-eat and processed food. The reports suggest that the total percentage of processed food is very low in developing countries like India. The non-processed food also adds to the wastage. A low level of processing does not mean that Indians consume less processed food. The notion is changing because of awareness about the nutritious effect of processed foods and fast life. Changes in the socioeconomic environment are leading to higher demand for processed food. Packaging centers, integrated cold chain facilities, value-added centers, and other supporting institutions have promoted processed food to reach its peak. There are several constraints of processing in developing countries. The food sector is challenged by poorly fixed assets, greater flexibility, higher efficiency, lower lead time, and sustainability (Sartal et al. 2019). Multidimensional and fundamental plans have been put up to shift the traditional food system to sustainable ones. The conventional food sector includes the sale of agricultural produce in the local market. Most of the small and marginal farmers sell their produce through the traditional channel, which leads to poor profitability (Ashfaq et al. 2019). With growing urbanization, conventional supply methods have increased overproduction and unorganized distribution systems. However, there has been a shift towards healthy food and sustainable farming. Research paper analyses grassroots initiatives that are taken up by a few individuals to find the solution to the existing challenges (Gernert et al. 2018).

Industry 4.0 comprises big data, IoT, and smart factories. Technology implementation leads to Industry 4.0 (Haseeb et al. 2019), which engages in developing capabilities to transform the business

[1] Symbioisis School of Economics, Symbiosis International (Deemed University), Pune, India.
[2] EM Normandie Business School, Metis Lab, France.
[3] Department of Management, Macquarie University, Sydney, Australia.
* Corresponding author: vgv1976@gmail.com

into digitalization. The extent of technology adoption can be seen at different phases of the supply chain, namely artificial intelligence, predictive analytics, data analytics capability, visualization, and transparency (Soosay and Kannusamy 2018). Industry 4.0 can simplify supply chain management, traceability, and product diversification. Artificial intelligence with different sensors, like E-nose and E-tongue, has wide applications in the food industry (Mavani et al. 2021). The industry is witnessing a highly competitive environment in the current era (Mavani et al. 2021). The Food Industry 4.0 has a significant contribution to sustainable development goals. Indeed, the food sector has been facing issues and challenges. Researchers have focussed on food processing, food wastage, and technology adoption in the food industry. However, there is a need for a review that can document the critical application of technology in Food Industry 4.0 and pave the way toward a sustainable food industry. Studies must focus on the linkage of technology in food to address the research gap. We derived research questions that are as follows:

1. How Industry 4.0 can result in a sustainable food system?

To answer the above research questions, we have derived our research objective from reviewing the applications of technology in sustainable Food Industry 4.0 and exploring Industry 4.0 in a sustainable food system. The study has undergone an extensive literature review of 72 research studies on Food Industry 4.0. The study has examined the role of technology in the food industry and has derived a sustainable roadmap for Food Industry 4.0. The objective of the research is to identify the use of information and technology in enhancing the sustainable practices of the food industry. The study is based on desk research presenting a review of the technology adoption in the food cold chain.

Due to the lack of technology, there have been post-harvest losses, storage losses, and food waste. The concept of Food Industry 4.0 has been reviewed. The chapter can make a significant contribution to sustaining the food sector. To proceed with our research objectives, we have undergone an extensive literature review on sustainable Food Industry 4.0. The authors have prepared a database for collecting the research on Food Industry 4.0. in the second stage. The first author compiled the database designed by the co-authors and removed any redundant research study. The final database was prepared to undergo an extensive literature review. The research chapter is divided into six sections. Section 2 focuses on the literature review on Food Industry 4.0, and Section 3 presents the research methodology of the study. Section 4 presents the results, followed by the implications of the study. Finally, the conclusion gives the sustainable roadmap for food industry 4.0.

2. Literature Review

The section presents a review of research studies on the sustainable Food Industry 4.0. Globalization has led to consistent growth in capital and goods for sustainability (Pisching et al. 2015). The food supply chain challenges can be overcome by advancing toward Industry 4.0 (Manavalan and Jayakrishna 2019; Beavo et al. 2021).

2.1 Food Industry in Developed Countries

The food industry has generated employment and value addition. European Commission has the potential for the sustainable food industry but lacks transparency. The Food Industry 4.0 comprises food quality, supply chain efficiency, food safety, food waste, and value chain analysis. North America is the second-largest food industry in the world. The country has adopted innovative warehousing, robotics, and automation. China is the third-largest food and drink producer, employing millions of workers from different countries. The countries like Japan and South Korea follow strict monitoring in the food industry. Australia and New Zealand aim to ensure food quality safety and reduce risks (Gustavsson et al. 2011). The agricultural trade increased between China and Korea. This led to the promotion of cold chain logistics and value-added services (Wang 2018). Developed countries have also faced severe problems handling extensive data on food waste every year.

2.2 Food Industry in Developing Countries

In developing countries, huge populations, growth, and natural diversity are associated with the food industry. This is due to a shift towards urbanization and creating a change towards changing food habits. Biodiversity and ecosystem are essentially crucial in Indian contexts. The food industry requires the support of digitalization, technology, analytics, and advanced technical support. Developing countries face different challenges concerning technology and digital transformation. Despite challenges in Food Industry 4.0, it has been found that Industry 4.0 has the opportunity to create a sustainable roadmap. A developing country is one of the contributors to food waste, and the prime cause for this is a lack of technology adoption (Negi and Anand 2018).

2.3 Food Industry 4.0

Food Industry 4.0 has led us to rely on technology and advance toward sustainability. Food products that are temperature-sensitive find a good place with technology. The application of artificial intelligence, machine learning, cloud technologies, and data analytics has led to Food Industry 4.0. The technology application has resulted in minimizing waste and monitoring perishable agriculture commodities. It is revealed that organizations should focus on digital technologies to advance toward Industry 4.0. The food industry is driven by technology, performance, economics, competitive policy, collaboration, strategy, procurement, customer engagement, real-time information, and zero errors (Manavalan and Jayakrishna 2019). Industry 4.0 is driven by IoT, smart factories, and big data, which leads to IT implementation resulting in sustainable performance (Haseeb et al. 2019). Therefore, the application of IoT, big data, and smart factories is essential for the food industry.

2.4 Food Supply Chain

The Food Supply Chain (FSC) consists of practices for preserving perishable food products from microbial spoilage and has potential barriers that cannot be ignored (Ada et al. 2019). Cultural, business, regulatory, technology, managerial, knowledge, and skills are the prominent barriers in the FSC. Administrative factors, like poor leadership, top management commitment, lack of collaboration, and ineffective labor, result in disruptions of the FSC as well. The technology limitations too act as a barrier to it. The food industry demands new professional skills and a workforce to handle technology and digital transformation. The skills alone cannot develop a sustainable roadmap, and the skills need to be supported by training and education provided to the workforce and other factors.

2.5 Food Cold Chain

The Food Cold Chain (FCC) can be sustained using the concept of people, process, and technology. The perishable food loss drives a concern for integrating people, processes, and technology. With the growing issues, the concept of blockchain technology has made a wide application in sustaining FCC. The blockchain can also look into trust, traceability, and accountability in FCC (Kayikci et al. 2020). Cold Chain implies a complex web of functions or activities in the supply chain that ensures temperature control for perishable commodities (Shabani et al. 2015). FCC is a cold chain that helps preserve the products from spoilage and increases shelf-life. FCC includes pre-cooling, refrigerator carriers, cold warehousing, and traceability tools. The FCC is mainly required for perishable agricultural products, like fruits, vegetables, milk, meat, fish, chicken, and other processed food products. With the increase in customer-focused products, most of the food demanded is ready-to-eat or processed food. These customer-focused products rely heavily on logistic and cross-functional drivers for competitive advantage. The logistic driver infrastructure, facility, and transportation need to maintain proper temperature control for the processed food products for FCC.

2.6 Sustainable Food Supply Chain

The sustainable FSC is composed of sustainable performance in terms of social, environmental, and economic (León Bravo et al. 2021). The environmental aspect confines green packaging and transportation. Financial aspect confines supporting the supply chain partners and sustainable sourcing. The social part means the health and safety of the stakeholders, community, and human rights in the FSC. The FSC is characterized by complexity in quality, seasonality, diversified produce, traceability, and different agriculture practices (Bellarby et al. 2008). The players in the FSC are responsible for managing the risks and uncertainties (Choirun et al. 2020).

The Food Industry 4.0 has been supported with a circular FSC. The circular FSC results in the reuse, reduction, and recycling of products, which may lead to waste (Ada et al. 2021). The circular FSC is essential for sustainability. The circular FSC requires the support of digital transformation. The circular economy seeking sustainable development is the central objective of improving disruptions and reducing food waste (Yazdani et al. 2021). Industry 4.0 is driven by technology, like big data and analytics, autonomous robots, simulation, cybersecurity, cloud augmented reality, autonomous robots, and horizontal and vertical integration (Hasnan and Yusoff 2018). Such technology emergence in developing countries is a challenge as the agriculture sector is the last sector looking for advanced technology. These high-tech systems need to be applied in the agriculture sector to sustain the FSC (Demir and Dincer 2019).

3. Application of Technology in the Food Industry

Industry 4.0 is an important area for the sustainable and affordable food sector. The application of blockchain technology and digital approach is evolving for overall development in the food industry. It leads to the use of big data and application on food and bioconversion to improve the sustainability of the environment (Panetto et al. 2020). Usage of Industrial Internet of Things (IIOT),Cyber Physical Systems (CPS), and Software as a service (SAAS) in various departments, such as Enterprise resource planning (ERP), manufacturing execution system (MES), and food quality assurance has led to sustainability. High-tech image processing systems and cameras provide "quality control" real-time data about the shape, color, and foreign bodies in the food, thus helping in early detection, less contamination, and safe food. RFID and QR Code helps in the traceability of food items. Augmented Reality (AR) assists in constructing real-time models and designing the plant, machinery, and process with the help of ARENA, Techno Matrix, TrackSYS, and FlexSim. The software helps to provide adequate cost planning and reduce product failures. Automation of repetitive tasks with specialized robots of gripping technology help in loading/unloading, assembling, packaging, piling, and sorting food items. Robots are often used as chefs. AR has reshaped the marketing segment and has provided direct information on their mobiles by cutting down the advertisement costs. It also helps to provide faster learning to the trainees. Finally, 3D technology has provided customers with the opportunity of placing customized orders as per their requirements. Machine learning, artificial intelligence, real-time data, the IoT, data visualization, cloud computing, and cyber-physical systems help in building a sustainable food industry (Savas et al. 2021).

4. Digital Transformation for Food Industry 4.0

Digital solutions are essential for complex industrial environments. They are required for decision-making skills and the future growth of the food industry. Critical new concepts like the IoT and the cyber-physical system play an essential role in digitalization, real-time monitoring, and future perspective. The technology resolves food industry challenges. The requirements through technology, smart logistics, and real-time tracking are met. The application of a decision support system helps identify, evaluate, and provide directions for improvement of the knowledge, identification of crop and plant diseases detection, etc. Internet manufacturing technology includes enterprise resource planning, food quality assurance, research and development, facilities management, and manufacturing execution system. Artificial intelligence helps perform human tasks and reduce disruptions in the FSC. Quality control is done using tools like industrial robotics and augmented reality to understand the supply chain better. The food traceability system and automation of repetitive tasks help get better practice and monitor production. The FSC requires professional skills and training to adopt digital technologies (Akyazi et al. 2020).

5. Sustainable Road Map for Food Industry 4.0

The food sector is often researched for better monitoring and sustaining the sector. World Bank and International Organizations have paid attention to the sustainable development of the food system. The Food Industry 4.0 approach refers to a complex evolution assisted by technological advances (Savastano et al. 2018). The food industry is the largest manufacturing sector in turnover, employment generation, and value addition in the European Union countries. It is classified into dairy, drink, meat, food, and bakery sectors. Industry 4.0 addresses business opportunities to understand digital applications and sustain the food supply chain. The contract farming concept is also strengthening the food sector. The readiness toward implementation of Food Industry 4.0 is driven by technology, organizational, and environmental context. The technology context talks about the IT infrastructure, organizational context about the top management support, firm size, employee disruptions, and employee skills. Environmental context refers to legal issues, competitors, security, and privacy (Ichsan et al. 2019). We need to focus on our IT infrastructure, top management support, and environment for a sustainable road map.

The food industry requires improved transportation and processing techniques (Mangla et al. 2019). There is a need for efficient tools and techniques like thermal characteristics, processing equipment, and information (Amjadi 2005). There has been a lack of processing techniques for perishable agriculture commodities. This results in post-harvest losses and wastage of food. The freshness and quality of food are often lost in the distribution channel (Tijskens and Polderdijk 1996). Every food requires a specific temperature and humidity to increase shelf life (Bhatnagar et al. 2019). The lack of temperature management results in problems with the Food.

5.1 Food Waste Reduction Roadmap

With time, it has been reported that the trend of data on food wastage and spoilage has been increasing. The per capita food waste of selected countries worldwide in 2017 is maximum in Australia with 361 kg per year, followed by the USA at 278 kilograms per year (Barua and Hossain 2021). Turkey, Spain, Japan, Germany, Mexico, Italy, Morocco, Portugal, Jordan, Canada, Tunisia, and France are the other countries that top the list of per capita food waste, contributing more than 100 kgs per year. This is due to improper handling of the food, post-harvest losses, storage losses, etc. Around one-third of the food is wasted globally, contributing to artificial greenhouse gas emissions. The food industry needs to set a waste reduction target by adopting sustainable development goals. The amount and type of food wastes need to be analyzed consistently so that appropriate action may be taken by using the data for planning and delivering the reduction actions. The concept of circular economy in the food industry is also an initiative for reusing and recycling food to minimize waste. The roadmap for food waste reduction is implementing targets, measure, and acting for all the food business sectors. There is a need to set a food waste reduction target by measuring and reporting the food surplus (Food Waste Reduction Roadmap Progress Report 2019). The roadmap for sustainable food industry 4.0 is shown in Fig. 1.

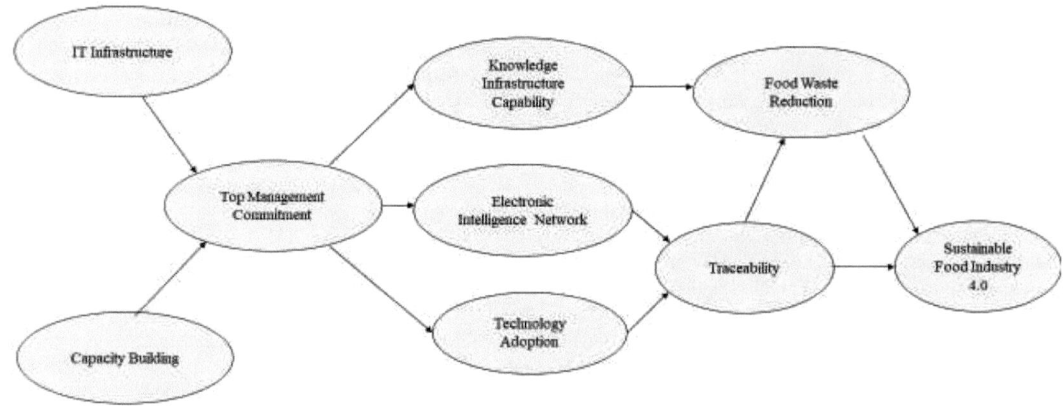

Figure 1. Roadmap for Sustainable Food Industry 4.0.

6. Conclusion

To create a sustainable and innovative food industry, there is a growing need to apply the IoT, data analytics, the e-food supply chain, big data, cloud computation, and cyber-physical systems. In turn, the FSC network is responsible for distributing goods and services to the final consumer. There should be data exchange, automation, and standard efficient food production (Ojo et al. 2018).

Planning is required at the national level for developing a sustainable roadmap for the food industry. Various guidelines have been formulated for sustaining the Food Industry 4.0 (Soosay and Kannusamym 2018). Industry 4.0 results in developing agile capabilities to meet the customers' requirements with responsiveness. High-quality, diversified produce, and food security can lead to a dynamic food system (Soosay and Kannusamy 2018). The drivers of Industry 4.0 can be applied, helping in providing sophisticated, effective communication and customer service. The adoption of technology should depend on the company's needs and capabilities for successful implementation. The central question is the economic viability among small and medium-scale enterprises in terms of investment and maintenance.

The food industry is one of the global economic industries driven by the demand for shorter product life cycles, cost reductions, and highly customized products. It is all about consistent digitization and linking all productive units in an interoperable environment. The potential of grassroots initiatives for pioneering sustainable transitions expands the options and scope of the food industry (Gernert et al. 2018). However, the supply chain operates in a complex manner. Amidst this complexity, blockchain technology has a broader scope in driving supply chain transparency and efficiency. With blockchain, the food supply chain helps in trust-building and better monitoring of the process (Manavalan and Jayakrishna 2019). Shortly, food security will be a crucial issue (de Amorim et al. 2019). Therefore, it is essential to start working on the roadmap for the sustainable food industry.

References

Ada, N., Kazancoglu, Y., Sezer, M.D., Ede-Senturk, C., Ozer, I. and Ram, M. 2021. Analyzing Barriers of Circular Food Supply Chains and Proposing Industry 4.0 Solutions. Sustainability 13(12): 6812.

Adelaja, A. and George, J. 2019. Effects of conflict on agriculture: Evidence from the Boko Haram insurgency. World Development 117: 184–195.

Akyazi, T., Goti, A., Oyarbide, A., Alberdi, E. and Bayon, F. 2020. A guide for the food industry to meet the future skills requirements emerging with industry 4.0. Foods 9(4): 492.

Ali, I., Arslan, A., Khan, Z. and Tarba, S.Y. 2021. The role of Industry 4.0 technologies in mitigating supply chain disruption: empirical evidence from the Australian food processing industry. IEEE Transactions on Engineering Management.

Ali, S.M., Moktadir, M.A., Kabir, G., Chakma, J., Rumi, M.J.U. and Islam, M.T. 2019. Framework for evaluating risks in the food supply chain: Implications in food wastage reduction. Journal of Cleaner Production 228: 786–800.

Amjadi, K. 2005. Integrating food hygiene into quantity food production systems. Nutrition & Food Science 35(3): 169–83.

Ashfaq, M., Hassan, S., Abbas, A., Razzaq, A., Mehdi, M., Ariyawardana, A. and Anwar, M. 2019. Critical issues at the upstream level in sustainable supply chain management of agri-food industries: Evidence from Pakistan's citrus industry. Sustainability 11(5): 1326.

Barua, P. and Hossain, N. 2021. Waste to energy: An overview by global erspective. Advanced Technology for the Conversion of Waste into Fuels and Chemicals, 1–49.

Batt, P.J. 2004. Incorporating satisfaction, trust, and power-dependence measures into an analysis of agribusiness supply chains. Agriproduct Supply-chain Management in Developing Countries 27–43.

Beavo, J.A., Golkowski, M., Shimizu-Albergine, M., Beltejar, M.C., Bornfeldt, K.E. and Ong, S.E. 2021. Phosphoproteomic analysis as an approach for understanding molecular mechanisms of cAMP-dependent actions. Molecular Pharmacology, 99(5): 342–357.

Bellarby, J., Foereid, B. and Hastings, A. 2008. Cool Farming: Climate impacts of agriculture and mitigation potential, https://eprints.lancs.ac.uk/id/eprint/68831/1/1111.pdf as retrieved on 15th November 2021.

Bhatnagar, A., Vrat, P. and Shankar, R. 2019. Multi-criteria clustering analytics for agro-based perishables in cold-chain. Journal of Advances in Management Research 16(4): 563–593.

Choirun, A., Santoso, I. and Astuti, R. 2020, April. Sustainability risk management in the agri-food supply chain: a literature review. In IOP Conference Series: Earth and Environmental Science (Vol. 475, No. 1, p. 012050). IOP Publishing.

Corallo, A., Latino, M.E. and Menegoli, M. 2018. From industry 4.0 to agriculture 4.0: a framework to manage product data in agri-food supply chain for voluntary traceability. International Journal of Nutrition and Food Engineering 12(5): 146–150.

Costa, C., Antonucci, F., Pallottino, F., Aguzzi, J., Sarriá, D. and Menesatti, P. 2013. A review on agri-food supply chain traceability by means of RFID technology. Food and Bioprocess Technology 6(2): 353–366.

de Amorim, W.S., Deggau, A.B., do Livramento Gonçalves, G., da Silva Neiva, S., Prasath, A.R. and de Andrade, J.B.S.O. 2019. Urban challenges and opportunities to promote sustainable food security through smart cities and the 4th industrial revolution. Land Use Policy 87: 104065.

Demir, Y. and Dincer, F.I. 2020. The Effects of Industry 4.0 on the Food and Beverage Industry. Journal of Tourismology 6(1): 133–145.

Energy & Environment, Waste Management. 2017. Per capita food waste of selected countries worldwide From https://www.statista.com/statistics/933059/per-capita-food-waste-of-selected-countries/ as retrieved on 15th December, 2020.

Food Waste Reduction Roadmap progress report 2019, https://wrap.org.uk /resources/report /food-waste-reduction-roadmap-progress-report-2019

Gernert, M., El Bilali, H. and Strassner, C. 2018. Grassroots initiatives as sustainability transition pioneers: Implications and lessons for urban food systems. Urban Science 2(1): 23.

Gustavsson, J., Cederberg, C., Sonesson, U., Van Otterdijk, R. and Meybeck, A. 2011. Global food losses and food waste, From https://www.madr.ro/docs/ind-alimentara/risipa_alimentara/presentation_food_waste.pdf, as retreived on 22nd September 2022.

Haseeb, M., Hussain, H.I., Ślusarczyk, B. and Jermsittiparsert, K. 2019. Industry 4.0: A solution towards technology challenges of sustainable business performance. Social Sciences 8(5): 154.

Hasnan, N.Z.N. and Yusoff, Y.M. 2018, November. Short review: Application areas of industry 4.0 technologies in food processing sector. In 2018 IEEE Student Conference on Research and Development (SCOReD) (pp. 1–6). IEEE.

Ichsan, M. and Dachyar, M. 2019, August. Readiness for Implementing Industry 4.0 in Food and Beverage Manufacturer in Indonesia. In IOP Conference Series: Materials Science and Engineering (Vol. 598, No. 1, p. 012129). IOP Publishing.

Kayikci, Y., Subramanian, N., Dora, M. and Bhatia, M.S. 2020. Food supply chain in the era of Industry 4.0: Blockchain technology implementation opportunities and impediments from the perspective of people, process, performance, and technology. Production Planning & Control 1–21.

León Bravo, V., Moretto, A. and Caniato, F. 2021. A roadmap for sustainability assessment in the food supply chain, British Food Journal 123(13): 199–220.

Manavalan, E. and Jayakrishna, K. 2019. A review of Internet of Things (IoT) embedded sustainable supply chain for industry 4.0 requirements. Computers & Industrial Engineering 127: 925–953.

Manavalan, E. and Jayakrishna, K. 2019. A review of Internet of Things (IoT) embedded sustainable supply chain for industry 4.0 requirements. Computers & Industrial Engineering 127: 925–953.

Mangla, S.K., Sharma, Y.K., Patil, P.P., Yadav, G. and Xu, J. 2019. Logistics and distribution challenges to managing operations for corporate sustainability: study on leading Indian diary organizations. Journal of Cleaner Production 238: 117620.

Mavani, N.R., Ali, J.M., Othman, S., Hussain, M.A., Hashim, H. and Rahman, N.A. 2021. Application of Artificial Intelligence in Food Industry—a Guideline. Food Engineering Reviews 1–42.

Negi, S. and Anand, N. 2018. Wastage and cold chain infrastructure relationship in Indian food supply chain: A study from farm to retail. In Supply Chain Management Strategies and Risk Assessment in Retail Environments (pp. 247–266). IGI Global.

Nguyen, G.N. and Sarker, T. 2018. Sustainable coffee supply chain management: a case study in Buon Me Thuot City, Daklak, Vietnam. International Journal of Corporate Social Responsibility 3(1): 1–17.

Ojo, O.O., Shah, S., Coutroubis, A., Jiménez, M.T. and Ocana, Y.M. 2018, November. Potential impact of industry 4.0 in sustainable food supply chain environment. In 2018 IEEE International Conference on Technology Management, Operations and Decisions (ICTMOD), pp. 172–177. IEEE.

Panetto, H., Lezoche, M., Hormazabal, J.E.H., Diaz, M.D.M.E.A. and Kacprzyk, J. 2020. Special issue on Agri-Food 4.0 and digitalization in agriculture supply chains-New directions, challenges and applications, https://pdf.sciencedirectassets.com/271439/1-s2.0-S0166361520X00029/1-s2.0-S0166361519311145/am.pdf.

Pisching, M.A., Junqueira, F., Filho, D.J.S. and Miyagi, P.E. 2015. Service composition in the cloud-based manufacturing focused on the industry 4.0. In Technological Innovation for Cloud-Based Engineering Systems: 6th IFIP WG 5.5/SOCOLNET Doctoral Conference on Computing, Electrical and Industrial Systems, DoCEIS 2015, Costa de Caparica, Portugal, April 13–15, 2015, Proceedings 6 (pp. 65–72). Springer International Publishing.

Sartal, A., Carou, D., Dorado-Vicente, R. and Mandayo, L. 2019. Facing the challenges of the food industry: Might additive manufacturing be the answer?. Proceedings of the Institution of Mechanical Engineers, Part B: Journal of Engineering Manufacture 233(8): 1902–1906.

Savastano, M., Amendola, C. and D'Ascenzo, F. 2018. How digital transformation is reshaping the manufacturing industry value chain: The new digital manufacturing ecosystem applied to a case study from the food industry. pp. 127–142. *In*: Network, smart and open. Springer, Cham.

Shabani, A., Torabipour, S.M.R. and Saen, R.F. 2015. A new super-efficiency dual-role FDH procedure: an application in dairy cold chain for vehicle selection. International Journal of Shipping and Transport Logistics 71(4): 426–456.

Soosay, C. and Kannusamy, R. 2018. Scope for industry 4.0 in agri-food supply chain. pp. 37–56. *In*: The Road to a Digitalized Supply Chain Management: Smart and Digital Solutions for Supply Chain Management. Proceedings of the Hamburg International Conference of Logistics (HICL), Vol. 25. Berlin: epubli GmbH.

Target. Measure. Act. The Food Waste Reduction Roadmap Progress Report 2019 https://wrap.org.uk/sites/default/files/2020-07/WRAP-Food-Waste- Reduction_ Roadmap_ Progress- Report-2019.pdf.

Tijskens, L.M.M. and Polderdijk, J.J. 1996. Generic model for keeping quality of vegetable produce during storage and distribution. Agricultural Systems 51(4): 431–52.

Wang, M. 2018. Impacts of supply chain uncertainty and risk on the logistics performance. Asia Pacific Journal of Marketing and Logistics, 30(3): 689–704.

Yazdani, M., Gonzalez, E.D.R.S. and Chatterjee, P. 2021. A multi-criteria decision-making framework for agriculture supply chain risk management under a circular economy context. Management Decision 59(8): 1801–1826.

5

Eco-friendly and Cost-effective Methods Applied to Sustainable Food Industries

Cennet Pelin Boyaci Gunduz

1. Introduction

Sustainability is meeting the needs of the present without compromising the ability of future generations to meet their own needs (UN Brundtland Report 1987). Today, the world population is over 7.7 billion and it is expected to increase to almost 8.5 billion in 2030, 9.7 billion in 2050, and nearly 10.9 billion around 2100 (SIStat 2021). As the world's population increases, major transformations are required to sustainably feed the growing population (FAO 2017a). Overproduction in many industries to meet the growing needs requires high amounts of natural resources; unfortunately, the effectiveness and continuity of production are affected by the scarcity of natural resources (Nikolaou et al. 2021). In the case of food, ensuring sustainable food production by preserving finite natural resources is a major challenge. Current food systems are unequal and fragile. The production and consumption of safe, nutritious, sufficient and affordable food, and preserving the ecosystems and natural resources on which food systems depend are very important. Therefore, it is an urgent need to make innovations that promote sustainability in the food industry, right from production to consumption by using land, water, and other inputs more efficiently, reducing fossil fuel usage to lead to a drastic cut of agricultural green-house gas emissions, shifting more sustainable diets, producing ingredients and materials from renewable sources, preventing food loss, and also reducing wastage at all stages of the food system (FAO 2017d).

Our current systems are facing complex and systemic challenges including the scarceness of natural resources, increase in demand for greater quality, quantity, and diversity of food, food loss and waste, hunger and food insecurity, unhealthy diets, climate change, pandemics, and ecosystem change (UNEP 2016a). The global food system needs to be redesigned to be more environmentally friendly, productive, sustainable and resilient, and also more inclusive of poor and marginalized populations. For that purpose, interconnected actions at the local, national, regional, and global levels are critically important (FAO 2018).

Adana Alparslan Turkes Science and Technology University, Faculty of Engineering, Department of Food Engineering, Adana, Turkey.
Email: cennetpelinboyaci@gmail.com

Sustainable development must balance social, economic, and environmental sustainability. From the industrial perspective, a sustainable company aims to reach the social, environmental, and economic objectives of sustainable development. Based on numerous scientific research, technological improvements should be implemented in the food industry. For that purpose, scientific research and technological innovations have been conducted worldwide to find sustainable solutions to the challenges in the food systems, and some companies have already shifted to more sustainable processes. In this chapter, sustainability in food systems and some eco-friendly and cost-effective processes applied in sustainable food industries will be discussed.

2. Sustainability in Food Systems

The food system is a complex system that contains sub-systems interacting with other key elements. It includes a wide range of actors and their interlinked value-adding activities included in the production, processing, packaging, distribution, preparation, consumption, and disposal as well as its impacts on the environment, health, and society (FAO 2018; von Braun et al. 2020). A sustainable food system is at the center of the United Nations Sustainable Development Goals, and it can be achieved when a positive value is generated along three dimensions simultaneously: social, economic, and environmental (FAO 2018). Especially, the COVID-19 pandemic has affected the entire food system and shows how fragile it is. It is very important to transform food systems to become more sustainable and resilient to future disturbances, shocks, and challenges (Boyaci-Gunduz et al. 2021). In 2021, the Food and Agriculture Benchmark, ranked by the World Benchmarking Alliance, assessed the most influential food and agriculture companies worldwide based on their contributions to transforming the global food system and by assessing companies from farm to fork on their environmental, nutritional, and social impact (World Benchmarking Alliance 2021). Global companies at the top of this ranking showed that they are meeting numerous expectations on a variety of topics. But, unfortunately, the overall average benchmark performance showed that the food and agriculture sector was not on track for transitioning to a sustainable food system. Actual transformation to more sustainable food systems could be achieved by shifting from individualized plans to more collective and integrated strategies with interdependent and combined objectives (De Bernardi and Azucar 2020). To facilitate the transition, it is very important to integrate research and sustainable processes in the food industry.

3. Eco-Friendly and Cost-Effective Methods in Food Industry

From the food industry point of view, companies should reach sustainable development goals from three pillars such as social, economic, and environmental based on the set science-based targets. A growing number of food companies have been integrating sustainability into their business strategy by considering a wide array of environmental, economic, and social factors. Some food companies have already redesigned food production, processing, and waste stages and applied some practices to meet environmental sustainability from the life cycle perspective. Each product has an impact on the environment, and life cycle assessment (LCA) is an internationally standardized environmental evaluation method that measures all possible environmental impacts of the product at each stage of its lifecycle (Despoudi 2021; Kloepffer 2008). LCAs are important for companies to understand the environmental impact of each step in their operations and provide an initial baseline on which to build sustainability progress (Chobani 2019). It is then possible to develop eco-friendly and cost-effective processes that can lead to sustainable production.

Many food companies have shifted to more productive and less wasteful production and processing methods to reach sustainable development goals. In that context, the usage of efficient water, energy, land, and other inputs and the application of processes using fewer resources are commonly applied solutions in the sustainable food industry. Decreasing greenhouse gas (GHG) emissions has paramount importance, and companies have started reducing fossil fuel usage to lead to a drastic cut in GHG emissions. Many global as well as some start-up companies have been trying to produce new foods with reduced environmental impact and carbon footprint and also use ingredients and materials that are produced from renewable sources. Bio-based ingredients from microorganisms through biotechnology and fermentation have been produced in the food industry. Reducing loss and waste at all stages, utilization of waste and by-products, using environmentally-friendly food packaging, and recycling food packaging materials are critically important. Moreover, shifting to more sustainable diets and improving digitalization at all stages of the food systems are among the innovations in the food industry.

3.1 Efficient Use of Energy, Water, Land, and Other Inputs

The water-food-energy nexus, which describes the complex and inter-related nature of our global resource systems, is central to sustainable development (FAO 2014a). Because food systems rely on natural resources, such as land, water, soil, biodiversity, minerals, and fossil fuels (UNEP 2016b). Agricultural production is the largest consumer of the world's freshwater resources. Worldwide, 72% of all water withdrawals are used by agriculture (UN-Water 2021a). The food industry accounts for around 30% of the world's total energy consumption (FAO 2011a). Food systems include the production, processing, packaging, distribution, preparation, consumption, and disposal; all of these steps use energy and result in released GHGs into the atmosphere (FAO 2018; von Braun et al. 2020). Therefore, the global food system makes a significant contribution to total anthropogenic GHG emissions. In that context, decreasing GHG emissions is critically important for the sustainable production of food companies.

3.1.1 Decreasing GHG Emissions

More than one-quarter of the energy used worldwide is consumed during food production and supply (UN-Water 2021b). The Joint Research Centre (JRC), the European Commission's science and knowledge service, developed a global food emission database (EDGAR-FOOD) that estimated GHG emissions involving all countries and sectors of the food system, from production to disposal, and reported that a third of global GHG emissions came from the food system for the years 1990 to 2015. The report highlighted that the food system emissions amounted to 18 Gt CO_2eq yr^{-1} in 2015, and the largest contribution was agriculture and land-use change activities (Crippa et al. 2021). Tubiello et al. (2021) estimated GHG emissions for the period 1990 to 2018 and reported that total GHG emissions from the food system were around 16 Gt CO_2eq yr^{-1} in 2018; three-quarters of these emissions were produced either within the farm gate or in food production activities, such as manufacturing, processing, transporting, and disposing of wastes.

GHG emissions from the food systems are carbon dioxide (CO_2), methane (CH_4), nitrous oxide (N_2O), and fluorinated gases (F-gases) (EDGAR-FOOD 2021; EPA 2021). Based on the data on GHG emissions from the food system in 2015, half of the GHG emissions belong to CO_2 (52%) and are followed by CH_4 (34%), N_2O (10%), and F-gases (2%) (Crippa et al. 2021). In food systems, CO_2 emissions are mainly linked to fossil fuel consumption for energy production at different stages of the food systems and land-use change, CH_4 emissions due to livestock, waste management and rice cultivation, N_2O emissions due to excessive fertilizer use, and F-gases mainly due to their use in the food industry for refrigeration (Baldwin 2015; EDGAR-FOOD 2021; EPA 2021).

Food systems rely on natural resources, such as land, water, biodiversity, soil, minerals, and fossil fuels, and are globally the dominant user of some of the sources. Therefore, it is critically important to build resource-smart food systems for sustainable development (UNEP 2016b). In that context, the global transition to net-zero emissions plays an important role in the industry. For that purpose, some companies in the food and beverage industry have already calculated their carbon footprints and developed eco-friendly solutions to become more sustainable, particularly when it comes to emissions (Food Institute 2020). From the environmental point of view, companies measure direct, indirect (scope 1 and 2) emissions and indirect value chain (scope 3) emissions, and then develop strategies such as increasing energy and water efficiency, sourcing renewable energy alternatives, reducing CH_4 from cattle digestion, practicing regenerative agriculture, and allowing digital technologies for decreasing the emissions. Measuring carbon footprint is the key point for companies to develop strategies to reduce emissions; for example, as an eco-friendly and powerful tool, blockchain technology could be used to trace CO_2 emissions and improve the transparency, accountability, and traceability of the companies' gas emissions based on the quantification of the carbon footprint through smart sensors compatible with Internet of Things (IoT) (Canorea 2021).

The world is transitioning to a zero-carbon economy and science-based targets provide certain pathways for companies to decrease GHG emissions (Science-Based Targets Initiative 2021). Global food and beverage companies, including Unilever, Nestle, Mars, Danone, PepsiCo, Mondalez International, General Mills, Coco-Cola, Kellogg's, and Kraft Heinz, set out a range of targets and actions to reach net-zero emissions no later than 2050.

3.1.2 Renewable Energy

The global food system is becoming more energy-intensive (Crippa et al. 2021). Demand for food production increases due to population growth, changes in consumption patterns worldwide, and increased demand in food production, whichf cause intensive energy demand for resources (Bajan et al. 2020; Bolandnazar et al. 2014 Pelletier et al. 2011). Many companies are looking for renewable and alternative energy resources to reduce GHG emissions. For example, Unilever has committed to reaching net-zero emissions covering scope 1, 2, and 3 emissions by 2039 and is planning to achieve a 100% renewable grid electricity target in 2020 (Unilever 2021a). To achieve 100% renewable electricity, the companies must source 100% of the electricity across their global operations with electricity generated from renewable sources, such as biomass (including biogas), solar, geothermal, water, and/or wind, either sourced from the market or self-produced (Center for Climate and Energy Solutions 2021). Different renewable sources are used depending on the company, and the place of the facility and global food companies increased their efforts to use renewable energy. Some other global RE100 food and beverage companies aiming for 100% renewable electricity in the upcoming years are Nestle, Pepsi Co, Danone, Heineken, General Mills, Starbucks, AB InBev, Mars, McCain Foods, Kellogg's, Nissin Foods, Keurig Dr. Pepper, and Coca-Cola European Partners (RE100 2021).

In the food and beverage industry, using renewable and alternative energy sources is eco-friendly since they typically have the least impact on the environment due to water efficiency and reduced or completely removed GHG emissions (Baldwin 2015). Renewable energy technologies are very important for companies to be sustainable since they deliver access to a secure, cost-effective, and environmentally sustainable supply of energy (IRENA 2015). The initial costs of installing renewable energy technologies may seem expensive, but these costs are decreasing (Inspire Clean Energy 2021) with a decrease in renewable energy technology costs. In fact, renewable energy provides an excellent opportunity for the mitigation of GHG emissions.

Over the past decade, renewable power generation costs have decreased significantly (IRENA 2020). The ratio of lifetime costs to lifetime electricity generation, levelized cost of energy (LCOE), is often used as a metric to measure and compare the costs of power generation (Tran and Smith 2018). According to the International Renewable Energy Agency (IRENA) renewable cost database—the global weighted average LCOE of utility-scale solar photovoltaics—concentrating on solar power, onshore wind, and offshore wind costs has decreased (IRENA 2020, 2021). Especially, the global weighted average LCOE of utility-scale solar photovoltaics fell by 85% in 10 years. For onshore and offshore wind, the global weighted average cost of electricity between 2010 and 2020 fell by 56% and 48%, respectively. Over the period 2010 to 2020, the global weighted average cost of electricity from concentrating solar power (CSP) fell to 68% (IRENA 2021). Environmental concerns, technological improvements, and government policies are the driving forces of the decreased renewable energy costs; with increasing economies of scale, further technological improvements, and more competitive supply chains, renewable energy costs could continue to decrease.

On the other hand, efficient use of not only water and energy but also land is important to be sustainable, and energy production technologies have varying land intensities (IRENA 2015). The transition to renewable energy sources is expected to increase the global competition for land, and this could result in some environmental impacts such as increasing biodiversity loss, water usage, or indirect land-use change emissions (Scheidel and Sorman 2012; van de Ven et al. 2021). Large-scale implementation of renewable energy can decrease GHG emissions and enhance resource efficiency, but it could affect biodiversity and ecosystem services (Gasparatos et al. 2017). Therefore, the land-use implications of growing energy demand and the potential negative impacts of renewable energy on biodiversity and ecosystems should be well understood (IRENA 2015; Trainor et al. 2016). Land-use efficiency should be improved with increased yields and technology advancements depending on the power generation technology without causing biodiversity loss.

3.1.3 Water and Energy Efficiency

Energy efficiency plays an important role to achieve sustainability in food process industries. There is a need to increase energy efficiency and clean energy solutions in agri-food systems (Semedo 2015). Energy efficiency could be achieved by shifting to new energy-smart technologies, efficient behavior, and alternative energy sources and processes that give the same with less energy (de Mello Santana and Bajay 2016). It could be achieved by enabling the efficient management of natural resources, using energy more efficiently, increasing the efforts to use renewable energy in the food systems, relying more on low-carbon energy systems, applying more eco-friendly farming methods that significantly improve yields in agricultural production, replacing fossil fuels with low-carbon resources and clean energy technologies, and reducing food losses and wastes along the food chain (FAO 2011a).

Different types of energy sources are required in the food industry, and energy consumption and emissions depend on the product type, the technique, and the form of energy used. Mainly, two types of energy are used—i.e, thermal energy and electric energy. Processing techniques include phase changes, such as evaporation, drying, freezing, and distillation, that are more energy-intensive compared to other processes (Morawicki and Hager 2014). Ladha-Sabur et al. (2019) calculated the energy usage within the food industry depending on the processes, product type, and transportation method based on the literature and reported that the thermal processes consumed large proportions of the total processing energy according to the energy figures. Heat production and subsequent chilling account for an important portion of energy since the precise temperature is an important parameter in the food industry to ensure optimum product. For that purpose, using energy-efficient refrigeration and heat pump solutions utilizing natural refrigerants is important for energy efficiency (GEA 2019).

For the food industry, it is very important to increase energy efficiency through the installation of cost-effective energy-saving technologies. For example, in the brewing industry, malting and wort production are the most energy-intensive operations (Morawicki and Hager 2014). Cordella et al. (2007) applied LCA methodology to the beer life cycle and the highest energy, and water consumptions were calculated at the malting and wort production stages. Kilning is the last stage of the malting process, which is applied to reduce the moisture of green malt by applying heated air that can reach 80–105°C at the final curing stage depending on the beer type. From a sustainable point of view, energy-efficient methods have paramount importance in sustainable food companies. For example, a heat pump technology was developed by GEA to upgrade the waste heat from the cooling plant and transfer it to heat the air used for the kilning process at the Interflour's Intermalt malting plant. Intermalt company declared that heat pump technology results in significant cost savings and reduced energy consumption for a sustainable production process (GEA 2021b). Similarly, energy efficiency is very important in the dairy industry, especially for thermal processes. Transfer of excess heat from the cooling plant to warm-up water for the milk pasteurization process and condensation of evaporated water to be reused in the dairy plant are among the eco-friendly methods applied to sustainable dairy industries (GEA 2021a).

Refrigeration is crucial for the food and beverage sector. Besides improving energy efficiency by smart alternatives to the electrical grid when powering refrigeration plants, it is also very important to use climate-friendly and cost-effective refrigerants (International Institute of Refrigeration 2015). Refrigerant gases are used as a heat exchange material in refrigerators, air conditioning systems, and freezers in the food and beverage industry (Martínez et al. 2021). F-gases that have very high global warming potentials (GWPs), relative to other GHGs, are mainly used as refrigerant gases (EPA 2021). F-gases have a severe and powerful impact on the climate (Sovacool et al. 2021). For controlling and preventing emissions from F-gases, legislation was adopted in different countries to restrict the use of F-gases as refrigerants in refrigeration systems. Many of them have been banned or phased out, and it was reported that the European Union F-gas emissions will be cut by two-thirds in 2030 compared to the 2014 levels as a result of the F-gas regulations (European Commission 2021; Garry 2020). For that purpose, switching from F-gas refrigerants to more environmentally-friendly, energy-efficient, future-proof, and natural refrigerants is very important for the sustainable food industry.

Switching to lower-GWP products is important for sustainable food and beverage companies to be environmentally friendly and also for energy efficiency (Martínez et al. 2021). There is a growing trend toward using natural refrigerants in the food industry, for both ecological and economic reasons. Natural refrigerants are made of substances that occur naturally in the environment and commonly used natural refrigerants, which include CO_2, ammonia, water, air, and hydrocarbons, such as propane, isobutane, and propylene (propene) (EIA 2018; Refrigerants Naturally 2021). They make a negligible or zero contribution to GHG emissions. Moreover, natural refrigerant systems could be more energy-efficient, consequently reducing indirect emissions. Refrigeration systems with natural refrigerants have the lowest lifetime costs when all system expenses are considered from installation and operation to disposal (GEA 2021c). Therefore, natural refrigerant systems have the potential to be more cost-efficient in the long term despite a slightly higher initial investment. In that context, natural refrigerants are very environmentally friendly and are considered less expensive alternatives compared to other synthetic refrigerants (Eurammon 2021; Refrigerants Naturally 2021). Refrigerants, Naturally! is an initiative of international companies taking action against global warming by switching harmful F-gases to eco-friendly natural refrigerants for a safe, reliable, and cost-effective cooling technology (UN 2017). Global companies, including Unilever, Coca-Cola, and PepsiCo, are among the partners of that initiative, and they put HFC-free refrigeration units into operation worldwide by retrofitting or replacing HFC refrigerants with low GWP refrigerants, such as hydrocarbons, ammonia, and CO_2 (Coca-Cola Company 2020; Unilever 2021a).

3.2 Sustainable Agricultural Practices

As reported in the previous section, the entire food system has a great contribution to total anthropogenic emissions. Especially, agriculture is one of the main contributors to climate change due to the high amount of GHG emissions produced on farm and agricultural land (Tubiello et al. 2021). For that purpose, sustainable applications in agricultural and food production have paramount importance. According to the Food and Agriculture Organization of the United Nations, to be sustainable, agriculture must fulfill the needs of present and future generations for its products and services. At the same time, profitability, environmental health, and social and economic equity must be enabled. From the environmental point of view, sustainable agriculture must develop healthy ecosystems and support the efficient use of land, water, and natural resources through sustainable agricultural practices that use technology, research, and development (FAO 2014b).

In the food industry, some companies have developed sustainable sourcing commitments and codes for the application of the best agronomic management practices for advancing the environmental and socio-economic performance of farms in their supply chains (Lam et al. 2021). Many food companies have already taken steps for regenerative agriculture. Unilever reported the regenerative agricultural practices, such as allowing efficient use of resources, ensuring crop production with high yield and nutritional quality, nourishing the soil and decreasing adverse effects on soil fertility, restoring and regenerating the land, increasing water and air quality and biodiversity, optimizing the usage of renewable resources while minimizing the use of non-renewable resources, and capturing carbon (Unilever 2017; 2021c). For Danone, regenerative agriculture includes farming practices that protect soil, water, and biodiversity and respect animal welfare. For that purpose, practices increasing soil quality, managing manure properly, protecting biodiversity, and managing and monitoring water usage in terms of quantity and quality are among the categories assessed (Danone 2021a, b). Dairy and livestock farming is responsible for an important portion of the emissions generated during sourcing the ingredients at Nestle. The company reported some of the sustainable actions for dairy and livestock farming such as feeding livestock with more sustainable feed, thus making farms more productive through better herd management, cutting down CH_4 produced by animals during digestion through nutrition changes, and caring for grassland to store more carbon by using regenerative agricultural practices and organic fertilizers (Nestle 2021).

From environmental and economical points of view, sustainable agriculture stands on the management of land, resources, and ecosystem services (Shelef et al. 2018). Sustainable agriculture land management includes categories, such as nutrient management, agronomic practices, agroforestry, soil and water conservation, integrated livestock management, restoration and rehabilitation, tillage and residue management, and sustainable energy and integrated pest management (Wekesa and Jönsson 2014). For resource management, it is very important to decrease inputs and utilize environmentally friendly and cost-effective renewable sources. Also, maintaining ecosystem services, biodiversity and the livelihood of animals is important for sustainable agriculture (Shelef et al. 2018).

Unfortunately, soil biodiversity declined due to the intensification, expansion, and mechanization of agriculture, and eco-friendly agricultural management practices should be applied to improve habitat quality for beneficial organisms (Uden 2012). Moreover, agricultural soils are among the main anthropogenic source of N_2O since synthetic nitrogen (N) fertilizers significantly contribute to GHG emissions (Chai et al. 2019; Van Groenigen et al. 2010). Good agronomic practices, less resource usage, precision farming, usage of more productive crop varieties, and decreasing N-fertilizer usage by precision agriculture are among the mitigation practices for sustainable agriculture (Bakken and Frostegård 2020; Lam et al. 2021).

It is very important to find solutions to unsustainable practices in agricultural production. Crop production could increase from 60% to 100% by the year 2050 to meet the nutritional needs of the world population of 9–10 billion. In that context, crop production systems with higher yield and nutritional content and, at the same time, with a lower effect on the environment are necessary (Delgado et al. 2019). Healthy ecosystems are connected with sustainable agricultural production, and farmers should be aware of maintaining a good ecological status on their land for their long-term productivity. For that purpose, economic interests should be balanced with eco-friendly methods to maintain a profitable, productive, and sustainable agri-food sector (Uden 2012). Gomes and Reidsma (2021) investigated the awareness of Dutch arable farmers about soil management approaches to minimize soil CO_2 and N_2O emissions in agricultural production in the Netherlands, where intensive farming practices take place; they also reported that it was very important to motivate farmers toward farming systems with healthy soils (Gomes and Reidsma 2021). Some GHGs mitigation approaches for healthy soils and sustainable agriculture include crop rotations, inclusion of cover crops, implementation of no-tillage or other conservation tillage systems to reduce carbon losses, increasing nitrogen use efficiency by matching applied fertilizer quantity and plant demands, application of bio-fertilizers, addition of organic inputs, improved grazing land management, more efficient and accurate water usage, improved irrigation, valorization of organic waste, usage of treated wastewater for agriculture, genotype crop improvement, new crops development, such as using resistant crops to stresses, enhancing tolerance to abiotic and biotic stresses using advanced breeding and biotechnology approaches, application of integrated pest management methods by displacements of polluting, and high energy-consuming chemicals and usage of natural enemies to reduce pests and diseases (Achmon et al. 2016; Delgado et al. 2019; Gomes and Reidsma 2021; Lopes et al. 2011; Oberč and Arroyo Schnell 2020; Ogle et al. 2019; Paustian et al. 2019; Shelef et al. 2018; Thierfelder et al. 2016; Van Groenigen et al. 2010; Witzgall et al. 2021).

3.3 Production of New Foods and Use of New Ingredients With Reduced Environmental Impact and Carbon Footprint

3.3.1 Cellular Agriculture

Cellular agriculture is an innovative approach to sustainable food production. It is defined as the utilization of cell cultures for the production of agricultural products and includes tissue engineering and fermentation-based approaches (Rischer et al. 2020; Stephens et al. 2018). Production of various food products and ingredients by cultivating cells in fermenters instead of traditional agricultural systems has been gaining importance in sustainable food industries. For example, the production of coffee cells in a bioreactor through cellular agriculture was recently reported by VTT scientists from Finland as an innovation that can help to make coffee production more sustainable (VTT Technical Research Centre of Finland 2021).

Cell-based meat has been an emerging field of cellular agriculture (Stephens et al. 2018). It could provide several benefits including environmental sustainability, animal welfare, and human health (Mattick 2018). Cell-based meats are also referred to as "cultured meat", "lab-grown meat", "in-vitro meat", "cultivated meat", "cellular meat", or "clean meat", which is grown from animal stem cells by the applications of tissue engineering techniques and research and development in that field are proceeding fast (Santo et al. 2020). Cell-based meat production is mainly at the laboratory or pilot scale. But the research and innovations in the cultivated meat, chicken, seafood, milk, or supporting technologies field have been proceeding so fast to shift cultivated meat into commercial-scale production (GFI 2020). It is a sustainable solution for the meat industry, but it needs to make significant progress in the price and availability of growth factors to become cost-competitive in the future for the meat industry (Watson 2021b). It was estimated that cell culture media accounts for an important part of production cost; 99% of that cost was due to the price of growth factors (Ben-Arye and Levenberg 2019; Specht 2020). In that context, reducing the cost of growth factors and developing media formulations from animal-free components are critically important. With increasing research to develop new methods and ingredients, the production of cell-based meat at an industrial scale could be enabled in the near term.

3.3.2 Plant-Based Meat Alternatives

Alternative proteins obtained from alternative sources, such as plants, microorganisms, and insects, have been gaining significant attention for the development of eco-friendly food product formulations that have a lower environmental footprint, and these also offer resources to feed a growing world population (Grossmann and Weiss 2021). Meat is an important source of protein and bioavailable micronutrients (Wyness 2016). However, because of increasing consumer awareness and environmental concerns among the public, the development and production of novel alternatives to meat from farmed animals have been increasing (van der Weele et al. 2019). For example, cell-based meat is popular as mentioned in the previous section.

Environmental concerns related to the livestock sector mainly arose from the high level of GHG emissions, land, and water use during the production of meat products. This is due to meat and dairy products in the food industry representing a major part of anthropogenic GHG emissions (Gaillac and Marbach 2021). Xu et al. (2021) stated that global GHG emissions from animal-based food products are twice those of plant-based foods. As they reported, GHG emissions from the production of food were found to be $17{,}318 \pm 1{,}675\,\text{Tg}\ CO_2\text{eq}\,\text{yr}^{-1}$ with 57% of these emissions corresponding to the production of animal-based food, including livestock feed and beef (25%) that were the largest contributing animal-based commodities. Emissions from the livestock sector are mainly related to enteric fermentation, feed production, manure management, and energy consumption (FAO 2017c; Gerber et al. 2013). Enteric fermentation accounts for the CH_4 produced as a part of the digestive process of ruminant animals. Manure acts as a source of both CH_4 and N_2O, and different manure management systems can lead to different emission levels. Moreover, CO_2 and N_2O emissions are produced from feed production processing and transport; also, CO_2 emissions are generated because of energy consumption along the entire supply chain (Gerber et al. 2013; Grossi et al. 2018). Besides energy, water usage is very high in the production of meat products. FAO (2017b) reported that a kilogram of beef takes up to 15 tonnes of water to produce, while a kilogram of cereal needs 1–3 tonnes of water to grow. Also, animal-sourced foods have large land footprints because of the consumption of more food macronutrients by animals than they produce (Hayek et al. 2021; Shepon et al. 2018). As a result of the high carbon, water, and land footprints of meat products, also health and animal-welfare considerations, there is an increasing demand by consumers for diets that are associated with lower environmental impacts (McClements et al. 2021).

Various non-animal protein-rich sources, such as plants, fungi, algae, or cultured meat, have been used as novel alternatives in the production of meat alternatives (McClements et al. 2021; van der Weele et al. 2019). Proteins, fats, carbohydrates, and other substances obtained from non-animal sources, which are physically, enzymatically, or biologically structured to mimic muscle tissue, are used in the production of meat analogs (McClements et al. 2021). Especially, plant-based and cell-based alternatives to meat from farmed animals have been increased. As written in the previous parts, research and development in cellular agriculture to produce cell-based meats are proceeding fast (Santo et al. 2020).

There are some companies and also innovative start-ups that have started research on meat alternatives by using technology to engineer meat in laboratories or produce it from plant-based products (CB Insights 2021; Plant-Based-Meat Global Market Report 2021). For example, Impossible Foods is creating plant-based replacements for meat products that are more sustainable. As it was declared, the Impossible Burger generates 89% fewer GHG emissions, while requiring 96% less land (viable habitat) and 87% less fresh water compared to a traditional beef burger (UNFCCC 2019).

During the production of plant-based alternatives, it is critically important to generate the desired appearance, flavor, texture, nutrition, and functionality of those kinds of foods using sustainable, affordable, and healthy plant-based ingredients (McClements and Grossmann 2021). Various ingredients derived from pulses, grains, root vegetables, oils, and other plants and mycoprotein derived from fungi or seaweed, algae, and microalgae for seafood analogs are used in the production of plant-based alternatives to catch the approximate texture, flavor, and/or nutrient profiles of farmed meat. Especially, soy, wheat, or pea protein isolates are commonly used (McHugh and Avena-Bustillos 2019; Santo et al. 2020). Methods for transforming plant-based ingredients into meat products involve a variety of processes, such as stretching, press forming, kneading, layering, folding, 3D printing, extrusion and shear-cell processing, and different plant-based foods with different forms, textures, and nutritional quality are produced with those methods. Among them, extrusion cooking and shear cell technology are commonly applied (GFI 2021; McHugh and Avena-Bustillos 2019; Rubio et al. 2020). As reported recently, the high moisture extrusion process could offer a cost-effective technology for the large-scale production of meat analogs (Bakhsh et al. 2021; Sun, Ge et al. 2021).

From the cost-competitive point of view, plant-based alternatives are getting cheaper; but they still have a higher price compared with regular meat products (Piper 2021; Plant-Based-Meat Global Market Report 2021). Plant-based meat production requires less grain, water, and energy compared to farmed meat production; however, the meatless alternatives are made on a smaller scale, which results in farmed meat products remaining affordable to consumers compared to plant-based meat alternatives (Rosenberg 2020). Their small-scale production, research and development costs, infrastructure, and processing costs of making plant-based meat products with the same texture as animal meat could be the reason for their high prices (Specht 2019). However, it is expected that with increasing demand and utilization of the most efficient technologies, they will become more cost-efficient during their production (Cohen 2021). Pandemics affected consumer preferences because of the vulnerabilities of the meat supply chain and also viral outbreaks in slaughterhouses, and this has increased the demand for plant-based proteins (Bryant and Sanctorum 2021; Ho 2020b; Smithers 2020). Another reason for the high costs of plant-based meat alternatives is local production at limited facilities, and this small-scale production increases the prices because of the supply chain costs. It is very important to shorten supply chains, lower costs, and ultimately bring more accessible and affordable plant-based products to consumers (Ho 2020b). For example, Plant & Bean is a company producing plant-based alternatives and aims for large-scale production at various locations to significantly shorten supply chains for significant cost-savings. Moreover, the company makes collaborations to develop technology to reduce the cost of plant-based meat ingredients, such as the development of computational breeding technology to reach a 50% decrease in the cost of peas and beans, optimization of protein extraction at a significantly lower cost and improving extrusion technology to produce higher machine efficiencies, lower energy use, and better texture in plant-based meat alternatives (Ho 2020b; Plant and Bean 2021).

3.3.3 Dairy Alternatives

Beef meat and cattle milk are responsible for the most emissions in the livestock sector (FAO 2017c). Therefore, the dairy industry aims to be more sustainable. With increased innovative and modern farming practices, the environmental impact of dairy milk has decreased significantly compared to previous years (Capper and Cady 2019). Most of the energy in the dairy industry is used for heating and cooling during pasteurization and processing raw milk into various dairy products and also for equipment sterilization, lighting, and transport. In that context, energy savings by heat pump technology and renewable energy sources that use less water by treating and reusing wastewater or allow the usage of milk concentrate for some types of dairy products to reduce transport emissions are among the applied methods in the sustainable dairy industry (Gillman 2016). However, because of animal welfare, environmental concerns, health reasons, and consumer preferences, plant-based dairy alternatives have been gaining attention as promising alternatives. The most known plant-based dairy alternatives that are produced from cereal, legume, nut, seed, and pseudo-cereals include almond, coconut, cashew, soy, oat, peanut, hazelnut, rice, sesame, flaxseed, and walnut milk substitutes (Sethi et al. 2016).

Plant-based milk substitutes are designed to have a similar appearance, taste, composition, and structure as cow's milk to be used as dairy alternatives in similar applications (McClements et al. 2019). It is important to simulate the desired quality attributes of dairy milk, such as colloidal texture. Fat globules and casein micelles contribute to dairy milk's creamy texture and appearance (Jukkola and Rojas 2017). Releasing oil bodies and colloidal matter by breaking down the structures of plants or production of simulated fat globules using plant-based ingredients are among the applied technologies for plant-based substitutes (Do et al. 2018; McClements 2020; McClements et al. 2019). Using organic solvents during the extraction of oils is costly and also is not eco-friendly. Therefore, more eco-friendly methods include wet processing by soaking in water to soften the plant, grinding to release the oil bodies, heating and homogenizing, or dry processing by drying and milling into flours (McClements 2020; McHugh 2018).

Plant-based dairy alternatives are generally accepted as sustainable and eco-friendly in terms of carbon emissions, land, water, and pesticide use compared to dairy products. However, different plants could have different growth and processing needs during their production. For example, according to the LCA analyses conducted for California, unsweetened almond milk as a plant-based dairy alternative showed almond milk's GWP is as low as other plant alternatives; however, fresh water consumption is much higher because of increasing irrigation water demand compared to other plant-based milk alternatives that are produced from largely rainfed crops. It is still a more sustainable option compared to dairy milk, but there is a need for high amounts of water and pesticide usage during almond milk production that may cause long-lasting effects on the environment (Winans et al. 2020). Similarly, rice milk has a large water footprint and is associated with higher GHG emissions than other plant-based milk alternatives (Bogueva and Marinova 2021). Soy or oat milk alternatives could be more eco-friendly choices in terms of water use. Since soybeans are legumes, they also fix nitrogen in the soil and enable decreasing the nitrogen fertilizers need (Held 2019; Röös et al. 2018).

On the other hand, oats are typically grown with glyphosate-based pesticides, which is important since the application of pesticides is not eco-friendly for biodiversity (Bogueva and Marinova 2021). Hemp, hazelnut, or pea milk alternatives are just a few other plant-based options. However, more researches and systematic comparisons of dairy and plant-based dairy alternatives are necessary to completely say it is eco-friendly and sustainable to a specific plant-based dairy alternative; indeed, it is difficult to draw solid conclusions when comparing results from different studies (Röös et al. 2018). Ultimately, many plant-based dairy alternatives are more eco-friendly compared to dairy milk, but evaluation of the environmental and biodiversity impacts of specific plants is very important. Research and innovations are progressing very fast in that field, not only in milk but also in dairy products, such as yogurt, cheese, and ice cream produced from various plant-based alternatives. Grossmann and McClements (2021) reported that the use of more environmentally sustainable ingredients compared to animal milk will most likely lead to more sustainable cheese analogs. For that purpose, biotechnology and fermentation companies have already started to develop ingredients to improve plant-based dairy products. Besides sustainability and environmental impacts, the nutrition profile and sensorial properties of plant-based dairy alternatives should be also similar to dairy milk and products. It is important to combine technologies, including plant-based, fermentation, and cell-based inputs, for sustainable production for plant-based alternatives (GFI 2020). Biotechnology companies have already started to develop ingredients through fermentation for plant-based products.

3.3.4 Developing Bio-Based Ingredients from Microorganisms Through Precious Fermentation

Fermentation has been used in food production since ancient times. From a biochemical perspective, it is defined as the anaerobic catabolism of an organic compound. But the role of fermentation has expanded to a much broader range of applications over the past century, and the term covers any microbial product that is produced by anaerobic or aerobic catabolism (Erten et al. 2016; Specht 2021). According to the current meaning, fermentation is the use of microorganisms to produce different fermented foods with desirable characteristics and compounds from small laboratory scale to large industrial-scale production (Nout 2014). A broad range of microbial species plays important roles in food and biotechnology industries not only for the production of fermented foods and beverages but also for aroma compounds, antimicrobial compounds, vitamins, pigments, and some food additives production (Erten et al. 2021). In that context, fermentation science has been started to be used as a sustainable tool in the food industry to produce more eco-friendly ingredients.

From a sustainable perspective, precision fermentation uses microorganisms to produce various biomolecules (EIT Food 2021b). EIT Food (2021b) defines precision fermentation as a sustainable, cost-effective, and environmentally friendly technology to transform the food sector since it does not require animals or fields. Precision fermentation uses microbial hosts to act as cell factories for the production of specific functional compounds, such as enzymes, natural pigments, vitamins, flavoring agents, fats, and proteins. Especially, proteins that have specific and improved features are produced through precision fermentation. There are some companies and start-ups conducting research on those kinds of functional ingredients to improve the sensory characteristics and functional attributes of plant-based products (Specht 2021). For example, Biospringer, a biotech unit under Lesaffre, has produced a new yeast protein, which is specifically developed for plant-based meat and cheese alternatives using the company's patented yeast fermentation technology. The product is created to separate the proteins that could achieve a clean taste profile as well as better textural qualities, and the company reported that the ingredient showed a good melt-behavior in vegan cheese and improved chewability in meat analogs (Ho 2020a). Other companies producing protein alternatives include egg proteins production by EVERY Company, dairy proteins by Perfect Day, and heme protein by Impossible Foods. The EVERY Company launched its animal-free egg product called ClearEgg which is produced by Komagataella phaffii yeast strain via precision fermentation without using animals (Watson 2021a).

Food colorants have common usage in the food industry to improve food quality and safety during processing, storage, and packaging stages, and sustainable production of bio-based food additives and colorants shows a growing trend (Sun et al. 2021). The natural pigments obtained from plants and microorganisms are good alternatives to synthetic food colorants. Nevertheless, compared to plants, microbial pigments are more stable and unaffected by seasonal variations. In addition, the usage of microbial culture for pigment production is sustainable and environmentally friendly since large-scale production with little or no impact on biodiversity and the environment (Rana et al. 2021). This is because a mass switch to natural extracts from fruit and vegetables for the production of natural colorants requires water and land resources and can have significant land, carbon, and water footprints. Therefore, the production of food colorants by fermentation is a sustainable and safe tool (Michail 2020). From a cost-effective point of view, microbial food colorants could be produced at relatively low costs (Sen et al. 2019). Another potential source of natural and sustainable food additives and colorants are by-products and wastes.

3.4 Recovery of Wastes and By-Products

According to FAO, one-third of food produced for human consumption is lost or wasted worldwide, which amounts to approximately 1.3 billion tons per year (FAO 2011b). This amount of food loss and waste is generated during the all food supply chain from initial production to final consumption. The utilization of waste as a renewable, plentiful, and low-cost source for the production of high value-added products has paramount importance for sustainable food industries (Doria et al. 2021).

Recovery of wastes and by-products of various food sectors include fruits, vegetables, alcoholic beverages, olive processing, coffee, dairy, animal and seafood industries that result in rich and low-cost sources of a variety of valuable compounds. Utilizing food-processing residues for the production of high value-added substances is very important for sustainable food industries. In that context, the valorization of agro-food industry wastes could be a cheap source of high value-added bio-compounds that can be used in various sectors (Nasti et al. 2021). However, it is very important to develop an economically feasible, sustainable, and safe recovery method from food waste into higher value-added compounds by a holistic approach as reported by Galanakis (2015). This also contributes to the circular economy approach, which is an economic system that substitutes the concept of end-of-life with reducing, reusing, recycling, and recovering materials in the processes in which products and services are produced and consumed (EIT Food 2021a; Kirchherr et al. 2017). The circular economy uses the by-products from one industrial process to feed into another and aims to create closed-loop processes by focusing on returning wastes as input back to the production processes (EIT Food 2021a; Nikolaou et al. 2021).

Conventional by-product disposal methods of soil fertilizers, animal feed, or biofuels could not provide the industry with enough economic encouragement to deal with environmental issues. Therefore, many innovative extraction technologies have been developed for the extraction of precious bioactive compounds from agri-food industry wastes and by-products. Traditional extraction techniques use a high volume of some toxic solvents and need long extraction times to obtain maximum performance, which results in an additional operating cost. In that context, more "green" technologies than conventional ones such as the elimination of toxic solvents, reducing operating costs with the maximization of yields, and using eco-friendly techniques were developed to become more eco-friendly and cost-effective (Pérez-Serrano et al. 2020). For example, supercritical fluid extraction, ultrasound-assisted extraction, pressurized hot water extraction, pulsed electric fields, or microwave-assisted extraction are among the investigated eco-friendly alternative extraction techniques for the recovery and valorization of high value-added compounds in the food industry (Barba et al. 2019).

Fruit and vegetable losses and wastes do not represent only the wasting of food products but also indirectly include the wasting of limited resources, such as land, water, fertilizers, chemicals, energy, and labor. Moreover, it contributes to environmental problems (Sagar et al. 2018). For that purpose, recovery and valorization of food and beverage wastes which are rich sources of potentially valuable bioactive compounds are very important. For example, the wine industry generates a high amount of waste and by-products, which could be potential sources of bioactive molecules including phenols, vitamins, and dietary fiber (Coelho et al. 2020). Therefore, the recovery of valuable compounds from grape processing by-products is an important sustainable solution for food industries (Barba et al. 2019; Maroun et al. 2017). In addition, the valorization of coffee by-products pays considerable attention since the coffee processing chain results in high volumes of waste and coffee by-products that are rich in carbohydrates, proteins, pectins, and bioactive compounds like polyphenols as cheap renewable resources (Murthy and Madhava Naidu 2012; Nasti et al. 2021). Especially spent coffee grounds and coffee silverskins are two high-volume industrial by-products rich in precious compounds and the recovery yield and purity of those compounds depend on the selected extraction method and operating conditions (Melo et al. 2017). Some recycling applications of spent coffee grounds in various industrial sectors, such as compost, soil conditioner, fertilizers, or bio-based additives have been reported (Girotto et al. 2018; Ronga et al. 2016). For the sustainable food industry, not only the usage of wastes for different purposes, but also the valorization of coffee biowaste within a circular economy approach and the application of eco-friendly green technologies have paramount importance (La Scalia et al. 2021).

Spent coffee grounds contain large amounts of natural antioxidants, lipids, protein, carbohydrates, and other valuable components, and they are also resistant to thermal food processing and digestion process (Martinez-Saez et al. 2017; Muangrat and Pongsirikul 2019). Therefore, they can be incorporated into some bakery products, and this results in a value-added opportunity for coffee by-product utilization at a very low cost (del Castillo et al. 2017; Martinez-Saez et al. 2017). Nasti et al. (2021) reported an eco-sustainable method for the extraction of coffee silverskin by applying a supercritical fluid extraction technique that guarantees a waste-free, safe, and low-cost extraction process. Supercritical fluid extraction is an environmentally friendly technology that is used for the extraction of bioactive compounds from natural sources, such as plants, food by-products, algae, and microalgae. The high capital investment cost is required in the supercritical fluid extraction process; but according to the cost analysis, the high investment for the apparatus is compensated by various benefits including low CO_2 cost and batch times, higher recovery and quality of desired compounds, and absence of pollution treatment costs (Nasti et al. 2021).

Production of high value-added substances in an eco-friendly and cost-effective way is very important for sustainable food industries. Therefore, much research has been conducted and various procedures have been developed to reduce and manage food wastes with their co- and by-products for valorization and recovery purposes. Furthermore, patented and commercialized applications for full-scale production have paramount importance for the implementation of recovery strategies (Galanakis et al. 2015). Because the concept of using waste to make useful products is a key pillar of the circular economy and the creation of circular economy models for foods generates various advantages, including cost benefits (Ellen MacArthur Foundation 2021a). Many food companies have already developed processes for waste reducing, reusing, recycling, and recovering. Besides, global food companies, many start-up companies aim to achieve zero waste for disposal by focusing on new food development from wasted ingredients of farms and food processing plants. For that purpose, upcycled food companies use food processing by-products by transforming them into new value-added products to naturally promote the circularity movement. For example, ReGrained transforms the spent grain that includes protein, fiber, and micronutrients and remains as waste in the brewing process into a flour called "SuperGrain+", which is then incorporated into snack bars (Peters 2019). Similarly, the Coffee Cherry Company has developed and patented an innovative process that converts coffee fruits into flour that can be added to bread, cakes, muffins, noodles, and pasta (IFST 2021). Planetarians is a plant-based ingredient start-up that developed a zero-waste technology by transforming unavoidable by-products from the vegetable oil, brewing, and distilling industries into additional protein through solid-state fermentation by fungi (Askew 2021; Planetarians 2021). This method is very cost-effective compared to producing meat by rearing livestock and also cheaper than other alternative sources of protein, such as soy concentrate or pea isolate (Ellen MacArthur Foundation 2021a). As discussed by Finn and Roversi (2021), upcycling food byproducts and food waste can play in achieving a more circular and sustainable food economy.

3.5 Sustainable Packaging Alternatives

Food packaging is an integral part of food systems since it maintains food safety and also food quality during the processing steps of transport, distribution, and storage and enables the preservation of foods, thereby reducing food loss and waste. Therefore, the packaging design and selection of food packaging materials have paramount importance for the food industry (Otto et al. 2021). Companies are transforming to more sustainable and eco-friendly packaging materials because of increased awareness of environmental concerns and governmental regulations. Moreover, from a circular economy point of view, the reduction, reusing, recycling, and redesign of packaging materials have the potential to decrease the environmental impact of food packaging (Geueke et al. 2018).

Packaging materials cause a high amount of waste and pollution in the environment. In that context, besides the development of sustainable and eco-friendly packaging materials, the collection and sorting of packaging materials for recycling, composting, reusing, waste-to-energy processing, and also reducing material and resource use have paramount importance for sustainability in the packaging value chain (Boz et al. 2020). Conventional packaging materials, such as various forms of plastics, are considered as not environmentally friendly because of the usage of non-renewable sources, low recycling rates, and non-biodegradable forms of the majority of them (Trajkovska Petkoska et al. 2021). In recent years, both scientific research and innovations in the food industry have been progressing on sustainable packaging alternatives. Edible packaging is regarded as a sustainable and biodegradable alternative in the food packaging field, and research have been conducted on edible packaging that uses sustainable and biodegradable materials that are applied as a consumable wrapping or coating around the food, which generates no waste. Various bio-based materials including polysaccharides, proteins, lipids, or composites, were developed for food packaging alone or in combination (Trajkovska Petkoska et al. 2021). The usage of natural materials generated from renewable sources is important for eco-friendly packaging. For example, natural fiber-reinforced polylactic acid (PLA) composite is investigated as an eco-friendly packaging alternative since biodegradable natural/PLA composites are recyclable green materials because of production from renewable sources (Ilyas et al. 2021).

Distinct properties of each packaging material could result in different characteristics. Besides being sustainable and eco-friendly, food safety, convenience, and also cost-effectiveness should be taken into consideration in food packaging design. Compared to petroleum-based fibers, natural fibers are composed of cheap resources and are more environmentally friendly (Anuar et al. 2020). The incorporation of plant-based food waste or by-products raw materials into biodegradable packaging for enhancement of packaging performance is among the other novel approaches (Zhang and Sablani 2021). Various innovative and sustainable packaging solutions have been developed in the food industry worldwide. China dairy heavyweight Mengniu has started to use post-consumer recycled resin-based shrink film as the second packaging, which is a sustainable approach due to the recycling of plastic waste and also reduced GHG emissions. The company reported that despite the current higher cost, with an increased collection of post-recycled resin, it could become more cost-effective (Neo 2021).

From a circular economy perspective, reusing and reprocessing plastic is critically important. Plastic has a low carbon footprint compared to other materials, but single-use plastic is an important source of waste (Unilever 2021b). A study conducted in Singapore reported that some types of plastic bags could have a lower environmental footprint than some forms of packaging materials and instead of single-use, multiple reuses of plastics could minimize the environmental footprint (Ahamed et al. 2021). Global food and beverage companies have already set targets to reduce the total amount of virgin plastic usage and transform plastic packaging design into a more reusable, recyclable, or compostable form for achieving a circular economy for plastic. Moreover, targets include weight reduction, sustainable sourcing, and also packaging design for improved recovery (Boz et al. 2020; Unilever 2021b). For example, Unilever aims to keep plastic in use for as long as possible in a circular loop system by collecting, processing, and repeatedly reusing it (Unilever 2021b). The company aims to reduce the total amount of virgin plastic used, replace virgin plastic with post-consumer recycled plastic, and only use reusable, recyclable, or compostable plastic packaging by 2025 (Ellen MacArthur Foundation 2021b; Unilever 2019). Moving to plant-based tea bags, 100% recyclable bottles or recyclable paperboard-based rings for beverage cans are among the sustainable packaging alternatives in the food industry (Arthur 2020; Food Drink Europe 2021). Therefore, together with packaging alternatives, waste management and process development for re-use purposes are sustainable approaches in food packaging.

4. Conclusions

In this chapter, some eco-friendly and cost-effective processes applied in sustainable food industries have been discussed. Innovative technologies provide sustainable food solutions with increased productivity and reduced pressures on resources. As emphasized above parts, interconnected actions are necessary to transform current food systems to make them more sustainable and resilient. In recent years, sustainability has been among the food quality parameters by many consumers and producers. Therefore, consumer awareness could be a driving force in the progress of sustainable methods in the food industry. On the other hand, food safety is an integral part of sustainable food systems, and it should be a high priority for the sustainable production and distribution of safe and nutritious food. In addition, the application of digitalization in the food systems, such as the application of artificial intelligence, blockchain, the IoT, and big-data analytics contribute to the traceability of food systems. Blockchain technology has emerged as a new concept in the field of food systems to improve food safety, increase efficiency in food processing, and improve transparency, traceability, and security during storage and food supply chains in the sustainable food industries.

In conclusion, overproduction in many industries to meet the growing needs requires high amounts of natural resources. In that context, transforming to more sustainable and resilient food systems is critically important. Undoubtedly, the application of eco-friendly and cost-effective methods in the sustainable food industries will be the core for the continuity of the food systems.

References

Achmon, Y., Harrold, D.R., Claypool, J.T., Stapleton, J.J., VanderGheynst, J.S. and Simmons, C.W. 2016. Assessment of tomato and wine processing solid wastes as soil amendments for biosolarization. Waste Manag 48: 156–164.

Ahamed, A., Vallam, P., Iyer, N.S., Veksha, A., Bobacka, J. and Lisak, G. 2021. Life cycle assessment of plastic grocery bags and their alternatives in cities with confined waste management structure: A Singapore case study. J. Clean. Prod. 278: 123956.

Anuar, H., Apandi, S., Rashid, S., Ali, F., Buys, Y., Manshor, M.R. et al. 2020. The effects of supercritical carbon dioxide on the degradation and antimicrobial properties of PLA biocomposite IIUM Eng. J. 21: 212–227.

Arthur, R. 2020. Coca-Cola introduces recyclable paperboard-based rings on can multipacks, https://www.foodnavigator-asia.com/Article/2020/08/11/Coca-Cola-introduces-recyclable-paperboard-rings-on-multipacks, Accessed 23 December 2021.

Askew, K. 2021. Single ingredient plant-based meat start-up launches fund raise: "We have done the hard work. Now we feel ready to move further', https://www.foodnavigator.com/Article/2021/09/01/Single-ingredient-plant-based-meat-start-up-launches-fund-raise-We-have-done-the-hard-work-Now-we-feel-ready-to-move-further, Accessed 23 October 2021.

Bajan, B. Mrówczyńska-Kamińska, A. and Poczta, W. 2020. Economic Energy Efficiency of Food Production Systems. Energies 13(21): 5826.

Bakhsh, A., Lee, S.-J., Lee, E.-Y., Hwang, Y.-H. and Joo, S.-T. 2021. Traditional Plant-based Meat Alternatives, Current, and Future Perspective: A Review. Journal of Agriculture & Life Science 55: 1–11.

Bakken, L.R. and Frostegård, Å. 2020. Emerging options for mitigating N_2O emissions from food production by manipulating the soil microbiota. Curr Opin Environ Sustain 47: 89–94.

Baldwin, C.J. 2015. The 10 Principles of Food Industry Sustainability. John Wiley & Sons, UK.

Barba, F., Gómez, B., Denoya, G., Brncic, M., Rimac, S., Lorenzo, J.M. et al. 2019. Green Extraction of Nutritional and Antioxidant Valuable Compounds from Wine By-Products, pp. 201–214. *In*: Barba, F.J., Roselló-Soto, E., Brnčic, M. and Lorenzo, J.M. (eds.). Green Extraction and Valorization of By-Products from Food Processing. CRC Press, New York, USA.

Ben-Arye, T. and Levenberg, S. 2019. Tissue Engineering for Clean Meat Production. Front Sustain Food Syst 3: 46.

Bogueva, D. and Marinova, D. 2021. Which plant-based milk is better for the planet? This is what the science says, https://ideas.ted.com/which-plant-based-milk-is-best-for-the-planet/, Accessed 15 November 2021.

Bolandnazar, E., Keyhani, A. and Omid, M. 2014. Determination of efficient and inefficient greenhouse cucumber producers using Data Envelopment Analysis approach, a case study: Jiroft city in Iran. J. Clean. Prod. 79: 108–115.

Boyaci-Gunduz, C.P., Ibrahim, S.A., Wei, O.C. and Galanakis, C.M. 2021. Transformation of the Food Sector: Security and Resilience during the COVID-19 Pandemic. Foods 10(3): 497.

Boz, Z., Korhonen, V. and Sand, C. 2020. March 2020. Consumer Considerations for the Implementation of Sustainable Packaging: A Review. Sustainability 12: 2192.

Bryant, C. and Sanctorum, H. 2021. Alternative proteins, evolving attitudes: Comparing consumer attitudes to plant-based and cultured meat in Belgium in two consecutive years. Appetite 161: 105161.

Canorea, E. 2021. Blockchain Technology To Trace CO2 Emissions and Neutralize Carbon Footprint, https://www.plainconcepts.com/blockchain-carbon-emissions-footprint/#:~:text=One%20of%20the%20great%20contributions,smart%20sensors%20compatible%20with%20IoT.&text=This%20chain%20transforms%20carbon%20emissions,exchange%20to%20offset%20their%20footprint., Accessed 12 January 2022.

Capper, J.L. and Cady, R.A. 2019. The effects of improved performance in the U.S. dairy cattle industry on environmental impacts between 2007 and 2017. Anim. Sci. J. 98: 1.

CB Insights. 2021. Our Meatless Future: How The $2.7T Global Meat Market Gets Disrupted, 09 August 2021, https://www.cbinsights.com/research/future-of-meat-industrial-farming/, Accessed 14 October 2021.

Center for Climate and Energy Solutions. 2021. Renewable Energy, https://www.c2es.org/content/renewable-energy/, Accessed 17 September 2021.

Chai, R., Ye, X., Ma, C., Wang, Q., Tu, R., Zhang, L. et al. 2019. Greenhouse gas emissions from synthetic nitrogen manufacture and fertilization for main upland crops in China. Carbon Balance Manag 14(1): 20.

Chobani. 2019. Chobani 2019 Sustainability Report, https://assets.ctfassets.net/3s6ohrza3ily/5Bry9RmMqnd4dF0Yxr8Vy/bbc8cc7867a831c569b355169325354e/COMP_2019_Sustainability_Project_v17.pdf, Accessed 29 September 2021.

Coca-Cola Company. 2020. The Coca-Cola Company - Climate Change 2020, https://www.coca-colacompany.com/content/dam/journey/us/en/policies/pdf/sustainability/2020-cdp-climate-response.pdf, Accessed 28 September 2021.

Coelho, M.C., Pereira, R.N., Rodrigues, A.S., Teixeira, J.A. and Pintado, M.E. 2020. The use of emergent technologies to extract added value compounds from grape by-products. Trends Food Sci. Technol. 106: 182–197.

Cohen, M. 2021. Impossible Foods, Beyond Meat battle to achieve price parity with real meat, https://www.cnbc.com/2021/08/25/impossible-foods-beyond-meat-battle-price-parity-with-real-meat.html, Accessed 15 October 2021.

Cordella, M., Tugnoli, A., Spadoni, G., Santarelli, F. and Zangrando, T. 2007. LCA of an Italian lager beer. Int. J. Life Cycle Assess 13(2): 133.

Crippa, M., Solazzo, E., Guizzardi, D., Monforti-Ferrario, F., Tubiello, F.N. and Leip, A. 2021. Food systems are responsible for a third of global anthropogenic GHG emissions. Nat. Food 2(3): 198–209.

Danone, 2021a. Danone's Regenerative Agriculture Scorecard, https://www.danone.com/content/dam/danone-corp/danone-com/about-us-impact/policies-and-commitments/en/2021/Danone-regenerative-agriculture-2021-scorecard.pdf, Accessed 05 October 2021.

Danone, 2021b. Regerenative agriculture for a regenerative future, June 2021, https://www.danone.com/content/dam/danone-corp/danone-com/about-us-impact/policies-and-commitments/en/2021/Danone-regenerative-agriculture-2021.pdf, Accessed 05 October 2021.

De Bernardi, P. and Azucar, D. 2020. Innovation in Food Ecosystems Entrepreneurship for a Sustainable Future. Springer, Switzerland.

de Mello Santana, P.H. and Bajay, S.V. 2016. New approaches for improving energy efficiency in the Brazilian industry. Energy Reports 2: 62–66.

del Castillo, M.D., Iriondo-DeHond, A., Martinez-Saez, N., Fernandez-Gomez, B., Iriondo-DeHond, M. and Zhou, J.-R. 2017. Applications of recovered compounds in food products. pp. 171–194. *In*: Galanakis, C.M. (ed.). Handbook of Coffee Processing By-Products: Sustainable Applications Academic Press, UK.

Delgado, J.A., Short, N.M., Roberts, D.P. and Vandenberg, B. 2019. Big Data Analysis for Sustainable Agriculture on a Geospatial Cloud Framework. Front Sustain Food Syst 3: 54.

Despoudi, S. 2021. Green supply chain pp. 35–61. *In*: Galanakis, C. (ed.). The Interaction of Food Industry and Environment. Academic Press, UK.

Do, D.T., Singh, J., Oey, I. and Singh, H. 2018. Biomimetic plant foods: Structural design and functionality. Trends Food Sci Technol 82: 46–59.

Doria, E., Boncompagni, E., Marra, A., Dossena, M., Verri, M. and Buonocore, D. 2021. Polyphenols Extraction From Vegetable Wastes Using a Green and Sustainable Method. Front Sustain Food Syst 5: 342.

EDGAR-FOOD, 2021. EDGAR - Emissions Database for Global Atmospheric Research, A global emission inventory of GHGs from the food systems, https://edgar.jrc.ec.europa.eu/edgar_food#ack, Accessed 06 September 2021.

EIA, 2018. Environmental Investigation Agency-Summary: Technical Report on Energy Efficiency in HFC-free Supermarket Refrigeration https://eia-international.org/wp-content/uploads/Energy-efficiency-in-HFC-free-supermarket-refrigeration-summary.pdf, Accessed 27 September 2021.

EIT Food, 2021a. Future Learn, Circular Business Models for Sustainable Urban Food Systems.

EIT Food, 2021b. Precision Fermentation: From Biotechnology to Sustainable Nutrition, Bringing precision fermentation to the mass market, https://www.eitfood.eu/projects/precision-fermentation-from-biotechnology-to-sustainable-nutrition, Accessed 6 December 2021.

Ellen MacArthur Foundation, 2021a. Alternative meat made from food by-products: Planetarians, https://ellenmacarthurfoundation.org/circular-examples/planetarians, Accessed 23 October 2021.

Ellen MacArthur Foundation, 2021b. Global Commitment 2021 Signatory Report, Unilever Packaging producers and users, https://ellenmacarthurfoundation.org/global-commitment/signatory-reports/ppu/unilever, Accessed 22 December 2021.

EPA, 2021. EPA United States Environmental Protection Agency Global-Greenhouse Gas Emissions Data, https://www.epa.gov/ghgemissions/global-greenhouse-gas-emissions-data, Accessed 10 September 2021.

Erten, H., Boyaci Gunduz, C.P. and Ağirman, B. 2021. Introduction to Yeasts Genera Important in Food Biotechnology, pp. 33–101. *In*: Wu, Q.-S., Zou, Y.-N., Zhang, F., Shu, B. (eds.). An Introduction to Microorganisms. Nova Science Publishers, New York, USA.

Erten, H., Boyaci Gunduz, C.P., Ağirman, B. and Cabaroğlu, T. 2016. Fermentation, Pickling, and Turkish Table Olives, pp. 209–230. *In*: Hui, Y.H., Evranuz, E.Ö., Bingöl, G., Erten, H., Jaramillo-Flores, M.E. (eds.). Handbook of Vegetable Preservation and Processing. CRC Press, Boca Raton, FL, USA.

Eurammon, 2021. Fresh for the future—natural refrigerants in the food industry, https://www.eurammon.com/natural-refrigerants/background-articles/10-background-articles/176-fresh-for-the-future-natural-refrigerants-in-the-food-industry, Accessed 27 September 2021.

European Commission, 2021. EU legislation to control F-gases, https://ec.europa.eu/clima/policies/f-gas/legislation_en, Accessed 23 September 2021.

FAO, 2011a. Food and Agriculture Organization of the United Nations, Energy-Smart" Food for People and Climate, http://www.fao.org/3/i2454e/i2454e.pdf, Accessed 08 September 2021.

FAO, 2011b. Food and Agriculture Organization of the United Nations, Food and Global food losses and food waste – Extent, causes and prevention. Rome, https://www.fao.org/3/i2697e/i2697e.pdf, Accessed 19 October 2021.

FAO, 2014a. Food and Agriculture Organization of the United Nations, The Water-Energy Food Nexus-A new approach in support of food security and sustainable agriculture, http://www.fao.org/3/bl496e/bl496e.pdf, Accessed 09 September 2021.

FAO, 2014b. Food and Agriculture Organization of the United Nations. Sustainable agriculture. Post 2015 and SDGs. Nourishing People, Nurturing the Planet, FAO and the post-2015 development agenda issue papers, March 2014., http://www.fao.org/fileadmin/user_upload/post-2015/14_themes_Issue_Papers/EN/Full_set_Englishrev_01.pdf, Accessed 05 October 2021.

FAO, 2017a. Food and Agriculture Organization of the United Nations. The future of food and agriculture—Trends and challenges. Rome.

FAO, 2017b. Food and Agriculture Organization of the United Nations. Water for Sustainable Food and

Agriculture A report produced for the G20 Presidency of Germany. Rome., https://www.fao.org/3/i7959e/i7959e.pdf, Accessed 12 October 2021.

FAO, 2017c. GLEAM 2.0 - Assessment of greenhouse gas emissions and mitigation potential, http://www.fao.org/gleam/results/en/, Accessed 13 September 2021.

FAO, 2017d. World's future food security "in jeopardy" due to multiple challenges, report warns., http://www.fao.org/news/story/en/item/471169/icode/, Accessed 21 August 2021.

FAO, 2018. Food and Agriculture Organization of the United Nations. Sustainable food systems-Concept and framework.

Food Drink Europe, 2021. Four new ways Unilever is rethinking plastic packaging, https://www.fooddrinkeurope.eu/industry-action/four-new-ways-unilever-is-rethinking-plastic-packaging/, Accessed 14 September 2021.

Food Institute, 2020. Major Food Companies aim to reduce emissions, 22 September 2020, https://foodinstitute.com/focus/major-food-companies-aim-to-reduce-emissions/, Accessed 17 September 2021.

Gaillac, R. and Marbach, S. 2021. The carbon footprint of meat and dairy proteins: a practical perspective to guide low carbon footprint dietary choices. J. Clean. Prod. 321: 128766.

Galanakis, C.M. 2015. The Universal Recovery Strategy, pp. 59–81. *In*: Galanakis, C.M. (ed.). Food Waste Recovery Processing Technologies and Industrial Techniques. Elsevier, UK.

Galanakis, C.M., Martinez-Saez, N., Castillo, M.D.d., Barba, F.J. and Mitropoulou, V.S. 2015. Patented and Commercialized Applications, pp. *In*: Galanakis, C.M. (ed.). Food Waste Recovery Processing Technologies and Industrial Techniques. Academic Press, Elsevier, London, UK.

Garry, M. 2020. Cooling in 2021: Climate Change, Biden Administration to Spark NatRef Adoption, 16 December 2020, https://hydrocarbons21.com/articles/9865/cooling_in_2021_climate_change_biden_administration_to_spark_natref_adoption, Accessed 12 September 2021.

Gasparatos, A., Doll, C.N.H., Esteban, M., Ahmed, A. and Olang, T.A. 2017. Renewable energy and biodiversity: Implications for transitioning to a Green Economy. Renew. Sust. Energ. Rev. 70: 161–184.

GEA, 2019. SEnS-GEA's Sustainable Energy Solutions, https://www.gea.com/en/articles/sens/index.jsp, Accessed 14 September 2021.

GEA, 2021a. Efficiency up, carbon footprint down with GEA Sustainable Energy Solutions, https://www.gea.com/en/customer-cases/aurivo-reduces-fossil-fuel-consumption.jsp, Accessed 08 November 2021.

GEA, 2021b. Going with the grain—GEA brings eco improvements to Intermalt, https://www.gea.com/en/customer-cases/Intermalt-partners-with-gea.jsp, Accessed 08 November 2021.

GEA, 2021c. What are natural refrigerants?, https://www.gea.com/tr/articles/natural-refrigerants/natural-refrigerants-climate-neutral.jsp, Accessed 28 September 2021.

Gerber, P.J., Steinfeld, H., Henderson, B., Mottet, A., Opio, C., Dijkman, J. et al. 2013. Tackling climate change through livestock: a global assessment of emissions and mitigation opportunities. Food and Agriculture Organization of the United Nations (FAO), Rome., http://www.fao.org/3/a-i3437e.pdf, Accessed 13 October 2021.

Geueke, B., Groh, K. and Muncke, J. 2018. Food packaging in the circular economy: Overview of chemical safety aspects for commonly used materials. J. Clean. Prod. 193: 491–505.

GFI, 2020. Good Food Institute-2020 State of the Industry Report Cultivated Meat, https://gfi.org/wp-content/uploads/2021/04/COR-SOTIR-Cultivated-Meat-2021-0429.pdf, Accessed 9 November 2021.

GFI, 2021. Good Food Institute-The science of plant-based meat, https://gfi.org/science/the-science-of-plant-based-meat/, Accessed 18 October 2021.

Gillman, S. 2016. Green milk – making dairy more environmentally friendly, Horizon-The EU Research and Innovation Magazine., https://ec.europa.eu/research-and-innovation/en/horizon-magazine/green-milk-making-dairy-more-environmentally-friendly, Accessed 12 November 2021.

Girotto, F., Pivato, A., Cossu, R., Nkeng, G.E. and Lavagnolo, M.C. 2018. The broad spectrum of possibilities for spent coffee grounds valorisation. J. Mater. Cycles Waste Manag. 20(1): 695–701.

Gomes, A. and Reidsma, P. 2021. Time to Transition: Barriers and Opportunities to Farmer Adoption of Soil GHG Mitigation Practices in Dutch Agriculture. Front Sustain Food Syst. 5: 335.

Grossi, G., Goglio, P., Vitali, A. and Williams, A.G. 2018. Livestock and climate change: impact of livestock on climate and mitigation strategies. Anim Front 9(1): 69–76.

Grossmann, L. and McClements, D.J. 2021. The science of plant-based foods: Approaches to create nutritious and sustainable plant-based cheese analogs. Trends. Food Sci. Technol. 118: 207–229.

Grossmann, L. and Weiss, J. 2021. Alternative Protein Sources as Technofunctional Food Ingredients. Annu. Rev. Food Sci. Technol. 12: 93–117.

Hayek, M.N., Harwatt, H., Ripple, W.J. and Mueller, N.D. 2021. The carbon opportunity cost of animal-sourced food production on land. Nature Sustainability 4(1): 21–24.

Held, L.E. 2019. Which Plant-Based Milk is Best for the Environment?, https://foodprint.org/blog/which-plant-based-milk-is-best-for-the-environment/, Accessed 16 November 2021.

Ho, S. 2020a. Biospringer Debuts Yeast Protein For Plant-Based Cheese & Meat That Combats 'Off-Notes & Off-Tastes', 10 December 2020. https://www.greenqueen.com.hk/biospringer-debuts-yeast-protein-for-plant-based-cheese-meat-that-combats-off-notes-off-tastes/, Accessed 7 December 2021.

Ho, S. 2020b. Plant & Bean To Open Europe's Biggest Plant-Based Meat Production Facility With Plans For U.S. & Asia, 10 December 2020., https://www.greenqueen.com.hk/plant-bean-to-open-europe-biggest-plant-based-meat-production-facility/, Accessed 17 October 2020.

IFST, 2021. Upcycling to a circular food system. Institute of Food Science and Technology, Features, Steven M. Finn&Sara Roversi. Food Sci. Technol. 35(3): 48–51.

Ilyas, R.A., Sapuan, S.M., Harussani, M.M., Hakimi, M.Y.A.Y., Haziq, M.Z.M., Atikah, M.S.N. et al. 2021. Polylactic Acid (PLA) Biocomposite: Processing, Additive Manufacturing and Advanced Applications. Polymers 13(8): 1326.

Inspire Clean Energy, 2021. Is Renewable Energy Cheaper?, https://www.inspirecleanenergy.com/blog/clean-energy-101/is-renewable-energy-cheaper, Accessed 21 September 2021.

International Institute of Refrigeration, 2015. The Role of Refrigeration in the Global Economy. 29th Informatory Note on Refrigeration Technologies. https://sainttrofee.nl/wp-content/uploads/2019/01/NoteTech_29-World-Statistics.pdf, Accessed 24 September 2021.

IRENA, 2015. The International Renewable Energy Agency (IRENA)-Renewable Energy in the Water, Energy & Food Nexus, January 2015, https://www.irena.org/-/media/Files/IRENA/Agency/Publication/2015/IRENA_Water_Energy_Food_Nexus_2015.pdf, Accessed 17 September 2021.

IRENA, 2020. The International Renewable Energy Agency (IRENA) Renewable Power Generation Costs in 2019, https://www.irena.org/-/media/Files/IRENA/Agency/Publication/2020/Jun/IRENA_Power_Generation_Costs_2019.pdf, Accessed 20 September 2021.

IRENA, 2021. The International Renewable Energy Agency (IRENA) , Renewable Power Generation Costs in 2020, https://www.irena.org/-/media/Files/IRENA/Agency/Publication/2020/Jun/IRENA_Power_Generation_Costs_2019.pdf, Accessed 20 September 2021.

Jukkola, A. and Rojas, O.J. 2017. Milk fat globules and associated membranes: Colloidal properties and processing effects. Adv. Colloid. Interface Sci. 245: 92–101.

Kirchherr, J., Reike, D. and Hekkert, M. 2017. Conceptualizing the circular economy: An analysis of 114 definitions. Resour Conserv. Recycl. 127: 221–232.

Kloepffer, W. 2008. Life cycle sustainability assessment of products (with Comments by Helias A. Udo de Haes, p. 95). Int J. Life Cycle Assess. 13, 2: 89.

La Scalia, G., Saeli, M., Miglietta, P.P. and Micale, R. 2021. Coffee biowaste valorization within circular economy: an evaluation method of spent coffee grounds potentials for mortar production. Int. J. Life Cycle Assess. 26(9): 1805–1815.

Ladha-Sabur, A., Bakalis, S., Fryer, P.J. and Lopez-Quiroga, E. 2019. Mapping energy consumption in food manufacturing. Trends. Food Sci. Technol. 86: 270–280.

Lam, W.Y., Sim, S., Kulak, M., van Zelm, R., Schipper, A.M. and Huijbregts, M.A.J. 2021. Drivers of variability in greenhouse gas footprints of crop production. J. Clean. Prod. 315: 128121.

Lopes, C.M., Santos, T.P., Monteiro, A., Rodrigues, M.L., Costa, J.M. and Chaves, M.M. 2011. Combining cover cropping with deficit irrigation in a Mediterranean low vigor vineyard. Sci. Hortic. 129(4): 603–612.

Maroun, R., Rajha, H., Vorobiev, E. and Louka, N. 2017. Emerging Technologies for the Recovery of Valuable Compounds From Grape Processing By-Products. pp. *In*: Galanakis, C.M. (ed.). Handbook of Grape Processing By-Products Sustainable Solutions. Academic Press, Elsevier, London, UK.

Martinez-Saez, N., García, A.T., Pérez, I.D., Rebollo-Hernanz, M., Mesías, M., Morales, F.J. et al. 2017. Use of spent coffee grounds as food ingredient in bakery products. Food Chem. 216: 114–122.

Martínez, C.I.P., Poveda, A.C. and Barreto, D.R.T. 2021. Diagnosis and baseline refrigerant use in the Colombian meat, dairy, and fruit and vegetable industries. Int. J. Refrig. 131: 448–458.

Mattick, C.S. 2018. Cellular agriculture: The coming revolution in food production. Bull. At. Sci. 74(1): 32–35.

McClements, D.J. 2020. Development of Next-Generation Nutritionally Fortified Plant-Based Milk Substitutes: Structural Design Principles. Foods 9(4): 421.

McClements, D.J. and Grossmann, L. 2021. A brief review of the science behind the design of healthy and sustainable plant-based foods. NPJ Sci. Food. 5(1): 17.

McClements, D.J., Newman, E. and McClements, I.F. 2019. Plant-based Milks: A Review of the Science Underpinning Their Design, Fabrication, and Performance. Compr. Rev. Food Sci. Food Saf. 18(6): 2047–2067.

McClements, D.J., Weiss, J., Kinchla, A.J., Nolden, A.A. and Grossmann, L. 2021. Methods for Testing the Quality Attributes of Plant-Based Foods: Meat- and Processed-Meat Analogs. Foods 10(2): 260.

McHugh, T. 2018. How Plant-Based Milks Are Processed. Food Technology Magazine 72: 12.

McHugh, T. and Avena-Bustillos, R. 2019. How Plant-Based Meat and Seafood Are Processed. Food Technology Magazine 73: 10.

Melo, M., Silvestre, A., Portugal, I. and Silva, C. 2017. Emerging technologies for the recovery of valuable compounds from coffee processing by-products. pp. 141–169. *In*: Galanakis, C.M. (ed.). Handbook of Coffee Processing By-Products: Sustainable Applications Academic Press, UK.

Michail, N. 2020. The search for natural and nature-friendly food colours, 25 September 2020, https://insights.figlobal.com/new-product-development/search-natural-and-nature-friendly-food-colours, Accessed 7 December 2021.

Morawicki, R.O. and Hager, T. 2014. Energy and Greenhouse Gases Footprint of Food Processing. pp. 82–99. *In*: Alfen, N.K.V. (ed.). Encyclopedia of agriculture and food systems. Academic Press, Elsevier, UK.

Muangrat, R. and Pongsirikul, I. 2019. Recovery of spent coffee grounds oil using supercritical CO2: Extraction optimisation and physicochemical properties of oil. CYTA J. Food 17(1): 334–346.

Murthy, P.S. and Madhava Naidu, M. 2012. Sustainable management of coffee industry by-products and value addition—A review. Resour. Conserv. Recycl. 66: 45–58.

Nasti, R., Galeazzi, A., Marzorati, S., Zaccheria, F., Ravasio, N., Bozzano, G.L. et al. 2021. Valorisation of Coffee Roasting By-Products: Recovery of Silverskin Fat By Supercritical CO2 Extraction. Waste Biomass Valorization 12(11): 6021–6033.

Neo, P. 2021. Film stars: Mengniu first China food firm to use recycled resin packaging as sustainability demand rise. 6 September 2021, https://www.foodnavigator-asia.com/Article/2021/09/06/Film-stars-Mengniu-first-China-food-firm-to-use-recycled-resin-packaging-as-sustainability-demands-rise, Accessed 20 December 2021.

Nestle, 2021. Accelerate, Transform, Regenerate: Nestlé's Net Roadmap, February 2021, https://www.nestle.com/sites/default/files/2020-12/nestle-net-zero-roadmap-en.pdf.

Nikolaou, I.E., Jones, N. and Stefanakis, A. 2021. Circular Economy and Sustainability: the Past, the Present and the Future Directions. CIES 1(1): 1–20.
Nout, M.J.R. 2014. Food Technologies: Fermentation. pp. 168–177. *In*: Motarjemi, Y., Moy, G.G. and Todd, E.C. (eds.). Encyclopedia of Food Safety, Volume 1. Academic Press, San Diego, CA, USA.
Oberč, B.P. and Arroyo Schnell, A. 2020. Approaches to sustainable agriculture. Exploring the pathways towards the future of farming. Brussels, Belgium: IUCN EURO.
Ogle, S.M., Alsaker, C., Baldock, J., Bernoux, M., Breidt, F.J. and McConkey, B. et al. 2019. Climate and Soil Characteristics Determine Where No-Till Management Can Store Carbon in Soils and Mitigate Greenhouse Gas Emissions. Sci. Rep. 9(1): 11665.
Otto, S., Strenger, M., Maier-Nöth, A. and Schmid, M. 2021. Food packaging and sustainability – Consumer perception vs. correlated scientific facts: A review. J. Clean Prod. 298: 126733.
Paustian, K., Larson, E., Kent, J., Marx, E. and Swan, A. 2019. Soil C Sequestration as a Biological Negative Emission Strategy. Front. Clim. 1(8).
Pelletier, N., Audsley, E., Brodt, S., Garnett, T., Henriksson, P., Kendall, A. et al. 2011. Energy Intensity of Agriculture and Food Systems. Annu. Rev. Environ. Resour. 36(1): 223–246.
Pérez-Serrano, M., Landete-Castillejos, T., Roca-Pérez, L., Brnčić, M., Rimac-Brnčić, S., Lorenzo, J.M. et al. 2020. Green Extraction, pp. *In*: Barba, F.J., Roselló-Soto, E., Brnčić, M. and Lorenzo, J.M. (eds.). Green Extraction and Valorization of By-Products from Food Processing. CRC Press, Taylor & Francis Group, Boca Raton, FL, USA.
Peters, A. 2019. Everything you need to know about the booming business of fighting food waste, https://www.fastcompany.com/90337075/inside-the-booming-business-of-fighting-food-waste, Accessed 16 December 2021.
Piper, K. 2021. Impossible Foods' plant-based meat just got closer to the price of regular meat, https://www.vox.com/future-perfect/2021/2/2/22260454/impossible-foods-burger-plant-based-meat, Accessed 18 October 2021.
Planetarians, 2021. Next Gen Plant Based Meat, https://www.planetarians.com/, Accessed 25 October 2021.
Plant-Based-Meat Global Market Report, 2021. Plant-Based-Meat Global Market Report 2021: COVID-19 Growth and Change to 2030. The Business Research Company, May 2021.
Plant & Bean, 2021. Our Technology Focus, https://plantandbean.com/, Accessed 17 October 2021.
Rana, B., Bhattacharyya, M., Patni, B., Arya, M. and Joshi, G.K. 2021. The Realm of Microbial Pigments in the Food Color Market. Front Sustain Food Syst 5: 54.
RE100, 2021. RE100 Members, https://www.there100.org/re100-members, Accessed 17 September 2021.
Refrigerants Naturally, 2021. Natural refrigerants-Reducing greenhouse gas emissions, https://www.refrigerantsnaturally.com/natural-refrigerants/, Accessed 27 September 2021.
Rischer, H., Szilvay, G.R. and Oksman-Caldentey, K.-M. 2020. Cellular agriculture—industrial biotechnology for food and materials. Curr. Opin. Biotechnol. 61: 128–134.
Ronga, D., Pane, C., Zaccardelli, M. and Pecchioni, N. 2016. Use of Spent Coffee Ground Compost in Peat-Based Growing Media for the Production of Basil and Tomato Potting Plants. Commun. Soil. Sci. Plant Anal. 47(3): 356–368.
Rosenberg, L. 2020. Plant-Based Meat Prices Are About to Drop—but Why Was It Expensive in the First Place? https://www.greenmatters.com/p/why-is-plant-based-meat-so-expensive, Accessed 18 October 2021.
Röös, E., Garnett, T., Watz, V. and Sjörs, C. 2018. The role of dairy and plant based dairy alternatives in sustainable diets. Swedish University of Agricultural Sciences, the research platform Future Food, SLU Repro, Uppsala.
Rubio, N.R., Xiang, N. and Kaplan, D.L. 2020. Plant-based and cell-based approaches to meat production. Nat. Commun. 11(1): 6276.
Sagar, N.A., Pareek, S., Sharma, S., Yahia, E.M. and Lobo, M.G. 2018. Fruit and Vegetable Waste: Bioactive Compounds, Their Extraction, and Possible Utilization. Compr. Rev. Food Sci. Food Saf. 17(3): 512–531.
Santo, R., Kim, B.F., Goldman, S.E., Dutkiewicz, J.A.N., Biehl, E., Bloem, M.W. et al. 2020. Considering Plant-Based Meat Substitutes and Cell-Based Meats: A Public Health and Food Systems Perspective. Front Sustain Food Syst. 4: 134.
Scheidel, A. and Sorman, A.H. 2012. Energy transitions and the global land rush: Ultimate drivers and persistent consequences. Glob. Environ. Change 22(3): 588–595.
Science Based Targets Initiative. 2021. What are 'science-based targets'? https://sciencebasedtargets.org/how-it-works, Accessed 28 September 2021.
Semedo, M.H. 2015. Preface-Opportunities for agri-food chains to become energy-smart, http://www.fao.org/3/ai5125e.pdf, Accessed 14 September 2021.
Sen, T., Barrow, C.J. and Deshmukh, S.K. 2019. Microbial Pigments in the Food Industry—Challenges and the Way Forward. Front. Nutr. 6(7).
Sethi, S., Tyagi, S.K. and Anurag, R.K. 2016. Plant-based milk alternatives an emerging segment of functional beverages: a review. J. Food. Sci. Technol. 53(9): 3408–3423.
Shelef, O., Fernández-Bayo, J.D., Sher, Y., Ancona, V., Slinn, H. and Achmon, Y. 2018. Elucidating Local Food Production to Identify the Principles and Challenges of Sustainable Agriculture, pp. 47–81. *In*: Galanakis, C.M. (ed.). Sustainable Food Systems from Agriculture to Industry. Academic Press, UK.
Shepon, A., Eshel, G., Noor, E. and Milo, R. 2018. The opportunity cost of animal based diets exceeds all food losses. Proceedings of the National Academy of Sciences 115(15): 3804–3809.

SIStat, 2021. Statistical Office of the Republic of Slovenia-Population in the grip of the COVID-19 pandemic., https://www.stat.si/StatWeb/en/News/Index/9566, Accessed 19 August 2021.

Smithers, R. 2020. UK demand for new vegan food products soars in lockdown. 25 July 2020, https://www.theguardian.com/lifeandstyle/2020/jul/25/uk-demand-for-new-vegan-food-products-soars-in-lockdown, Accessed 17 October 2021.

Sovacool, B.K., Griffiths, S., Kim, J. and Bazilian, M. 2021. Climate change and industrial F-gases: A critical and systematic review of developments, sociotechnical systems and policy options for reducing synthetic greenhouse gas emissions. Renew Sust Energ Rev 141: 110759.

Specht, L. 2019. Why plant-based meat will ultimately be less expensive than conventional meat. Good Food Institute., https://gfi.org/blog/plant-based-meat-will-be-less-expensive/, Accessed 18 October 2021.

Specht, L. 2020. An analysis of culture medium costs and production volumes for cultivated meat. The Good Food Institute., https://gfi.org/wp-content/uploads/2021/01/clean-meat-production-volume-and-medium-cost.pdf, Accessed 11 January 2022.

Specht, L. 2021. Good Food Institute-The science of fermentation, https://gfi.org/science/the-science-of-fermentation/, Accessed 7 December 2021.

Stephens, N., Di Silvio, L., Dunsford, I., Ellis, M., Glencross, A. and Sexton, A. 2018. Bringing cultured meat to market: Technical, socio-political, and regulatory challenges in cellular agriculture. Trends. Food Sci. Technol. 78: 155–166.

Sun, C., Ge, J., He, J., Gan, R. and Fang, Y. 2021. Processing, Quality, Safety, and Acceptance of Meat Analogue Products. Engineering 7(5): 674–678.

Sun, L., Xin, F. and Alper, H.S. 2021. Bio-synthesis of food additives and colorants-a growing trend in future food. Biotechnol. Adv 47: 107694.

Thierfelder, C., Rusinamhodzi, L., Setimela, P., Walker, F. and Eash, N.S. 2016. Conservation agriculture and drought-tolerant germplasm: Reaping the benefits of climate-smart agriculture technologies in central Mozambique. Renew Agric. Food Syst. 31(5): 414–428.

Trainor, A.M., McDonald, R.I. and Fargione, J. 2016. Energy Sprawl Is the Largest Driver of Land Use Change in United States. PLOS ONE 11(9): e0162269.

Trajkovska Petkoska, A., Daniloski, D., D'Cunha, N.M., Naumovski, N. and Broach, A.T. 2021. Edible packaging: Sustainable solutions and novel trends in food packaging. Food Res. Int. 140: 109981.

Tran, T.T.D. and Smith, A.D. 2018. Incorporating performance-based global sensitivity and uncertainty analysis into LCOE calculations for emerging renewable energy technologies. Appl. Energy 216: 157–171.

Tubiello, F.N., Rosenzweig, C., Conchedda, G., Karl, K., Gütschow, J. and Xueyao, P. et al. 2021. Greenhouse gas emissions from food systems: building the evidence base. Environ Res Lett 16, 6: 065007.

Uden, G.v. 2012. European Learning Network on Functional AgroBiodiversity (ELN-FAB). Functional agrobiodiversity Nature serving Europe's farmers., https://ec.europa.eu/environment/nature/natura2000/platform/documents/functional_agrobiodiversity_eln-fab_publication_en.pdf, Accessed 07 October 2021.

UN-Water, 2021a. Summary Progress Update 2021: SDG 6—water and sanitation for all. Version: July 2021., file:///C:/Users/User/Downloads/SDG-6-Summary-Progress-Update-2021_Version-July-2021.pdf, Accessed 08 September 2021.

UN-Water, 2021b. United Nations-UNWATER-Water, Food and Energy, https://www.unwater.org/water-facts/water-food-and-energy/, Accessed 07 September 2021.

UN, 2017. United Nations Sustainable Development Goals Partnerships Platform-Refrigerants, Naturally!, file:///C:/Users/User/Downloads/SDG-6-Summary-Progress-Update-2021_Version-July-2021.pdf, Accessed 23 September 2021.

UN Brundtland Report, 1987. United Nations-Report of the World Commission on Environment and Development: Our Common Future.

UNEP, 2016a. Food and food waste, https://www.unep.org/explore-topics/resource-efficiency/what-we-do/sustainable-lifestyles/food-and-food-waste, Accessed 30 August 2021.

UNEP, 2016b. Food systems and natural resources, file:///C:/Users/User/Downloads/-Food_systems_and_natural_resources-2016. Food_Systems_and_Natural_Resources.pdf1.pdf, Accessed 09 September 2021.

UNFCCC, 2019. United Nations- Climate Change. Impossible Foods: Creating Plant-Based Alternatives to Meat | Singapore, Hong Kong, USA, Macau, https://unfccc.int/climate-action/momentum-for-change/planetary-health/impossible-foods, Accessed 14 October 2021.

Unilever, 2017. Sustainable Agriculture Code 2017, https://assets.unilever.com/files/92ui5egz/production/555201fc976d5389fbdf9845099c15b61d1ceb53.pdf/sustainable-agriculture-code--sac---2017.pdf, Accessed 04 October 2021.

Unilever, 2019. Unilever announces ambitious new commitments for a waste-free world, https://assets.unilever.com/files/92ui5egz/production/489410442380812907bc3d97be02ccda1a44ab4b.pdf/Regenerative-Agriculture-Principles-and-Implementation-Guide-April-2021.pdf, Accessed 22 December 2021.

Unilever, 2021a. Climate Transition Action Plan, 22 March 2021, https://www.unilever.com/Images/unilever-climate-transition-action-plan-19032021_tcm244-560179_en.pdf, Accessed 16 September 2021.

Unilever, 2021b. Rethinking plastic packaging, https://www.unilever.com/planet-and-society/waste-free-world/rethinking-plastic-packaging/, Accessed 14 September 2021.

Unilever, 2021c. The Unilever regenerative agriculture principles with implementation guides 2021., https://assets.unilever.com/files/92ui5egz/production/48941044238081 2907bc3d97be02ccda1a44ab4b.pdf/Regenerative-Agriculture-Principles-and-Implementation-Guide-April-2021.pdf, Accessed 05 October 2021.
van de Ven, D.-J., Capellan-Peréz, I., Arto, I., Cazcarro, I., de Castro, C., Patel, P. et al. 2021. The potential land requirements and related land use change emissions of solar energy. Sci. Rep. 11(1): 2907.
van der Weele, C., Feindt, P., Jan van der Goot, A., van Mierlo, B. and van Boekel, M. 2019. Meat alternatives: an integrative comparison. Trends Food Sci Technol 88: 505–512.
Van Groenigen, J.W., Velthof, G.L., Oenema, O., Van Groenigen, K.J. and Van Kessel, C. 2010. Towards an agronomic assessment of N2O emissions: a case study for arable crops. Eur. J. Soil Sci. 61(6): 903–913.
von Braun, J., Afsana, K., Fresco, L., Hassan, M. and Torrero, M. 2020. Food Systems—Definition, Concept and Application for the UN Food Systems Summit. United Nations Food Systems Summit 2021 Scientific Group.
VTT Technical Research Centre of Finland, 2021. Sustainable coffee grown in Finland—the land that drinks the most coffee per capita produces its first tasty cup with cellular agriculture, https://www.vttresearch.com/en/news-and-ideas/sustainable-coffee-grown-finland-land-drinks-most-coffee-capita-produces-its-first, Accessed 11 November 2021.
Watson, E. 2021a. Clara Foods rebrands as The EVERY Company, launches "nearly invisible" animal-free egg protein (without chickens), 05 October 2021, https://www.foodnavigator-usa.com/Article/2021/10/05/Clara-Foods-rebrands-as-The-EVERY-Company-launches-nearly-invisible-animal-free-egg-protein-without-chickens?utm_source=newsletter_daily&utm_medium=email&utm_campaign=05-Oct-2021&cid=DM982320&bid=1728361935, Accessed 7 December 2021.
Watson, E. 2021b. When will cell-culture meat reach price parity with conventional meat?, https://www.foodnavigator-usa.com/Article/2021/03/15/When-will-cell-cultured-meat-reach-price-parity-with-conventional-meat?utm_source=copyright&utm_medium=OnSite&utm_campaign=copyright, Accessed 23 October 2021.
Wekesa, A. and Jönsson, M. 2014. Sustainable Agriculture Land Management, https://viagroforestry.org/app/uploads/2018/11/Vi_SALM-Training-Manual_Chapter-1-to-3.pdf, Accessed 09 October 2021.
Winans, K.S., Macadam-Somer, I., Kendall, A., Geyer, R. and Marvinney, E. 2020. Life cycle assessment of California unsweetened almond milk. Int J Life Cycle Assess 25(3): 577–587.
Witzgall, K., Vidal, A., Schubert, D.I., Höschen, C., Schweizer, S.A., Buegger, F. et al. 2021. Particulate organic matter as a functional soil component for persistent soil organic carbon. Nat. Commun. 12(1): 4115.
World Benchmarking Alliance, 2021. 2021 Food and Agriculture Benchmark-September 2021, https://www.worldbenchmarkingalliance.org/publication/food-agriculture/, Accessed 28 September 2021.
Wyness, L. 2016. The role of red meat in the diet: nutrition and health benefits. Proceedings of the Nutrition Society 75(3): 227–232.
Xu, X., Sharma, P., Shu, S., Lin, T.-S., Ciais, P., Tubiello, F.N. et al. 2021. Global greenhouse gas emissions from animal-based foods are twice those of plant-based foods. Nat Food 2(9): 724–732.
Zhang, H. and Sablani, S. 2021. Biodegradable packaging reinforced with plant-based food waste and by-products. Curr Opin. Food Sci. 42: 61–68.

Part II

Conventional Technologies in the Food Industry

6

Sterilization Methods

Jeanne Moldenhauer

1. Background: Microbiological Control Strategies in the Manufacture of Foods Products

A key consideration in the manufacture of regulated products is determining the desired quality attributes for the product and providing manufacturing processes that can produce these products with the desired quality attributes. There are many different finished product quality attributes that can be important in food products. They are generally known as organoleptic. Typically, they include color, shape, size, smell or odor, taste, or texture. Others include nutritional profile and wholesomeness of the food (Anonymous 2016) Many of the attributes are sensory, based upon customer satisfaction. For sterile products, the absence of microorganisms is essential (Moldenhauer 2014). Manufacturing supplies and equipment should be cleaned and sanitized prior to use in the manufacturing area. Most regulatory agencies have defined expectations defining the type of cleaning or sanitization required.

2. Defining Sterilization

Sterilization processes are used to destroy or eliminate the microorganisms that are present on the surfaces or materials and those within the materials. All sterilization processes are intended to result in materials or surfaces that have become sterile or free from viable microorganisms. While this dictionary definition of sterility sounds good, in the real world, we do not have a way to routinely test or evaluate for the absence of all viable microorganisms in all samples (Agalloco 2008; Moldenhauer 2014; 2019).

When products are labeled sterile, the common assumption is that the vendor is claiming to meet the dictionary definition of sterility. This is further complicated by the fact that most sterile products are released using a national standard—the compendial sterility test methodology. It is common to assume that passing this test indicates that there are no viable organisms present. However, in 1956, Bryce identified two critical limitations of the compendial sterility test method. His first concern was that the viable organisms present in the test sample can only be cultivated if they are able to do so under specific test conditions. Secondly, the number of samples tested is so small (it is a destructive test) that at best it can only be used to provide a gross estimate of the "sterility" of the test sample (Bryce 1956; Moldenhauer 2014; 2019).

Knudsen (1949) added to Bryce's comments to indicate that the sample size is not of a statistically significant population to accurately estimate sterility. As such, the compendial sterility

Excellent Pharma Consulting, Inc., Lake Villa, IL.
Email: jeannemoldenhauer@yahoo.com

test is a poor indicator of test sample sterility. The ability to accurately detect contamination with a 95% confidence level is about 15% when using this test method. The probability of microorganisms surviving is 10^{-1} or 10%. While it is possible to improve these numbers by increasing the sample size, most have not considered this type of increase, because the test cost would increase. Since this is a destructive test, the test samples used in a statistically valid sample plan would be costly and possibly wasteful (DeSantis 2008; Moldenhauer 2014; 2019).

The sterility test method would be more reliable if one were able to detect single microorganisms, in-line, and without destroying the sample. At such a time when a method like this is available for commercial use, the validity of the test may be greatly improved (Sutton and Moldenhauer 2004; Moldenhauer 2014; 2019).

Sterilization processes are utilized to eliminate the microbial contamination that may be present. Scientists have carefully researched the destruction of microorganisms (Agalloco 2008). Defining when a cell is dead is a complex issue. In fact, no single definition exists of when a cell is dead. Many use the definition that the cell is unable to continue to replicate, but microbial cells can have long dormant phases. Some cells can form spores that are resistant to adverse conditions and exist in the spore state for very long periods. Some have even reported spores that have existed for hundreds of years. But when the cell is exposed to favorable growth conditions, these spores can be revived and replicated (Setlow 2009). Some other microorganisms that are not able to form spores can form viable but non-culturable (VBNC) cells. Oliver et al. (1995) states current estimated number of microorganisms in the environment are much lower than the real number of cells present. One of the problems in estimating counts is the capability to recover microorganisms, which is dependent upon the culture methods that are used (Moldenhauer 2019).

The use of a sterilization process is to eliminate the viable microorganisms present to a specified level of sterility assurance (SAL), which is also called the probability of a non-sterile unit (PNSU). Studies are performed to show that the sterilization process is effective. Generating the documentation to show effectiveness and control of the process is called validation. Good science and engineering practices are used to assess the functions and capabilities of the sterilization process (Moldenhauer 2014; 2019).

Most regulatory agencies have an expectation for sterilization processes to be validated. The definition of validation may not be identical in all documents, but all of them include requirements to provide documented evidence of the system's performance within specified parameters. Agalloco (1993) provides a definition that is comprehensive and descriptive:

> "Validation is a defined program which in combination with routine production methods and quality control techniques provides documented assurance that a system is performing as intended and/or that a product conforms to its pre-determined specifications. When practiced in a "life cycle" model it incorporates design, development, evaluation, operational and maintenance considerations to provide both operational benefits and regulatory compliance" (Moldenhauer 2014; 2019).

3. What is the Difference Between Sterilization and Sanitization?

Sanitization is defined as the destruction of microorganisms, which may or may not be pathogenic, on surfaces using chemicals or heat. Sanitizing destroys the microorganisms; however, it may or may not achieve the same lethality offered by sterilization cycles. When using either of these terms, it is useful to clearly define the intended meaning within your documentation system as some individuals would do (Moldenhauer 2013; 2014; 2019).

Some companies use the term sanitization when describing processes used to reduce bioburden, without needing all the validation and documentation associated with a sterilization cycle. For example, the microorganisms may be destroyed, but the cycle is neither designed nor expected to meet the same levels of sterility assurance as a validated sterilization cycle (Moldenhauer 2013, 2014, 2019).

4. Key Concepts for the Validation of Sterilization Processes

Sterilization makes good business sense whether it is required for regulatory requirements or not. There are many things that can be learned during validation—e.g., what parameters are critical to a valid process, having employees learn to correctly execute the process, and how to optimize the process.

There are a variety of fundamental concepts of validation that are applicable when validating sterilization processes. Validation protocols should be based on good science and engineering practices. The scientific method is used in designing and executing validation studies, typically documented in a protocol or series of protocols. This involves the assessment of the critical performance aspects of the sterilizer and the development of methods to evaluate this performance. The experiments are designed based on the hypothesis developed (e.g., sterilization will be accomplished to a specific sterility assurance level). Sterility assurance levels for food products are typically between 10^{-3} and 10^{-6}. (The sterility assurance level is the probability of a non-sterile unit.) The testing is conducted to assess whether the hypothesis developed is true or not. The observations made and the results obtained led to the development of a conclusion. The conclusion identifies whether the testing is acceptable (Agalloco 2008; Moldenhauer 2013; 2014; 2019).

There is the expectation that testing will be conducted to verify that all critical parameters are met. This is accomplished using test equipment, which is properly calibrated or qualified as appropriate. For calibrated equipment, the accuracy should be traceable to established national or international standards (Agalloco 2008; Moldenhauer 2013; 2014; 2019).

Another expectation is for the reproducibility of the system's performance (consistency of performance). In the case of sterilization systems, it is expected that the performance is evaluated within a specific sterilization cycle as well as across a series of cycles (Agalloco 2008) For sterilization, this includes a minimum of three consecutive acceptable cycles as part of the initial validation. An assessment should be conducted to determine the number of studies that should be performed to ensure appropriate testing has been conducted to demonstrate the consistent and reliable performance of the system. For example, if you are qualifying both minimum and maximum load configurations, it may be necessary to perform three studies each for the minimum and maximum load configuration. (Moldenhauer 2013; 2014; 2019).

Worst-case conditions are evaluated during the validation studies. Among the worst-case conditions for sterilization, it is common to test at or below the minimum allowable conditions for exposure time and/or temperature. There is a trend in some companies to just run the validation studies at the nominal production cycles; however, in the event of a cycle that is at the lower acceptable specification limit, one may have no data to support the efficacy of the cycle (Moldenhauer 2013; 2014; 2019).

There is a concept that has developed over the years for sterilization: "the bugs don't lie". While the source of this statement has been attributed to several different individuals, it is widely accepted in the regulated industry. This statement indicates that the microorganisms used as part of the test system evaluate the performance of the sterilizer process. Whether they live or die reflects all the conditions to which there were exposed during the sterilization cycle. For example, it is possible to obtain thermal data that indicates an acceptable cycle was delivered to a sterilizer load, and yet they have biological indicator results that show that an unacceptable condition occurred. Over the years, many have thought "there is something wrong with the biological indicators", when the expected sterilization process was not delivered to all the areas where the biological indicators were placed within the load (Moldenhauer 2013; 2014; 2019).

Validation of the sterilizer once is not enough. It must also be maintained in a validated state, including maintenance of the equipment, software, support systems, etc., so that the validation state is maintained. There should be controls established to maintain the system in a state of control. Procedures should be established to monitor and evaluate changes made to the system, i.e. it should be part of a change control system. When changes are made, they should be reviewed to assess the impact on the validated state of the system. Preventative maintenance should be conducted to ensure that the system is maintained in "good" operating conditions. Access to the system should be limited to appropriately qualified and trained individuals. It is typically not sufficient to only validate the system when changes are made. For most regulated industries, there is an expectation that periodic validation or qualification of the sterilization system is also conducted (Moldenhauer 2013; 2014; 2019).

5. What Regulatory Requirements Must be Met for Sterilization Validation?

Since sterilization is a critical operation, there are many regulatory expectations for the validation of these processes. Within these regulatory expectations, there may be differences in how the validation should be conducted and what parameters are most important in the validation. It is important to understand whether there are regulatory requirements for sterilization that must be met for your affected product and the countries in which the product is sold (Moldenhauer 2013; 2014; 2019).

Additionally, organizations like the Parenteral Drug Association (PDA) have issued technical reports that reflect industry standards for validation, while written for drug and device applications, the science in these documents is sound and useful in designing sterilization cycles. Among the PDA documents governing sterilization are (Moldenhauer 2014; 2019):

- PDA Technical Report 1, Revised 2007, (TR 1) Validation of Moist Heat Sterilization Processes Cycle Design, Development, Qualification, and Ongoing Control (PDA Technical Report Number 1, 2007)
- PDA Technical Report 3 (TR3) Validation of Dry Heat Sterilization and Depyrogenation Cycles
- PDA Technical Report No. 3 (Revised 2013) Validation of Dry Heat Processes Used for Sterilization and Depyrogenation (PDA, 2013)
- PDA Technical Report 7 (TR7) Depyrogenation (PDA Technical Report Number 7, 1981)
- PDA Technical Report 48 (TR48) Moist Heat Sterilizer Systems: Design, Commissioning, Operation, Qualification, and Maintenance. (PDA Technical Report Number 48, 2010)
- PDA Technical Report No. 61 Steam In Place (PDA Technical Report Number 61, 2013b)

The International Organisation for Standardisation (ISO) has also published a variety of documents for sterilization processes. In addition to sterilization, they have documents for calibration and biological indicators. These documents have also predominantly been written for healthcare products but have good science that can be used. Some of the ISO documents include (Moldenhauer 2014 and 2019):

- ISO 13408-2006(r)2012 Aseptic processing of health care products—Part 5: Sterilization in place.
- ISO 14160: 2011 Sterilization of health care products—Liquid chemical sterilizing agents for single-use medical devices utilizing animal tissues and their derivatives—Requirements for characterization, development, validation and routine control of sterilization processes for medical devices.
- ISO 17665-1:2006 Sterilization of health care products—Moist heat—Part 1: Requirements for the development, validation and routine control of a sterilization process for medical devices.
- ISO 17665-3:2013 Sterilization of health care products—Moist heat—Part 3: Guidance on the designation of a medical device to a product family and processing category for steam sterilization.
- ISO 20857:2010 Sterilization of health care products—Dry heat—Requirements for the development, validation and routine control of a sterilization process for medical devices.
- ISO 25424:2009 Sterilization of medical devices—Low temperature steam and formaldehyde—Requirements for development, validation and routine control of a sterilization process for medical devices.

- ISO 11137 Sterilization of health care products package.
- Note: For foods, there are standards for food safety and management that may have relevant information.

The American National Standard Institute (ANSI) has also identified some sterilization documents that include good science and useful information. (Moldenhauer 2014, 2019):

- ANSI/AAMI/ISO 13408-5:2006/(R)2012. Aseptic processing of health care products – Part 5: Sterilization in place (ANSI 2012)
- ANSI/AAMI/ISO TIR 17665-2: 2009. Sterilization of health care products—Moist heat—Part 2: Guidance on the application of ANSI/AAMI/ISO 17665-1

The USP has also generated several documents on sterilization (Moldenhauer 2014, 2019)

- <1211> Sterilization and Sterility Assurance of Compendial Articles (USP 2013)
- <1229> Sterilization of Compendial Articles (USP 2013)
- <1229.1> Steam Sterilization by Direct Contact (USP 2013)
- <1229.2> Moist Heat Sterilization of Aqueous Liquids (USP 2013)
- <1229.3> Monitoring of Bioburden (USP 2013)
- <1229.4> Sterilizing Filtration of Liquids (USP 2013)
- <1229.5> Biological Indicators for Sterilization (USP 2013)
- <1229.6> Liquid Phase Sterilization (USP 2013)
- <1229.7> Gaseous Sterilization (USP 2013)
- <1229.8> Dry Heat Sterilization (USP 2013)
- <1229.9> Physicochemical Integrators and Indicators for Sterilization (USP 2013)
- <1229.10> Radiation Sterilization (USP 2013)
- <1229.11> Vapor Phase Sterilization (USP 2013)
- <1229.12> New Sterilization Methods (USP 2013)
- <1229.13> Sterilization in Place (USP 2013)
- <1229.14> Sterilization Cycle Development (USP 2013)
- <1229.15> Sterilizing Filtration of Gases (USP 2013)

All the documents from regulatory agencies and industry are living documents. They change over time to reflect current expectations for sterilization validation. Some of these documents have conflicting requirements or different requirements than other documents, so it is always important to check the most recent issuance of the document being used (Moldenhauer 2014; 2019).

This list is not all-inclusive by any means. As such, you should be clear in your documents regarding the regulations applicable to you.

6. Types of Sterilization Processes

There are numerous types of sterilization processes: sterilizing filtration, radiation sterilization, gas sterilization, vaporized gas sterilization (decontamination), liquid phase sterilization, dry heat sterilization (and possibly depyrogenation), and moist heat sterilization. Sterilizing filtration and moist and dry heat sterilization are most commonly used in pharmaceutical products, although they are used in some food and drinks; meanwhile, gas sterilization and radiation sterilization are most commonly used in the medical device industry. Liquid phase sterilization (chemical sterilization) has not been widely used in the past in pharmaceuticals but has increasing applicability with the need to sterilize cell cultures and tissue cultures in the biotechnology industry. This type of sterilization is also routinely used in the food industry—e.g., ozonated water (Moldenhauer 2014; 2019).

6.1 Sterilizing Filtration Frequently Misnomered as Aseptic Processing

Sterilizing filtration involves the filling of containers with ingredients that have been sterilized or are sterilized during the process (i.e., through a sterilizing filter). This is an extremely critical process as the filled container is not further processed in any other sterilization cycle. All the components are pre-sterilized and assembled in an aseptic environment using an aseptic technique. This type of sterilization process is dependent upon the physical removal from the material to be sterilized and not on the destruction of the microorganisms present in the item. This section of the chapter specifically refers to a subset of aseptic processing and sterilizing filtration (Moldenhauer 2014; 2019). In recent years, this type of process has been used with foods and drinks to improve the taste and extend the shelf-life.

The filtration process can be used and validated to manufacture sterile products as defined by the USP <1229> Sterilization of Compendial Articles (USP 36–NF31, 1229) This definition does not address the removal of viral contaminants (Moldenhauer 2014; 2019). There are several different factors that can affect the filtration process. When a filter is assessed for the ability to be used in a sterilization process, filter manufacturers frequently ask numerous questions that are taken into consideration in the filter validation like those aspects discussed in USP <1229.4> (USP 2013; Moldenhauer 2014; 2019):

- Type and number of microorganisms present in or on the item being sterilized
- Properties of the liquid being filtered—e.g., whether it is aqueous or oil-based or the entire liquid is soluble, the chemistry of the liquid, viscosity of the liquid, its surface tension, pH, osmolarity, ionic strength, and temperature
- The design of the filter—e.g., whether it is a flat disk, a pleated filter, or it has tortuous paths or non-tortuous paths, the area available for filtration, the nominal pore size, and the thickness of the filter.
- The material used to manufacture the filter—e.g., mixed esters of cellulose, cellulose acetate, cellulose nitrate, polysulfone, and polyvinylidene fluoride [PVDF].
- The process parameters for filtration—e.g., the temperature, flow rate, volumes to be filtered, the filtration time, the differential pressure, and pressure pulsations.

These types of parameters are so important, that the FDA requires that changes affecting many of these parameters be evaluated for their effect on the validation of the process (FDA 2004; Moldenhauer 2014, 2019). The filter selected for the process is critical. As such, it is important to have a reliable supplier (USP <1229.4>, 2013; Moldenhauer 2014; 2019).

The end-user of the filter has many responsibilities in the sterilization process; one of which is to validate the effectiveness of the filter to sterilize the liquid processed through it. The user should define the allowable and validated parameters for the process and then establish controls to ensure that operations take place within the validated process. Some of the controls to be established are the requirements for filter integrity testing, including both how and when it will be performed. There also should be requirements that specify what solution is evaluated in the test—e.g., the actual product, placebo, or other solutions. Additionally, they should have microbiological controls established to ensure that the bioburden level of the process does not exceed the capabilities of the filter (Moldenhauer 2014 and 2019).

Specialized clean rooms are used for the sterilizing filtration process (ISO 5/Grade A). Very tight standards exist for the particulate and viable microorganism levels that are allowed in these areas (Moldenhauer 2014). Validation of the process for sterilizing filtration of liquids is achieved by the successful performance of process simulation tests, which are also called media fills. Media fills are conducted after the completion of validation of the supporting systems, such as equipment sterilization, room and area cleaning and sanitization, qualification of disinfectants, environmental monitoring, validation of HVAC systems, gowning qualification, and so forth. An important pre-requisite is sufficient operator training to clearly understand acceptable aseptic behavior in a clean room. This is critical to maintaining the aseptic area (Moldenhauer 2014; 2019).

It can be difficult for clean room personnel to comprehend the impact of a quick movement in the room on the contamination level. One of the concerns is how slow should the movements be made to protect the environment. Some of the newer rapid microbiology technologies for environmental monitoring have the capability to measure viable microorganisms in real time. One of these technologies even has a webcam associated with the product to record the actions taking place, especially in the event of a deviation being recorded. Allowing operators to mimic various actions they use and see/hear that it affects the counts and whether they stay within limits can be useful in training (Moldenhauer 2014; 2019).

Designing the validation studies to perform requires careful thought and planning. One intent of the studies is to show that various interventions that occur in normal production can be performed while maintaining the sterility of the solution filtered. One must also show that the process controls established are sufficient to produce a sterile liquid.

The sterilizing filter used should have a pore size of 0.2 µm or smaller. It needs to be validated to ensure that it can remove microbiological contaminants (excluding virus removal) from the liquid. Challenge studies are typically performed with Brevundimonas diminuta (ATCC19146) as this organism is very small and a good challenge for this filter pore size. The validation should include a challenge of at least 10^7 microorganisms per cm^2 of filtration area (FDA 2004; Moldenhauer 2014; 2019).

Unlike many of the other types of sterilization processes available, you cannot calculate the sterility assurance level (SAL) of an aseptic filtration process. Many companies claim it has a SAL of 10^{-3}, but there are no valid calculations to do this. When we run process simulation runs (media fills) to validate the process, we only get a snapshot in time of the contamination present in the process. This is different from a SAL calculation where you are determining the probability of a non-sterile unit in the process. Lack of contamination on one day does not give any mathematical probability of whether contamination will occur on a subsequent day (Moldenhauer 2014; 2019).

Sterilizing aseptic filtration is the preferred method of sterilization for liquids that cannot withstand the rigors of other sterilization cycles. It does not add heat to the product and rarely, if ever, negatively affects the stability of the product (Moldenhauer 2014).

6.2 Chemical Sterilization Methods—Also Known as Liquid Phase Sterilization

Chemical sterilization methods may be useful in situations where the item to be sterilized cannot withstand the heat associated with many other sterilization methods. Additionally, some items cannot be subjected to sterilizing filtration, e.g., thick solutions and items with chunky components. Chemical sterilization methods are also useful in re-sterilizing medical devices, sterilizing in place lines within the equipment, decontaminating equipment and drains, and others. Several of these methods can also be used to remove biofilm. Chemical sterilization methods can be useful for these types of situations. Chemical sterilization involves subjecting the item to exposure to a chemical sterilant like the following (Moldenhauer 2014; 2019):

- Aldehydes: glutaraldehyde and formaldehyde. Note: formaldehyde and glutaraldehyde are accepted more often in countries outside the USA.
- Acids: peracetic acid, nitric acid, and sulfuric acid.
- Bases: sodium hydroxide and potassium hydroxide.
- Oxygenating compounds: hydrogen peroxide, ozone (typically used as ozonated water), and chlorine dioxide. Note: for many foods and cell culture operations, ozonated water is used to achieve sterilization.
- Halides: sodium hypochlorite and chlorine.

Liquid phase sterilization is a type of chemical sterilization where the item is submersed or rinsed in the chemical solution. When these processes are conducted under the specified conditions, it is possible to eliminate bacterial and fungal contaminants (USP <1229.6>, 2013b; Moldenhauer 2014; 2019). In addition to the solutions mentioned in USP <1229.6>, newer disinfecting solutions have been identified that use polarity or electrical charges to inactivate microorganisms. Some of these are promising for this type of sterilization as these methods are typically bactericidal, sporicidal, fungicidal, and viricidal. They can also be used to eliminate prions.

A key consideration in using chemical sterilants is whether the article to be sterilized is chemically compatible with the chemical sterilant. Many chemical sterilants can pose risks to the individuals working with them. As such, information on the safety issues associated with the sterilant is very important. In most cases, the vendor of the sterilant can provide this information. There are also many published articles on material compatibility with chemical sterilants (Moldenhauer 2014; 2019).

An important consideration with using liquid phase sterilization is how the item will be removed and handled after being in the chemical sterilant. It is critical that it is not recontaminated as it may become unsuitable for use following the sterilization process. This is extremely important when handling re-usable medical devices—i.e., the maintenance of sterility. Liquid phase sterilants are being evaluated for use in grocery store sprayers (over vegetables and fruits) to aid in stopping contamination and prolonging the product shelf-life.

The key factors to consider in the effectiveness of the sterilization cycle are the concentration of the sterilant and temperature, although the pH can be a factor for some sterilants, the extent of mixing (if used), and soil (cells or cellular debris) may also be important. During validation, it is common to use Bacillus atrophaeus ATCC 9372 or Bacillus subtilis ATCC 6633 as the biological indicator since they are representative of worst-case bioburden isolates for this type of process. The indicators are directly inoculated upon the surface to be sterilized. The placement of the biological indicators should be determined by evaluating the most difficult areas for the sterilant to reach (USP <1229.6>, 2013b; Moldenhauer 2014; 2019).

Validation may be conducted using a half-cycle approach or the bracketing method. The premise of this approach is that a cycle is determined (exposure dwell time) that provides the total kill of the biological indicator (starting population of 10^6). The exposure dwell time for routine production cycles is twice as much, resulting in a sterility assurance level or probability of a non-sterile unit of 10^{-6}. So, if you could validate a 10-minute cycle to yield the total kill of the biological indicator at 10^6 population, routine production would use a 20-minute cycle to provide a SAL of 10^{-6} (USP <1229.6>, 2013b; Moldenhauer 2014 and 2019).

Another method that can be used for validation is the bracketing approach. In this method, the worst-case parameters for over-processing and under-processing are validated, allowing the use of a cycle between these two sets of parameters. It is based upon the needs for the materials in the process and the bioburden. It is possible then for the user to determine the death rate for both extreme conditions tested. Short exposure times are common because of the high inactivation rate of the sterilants. As such, one may have difficulty selecting cycles with recoverable biological indicators (USP <1229.6>, 2013; Moldenhauer 2014; 2019).

The equipment used in these types of processes is subjected to the typical installation qualification and operational qualification. The loading of the equipment for the sterilization process is important. Chemical sterilants are surface sterilants, so the parts to be sterilized must contact the sterilizing agent. This can be aided by the use of mixing or recirculation during the sterilization process. For some of the water-based sterilants, the flow rate of the solution may also be an important parameter. It is common to define a maximum load per sterilizing chamber as it represents the maximum surface area to be sterilized (USP <1229.6>, 2013b; Moldenhauer 2014).

Successful validation is accomplished by performing sufficient replicates to show uniformity and reproducibility. During these cycles, parameters include biological indicator kill, sterilant concentration, exposure time, agitation or recirculation rates, and other key physical parameters (USP <1229.6>, 2013; Moldenhauer 2014; 2019).

Unlike some other forms of sterilization, it is important to ensure that all the chemical sterilant is removed or inactivated following sterilization. This process must maintain the item in a sterilized manner. This should be evaluated from the completion of sterilization through placement into a sterile sealed container (USP <1229.6, 2013; Moldenhauer 2014; 2019).

6.3 Gas Sterilization

Many types of instruments and medical devices cannot withstand the sterilization conditions associated with moist and dry heat sterilization. Gas sterilization is most commonly used with these types of equipment. There have been concerns about the environmental impact of ethylene oxide and concerns about its carcinogenicity as well. As such, many gas sterilization processes use ethylene oxide (EtO); however, there are other gases that are being used today, e.g., chlorine dioxide, mixed oxides of nitrogen (nitrogen dioxide), and ozone (O_3). There are some sterilants that exist in both a liquid and gas phase at the used temperatures, like hydrogen peroxide, peracetic acid, and paraformaldehyde. Some of these types of sterilants are discussed separately in the section entitled 'Vapor Sterilization' (Agalloco 2008; USP <1229.7>, 2013c; Moldenhauer 2014; 2019).

The key parameters of gas sterilization processes include concentration, relative humidity, pressure, and temperature. Ethylene oxide (EtO) has the unique ability to penetrate through polymers, papers, and other materials. As such, it is frequently used for terminal sterilization of medical devices, even in their final packages. There are other gases that may be used for these kinds of processes (USP <1229.7>, 2013c; Moldenhauer 2014; 2019). Gases typically do not penetrate containers like glass or hard plastics. If the part inside a solid container needs to be sterilized, it may not be the best type of process to use.

True gases are used allowing for single-point monitoring during operation to determine lethality. In some cases, more than one type of gas may be used concurrently (mixing). A difference between gas sterilization and other processes is that it does not condense during operation. The cycles must be developed to ensure sterility and maintain the quality attributes of the product (USP <1229.7>, 2013c; Moldenhauer 2014; 2019).

Other important parameters for this type of cycle include the immediate effects of the gas on the item being sterilized, the residual sterilants on the item, the by-products of the sterilants, and potential chemical reactions. Some chemical sterilants can be very corrosive, so this needs to be taken into consideration as well (USP <1229.7>, 2013c; Moldenhauer 2014; 2019).

6.3.1 Ethylene Oxide (EtO)

This type of process is used to sterilize items that are sensitive to temperatures greater than 60°C and/or cannot be radiation sterilized. Some typical items include plastics, options, and electrics (Anonymous 2013). In ethylene oxide (EtO) sterilization, the strong alkylating agent destroys microorganisms—primarily the cell DNA. This type of destruction follows first-order kinetics. It is dependent upon the concentration, humidity, and temperature (USP <1229.7>, 2013c; Moldenhauer 2014, 2019).

A typical EtO cycle operates between 30°C and 60°C, with a relative humidity greater than 30%, a gas concentration between 200 and 800 mg/L, and a typical cycle time of at least three hours. EtO has good penetrability and is very effective, killing viruses, bacteria and bacterial spores, and fungi. It is compatible with a variety of materials, even with more than one sterilization treatment. On the negative side, it is flammable, toxic, and a carcinogen. It even has the potential to be teratogenic (Anonymous 2013; Moldenhauer 2014; 2019).

Typical sterilization cycles include preconditioning, rehumidification in the chamber, air removal, the gas exposure period, removal of gas from the sterilizer, and post-sterilization aeration to remove toxic residues. These cycles are widely used for medical devices. (Anonymous 2013; USP <1229.7>, 2013c; Moldenhauer 2014; 2019).

EtO is different from heat-based cycles; in that, the gas is added at the beginning of the cycle with minimal additions of gas—i.e., only to maintain pressure. This gas can be explosive at concentrations greater than 3% volume in the air (USP<1229.7>, 2013c). Validation and routine biological monitoring is conducted using Bacillus atrophaeus (formerly called Bacillus subtilis variety niger and was also known as Bacillus globigii) (USP <1229.7>, 2013c; Moldenhauer 2014, 2019).

6.3.2 Ozone (O_3)

Ozone is used in a variety of settings for food, medical devices, and pharmaceuticals. It is used to sterilize water and air. It is also used as a surface disinfectant and sterilant. Unlike many disinfectants, it can oxidize most organic matter. On the negative side, it can be toxic and unstable (i.e., it has a short half-life), so it needs to be produced on-site (Anonymous 2013; Moldenhauer 2014; 2019).

Ozone is effective on a wide range of pathogens, including prions. It does not need the use of hazardous chemicals to be effective. It is manufactured using medical-grade air. Following the use of sterilization, the waste material is converted back to oxygen (Anonymous 2013; Moldenhauer 2014; 2019).

Ozone is a strong oxidizing agent. It is manufactured by passing a stream of oxygen or air through a high-voltage electrical field. It is lethal in concentrations from 2% to 10% in the air. It works best when the relative humidity is above 80%. Its degradation products are oxygen, in the presence of moisture and metals. Unfortunately, it does not penetrate as well as EtO (USP <1229.7>, 2013C; Moldenhauer 2014; 2019).

A typical sterilization process includes humidification, injection, exposure, and ventilation to remove the ozone following the completion of the cycle. Biological challenges typically utilize Geobacillus stearothermophilus (USP <1229.7>, 2013c; Moldenhauer 2014; 2019).

6.3.3 Nitrogen Dioxide (NO_2)

This type of gas can be used as a rapid and effective sterilant. It is effective against many microorganisms including bacteria, viruses, and spores. This gas can be used in an enclosed environment at room temperature and ambient pressure. Cells are inactivated by the degradation of DNA. It is effective at low concentrations of gas. It has a low boiling point of 21°C at sea level. As such, there is a high saturated vapor pressure at ambient temperature. With the low boiling point and the high saturated vapor pressure, condensation does not occur on the items subjected to the sterilization process, and no post-sterilization aeration is required. It is compatible with many medical-grade materials and adhesives. Geobacillus stearothermophilus is used as the biological indicator for this process. (Anonymous 2013; Moldenhauer 2014; 2019).

6.3.4 Chlorine Dioxide (ClO_2)

Another gas used for sterilization is chlorine dioxide. This type of gas is metastable meaning that it is stable, provided it is subjected to no more than small disturbances. Due to the difficulty in maintaining stability, it must be generated when needed for use. It is not a carcinogen nor is it flammable. It works at ambient conditions but does not penetrate as well as EtO (USP <1229.7>, 2013c; Moldenhauer 2014; 2019).

"Chlorine dioxide is a chemical compound with the formula ClO_2. This yellowish-green gas crystallizes as bright orange crystals at −59°C. As one of several oxides of chlorine, it is a potent and useful oxidizing agent used in water treatment and bleaching" (Moldenhauer 2014; 2019).

Typical sterilization cycles include preconditioning, conditioning dwell period, charge, exposure, and aeration. The biological indicator of choice is Bacillus atrophaeus (USP <1229.7>, 2013C;

Moldenhauer 2014; 2019). Eylath et al. (2003) describe the successful use of this technology when sterilizing an aseptic filling line isolator. The authors reported that the chlorine dioxide distribution was excellent in this study with variability throughout the isolator and its associated ducts measured at less than 0.2 mg/L at an exposure concentration of 5 mg/L."

Lowe et al. (2013) found gaseous ClO_2 effective for sterilization in hospital rooms, when it was tested against common contaminants for nosocomial infections. In this study, concentrations of ClO_2 up to 385 ppm were safely maintained in a hospital room with enhanced environmental controls.

6.3.5 Validating Gas Sterilization Processes

The most common method of validation is the half-cycle approach. It is a variety of overkill approaches. In this cycle type, you determine what is required to kill all the biological indicators (10^6 spores) for validation and routinely double the cycle for routine production, corresponding to a PNSU of 10^{-6}. There have been other approaches to validation developed, including the Gillis and Mosley Parametric Evaluation and a Bracketing Approach (USP <1229.7>, 2013c; Moldenhauer 2014; 2019).

Equipment qualification is conducted like other sterilization equipment. It should include the requirements for all phases of sterilization from the prehumidification steps through any post-cycle aeration or other phases. Completion of this qualification should show that the equipment is installed and operating as intended.

Empty Chamber Distribution Study: Many still perform this type of study to show that the methods used for the introduction of gas and humidity are consistent and uniform. It may also provide useful information on the routine monitoring positions to use. No biological indicators are utilized (USP <1229.7>, 2013; Moldenhauer 2014).

Component and Load Mapping: This type of study is typically not included for gas sterilization, as the conduct of the study would alter the gas and humidity parameters. Biological indicators (passive) or control devices may be placed within load items that are expected to be the hardest for gas and humidity penetration to evaluate these parameters (USP <1229.7>, 2013c; Moldenhauer 2014).

Biological indicators: Bacillus atrophaeus (ATCC 9372) is used for all but ozone sterilization. For ozone Geobacillus stearothermophilus (ATCC 12980 or 7953) is the indicator of choice (USP <1229.7>, 2013c; Moldenhauer 2014).

Process Confirmation—Microbiological Challenge Studies: Combination studies are performed using physical, chemical, and microbiological challenges concurrently. The sensors are placed in the chamber, while the biological challenges are located within the load items. Replicate studies are required to show reproducibility and efficacy (USP <1229.7>, 2013c; Moldenhauer 2014).

Routine Process Control: Once validated, care should be taken to maintain the sterilizer in a qualifying condition. This includes things like change control, the establishment of procedures for operation and control, performing periodic evaluations of performance, routine calibration, and so forth.

6.4 Vapor Sterilization

Various types of vapor can be used for sterilization including hydrogen peroxide, formaldehyde (not widely accepted for use in the USA, but is used in the EU), and peracetic acid. Some also choose to vaporize ozone gas. In many cases, they are used and claimed for decontamination rather than sterilization. The chemical vapors have limited volatility at room temperature in the air. Raising the temperature of the inlet stream is used for hydrogen peroxide and formaldehyde to increase efficiency. It is mixed at temperatures above the boiling point, and the vapor and air are mixed. The hot "mixed" air is mixed further in the sterilizer chamber. Unfortunately, as the hot air contacts the colder surfaces in the chamber, condensation occurs. Aggressive mixing is needed in the vessel to achieve uniformity. The system lends itself to high relative humidity but care must be taken to achieve uniform temperature ranges across the chamber in order to get uniform humidity. Some variation always occurs with the inlet stream being very hot. It is also likely to obtain differences in concentration on the surface of water and vapor. Minimal penetration exists for the vapors. The use of vapors for sterilization is intrinsically more difficult than true gas sterilization processes (Agalloco 2008; Moldenhauer 2014; 2019).

6.4.1 Hydrogen Peroxide (H_2O_2)

In a hydrogen peroxide cycle, the first step is to remove the humidity from the chamber. This is accomplished using low humidity air circulation, followed by brief conditioning to increase the hydrogen peroxide concentration rapidly. The "exposure dwell period" is next. During this period, the hydrogen peroxide concentration is maintained by continuous addition or periodic injections. Following exposure, the aeration phase takes place using low humidity air to re-evaporate the hydrogen peroxide from the surfaces (Agalloco 2008; Moldenhauer 2014; 2019).

Hydrogen peroxide is a strong oxidizing agent, which makes it effective against a wide range of pathogens. It can be dangerous as concentrations are higher than 10% w/w. The cycle time for hydrogen peroxide is much lower than for EtO. The strong oxidizing power of hydrogen peroxide can also be a negative, as it can have issues with material compatibility. According to the information for the Sterrad system, paper products cannot be sterilized with this method due to cellulostics, a process where all the hydrogen peroxide is absorbed by the paper product. Additionally, this vapor does not have the penetrating power of EtO (Anonymous 2013; Moldenhauer 2014; 2019).

There are safety concerns for personnel, which also must be addressed with the use of hydrogen peroxide. The vapor is frequently used to sterilize large areas and sealed areas, like clean rooms and the inside of aircraft (Anonymous 2013; Moldenhauer 2014; 2019).

6.4.2 Peracetic Acid

Sterilization is achieved by misting this product and water throughout the chamber. Wetted surfaces are not sterilized. This necessitates the use of care in arranging the items in the load and the injection site. The items are held for a defined period, and the liquid is evaporated using how humidity air. It can be an effective surface sterilant (Agalloco 2008; Moldenhauer 2014; 2019).

6.4.3 Validating Vapor Sterilization Processes

In vapor sterilization, it is not as easy to measure concentration, humidity, and temperature (though not as important a parameter) as in gas sterilization. The key is to have reproducible attainment of the combined parameters. Cycles are performed to show reproducibility using a resistant biological indicator. One can use fraction negative cycles to show the resistance of the biological indicator to the process (Agalloco 2008; Moldenhauer 2014; 2019).

Most use the half-cycle approach for the qualification of these processes. The cycle is set at 8 to 9 times the D-value, which allows for an appropriate safety factor to be established. The routine cycle is twice this length. All biological indicators must be inactivated (Agalloco 2008; Moldenhauer 2014; 2019).

6.5 Radiation Sterilization

There are a variety of radiation processes that can be used, but most utilize gamma rays or electron beams. X-rays and ultraviolet light are also used. The effect of the radiation on the items being sterilized should be assessed. The radiation is used to destroy the microorganisms present. Radiation sterilization has several advantages including simplicity, lack of mechanical complexity, reproducibility, and overall efficiency (USP <1229.10>, 2013; Moldenhauer 2014; 2019).

The key concept of the sterilization process is the amount of radiation (dose) that is absorbed by the item. This dose can be precisely measured. ISO 11137-1 (sterilization of health care products) describes the various ways to determine the dose that should be used. They have methods called, Method 1, Method 2A, Method 2B, and Method VDmax. These methods have different testing approaches and a number of times to be used for testing. These differences are based on the assumptions regarding bioburden. Biological indicators are not used since accurate correlations exist between the measurement of the dose and the destruction of a variety of microbes. The measurement process—dosimetry—is used to assess the lethality achieved in the process. It is measured in kGy. It is directly related to the microbial lethal effects due to radiation, like the F_0 calculation in moist heat sterilization. Dosimeters can be used on the items being sterilized to assess process performance (USP <1229.10>, 2013; Moldenhauer 2014; 2019).

6.5.1 Gamma Sterilization

Gamma sterilization is conducted in a specialty facility that is tightly controlled. In this facility, the items to be sterilized are subjected to radiation from Cobalt 60 (Co60). This type of radiation allows for a uniform dose to be delivered. Gamma rays are emitted from the Co60 during the decay process where it becomes Ni60. This specific isotope has a half-life of 5.25 or it loses about 12% of its radioactivity each year. Due to this decay process, the cycle requires regular adjustments to ensure that the required dose is delivered. This may necessitate adding additional Co60 in the process to maintain the appropriate dose (USP <1229.10>, 2013; Moldenhauer 2014; 2019). The items being sterilized by irradiation do not become radioactive (Anonymous 2013). However, the radiation may affect the properties of the material (Bharati et al. 2009; Moldenhauer 2014; 2019).

6.5.2 X-ray Sterilization

Bremsstrahlung or high-energy (high power) X-rays are used in medical device processes for sterilization. They are useful since large packages and pallet loads can be sterilized using this technology. The ability to penetrate the package is very good when using loads of low-density packages. It has good dose uniformity ratios (Anonymous 2013; Moldenhauer 2014; 2019).

This type of sterilization is an electricity-based process. Neither chemical nor radioactive material is required. An X-ray machine is used to generate X-rays. Fortunately, this machine can be turned on when in use and turned off when not in use. It also does not require the various protections needed for other types of radiation sterilization, e.g., shielding (Anonymous 2013; Moldenhauer 2014; 2019).

X-rays are a result of colliding accelerated electrons with a dense material called the target. Materials like tantalum and tungsten may be used. The conversion requires a higher level of electrical energy than some other systems (Anonymous 2013; Moldenhauer 2014; 2019).

The photons generated in this type of process are highly penetrative. The materials to be sterilized are scanned with X-ray photons. If the systems are maintained appropriately, constant sterilization is then delivered (USP <1229.10>, 2013; Moldenhauer 2014; 2019).

6.5.3 Electron-Beam Sterilization (E-Beam Sterilization)

This type of sterilization provides a significantly higher dose rating than gamma or X-rays. As such, a lower exposure time is required, which in turn reduces the potential for product degradation. Electron beams do not have the penetrative efficiency of gamma or X-rays. Shielding, typically concrete, is used to protect the surrounding environment and personnel (Anonymous 2013; Moldenhauer 2014; 2019).

Sterilizers for this technology are available for both batch and continuous processes. Items to be sterilized are scanned with the electron beam within a defined radiation field. When appropriately controlled and operated, the systems deliver a constant dose over time. The higher dose rate and the absence of a local radioactive source make this type of sterilization attractive. The systems themselves come in small sizes that may be operated by the end user (USP <1229.10>, 2013; Moldenhauer 2014; 2019).

Care must be taken in developing the sterilization cycles as the penetration is much lower than obtained with photons. This higher dose rate can also lead to the items experiencing higher temperatures than in a gamma cycle. Dose mapping is very important to guarantee that items are properly sterilized (USP <1229.10>, 2013; Moldenhauer 2014; 2019).

6.5.4 Validating Radiation Sterilization Processes

A key consideration in the development of a radiation sterilization cycle is the effective radiation dose for the times to be sterilized while showing that this dose does not adversely affect the item's key attributes. Both the minimum and maximum allowable dose should not cause adverse effects on the item (USP <1229.10>, 2013; Moldenhauer 2014; 2019).

ISO standards identify various types of dose-setting methods, which are Method 1, Method 2A, Method 2B, and Method VDmax. Selecting the right method to use is dependent upon the production batch size, the bioburden present, and the sensitivity of the item's sensitivity to the type of radiation used (USP <1229.10>, 2013; Moldenhauer 2014; 2019).

- Radiation Dose: As the ionizing radiation penetrates an object, it deposits energy. The energy that is absorbed from radiation exposure is called a dose.
- Dose Setting: Using Method 1, the dose is compared to the radiation resistance of the bioburden. Typical D-value or resistance studies are not determined but rather a standard radiation resistance is used from the literature. The belief is that standard resistance is a greater challenge than routine bioburden. The site performs a verification dose study and determines the appropriate dose from a table. This is similar to the VDmax method, requiring bioburden and dose verification testing, but it is dependent upon low bioburden in the item to be sterilized. The assumption is <1,000 cfu for a 25 kGy sterilization dose and 0.1–1.5 cfu's per item for a 15 kGy sterilization dose. Method 2 is more complex. It uses a series of incremental dose exposures to establish a dose. The intent is that 1 sample out of 100 if treated with that dose will be non-sterile. This dose determination process provides a way to determine the dose. The intent is to determine the dose that will result in 1 sample out of 100 non-sterile samples (USP <1229.10>, 2013; Moldenhauer 2014; 2019).
- Compatibility of Materials: Following the determination of the required dose, a corresponding maximum dose is established. This dose represents the highest dose that is likely to be seen in a sterilization cycle, adding a safety factor, and based upon the effects on the materials in the item both immediate and long term (USP <1229.10>, 2013, Moldenhauer, 2014 and 2019).
- Dose Verification: During cycle development and dose setting the methods are dependent upon the bioburden approach. Cycle efficacy depends upon appropriate control of pre-sterilization bioburden and periodic monitoring of the process' effect on bioburden. The dose setting is based on the natural resistance of the bioburden and extrapolates to ensure a PNSU of 1×10^{-6}.
- Equipment Qualification: Initial qualification and periodic requalification studies are performed to ensure appropriate equipment performance. The key parameters in qualification differ based on the type of sterilization process. For gamma sterilization, key parameters are equipment controls and the parameters for the system's capability. E-beam and X-ray sterilization utilize controls for scan speed, source intensity, and system timers (USP <1229.10>, 2013; Moldenhauer 2014; 2019).
- Empty Chamber Dose Mapping: This type of study, while not required, provides a useful baseline of performance (USP <1229.10>, 2013; Moldenhauer 2014; 2019).
- Load Mapping: The purpose of this study is to determine the appropriate arrangement of items in the load that minimizes the dose variation across items. Several dosimeters are used in both

internal and external positions. This type of study can also be used to determine the effects of the maximum sterilization dose.

- Biological Indicators: The use of biological indicators is contraindicated as the dosimeters are more reliable, reproducible, and robust (USP <1229.10>, 2013; Moldenhauer 2014; 2019).
- Dosimetry: Cycle development and qualification require the use of periodically calibrated dosimeters (USP <1229.10>, 2013).
- Process Confirmation: Replicate studies are performed to show reproducible acceptable lethality based on the results from the dosimeters. They should be strategically placed across the material in the process (USP <1229.10>, 2013; Moldenhauer 2014; 2019).
- Routine Process Control: Controls must be established to ensure that the systems are maintained and controlled within the validated sterilization parameters.

6.6 Moist Heat Sterilization

Before addressing the topic of moist heat sterilization, it is important to differentiate between moist and dry heat. Moist heat sterilization is defined as a sterilization process that uses steam under pressure, as is conducted in an autoclave. The steam is considered saturated when the steam and the water are in equilibrium. A closed container of water, when heated to temperatures above 100°C, will demonstrate an increase in the steam pressure. As the temperature rises, so does the pressure. There are steam tables that can be utilized to show the relationship between temperature and pressure. When the steam is saturated, raising the pressure using external means will result in the steam starting to condense back into the water. Superheated steam is defined as steam whose temperature, at a given pressure, is higher than that indicated by the equilibration curve for the vaporization of water. The thermal energy (heat) delivered by different heating media using different types of moist heat sterilization processes, like saturated steam, air/steam mixtures, or superheated water, at the same temperature is significantly different (PDA Technical Report Number 1, 2007; Moldenhauer 2013; 2014).

With moist heat sterilization, the microorganisms present are destroyed by coagulating and denaturing the cell's enzymes and structural proteins. Typical sterilization cycles require that the exposure time is at 121°C in the range of 15 to 30 minutes for the spores most resistant to the sterilization process. These processes can be performed at other temperatures that yield equivalent lethality to those at a specified reference temperature (Moldenhauer 2013; 2014; 2019).

Moist heat sterilization processes are used to sterilize equipment in support of the sterile processing of products, cleaning supplies used in some clean room operations, primary packaging materials, terminal sterilization of liquid-filled containers, and so forth (Moldenhauer 2013; 2014; 2019).

The use of a steam sterilizer does not ensure that all parts of the load are subjected to moist heat. It is not sufficient to supply moist heat (steam) to the sterilizer. In steam sterilization, the moist heat must contact the items being sterilized. Some items make it difficult or impossible for all the air to be removed from them as well as allowing steam to penetrate. In these situations, they may be achieving sterilization through dry heat only. Some examples include filter housings, long tubing, and tightly closed containers. It is important to ensure that the items being sterilized are in contact with the sterilizing media that you intend to use (Moldenhauer 2013; 2014; 2019).

6.7 Dry Heat Sterilization

Dry heat sterilization refers to the use of hot air that is free of water vapor or has very little water vapor. In this type of process, the moisture present has little effect on the sterilization process. Dry heat sterilization has been used for many years. It is believed to be less complicated than moist heat sterilization. Sterilization is achieved by using much higher temperatures for much longer periods. Pharmaceuticals often use temperatures exceeding 250°C; however, some guidance indicates that one may use temperatures greater than 160°C. Dry heat kills microorganisms by the destructive oxidation of essential cell constituents. There are bacterial spores that are resistant to dry heat sterilization, although they may not be used in the validation of these processes. Frequently endotoxin challenges are conducted concurrently with the validation of these cycles. If used appropriately, the endotoxin challenges eliminate the need to use biological indicators since the endotoxin challenge is more severe than the resistance of the biological indicator. Dry heat sterilization processes are used for glassware and some equipment that can withstand higher temperatures for the prolonged periods required (Moldenhauer 2013; 2014; 2019).

6.8 Pasteurization

While this chapter is intended to describe sterilization methods, it is well-known that pasteurization is used as a method of eliminating microorganisms in food processing. As such, this section will explain the differences between sterilization and pasteurization. Pasteurization is used to heat an item (usually liquid foods) to a specific temperature in order to limit the growth of microorganisms in food. Sterilization, however, is a process that inactivates all microorganisms from an item. Both pasteurization and sterilization are utilized to increase the shelf-life of a product, most often a liquid. While the processes may both use heat treatment, they are not equivalent in their methods or results.

Pasteurization heats liquids to a specific temperature. The specified heating temperature is less than the boiling temperature. The heating is conducted rapidly and is followed by rapid cooling and sealing. The process is named after Louis Pasteur, who designed the process. Originally, it was used to keep beer and wine from souring. Today, it is widely used in dairy products (Anonymous 2022b).

A typical liquid process of pasteurization includes boiling the product to a temperature below the boiling temperature and then rapidly cooling it. Milk is often heated to 145°F (63°C) for 30 minutes and then rapidly cooled. At least two types of heating processes are used for milk under pasteurization. High temperature short time (HTST) involves the use of metal pipes that are surrounded by hot water and heated to 72°C (161°F) for 15 seconds. The milk is then cooled and sealed. Another process involves the use of ultra-high temperature (UHT) processing. This process keeps the milk at a temperature of 138°C (280°F) for a minimum of two seconds. The HSTH process yields milk with a refrigerated shelf life of two to three weeks, while UHT extends the shelf-life of milk to approximately two to three months (Anonymous 2022b). While sterilization claims are made in some countries, in the US these types of processes would include labeling that says pasteurization.

Table 1. Comparisons between pasteurization and sterilization (as used in other types of applications, e.g., pharmaceuticals, medical devices, and biological products).

Topic	Pasteurization	Sterilization
Definition	Pasteurization refers to the process of heating a liquid for a controlled time to reduce microbial growth.	Sterilization refers to the elimination of all viable microorganisms in a substance, e.g., liquids, solids, packages, and the like.
Processing Methods	Uses heat and may include rapid cooling	Uses various processes to achieve the killing all viable microorganisms.
Varieties of Methods	HTST and UHT	Aseptic Processing, Moist Heat Sterilization, Radiation Sterilization, Gaseous Sterilization, Liquid Phase Sterilization, and Vapor Phase Sterilization.
Effects on the Taste of Food	Remains the same	May alter the taste of food.
Effectiveness	Eliminates some microbiological organisms present	Eliminates microorganisms and spores.
Applicability	Liquids and foods	Pharmaceuticals, food, medical devices, biological products, medical purposes, cleaning, and so forth.

6.8.1 Sterilizer Configurations

For moist heat, sterilizers there are both batch configurations and continuous sterilizers available. Few pharmaceutical companies choose to use continuous sterilizers; however, they are often used in the food industry. Most pharmaceutical sterilizers used are operated as a batch unit—i.e., the material is moved to the sterilizer, processed in the sterilizer, and following sterilization is moved to the next production step. Batch sterilizers are available in a variety of sizes from bench-top sized to laboratory sized to units that handle two or more sterilizer trucks or pallets of loads. Some sterilizers are the size of railroad cars. Batch sterilizers can be used for any type of product that can be sterilized with moist heat (Moldenhauer 2013; 2014; 2009).

Sterilizers can be configured to be specific for one type of sterilization process or may be able to perform more than one type of sterilization process (multi-function sterilizers). Some of the typical sterilization processes available are saturated steam, steam-air mixtures, air overpressure cycles, and those which use superheated water (e.g., water-spray, raining water, water immersion, and rotary sterilization cycles) (PDA Technical Report Number 1, 2007; Moldenhauer 2014; 2019).

6.8.1.1 Determining Worst-Case Conditions

Overkill cycles deliver the most heat to the product and have the highest safety factor of sterility assurance. For products, they can also have the highest level of impact on product stability. Typically, they are routinely used for the sterilization of porous loads, hard goods, and some terminal sterilization activities when the item is not heated sensitive. The worst-case overkill cycle may be determined by lowering the sterilizing temperature by 1°C or more and reducing the exposure time by a few minutes from the set-point conditions—i.e., either at the minimum allowable conditions or below these conditions (sub-lethal). Alternatively, if one uses a load probe one can execute the study with a lower F_0 than the values routinely allowed. Following the approach used in gaseous sterilization, some use the half-cycle approach for validation, where the sterilization time used in the performance qualification is doubled for the routine production cycle. This approach is only useful if the excessive heat will not affect the product. The cycles developed for an overkill process are expected to destroy the entire microbiological indicator challenge (Agalloco 2008; Moldenhauer 2013; 2014; 2019).

Product-specific cycles are used most often for terminal sterilization of large-volume products. In the USP chapter, for moist heat sterilization of liquids, these types of cycles are described as bioburden-based approaches or biological indicator-bioburden-based approaches (USP <1229.2>, 2012b). Validation is performed at the minimum conditions allowed for the acceptable release of the product. These requirements may require the destruction of the entire biological indicator or require a specific biological indicator log reduction or $F_{Biological}$. Additionally, there are requirements established for the maximum heat delivered to the product, typically as either specified upper time/temperature requirements and/or a provision for the maximum allowable F_0 (the maximum allowable lethality delivered to the product). The maximum lethality does not aid in the sterilization of the product, rather it is related to the effect of the heat on the stability of the product (shelf-life). Resistant bioburden is an integral part of determining the lethality required. Therefore, these types of cycles require an on-going, vigilant evaluation of bioburden recovered in the plant, including an assessment of its heat resistance (Moldenhauer 2013) Note: in the USP monograph (USP <1229.2>, 2013), maximum F_0 is only required for moist heat sterilization of liquids (Moldenhauer 2014; 2019).

6.8.1.2 Load Configurations

The load size and the load arrangement are also important factors in the efficacy of the sterilization cycle. In porous load sterilization, it is important that the items be placed on the sterilizer carts or pallets in such a way that they do not adversely affect the sterilization of other items in the load e.g., not dripping condensate from one item onto another item. It is common to have a fixed loading pattern, which identifies the specific locations where items are to be placed. Heavier items are placed on the bottom of the load. Fixed loading patterns are also used for liquid loads. The containers may be placed into trays or on sterilizer carts in a defined arrangement. It is common to have detailed drawings or photographs, which clearly delineate the loading arrangements to be used (Moldenhauer 2013; 2014; 2019).

Following sterilization, items may be removed and transported to other areas for use or may not be used immediately, provisions are made to ensure that the item sterilized maintains its sterile state. This may include wrapping the item with sterilizer wrap or placing it into sterilizer bags or pouches prior to the sterilization process. When this is done, the methods should be established so that the wrapping or pouching process does not adversely affect the air removal or steam penetration (Moldenhauer 2013; 2014; 2019).

During cycle development, one should determine which items heat up faster than others and which are the slowest to heat up. Within specific items, studies are performed to evaluate the hardest-to-heat area of the item, which is where the thermocouples are placed. This may be accomplished by placing multiple probes or biological indicators into the item. If probes are used, the heating characteristics of the item are mapped, and the lowest probe value is where the thermocouples are placed in subsequent studies. Alternatively, biological indicators may be used to assess where the lowest level of kill is achieved and then this is the probing location. With biological indicators, it is common to use sub-lethal cycles to ensure the growth of the indicators after sterilization (Moldenhauer 2014; 2019).

One may validate both the minimum and maximum allowable loading configuration. However, the PDA's Technical Report Number 1 (2007) indicates that neither of these configurations may be the worst-case (Moldenhauer 2013; 2014; 2019).

6.8.1.3 Cycle Monitoring

Routine monitoring parameters for moist heat sterilization cycles include come-up time, exposure time and temperature, cooling time or final temperature, water temperature (for superheated water cycles), fan speed (for fan cycles), pressure, time at which air is added to the cycle (air overpressure), and so forth (Moldenhauer 2013; 2014; 2019).

6.8.1.3.1 Sterilization and The Way Microorganisms Die

A significant amount of research has been conducted studying the kinetics of microbial death. They found that the process of destroying or sterilizing the cells occurs at a defined and consistent rate that is dependent upon the variables that affect the reaction rate, as shown in Fig. 1. The D-value is determined by plotting the number of viable cells (Survivors) on the y-axis and the exposure time on the x-axis. A best-fit line is drawn through the points. The line can be reviewed graphically to determine the time it takes to reduce the microbial population by one-log at a specified set of sterilization conditions (Agalloco 2008; Moldenhauer 2014). However, today most people just use the calculator to determine the D-value using the equations from PDA Technical Report Number 1 (2007).

This relative relationship has led to the development of several calculations that can be used to model microbial destruction and the probability of sterility associated with a sterilization cycle. One of the most effective types of sterilization is moist heat (steam) under pressure. To use this methodology, the item to be sterilized must be able to withstand heat and moisture without damage. It is reported that most microbes grow in the range of -50°C to 80°C, although some form spores at temperatures outside of these ranges (DeSantis 2008; Moldenhauer 2013; 2014; 2019).

Sterilization applications utilize the understanding of how microorganisms die to predict the probability of a survivor in a sterilization process. This is expressed as PNSU or the probability of a non-sterile unit. For moist heat sterilization, the expected PNSU is less than 10^{-6}, which means that there is no more than a one in a million probability of a non-sterile unit. Not all sterilization processes

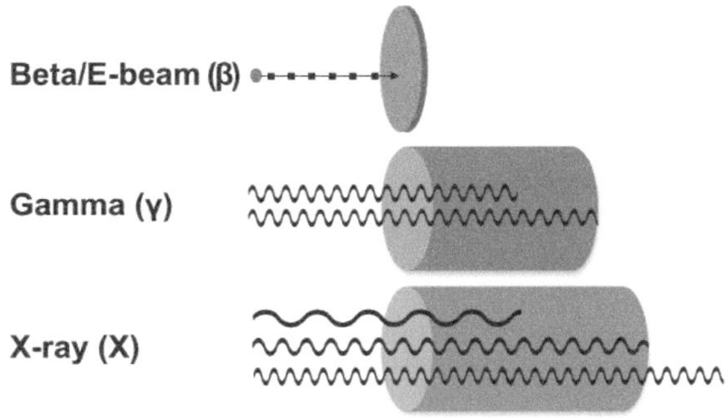

Figure 1. Penetration Differences for Different Types of Radiation. Figure courtesy of Wikipedia (Anonymous 2013).

predict the probability of non-sterility. For instance, an aseptic process validation measures the rate of contamination present at the time the process simulation test was performed but does not provide a probabilistic measure of sterility assurance (Moldenhauer 2013; 2014; 2019).

Copious information is available that indicates that microbial death can be described as a first-order chemical reaction. This implies that the death is a single-molecule reaction, probably because of a denaturization of a specific critical molecule in the cell (DeSantis 2008; Moldenhauer 2014; 2019). There are numerous publications available on the D-values associated with spores and food products.

6.8.1.3.2 Biological Indicators

Some types of bacteria (Gram-positive rods) can form structures that are resistant to adverse conditions called spores as a defensive survival method. These are different from fungal spores (mold spores) that are manufactured as reproductive bodies. The genus and species of bacteria resistant to a specific sterilization process may be different across sterilization processes. Additionally, the same bacterial spore may have different levels of resistance to varying sterilization processes. The level of resistance to a sterilization process is expressed as the D-value. D-values are specific for a sterilization medium and a specific temperature, and it is defined as the amount of time in minutes that is required at a specific reference temperature to reduce the microbial population by one log. The higher the D-value at a specific set of sterilizing conditions, the more resistant the organism is to the sterilization process (Moldenhauer 2013; 2014; 2019).

6.8.1.3.3 D-values

Geobacillus stearothermophilus spores ($D_T \sim 1.5$–6.0 minutes in various parenteral solutions) are considered one of the most resistant spores for moist heat sterilization. As such, many regulatory documents specify the use of these spores in the validation of moist heat sterilization cycles. Some other organisms have been identified that have sufficient resistance to be appropriate challenges for validation of moist heat sterilization cycles including Bacillus subtilis 5230 ($D_T \sim 0.5$ minute in 0.9% Saline), Bacillus smithii (Formerly Bacillus coagulans, FRR B666) ($D_T \sim 1.5$ minute in Sterile Water), and Clostridium sporogenes ATCC 7955 ($D_T \sim 0.3$–7.0 minutes in various parenteral solutions) (Sadowski 2009; Moldenhauer 2011). It is important to know the regulatory expectations in your country when selecting the appropriate biological indicator challenge to use (Moldenhauer 2013; 2014; 2019).

A variety of spores are also available for dry heat sterilization. Typically, spores of Bacillus subtilis have been used. Some have successfully used Geobacillus stearothermophilus. Typically, the most resistant of spores will have a D-value of 6–10 minutes at 170°C. If the cycle is used to depyrogenate and is run at a much higher temperature, the microbial D-values may be only a few seconds (Anonymous 2011; Moldenhauer 2013; 2014; 2019).

Selection of the biological indicator to use includes consideration of many things including an organism that is not pathogenic; an organism that is easy to cultivate; an organism that can be cleaned, harvested, and grown with relative ease; an organism more resistant to the sterilization process than the items being sterilized; an organism which is stable; an organism which is not inhibited by the items it will be used to test; and an organism which provides reproducible results when tested at a specific temperature (Sadowski 2009; Moldenhauer 2013; 2014; 2019).

For moist heat sterilization cycles, the biological indicator to be used in the process is based upon the sterilization cycle-design model (approach) used. Most companies using an overkill model choose to use Geobacillus stearothermophilus. When product-based models are used, other acceptable biological indicators are frequently used (Sadowski 2009; Moldenhauer 2013; 2014; 2019).

6.8.1.3.4 Z-Values

Another important term when using biological indicators is the z-value. This term refers to the temperature dependence on the D-value. It represents the number of degrees of temperature that are required to achieve a ten-fold change in the D-value. For example, if the z-value is 10°C, changing the sterilizing temperature from 120°C to 130°C will yield a ten-fold change in the D-value. As such, a much shorter sterilization time will be required at the higher temperature. The z-value tends to be constant across a broader range of temperatures. Geobacillus stearothermophilus spores have a typical z-value of 10°C for temperatures between 100°C to 135°C. Bacillus subtilis spores used for dry heat have a typical z-value of 20°C in the range of 150°C to 190°. Dry heat cycles are used to concurrently depyrogenate using temperatures of 225°C to 300°C; the z-values used are between 45°C to 53°C (Agalloco 2008; Moldenhauer 2013; 2014; 2019).

Understanding the z-value can aid in creating a higher temperature sterilization cycle to sterilize a product quicker than a routine cycle with less product degradation.

6.8.1.3.5 F-Value

The F-value or thermal death time is used along with the temperature data and z-value to estimate the lethality for moist and dry heat sterilization cycles (Agalloco 2008). It is the time in minutes required to deliver a sterilization cycle equivalent to that of the reference temperature T. For most moist heat sterilization cycles the reference temperature is 121°C and for dry heat sterilization cycles, it may be 160°C or 250°C (Moldenhauer 2013; 2014; 2019).

The reference temperature benchmarks the cycle in terms of the heat delivered at the specified temperature. The F-value is specific for a defined reference temperature. Equation 1 provides the equation for the F-value (PDA 2007; De Santis 2008; Moldenhauer 2013; 2014; 2019).

This formula approximates to:

$$F = \Sigma\, 10^{(T_0 - T_{ref})/z}\, \Delta t \tag{1}$$

where,

T_0	=	Temperature within the item being heated
T_{ref}	=	Reference temperature
z	=	Z-value of the challenge organism
dt	=	The change in time (minutes)
Δt	=	The chosen time interval (minutes)

For moist heat sterilization, when the reference temperature is 121.1°C and the z-value is 10°C, the F-value is called F_0. For dry heat sterilization, many use the term F_H, where the reference temperature is dependent upon whether the cycle is used for sterilization or depyrogenation. Some companies use F_P designation for depyrogenation (Moldenhauer 2014; 2019).

Figure 2 provides an example of a microbial survivor curve for moist heat sterilization. Although this example is for moist heat, similar curves can be performed for dry heat. In this figure, the starting population of resistant microorganisms, N_0 is 10^6. The D-value at the specified reference

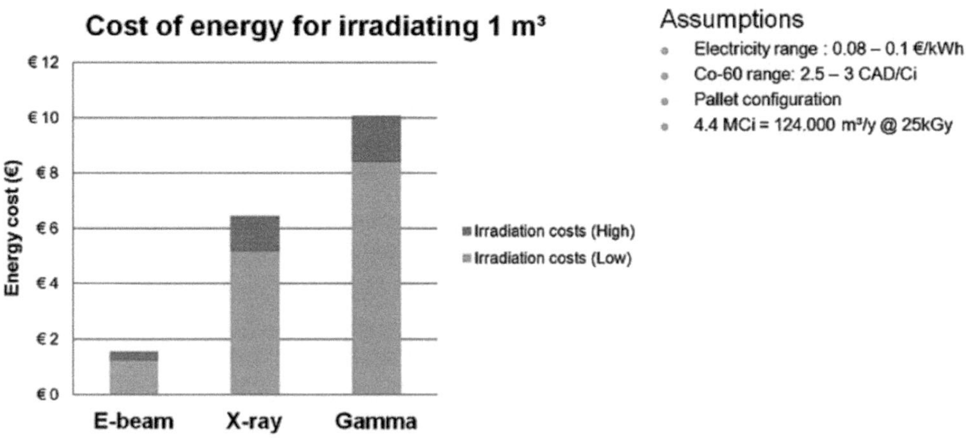

Figure 2. Cost of Irradiation Using Different Processes. Figure courtesy of Wikipedia (Anonymous 2013).

temperature is 2.5 minutes. The desired level for the probability of a non-sterile unit is 10^{-6}. The desired probability of survivors in this graph is designated as N_F or the final population (PDA Technical Report Number 1 2007; Moldenhauer 2013, 2014; 2019).

The microbial survival curve model is representative of first-order kinetics—i.e., the death kinetics of single-celled spores. While this model may not be representing all the different types of challenge systems, this is still the best model available. To properly use this model in microbial destruction calculations, it is important that the challenge organism is in a homogeneous culture and that the constant lethal stress or the equivalent lethal stress is applied to the challenge (PDA 2007; Moldenhauer 2013; 2014; 2019).

6.8.1.3.6 Probability of a Non-Sterile Unit (PNSU)

The probability of a non-sterile unit calculation is used to determine the likelihood of obtaining a non-sterile unit at the end of the sterilization cycle. Generally, it is accepted that an acceptable probability is 1 in a million or 10^{-6} for sterilization. Some assume this is a synonymous term with sterility assurance level (SAL). The exponent for the SAL should be a positive number rather than the negative exponent in a PNSU value. Many prefer the term PNSU since the values of the exponents are not ambiguous. The PNSU can be calculated by solving for B using Equation 2 (Agalloco 2008; Moldenhauer 2013; 2014; 2019):

$$\text{Log}_{10} \text{B} = \text{Log}_{10} \text{A} - (\text{F}_0/\text{D}_{121}) \quad (2)$$

where,

B = Probability of a Non-Sterile Unit

A = Number of Presterilization Bioburden Microorganisms

F_0 = The minimum F_0 observed in the cycle

D_{121} = The D-value of the bioburden microorganism at the reference temperature of 121°C.

Note: For cycles using other F-values or D-values, the values would be substituted for F_0 and D_{121}. (Moldenhauer 2014, 2019)

6.8.1.3.7 Commissioning or Engineering Studies

Once the sterilizer has been installed at the intended location, there are some activities conducted prior to initiating the formal validation of the system. Some of the activities that may be conducted include cleaning, adding lubricants, and so forth. It is also beneficial to execute some runs of the sterilizer to take a “look-see” at how the system is operating. The engineers frequently use this time to ensure that the system is sequencing as expected, no major alarms occur, and a general assessment of the system appears to be operating reliably so that validation can commence. This is performed to reduce the risk of failures during the actual validation process. Typically, these activities fall under the responsibility of the engineering/maintenance department (Moldenhauer 2013; 2019).

Commissioning of the equipment includes verifying that the sterilizer is performing sufficiently to proceed with validation. Additional information on this topic can be found in the PDA's Technical Report No. 48 Moist Heat Sterilizer Systems: Design, Commissioning, Operation, Qualification, and Maintenance (PDA 2010).

6.8.1.3.8 The Physical Characteristics of a Sterilization Cycle:

It is important to look at the physical parameters of the cycle like temperature, pressure, calibration, and other physical monitoring parameters. Temperature distribution studies and heat penetration studies are conducted using temperature monitoring devices, usually thermocouples. The thermocouples selected for use should be appropriate for the temperature range being monitored. Some companies choose to use resistance temperature detectors (RTDs) because they have a higher level of accuracy than thermocouples. However, thermocouples are less expensive, have a faster response time, and are made to withstand the "abuse" they get during validation studies. The thermocouples selected for use must be calibrated and accurate to a specified level for the temperature range in which they will be used (Agalloco 2008; Moldenhauer 2013; 2014; 2019).

Care should be taken when placing the thermocouples in the sterilizer load for validation. When placing temperature distribution probes, it is important that the probes are not in contact with other surfaces that may produce results that are not representative of the temperature in the sterilizer chamber. The placement of heat penetration probes also requires care. The probes should be placed within items in such a way that they do not block off the path of steam penetration to the item. Additionally, the placement of the probes must not create artificial openings where steam could penetrate the item in a way that is not possible in routine sterilization. For thermocouples that have wires which attach to the data recorder, one must also be sure to position the wires so that they do not become damaged by the handling of the load being placed into or out of the sterilizer. The operation of some sterilizers makes it difficult to utilize wired probes as part of the validation process, e.g., probes that operate through a tunnel or probes in sterilizers that rotate or agitate. There are probes that have been designed to wirelessly provide the temperature data to the data logger (Moldenhauer 2013; 2014; 2019).

Some sterilizers also utilize load probes. These probes are permanently installed in the sterilizer and monitor temperature in both validation and routine production cycles. When installed, the temperature values for these probes are also monitored during the validation cycle (Moldenhauer 2013; 2014; 2019).

For sterilizers that utilize recirculating water for cooling, thermocouples are also used to assess the temperatures attained in the water prior to the initiation of the cooling process. There is an expectation that the water is sterilized during the cycle so that the water used for cooling the product is also sterile (Moldenhauer 2013; 2014; 2019).

Sterilizer pressure is monitored in moist heat sterilization cycles. The pressure has lesser importance than temperature, but it is used to ensure that saturated steam is used during the exposure dwell period. There are some regulators who routinely check the temperatures observed on recording charts to the expected pressure readings from steam tables. Pressure can play a significant role in ensuring functional products at the completion of the cycle. For example, inadequate overpressure in a plastic cycle may result in deformed containers (Moldenhauer 2013; 2014; 2019).

The key instrumentation for monitoring or recording temperature and pressure data for the sterilizer should be calibrated in such a way that the calibration can be traced to a recognized national standard. Additionally, the thermocouples used in the validation should be calibrated before and after each study (Moldenhauer 2013; 2014; 2019).

It is possible to utilize a physical or chemical indicator and/or a physical/chemical integrator to monitor the sterilization cycle. There are a variety of indicators available for different types of sterilization. One of the simplest types of indicators is the use of autoclave tape, which changes color if the tape has been processed in the sterilizer. It does not provide information on the quality of the cycle and in many cases, it does not indicate whether a complete cycle was performed. Physical/chemical integrators assess the process conditions in the cycle, typically integrating the effects of multiple process parameters and providing some guidance on the cycle delivered. One such integrator provides a line showing whether a minimum F_0 value was achieved in the cycle. Neither indicators nor integrators are to be utilized in place of biological indicators. They do provide useful information, like indicating that a load has been processed through the sterilizer (Moldenhauer 2013; 2014; 2019).

6.8.1.3.9 Selecting an Approach for Sterilization Validation:

There are established models or approaches that can be used to determine the sterilizer validation criteria like minimum lethality that must be met. In the past, there were three models based on the thermal death time curve, an overkill model, a combined biological indicator bioburden-based model, and an absolute bioburden model. The key differences in these models are the amount of heat delivered to the product, the effect on the stability of the product, and the cost of the sterilization cycle (e.g., the utilities used). The Parenteral Drug Association (PDA) reissued its technical report on moist heat sterilization and converted the combined biological indicator bioburden-based model and the absolute bioburden model into a product-specific approach to cycle design (PDA 2007). The intent of each model or approach is to determine the lethality required to deliver the desired probability of a non-sterile unit (Moldenhauer 2013, 2014, 2019).

The semi-logarithmic survivor curve model can be used to determine the appropriate PNSU, which in most cases is 10^{-6}. This can be expressed as shown in Equation 3:

$$\text{Log } N_F = \text{-}F/D + \text{Log } N_0 \qquad (3)$$

In order to determine the lethality that is required for a specific sterilization cycle, this equation can be rearranged to solve for the F-value, as shown in Equation 4 (PDA 2007):

$$F = (\text{Log } N_0\text{-Log } N_F) \text{ x } D_T \qquad (4)$$

Where N_0 represents the starting population of the biological challenge organism, e.g., 10^6. The D-value is specific to the microorganism chosen as the biological indicator. N_F represents the final population of microorganisms, which is usually 10^{-6} or the desired probability of a non-sterile unit (Moldenhauer 2014 and 2019). An overkill cycle provides the most heat to the items being sterilized. It has the highest cost associated with the use of the sterilizer. This added heat has an impact on the stability of the item following sterilization. This type of cycle provides a high level of sterility assurance, regardless of the level of bioburden present and the heat resistance of that bioburden. It is assumed that the starting population of bioburden is 10^6 with a D_{121}°C of at least 1 minute and a z-value of 10°C. To have a PNSU of 10^{-6} would indicate a final biological population or N_F that is 10^{-6}. These values can be entered into the equation to determine the F-value—i.e., the expected lethality. This calculation is shown below using Equation 5 (PDA 2007; Moldenhauer 2014; 2019):

$$F_0 = D_{121^\circ C} \text{ x } (\text{Log } N_0 \text{ - Log } N_F) \qquad (5)$$

$$F_0 = 1.0 \text{ minute x } (\text{Log } 10^6 - \text{Log } 10^{-6}) = 12 \text{ minutes}$$

Within the definition of overkill in the PDA's Technical Report (PDA 2007), an overkill cycle must deliver a lethality of at least 12 minutes under the specified conditions. The PDA's Technical Report (PDA 2007) added to this definition stating:

A cycle designed with the overkill design approach can be defined as a sterilization cycle that is demonstrated to deliver an $F_{Biological}$ and $F_{Physical}$ of at least 12 minutes to the items being sterilized (Moldenhauer 2013; 2014).

These terms have been defined by PDA to mean (PDA 2007):

> $F_{Physical}$ is defined as the term used to describe the delivered lethality calculated based on the physical parameters of the cycle. The $F_{physical}$-value is calculated as the integration of the lethal rate (L) over time. The lethal rate is calculated per a reference temperature (T_{ref}) and z-value using the equation: $L = 10^{(T\text{-}Tref/z)}$. This typically refers to the data obtained from the temperature measurement devices, e.g., heat penetration probes.
>
> $F_{Biological}$ is defined as a term used to describe the delivered lethality measured in terms of the actual kill of microbiological organisms on or in a BI challenge system. The $F_{Biological}$ value is calculated as the D_T x LR, where D_T is the D-value of the BI system at the reference temperature, and T and LR are the actual log reduction of the BI population achieved during the cycle.

The change to add the requirements for $F_{Biological}$ to the definition of an overkill cycle was a change over previous definitions. Few companies before this time routinely monitored or assessed $F_{Biological}$ as part of their routine qualification requirements. For many companies that had cycled in place for many years, converting to this definition of an overkill cycle necessitated repeating or revising the cycle development studies previously conducted (Moldenhauer 2013; 2014; 2019).

There is a separate definition of overkill cycles within European regulations. In the European regulations and compendia for terminally sterilized dosage forms, an overkill cycle is specified as sterilization by moist heat at 121°C for at least 15 minutes (PDA 2007; Moldenhauer 2013; 2014; 2019).

It is common to expect that an overkill cycle model is used for the sterilization of equipment and components utilized in aseptic manufacturing operations. It is possible to use this type of sterilization model for some terminal sterilization activities, providing that the items being sterilized can withstand the cycle conditions (Moldenhauer 2013; 2014; 2019).

Overkill terminal sterilization cycles are too harsh for many products. Terminal sterilization is desired for products as a method to provide for user safety. Since many overkill cycles are too harsh, alternative models have been developed for those products with some thermal sensitivity, which are called product-specific models or approaches. (Older texts refer to these types of approaches as combined biological indicator bioburden-based or absolute bioburden-based cycles). With this type of approach, one must sufficiently reduce the microbial population to an acceptable level of sterility assurance while maintaining the product's attributes. In order to accomplish this, the models take into account the starting population of the microorganisms present (bioburden) and the heat resistance of those organisms, along with the provision of a safety factor. These values are utilized in the calculation to determine the minimum lethality. The use of this type of model requires the manufacturer to commit to ongoing monitoring of the product bioburden and its heat resistance. An example of the product-specific design model calculations is illustrated using Equation 5 (PDA 2007; Moldenhauer 2013; 2014; 2019).

Manufacturing Site Data for Bioburden indicated:

$N_0 < 25$ resistant organisms per unit of product

$D_{121^{\circ}C} < 0.3$ minutes

In order to add a safety factor, the following values were used:

$N_0 = 10^2$

$D_{121^{\circ}C} = 0.4$ minutes

The desired PNSU was determined to be 10^{-6}

$F_{121^{\circ}C} = (\text{Log } 10^2 - \text{Log } 10^{-6}) \text{ x } D_{121^{\circ}C} = 3.2$ minutes

The above example indicates that the model would deliver the desired PNSU using an F_0 of 3.2 minutes. However, many regulators would expect that additional precautions are established in support of cycles with F_0 less than 6 minutes, e.g., aseptic filling prior to terminal sterilization or other methods designed to ensure microbiological control (Moldenhauer 2013; 2014; 2019).

6.8.1.3.10 Validating Moist Heat Sterilization Systems

The equipment qualification refers to an assessment of the hardware/software used for the sterilization process. It does not address the actual sterilization cycle or the items being processed in the sterilizer. The equipment qualification includes verifying the proper installation of the equipment as well as verifying that it is operating correctly, prior to assessing its capability to sterilize the desired components, equipment, or product. These activities are conducted in the Installation Qualification (IQ), the Operational Qualification (OQ), or in a combined document that covers the Installation and Operation Qualification (I/OQ). Whether the protocols are combined or separate documents frequently is a function of how complex the equipment qualification process is. The same requirements apply in either case (Moldenhauer 2013; 2014; 2019).

- Installation Qualification (IQ): In some cases, companies contract with the vendor of the equipment to perform this testing. It is good practice for end-user personnel to participate in this activity, even when the vendor conducts the studies to gain knowledge of the system. The documentation generated becomes a technical handbook for the system as supplied to one's site. This information can be invaluable when determining at a later date whether replacement parts are or are not identical to those installed. It is at the end user's discretion whether the hardware and software are tested concurrently or as separate documents (Moldenhauer 2003). Some other considerations include (Moldenhauer 2013; 2014; 2019) verification that items match the purchase order, no damage is in shipping, all items are received, supporting documentation is received, and so forth.

  It is beneficial to make this document as comprehensive and detailed as possible. This document becomes the technical handbook for the sterilizer over time. It is especially useful when trying to assess the appropriateness of replacement parts and whether they are equivalent to the original installation (Moldenhauer 2013; 2014; 2019).
- Operational Qualification (OQ): It allows you to assess whether the system operates at the specified conditions, prior to adding the influence of the item to be sterilized into the mix. It may also be called Dynamic Equipment Qualification. Once you initiate sterilization of items within the unit, if a problem occurs it is not clear whether it is the sterilizer performance at fault or it affects the item in the sterilizer. The OQ of the hardware and the software or control system may be performed within a single document or in separate documents. The utilities used to support the sterilization system, e.g., clean steam, oil-free compressed air, and others, should be qualified either concurrently with the OQ or before the OQ is conducted (Moldenhauer 2013; 2014; 2019).
- Empty Chamber Temperature Distribution: Studies are conducted using temperature distribution probes, usually arranged in a fixed or geometric pattern, to evaluate the uniformity of heat delivered within the sterilizer chamber. For this study type, fixed patterns are used most often. This allows for a direct comparison of data from one study to another, which is useful when these studies are used following production deficiencies. Unless the heat is delivered to all areas of the sterilizer chamber, one cannot ensure that all of the items in the sterilizer will be appropriately sterilized in the cycle. During these studies, it is common to find that some of the sterilizer parameters may need to be modified, such as the bleeder valves, water spray nozzle directions, directional airflow, and so forth. The data obtained from these studies are also used as baseline data for evaluating the sterilizer's performance characteristics over the life of the sterilizer. Penetration probes are not typically used in these evaluations. It is important that the temperature distribution probes are properly located and that they do not come into contact with either the sterilizer trays or the sterilizer itself (Moldenhauer 2013; 2014; 2019).
- Software Evaluations: During the OQ, it is also common to evaluate several features of the control system including verification, that it sequences properly through the various stages of the

sterilization cycle, alarms trigger as designed, security features are enabled, all documentation required in the URS is present, stored records are retrievable, records are accurate and so forth (Moldenhauer 2014; 2019).

- Sterilization Cycle Development: After the OQ is complete, it is verified that the sterilizer is operating correctly, and cycle development studies are thus initiated. The requirements may differ somewhat for the sterilization of porous/hard good loads and those for terminal sterilization of liquids or dry heat sterilization in ovens or tunnels. The PDA's Technical Report Number 1 (PDA 2007) provides a decision tree that walks through the various evaluations that should be considered in assessing the process type to be used for moist heat sterilization. This document also indicates which parameters should be established for each different process type of cycle (Moldenhauer 2013; 2014; 2019).

Cycle development studies should be used to determine the physical parameters required to appropriately sterilize the component, equipment, or final product. This includes placing the item into a specified loading pattern, except for dry heat tunnels. Part of this process is to determine a range of acceptable parameters that may be met in performing the studies. The result of this process should be an item that is sterile and functional post-sterilization. Studies for cycle development should be formally documented (PDA 2007; Moldenhauer 2014; 2019).

- Pre-qualification Activities: To conduct cycle development studies, it is important to conduct some evaluations prior to the actual cycle development like the determination of the slowest-to-heat location and slowest-to-heat zone (of a liquid). Determination of the Slowest-to-Heat Location is important. Performing validation studies using the selected items is deciding where to place the penetration probes to measure the item's temperature accurately. Additionally, one should monitor the "cold spot" of the item to ensure that the entire item has seen the recorded temperature (at minimum). As such, it is necessary to perform studies that assess the heating characteristics of the item. These studies may also be called component-mapping studies. As such, the probes should be placed to measure the heat penetration in contact with the item. This information aids in the determination of the load come-up time. It is not required to do these studies in each sterilizer or even in a production sterilizer. It just should use the same type of sterilization process (PDA 2007; Moldenhauer 2013; 2014; 2019).

  Mapping studies are also performed for liquid-filled containers. It is assumed that these studies are not required for containers of 50 mL size or smaller because of the small size of the container. The studies usually look at different levels of solution within the container as illustrated in Fig. 3 (Moldenhauer 2014; 2019).

  D-value Studies should be conducted. They are affected by the size of the inoculum, type of organism, sporulation and growth media composition, pH, phase of spore maturity, and incubation temperature for recovery of organisms. The surface on which the biological indicator is placed can have an effect on the heat resistance of the biological indicator. For instance, placing a biological indicator on the surface of a porous rubber stopper can have increased heat resistance since the biological indicator is protected by the porosity of the stopper and the poor transfer of heat from the rubber. In general, we assume that placement on a glass or stainless-steel surface does not affect the D-value sufficiently to be of concern. Biological indicators placed on paper carriers can show different resistance based on the quality of the paper used as a carrier. For porous/hard good loads, one must evaluate the biological indicator's heat resistance when directly inoculated on the rubber stoppers being sterilized. Rubber inherently is a poor conductor of heat. Data has been published to show that heat resistance is affected by rubber composition. This effect varies based on the specific biological indicator used, the formulation of the stopper, and the types of treatments to which the stopper is subjected (Rubio and Moldenhauer 1995). As knowledge in this area has increased, concerns have arisen with other items that may change the biological indicator resistance. This has led to companies starting to

Figure 3. Example of a Milk Product Claiming Sterilization.[1]

[1]Downloaded from: image sterilized milk - Bing images on January 16, 2022/.

evaluate the heat resistance of tubing and other materials processed in the load (Moldenhauer 2013; 2014; 2019).

For liquid loads, the biological indicator should be placed in direct contact with the liquid to determine the D-value. The D-value can change based on the strain of the biological indicator used as well as the formulation of the liquid product. If a carrier is used for the biological indicator, the sterilization cycle design should take into account the D-value on the carrier as well as the D-value determined when in direct contact with the liquid, item, or component being evaluated (Moldenhauer 2013; 2014; 2019).

- Determination of the Selected Loading Configuration (Load Patterns): Typical considerations include the necessary capacity required to support production activities as well as ensure that the effectiveness of the sterilization process can be achieved. For instance, if a water spray cycle is used, the loading should allow for adequate drainage of the water that contacts the items being sterilized. When sterilizing hard goods, many of the items need to be wrapped in sterilization wrap to protect the sterile surfaces until they are used. If sterilization wrap is used, it should be non-shedding and provide a microbial barrier (keeping the item sterile). The amount of wrapping should not be excessive as this can reduce the effectiveness of air removal, steam penetration, and temperature penetration. Larger mass items should be placed on the lower level of sterilizer trucks in order to reduce the risk of wetting by condensation. Care should be taken to ensure that condensation from items minimizes drip onto other items. The configurations should be photographed or drawn to be able to reproduce the loading parameters (PDA 2007; Moldenhauer 2013; 2014; 2019).
- Sterilization of Porous/Hard Goods: Sterilization of porous items in a sterilizer load can be difficult. It is important to ensure that the air normally present in the load items is removed during the process to ensure that the desired heat is delivered to the items being sterilized. As such, the data generated must show that sufficient air is removed from the chamber and load prior to the exposure conditions of the cycle. Additionally, the air and non-condensable gases need to be removed from the steam supply provided to the sterilizer. It is also critical for saturated steam (which must have an appropriate dryness level) to be provided to the sterilizer. The number and type of studies required during this process may vary, depending on the site's knowledge of the items to be sterilized. Some of the typical parameters evaluated as part of this process include (PDA 2007; Moldenhauer 2013) jacket temperature, number and depth of

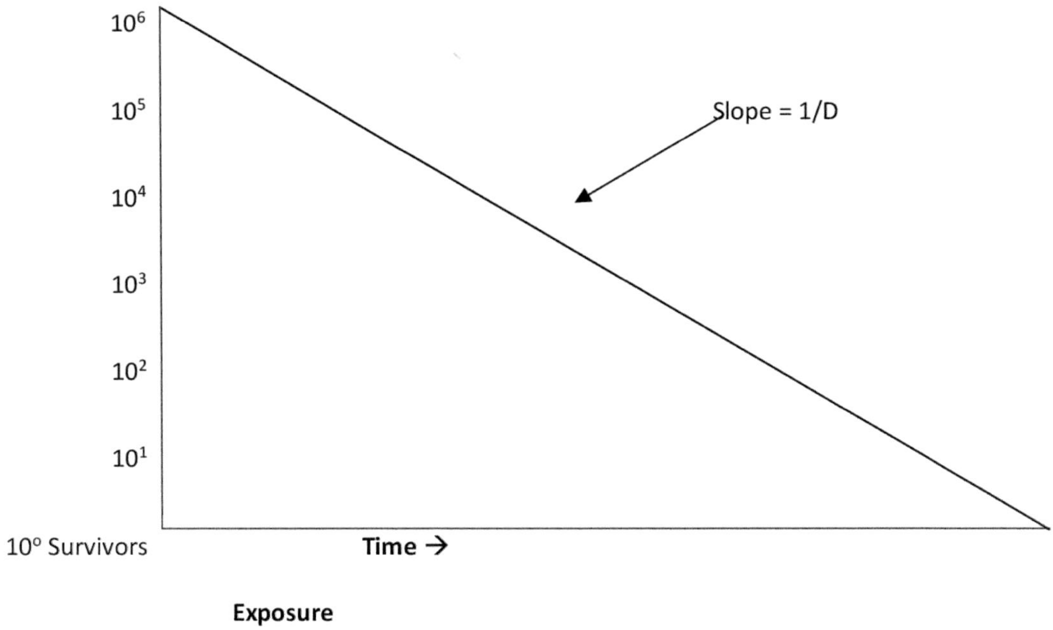

Figure 4. Microorganisms Die in a Logarithmic Fashion at a Defined Set of Parameters.

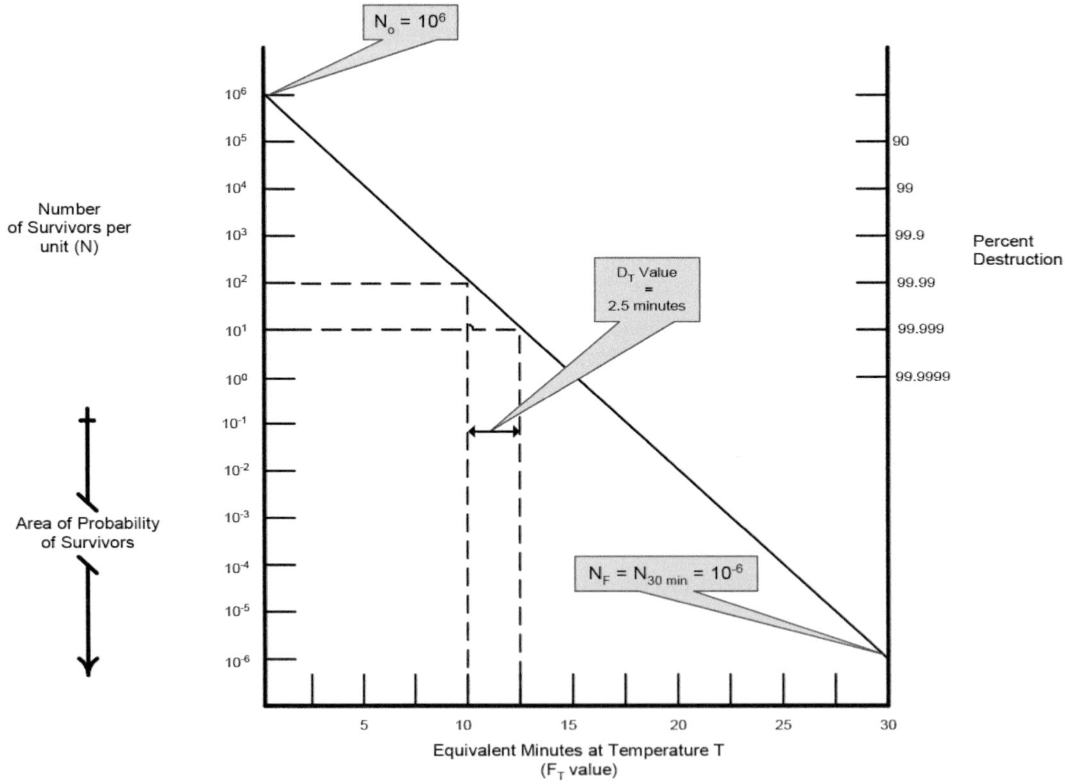

Figure 5. Microbial Survivor Curve.

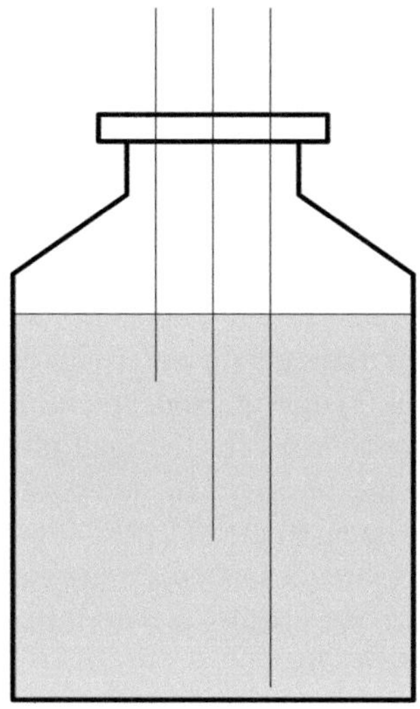

Figure 6. Example of Mapping in a Liquid-Filled Vial.

pre-vacuum conditions, steam pulse levels, time and conditions for chamber heat-up, exposure time and temperature, allowable time for cooling (exhaust phase) and drying, and air breaks. It is useful to standardize cycles to aid in reducing ongoing validation and re-validation costs (Moldenhauer 2014; 2019).

- Sterilization of Liquid Loads: In a liquid load, the liquid inside of the container is heated by the transfer of heat from the outside of the container. The water contains steam and provides the sterilant inside the container. Non-aqueous solutions, like suspensions and emulsions, require the use of a rotation or agitation cycle to keep the load in motion. Several types of sterilization processes may be used including saturated steam, overpressure cycles, steam air mixtures, and superheated water. For many liquid cycles, air removal is not performed; however, it is important to ensure that a system is used to ensure uniform heat distribution within the chamber. The containers in a loaded heat do not necessarily heat at the same rate. Items closer to the outside may heat more quickly than items in the center of the load. The concern is to be sure that the items receiving the least heat are sterile and that the items which receive the most heat are not adversely affected from a functional and stability point of view (PDA 2007; Moldenhauer 2013; 2014; 2019).
- Determining the Cycle Parameters to Use: The parameters to measure in a sterilization cycle are dependent upon the type of items to be sterilized, e.g., porous/hard goods loads, terminal sterilization loads, dry heat sterilization, or dry heat depyrogenation. In addition, one must also determine the criticality of these parameters. Some parameters are so important, frequently deemed critical parameters that when they are not met the sterilization cycle should be deemed unacceptable. Other parameters, while important may be mitigated, such as a pressure failure at some parts of the cycle might be mitigated by 100% inspection of the load for a type of physical defect. The PDA's Technical Report Number 1 provides a listing of process parameters to be considered for moist heat sterilization of porous loads including (PDA 2007). Jacket temperature and/or pressure, number and depth of vacuum pulses prior to exposure, level of the steam charge, number and requirements for positive pressure pulses, chamber come-up time, exposure time, exposure temperature set-point, temperature of the chamber drain during exposure, chamber temperature during exposure, allowable temperature range for the load probe during exposure, chamber pressure during exposure, minimum allowable F_0 during exposure, allowable time for the load to cool down after exposure, drying time after exposure, and the

rate for vacuum breaks. Depending upon the sterilizer configuration used, not all of the stated parameters are applicable (Moldenhauer 2013; 2014; 2019).

The PDA's Technical Report Number 1 provides a listing of process parameters to be considered for liquid-filled moist heat sterilization loads including (PDA 2007) jacket temperature and/or pressure, fan rotations per minute during the cycle, agitation rate during the cycle, water flow rate during the cycle, chamber water level prior to exposure, heat-up time prior to exposure, temperature heat-up rate prior to exposure, rate of temperature heat-up prior to exposure, pressure ramp-up rate prior to exposure, temperature set-point during exposure, exposure time, chamber pressure during exposure, redundant heating media temperature during exposure, load probe temperature(s) during exposure, minimum allowable F_0 for the load probe during exposure, minimum allowable F_0 for the load probe post-exposure, temperature cool-down rate post-exposure, pressure ramp-down rate post-exposure, load cool-down rate, and maximum allowable load probe accumulated F_0 post-exposure. Note that while PDA expresses this term post-exposure, some companies calculate the value accumulated only in the exposure period. The applicability of these parameters is determined by the type of sterilization process selected (Moldenhauer 2013; 2014; 2019).

- Performance Qualification (PQ): The purpose of performance qualification (PQ) studies is to demonstrate that the sterilization process consistently meets its pre-defined requirements. A minimum of three acceptable studies is required. If the minimum and maximum loads are performed, a minimum of three studies for each load configuration are performed. There are some provisions for bracketing of load configurations, e.g., qualification of the maximum and minimum container size may justify the elimination of testing for intermediate container sizes. The IQ and OQ should be successfully completed prior to initiating the PQ studies. Supporting systems like utilities necessary for the process should also have been completed by this time (Moldenhauer 2013; 2014; 2019).

  The PQ includes both physical and biological qualifications of the sterilizer load. It is common practice to perform the studies concurrently. For moist heat sterilization, the physical qualification data yields an $F_{Physical}$ value while the biological data yields an $F_{Biological}$ value. These values should be correlated to ensure that both the physical and biological parameters for the cycle design have been met. The studies should be performed under worst-case conditions (Moldenhauer 2013; 2014; 2019).

6.8.1.3.11 Challenges for Porous Load Cycles

Traditionally Geobacillus stearothermophilus is used as the challenge organism in this type of cycle. The typical cycle design utilizes an overkill approach. Data should be generated using the biological indicators directly inoculated onto the items or carriers may be used. Carriers are available in a variety of formats including paper, wire, stainless steel, aluminum, or other appropriate materials. The biological indicators should be placed in the most difficult to sterilize areas of the item. Some examples include within long tubing, areas where it is difficult to achieve air removal and/or steam penetration, within the pleats of filter cartridges, and so forth. When these studies are performed concurrently with physical qualification, care must be taken to ensure that the placement of thermocouples does not impact either the air removal or the steam penetration such that the biological challenge is seeing different conditions than if the probes were not in position. Biological challenge studies should be performed for both minimum and maximum load configurations (PDA 2007; Moldenhauer 2013; 2014; 2019).

6.8.1.3.12 Sterilizer Equivalence

Equivalence can be addressed with different types of evaluations. One consideration is the ability to use sterilization data from one sterilizer in the qualification of another sterilizer. While this has been accepted by some regulatory bodies, there is an expectation that the sterilizers use the same sterilization process, have the same physical size, construction and basic operation. It is expected, these sterilizers should be in the same general location, e.g., the same plant although maybe not in direct proximity of each other. In this scenario, each sterilizer would be tested to ensure that the operation of the sterilizer is within the expected norms prior to utilizing data from another sterilizer for qualification (Agalloco 2008; Moldenhauer 2013; 2014 and 2019).

6.8.1.3.13 Equivalence of Different Containers/Fill Volumes

Evaluating minor differences in container dimensions and/or in the fill volumes within a container can be accomplished by performing a side-by-side statistical evaluation of the heat penetration in both containers (or fill volumes). A test like a Student T-test may be used for this evaluation (Moldenhauer 2013; 2014; 2019)

6.8.1.3.14 Equivalence of Different Load Configurations

When evaluating changes to the load configurations for moist heat sterilization porous loads or hard good loads, the heat penetration data can be compared between the two load configurations in both the minimum and maximum loading arrangement. This evaluation may include a statistical evaluation of the data also (Moldenhauer 2013; 2014; 2019)

6.9 Dry Heat Sterilization

Since one log of endotoxin reduction requires substantially more heat than what is required for the inactivation of biological indicators, biological indicators are not typically required as the endotoxin reduction provides more challenge to the system that the destruction of the biological indicator (PDA Technical Report Number 1, 2007). Endotoxin is typically not an issue in food sterilization processes. Dry heat sterilizers are available in both batch and continuous configurations. Continuous configurations typically have a tunnel (Moldenhauer 2014 and 2019).

6.9.1 Dry Heat Ovens

Validation is conducted in a similar fashion to moist heat sterilizers. Studies are performed measuring temperature distribution, heat penetration, and biological (or endotoxin) challenges. The F_0 calculation for moist heat is not appropriate when using dry heat. For depyrogenation, in place of this calculation, the F_H calculation (some groups call this F_P) is used. It uses the same type of F_0 calculation, but the reference temperature is set at 250°C and the z-value is set at 46.4°C (PDA TR 3 (revised), 2013c). (Note that some authors prefer the use of a z-value of 47.6°C from the Tsuji and Lewis article.) If sterilization is the only objective of the cycle, spores of Bacillus atrophaeus are typically used as biological indicators in the cycle (Moldenhauer 2014; 2019). If the oven is used for sterilization only, the F_H calculation utilizes a reference temperature of 160°C and a z-value of 20°C (PDA TR3 (revised) 2013c).

6.9.2 Worst-Case Conditions

The oven validation studies are conducted using worst-case conditions. Typically, this is accomplished by reducing the exposure time and/or temperature (Moldenhauer 2014; 2019).

6.9.3 Load Size and Configuration

Dry heat sterilization is not as effective as moist heat sterilization since the hot air may not be able to account for differences in the loading. The loading configurations should be established and tightly controlled. It has been documented that in some cases minimum loads received lesser amounts of heat delivered than the corresponding maximum load (Agalloco 2008). The condensation of saturated steam delivers a greater amount of heat to the product than is available in a dry heat oven. Based on some of the reported variations, every load configuration should be validated. Some companies still try to use minimum and maximum loading (Moldenhauer 2013; 2014; 2019).

6.9.4 Monitoring of Cycles

Typical parameters monitored for dry heat sterilization cycles include exposure time and temperature, cooling time or final temperature, information on the blowers used, and so forth (Moldenhauer 2013; 2019).

6.9.5 Dry Heat Depyrogenation (and Sterilization) Tunnels

Tunnels allow for continuous dry heat sterilization with the item traveling on a belt through the sterilization process. They are validated in much the same way as a dry heat oven. Key validation parameters are the belt speed (which corresponds to the amount of time the item is subjected to the sterilization process) and the exposure set-point temperature. Biological indicators are not routinely used in these cycles since the intent is to perform depyrogenation. The endotoxin challenge units should be prepared directly inoculating the endotoxin onto the surface of the items to be processed in the tunnel. Data should be available to show the level of the challenge is sufficient to assess a minimum three-log reduction of endotoxin. Additionally, it is important to develop data on the recovery efficiency of the endotoxin method. For instance, it might look like a total reduction of endotoxin occurred when in reality the method was not appropriate to show the recovery of endotoxin from the surface of the item (Moldenhauer 2013; 2014; 2019).

6.9.6 Loading

Dry heat sterilization using a continuous process does not have the same type of loading configuration or arrangement as is possible with a batch sterilizer. The items are "pushed" into the sterilizer process on the conveyor belt. In most cases, those units at the beginning and trailing edges of the items typically represent the worst-case conditions (Agalloco 2008; Moldenhauer 2013; 2014; 2019).

It is common for many companies to bracket the loading qualification with the largest and smallest size item sterilized. Intermediate sizes are not routinely challenged. However, the company should have a plan ready to explain why the intermediate sizes do not require qualification.

6.9.7 Worst-Case Challenges

During validation, a company may lower the conveyor belt speed (exposure time) and/or the exposure temperature. Some companies qualify at nominal temperatures. When this occurs, they do not have the necessary data to show effective sterilization at the minimum allowable conditions unless nominal production is routinely performed at the minimum conditions (Moldenhauer 2014 and 2019).

6.9.8 Validating Dry Heat Sterilization Processes

There are both similarities and differences to other sterilization processes for validation.

- Equipment Qualification: The PDA's Technical Report Number 3 (revised) (2013) provides detailed instructions for equipment qualification. Some of the considerations at this stage include electrical program logic, verification of alarms, item interlocks overload, door interlocks, gasket integrity, air balance, vibration analysis, louver balance ability, louver position, gate balance ability, blower rotation, blower revolutions per minute (RPM), heater elements, room balance, temperature sensors, and belt speed. Other concerns include environmental qualification, uniformity of heating media, and empty chamber temperature distribution studies (Moldenhauer 2014 and 2019).

- Process Development: Overkill approaches are typically used for sterilization. The probability of a non-sterile unit is calculated using the following calculation (PDA TR 3 (revised) 2013; Moldenhauer 2014 and 2019):

  Log PNSU = $-F_H / D + \text{Log } N_0$

  Where: F_H = lowest F_H in the load

  D = D160°C value of the biological indicator

  N_0 = Initial population of the biological indicator

  If the F_H = 60 minutes, the initial population is 10^6 and the D-value at 160°C is 2 minutes, the calculation would be as follows:

  Log PNSU = $-60/2 + 6$

  Log PNSU = $-30 + 6$

  Log PNSU = -24

  PNSU = 10^{-24}

  There are allowances for a product-specific design approach, and a minimum PNSU of 10^{-6} (PDA TR 3 (Revised) 2013). During this phase of the project, the operating parameters are defined and typically include time and temperature during heat-up and cooling, pressure differentials, temperature during exposure, and exposure time (belt speed for tunnel systems). It should also include a specified PNSU. Studies should be performed to confirm the assumptions made (PDA TR 3 (revised) 2013; Moldenhauer 2014; 2019).

  For batch processes, loading patterns should be determined and based on data obtained in temperature distribution and heat penetration studies. Typically both types of the probe and the BI or endotoxin challenge are placed adjacent to each other. Consideration in these studies includes maximum variation in distribution probe measurements (probe to probe and within a probe), maximum difference in temperature between the probes and the controller setpoint, and minimum F_H (PDA TR3 (revised) 2013; Moldenhauer 2014; 2019).

  For tunnels, the worst-case load for each temperature should be determined, typically by calculating the throughput of items (mass/time), belt speed/item passage time, and the weight of the items in the load (capacity). After determining the worst-case parameters, the most mass container per unit area is selected, the belt speed is set to the maximum for any container and the tunnel is loaded at the maximum rate (PDA TR3 (revised) 2013; Moldenhauer 2014; 2019).
- Performance Qualification Studies: These studies should include physical qualification and biological qualification of the load (either biological indicators or endotoxin units) (PDA TR3 (revised) 2013; Moldenhauer 2014; 2019).
- Provisions are provided for process equivalency based on the chamber or tunnel size and configuration, airflow dynamics, temperature come-up and uniformity, conveyor speed, temperature equilibration time, and materials of construction. Other concerns are load configurations, temperature distribution, F_H values, biological inactivation, and the heat-up and cool-down rates (PDA TR3 (revised) 2013; Moldenhauer 2014; 2019).

7. Systems to Maintain Sterilizer Qualification

7.1 Personnel Training

Sterilizers are expensive and complex pieces of equipment. It is important that the individuals operating this equipment be thoroughly trained. Some companies have established certification programs for sterilizer operators' documentation requirements, e.g., SOPs and protocols. All of the activities surrounding the validation of sterilizers should be thoroughly documented. This may be in the form of protocols or standard operating procedures (SOPs) that govern the process (Moldenhauer 2014; 2019).

7.1.1 Summary Report

Final reports should be generated and approved at the completion of each phase of the validation testing. This report should be accurate, complete, and approved by those organizations that approved the protocol (Moldenhauer 2014; 2019).

7.1.2 Ongoing Activities

Once validated, it is important to maintain the sterilizer within its qualified parameters. As such, it is important to monitor the sterilizer's performance. Some of the evaluations should include a review of the cycle parameters during routine use, an evaluation of deviations that occur, a review of maintenance activities, and change control. Should a significant number of deviations occur, one must consider whether there are additional parameters that should be added for evaluation in routine qualification to better control the sterilizer's performance. Re-qualification should be conducted periodically as well as when deemed necessary by maintenance or other activities (Moldenhauer 2013; 2014; 2019).

8. Conclusion

Sterilization in any of the various forms described provides an important method to aid in the contamination control of a process. Within the types of sterilization, terminal sterilization of the product in its final closed container provides the highest level of control and prevention of contamination.

References

Agalloco, J. 1993. The Validation Life Cycle. Journal of Parenteral Science and Technology 47(3): 142–147.

Agalloco, J. 2008. Sterilization Process Validation. pp. 401–423. *In*: Microbiology in Pharmaceutical Manufacturing. Second Edition, Volume 1. Richard Prince, Ed. PDA/DHI. Bethesda, MD.

Amnon Eylath, Donald L. Wilson, David Thatcher, and Anthony Pankau. 2003. Successful Sterilization Using Chlorine Dioxide Gas part One: Sanitizing an Aseptic Fill Isolator. BIoProcess International. Downloaded from: file:///C:/DocLib/Literature/Chlorine%20Dioxide/Morrisey%20-%20Aseptic%20Isolator.pdf on January 4, 2018.

Annex 15 (2001) Final Version of Annex 15 to the EU Guide to Good Manufacturing Practice. Working Party on Control of Medicines and Inspections. Brussels. Downloaded from http://www.it-asso.com/gxp/eudralex_v21/contents/vol-4/pdfs-en/v4an15.pdf on July 15, 2011.

Anonymous (2011) Biological Process Validation of Dry-Heat Sterilization Cycles. In Pharmaceutical Validation Downloaded from http://pharmaceuticalvalidation.blogspot.com/2010/01/biological-process-validation-of-dry.html on September 11, 2011.

Anonymous (2011b) Validating Rapid Micro Methods. Downloaded from http://rapidmicromethods.com/files/validation.html on September 8, 2011.

Anonymous (2013) Sterilization (Microbiology). Downloaded from http://en.wikipedia.org/wiki/Sterilization_%28microbiology%29 on August 19, 2013.

Anonymous (2015) Finished Product Quality. HACCP Mentor/ Downloaded from: How to set finished product quality attributes - HACCP Mentor on January 10, 2022.

Anonymous (2022) Sterilization in Food Industry. AEEE Intec Downloaded from: Sterilization in food industry - Efficiency Finder (zero-emissions.at) on January 12, 2022.

Anonymous (2022b) Difference Between Pasteurization and Sterilization. Difference Between| Descriptive Analysis and Comparison. Downloaded from: Difference between Pasteurization and Sterilization | Pasteurization vs Sterilization on January 12, 2022.

ANSI (2012) ANSI/AAMI/ISO 13408-5: 2006/(R)2012. Aseptic Processing of Health Care Products – Part 5: Sterilization in Place. American National Standard. No city/state on report

ANSI/AAMI/ISO TIR 17665-2: 2009. Sterilization of Health Care Products – Moist Heat – Part 2: Guidance on the Application of ANSI/AAMI/ISO 17665-1. American National Standard. No city/state on report

Bharati, S., Soundrapandian, C., Basu, D. and Datta, S. 2009. Studies on a novel bioactive glass and composite coating with hydroxyapatite on titanium based alloys: Effect of γ-sterilization on coating. J. Eur. Ceram. Soc. 29(12): 2527–35.

Bryce, D.M. 1956. Tests for the sterility of pharmaceutical preparations; the design and interpretation of sterility tests. J Pharm. Pharmacol. 8: 561–572.

DeSantis, P. 2008. Steam Sterilization in Autoclaves. pp. 175–186. *In*: Agalloco, J. and Carleton, F.J. (eds.). Validation of Pharmaceutical Processes Third Edition. Infroma Healthcare USA, Inc. New York, NY.

EMEA (2000) Decision Trees for the Selection of Sterilisation Methods (CPMP/QWP/054/98). The European Agency for the Evaluation of Medicinal Products—Evaluation of Medicines for Human Use. Committee for Proprietary Medicinal Products (CPMP) London.

EN285 (2009) BS En 285: 2006+A2:2009 Sterilization—Steam Sterilizers—Large Sterilizers. European Committee for Standardization, Brussels.

EU GMPs (2008 revised) EU Guidelines to Good Manufacturing Practice Medicinal Products for Human and Veterinary Products Annex 1 (Corrected). EudraLex The Rules Governing Medicinal Products in the European Union. Volume 4. (Implementation by March 2010). European Commission. Brussels, Belgium.

FDA (1976) Proposed Current Good Manufacturing Practices for Large Volume Parenterals (LVP GMPs). Drafted 1976 and subsequently withdrawn 1993. Rockville, MD.

FDA (1994) Guidelines on the Principles of Process Validation. March. Rockville, MD.

FDA (1994b) Guidance for Industry for the Submission Documentation for Sterilization Process Validation in Applications for Human and Veterinary Drug Products.

FDA (1997) Hazard Analysis and Critical Control Point Principles and Application Guidelines (HACCP). National Advisory Committee on Microbiological Criteria for Foods. Available at: http://www.cfsan.fda.gov/~comm/nacmcfp.html .

FDA (2003) 21CFR§211.113b. Control of Microbiological Contamination. GMP Publications, Inc.

FDA (2004) Guidance for Industry: Sterile Drug Products Produced by Aseptic Processing—Current Good Manufacturing Practice. U.S. Health and Human Services. Food and Drug Association. (CBER) Office of Regulatory Affairs (ORA). Pharmaceutical CGMPs.

HTM-2010 (1994) Part 3: Validation and Verification. Her Majesty's Stationary Office.

ISO (2006) ISO 17665–1: 2006 Sterilization of Health Care Products—Moist Heat—part 1. Requirements for the development, validation and routine control of a sterilization process for medical devices. International Organisation for Standardisation. Geneva, Switzerland.

ISO (2007) ISO 11135:2007 Sterilization of health care products—ethylene oxide—part 1: requirements for development, validation, and routine control of a sterilization process for medical devices. International Organisation for Standardisation Geneva, Switzerland.

ISO (2008) ISO 10993-7 Biological evaluation of medical devices, part 7: ethylene oxide sterilization residuals. International Organisation for Standardisation. Geneva, Switzerland.

ISO (2012) ANSI/AAMI/ISO 13408-2006(r)2012 Aseptic Processing of Health Care Products—Part 5: Sterilization in Place. International Organisation for Standardisation. Geneva, Switzerland.

ISO (2013) ISO 17665–3: 2013 Sterilization of Health Care Products—Moist Heat—Part 3: Guidance on the Designation of a Medical Device to a Product Family and Processing Category for Steam Sterilization. International Organisation for Standardisation. Geneva, Switzerland.

Knudsen, L.F. 1949. Sample Size of Parenteral Solutions for Sterility Testing. J Amer Pharm Assoc. 38: 332–337.

Lowe, J., Gibbs, S. , Iwen, P., Smith, P. and Hewlett, A. 2013. Impact of Chlorine Dioxide Gas Steriliztaion on Nosocomial Organisms Viability in a Hospital Room. Int. J. Environ. Res. Public Health. 10(6): 2596–2605.

Miller, M.J., Claerbout, M.E., DeMaio, R.A., Farrington, J.K., Johnston, W.H., Knoy, D.K., Nicas, T.I., Rubino, M.J., Stickelmeyer, M.P. and Williams, K.L. 2008 Microbiological Control Strategies During the Development of Pharmaceutical Processes. *In*: Microbiology in Pharmaceutical Manufacturing Volume 1. Prince, R., Ed. PDA/DHI, Bethesda, MD. 237–282.

Moldenhauer, J. 2013. Validation of Moist and Dry Heat Sterilization. In Sterile Product Development: Formulation, Process, Quality and Regulatory Considerations. AAPS Advances in the Pharmaceutical Sciences Series 6. Kolhe, P., Shah, M., and Rathore, N. American Association of Pharmaceutical Sciences and Springer. New York, Heidelberg, Cordrecht, and London, pp. 535–574.

Moldenhauer, J. 1999. Contributing Factors to Variability in Biological Indicator Performance Data. PDA Journal of Pharmaceutical Science & Technology. 53(4): 157–162.

Moldenhauer, J. 2011. Personal Communication of Data for B. smithii.

Moldenhauer, J. 2014. Sterilization Processes. *In*: Madsen, R and Moldenhauer, J. (eds.). 2014. Contamination Control Volume 2. PDA/DHI Publishers. Bethesda, MD.

Moldenhauer, J. (ed.). 2019. Disinfection and Decontamination: A Practical Handbook. CRC Press, Taylor and Francis Group. Boca Raton, FL. ISBN-13-9780815379016.

Oliver, J.D., Hite, F., McDougald, D., Andon, N.L. and Simpson, L.M. 1995. Entry into, and resuscitation from, the viable but nonculturable state by Vibrio vulnificus in an estuarine environment. Appl. Environ. Microbiol. (July) 61(7): 2624–2630.

PDA (1980) PDA Technical Report 2 (TR2) Validation of Dry Heat Sterilization and Depyrogenation Cycles. PDA. Philadelphia, PA.

PDA (1981) PDA Technical Report 7 (TR7) Depyrogenation. Parenteral Drug Association. Bethesda, MD.

PDA (1998) PDA's Technical Report Number 27: Pharmaceutical Packaging Integrity. Parenteral Drug Association. Bethesda, MD.

PDA (2007) PDA Technical Report 1, Revised 2007, (TR 1) Validation of Moist Heat Sterilization Processes Cycle Design, Development, Qualification and Ongoing Control. Parenteral Drug Association. Bethesda, MD.

PDA (2010) Technical Report No. 48 Moist Heat Sterilizer Systems: Design, Commissioning, Operation, Qualification and Maintenance. Parenteral Drug Association. Bethesda, MD.

PDA (2010) Technical Report No. 48 Moist Heat Sterilizer Systems: Design, Commissioning, Operation, Qualification and Maintenance. Parenteral Drug Association. Bethesda, MD.

PDA (2013b) Technical Report No. 61 Steam In Place. Parenteral Drug Association. Bethesda, MD.

PDA (2013c) Technical Report No.3 (revised 2013) Validation of Dry Heat Processes Used for Depyrogenation and Sterilization. Parenteral Drug Association. Bethesda, MD.

PIC/s (2011) Recommendation on the Validation of Aseptic Processes. PI 007-06. Pharmaceutical Inspection Convention.

Rubio, S.L. and Moldenhauer, J. 1995. Effect of Rubber Stopper Composition, Preservative Pretreatment and Rinse Water Temperature on the Moist Heat Resistance of Bacillus stearothermophilus ATCC 12980. PDA Journal of Pharmaceutical Science & Technology 49(1): 29–31.

Sadowski, M. 2009. The Use of Biological Indicators in the Development and Qualification of Moist Heat Sterilization Processes. *In*: Biological Indicators for Sterilization Processes. Gomez, M. and Moldenhauer, J. (eds.). PDA/DHI Publishers. Bethesda, MD. 219–270.

Setlow, P. 2009. Bacterial Endospores: Mechanisms that Contribute to their Longevity and Resistance. *In*: Gomez, M. and Moldenhauer, J. Biological Indicators for Sterilization Processes. PDA/DHI. Bethesda, MD. 25–54.

Stamatis, D.H. 2003. Failure Mode and Effect Analysis, FMEA from Theory to Execution, 2nd Edition. ASQ Quality Press. Milwaukee, WI.

Sutton, S.V.W. and Moldenhauer, J. 2004 Towards an Improved Sterility Test. PDA Journal of Science and Technology 58(6): 284–286.

USP (2012) <1229> Sterilization of Compendial Articles. United States Pharmacopeia, USP 36-NF3, First Supplement. Pharmacopeial Convention. Rockville, MD.

USP (2012b) <1229.2> Moist Heat Sterilization of Aqueous Liquids. United States Pharmacopeia, USP 36-NF3, First Supplement. Pharmacopeial Convention. Rockville, MD.

USP (2013) <1129.4> Sterilizing Filtration of Liquids. Pharmacopeial Forum (PF Online) 39(2). *In*: Process Revision. New USP37-NF32 1S. downloaded from http://www.usppf.com/pf/pub/index.html on August 8, 2013.

USP (2013b) <1229.6> Liquid Phase Sterilization. Pharmacopeial Forum (PF Online) 39(4). *In*: Process Revision. New USP 37-NF32 2S. Dowloaded from http://www.usppf.com/pf/pub/index.html on August 8, 2013.

USP (2013c) <1229.7> Gaseous Sterilization. Pharmacopeial Forum (PF Online) 39(3). Downloaded from http://www.usppf.com/pf/pub/index.html on August 8, 2013.

USP (2013d) <1229.8> Dry Heat Sterilization. Pharmacopeial Forum (PF Online) 39(3). Downloaded from http://www.usppf.com/pf/pub/index.html on August 8, 2013.

USP (2013e) <1229.8> Radiation Sterilization. Pharmacopeial Forum (PF Online) 39(2). Downloaded from http://www.usppf.com/pf/pub/index.html on August 8, 2013.

Wikipedia (2018) Chlorine Dioxide. Downloaded from: https://www.bing.com/search?q=chlorine+dioxide+sterilization&qs=n&form=QBLH&sp=-1&pq=chlorine+dioxide+sterilization&sc=4-30&sk=&cvid=A72B2982C1FC4508A57DC534BF1F6B4A on January 4, 2018).

7

Electromagnetic Radiation

A New Approach to Sustainable Development in Food Sectors

Plachikkattu Parambil Akhila, Basheer Aaliya and
*Kappat Valiyapeediyekkal Sunooj**

1. Introduction

Sustainable food supply is a primary concern for food industries, governments, and international agencies worldwide. Food processors are under constant pressure to create technology that assures food safety, preserves nutritional value while improving the bio-accessibility of nutrients, is eco-friendly, and meets customer needs. Consumers desire more naturally produced and minimally processed foods as their awareness of nutrition and health has grown over the years (Pan et al. 2017). Furthermore, the growing population and insufficient cultivable area necessitate changes in current food processing systems (Arshad et al. 2021). Moreover, insects, microorganisms, and physiological deterioration cause roughly 25% of agricultural output loss globally (Singh and Singh 2019). Microbial contamination and pest infestation cause significant food losses during post-harvest storage, transportation, and marketing. Sprouting is the leading cause of vegetable crop losses, such as tubers and bulbs. Food deterioration and waste are also caused by the use of contaminated products and raw resources (Singh and Singh 2019). As a result, reducing food waste, developing sustainable preservation techniques, ensuring nutrient security, and managing food safety are critical elements for future requirements (Arshad et al. 2021). In both home and commercial food processing, electromagnetic energy is employed in one form or another to adjust food texture and flavour qualities, maintain microbiological safety, and extend shelf life (Mudgett 1988). The wavelengths of typical electromagnetic emissions range from 10^3 m (radio wave) to 10^{-12} m (γ -rays) (Moosekian et al. 2012). Ionising radiation like X-rays and gamma rays are among the highest frequencies of interest in the electromagnetic spectrum beyond the visible range. Ionising radiation effectively eradicates several common pathogenic organisms that cause foodborne illness, including *Listeria monocytogenes*, *E. coli*, *Vibrio* spp., and *Salmonella* spp. among others. Irradiation offers the benefit of allowing the food to be treated after being sealed in its final packaging, minimising or eliminating the chance of recontamination after this treatment (Borsa and Ph 2012). Although heat is not commonly thought of as electromagnetic energy, it is produced during the application of

Department of Food Science and Technology, Pondicherry University, Puducherry, 605014, India.
* Corresponding author: sunooj4u@gmail.com

longer wavelength and low-frequency radiation in the infrared range. It is used in several basic unit activities in conventional food processing (Mudgett 1988). In dielectric heating techniques, non-ionising radiation from electrical energy sources in the microwave and radio-frequency parts of the spectrum is also exploited (Mudgett 1988).

The application of electromagnetic radiation (EMR) has provided the food industry with exceptional prospects by producing high-quality products without compromising safety (Akhila et al. 2021). Food technologists, researchers, engineers, and the food industry are interested in this green technology since it fits with the concept of sustainable agriculture in a world with a growing population (Sommers 2012; Pan et al. 2017). This chapter provides a detailed description of electromagnetic radiation and its various application in the food sector, including radiation-dependent detection methods. Also, the chapter will analyse the sustainability of EMR in microbial safety and processing of food and agriculture produce.

2. Electromagnetic Technologies

Particles above absolute zero (–273°C) carry electromagnetic energy and release radiation with varying frequencies in the electromagnetic spectrum. Electrically charged atoms or molecules present in the matter oscillate sinusoidally varying electric and magnetic fields to produce a spectrum of electromagnetic frequency (Ramaswamy et al. 2012; Akhila et al. 2021). Electromagnetic waves are a part of everyday life, and we are exposed to them from various sources, including the Sun and cell phones (Moosekian et al. 2012). During the 18th and 19th centuries, electromagnetism flourished in science. In 1865, James Clerk Maxwell (1831–1879) described his groundbreaking theory about electromagnetic fields based on several findings in the realms of electricity and magnetism. Undoubtedly, none of those brilliant scientists could have forecast the significance of their findings in the emerging technology of the subsequent centuries (Barba et al. 2018). Gamma rays, X-rays, ultraviolet, visible light, infrared radiation, microwaves, and radio waves are all types of radiant energy that are generated in different ways and fall within the spectrum of electromagnetic radiation (EMR) (F. Lima et al. 2018). Figure 1 represent the application of electromagnetic radiation in food

Figure 1. Application of electromagnetic radiation in food industry.

sectors. The wavelength and penetration power affects the impact of radiation on various systems (F. Lima et al. 2018). Based on the penetration power, the electromagnetic spectrum can be divided into ionising and non-ionising radiation. This section will focus on the fundamentals of electromagnetic-based technologies, such as ionising, non-ionising radiation, and pulsed light (PL).

2.1 Ionising Radiation

Ionising radiation is currently the most popular type of radiation utilised in the treatment of foods. These radiations have the energy to displace electrons from the outer orbital shell of atoms, converting them to electrical charges known as ions (F. Lima et al. 2018). It is produced by unstable atoms with a surplus of energy, mass, or both, which can achieve stability by releasing these atoms or expelling the excess energy or mass they possess (Calado et al. 2014). The three ionising radiations used to irradiate food around the world are gamma irradiation (^{137}Cs and ^{60}Co), electron beam (from linear accelerators or other accelerating structures), and X-rays (from X-ray tubes or linear accelerators) (Pillai and Shayanfar 2017). However, these technologies differ in energy profiles, production techniques, shield demands, and regulatory environment (Pillai and Shayanfar 2017). Irradiation of food is a physical method of treatment that involves exposing a product, either in bulk or packaged, to precise doses of ionising radiation for sanitary, phytosanitary, or technological purposes. Irradiation has several benefits, including strong penetrating power and intensity, minimal quantifiable residues, minimal temperature rise, and only a few factors to manage (e.g., dose, temperature and dryness) (Supriya et al. 2014). A standard measurement called a gray (Gy) is used to measure the amount of energy deposited in a unit mass (J /kg). The dosages of ionising radiation used to treat food products typically vary from 1 to 44 kGy (Moosekian et al. 2012).

Irradiation of food has long been accepted as a dependable and safe approach for preserving food and improving its hygienic quality and nutritional content. Food irradiation has been proven to have a wide range of beneficial effects, including sprouting inhibition in potatoes and onions, insect disinfestation in grains and fruits, delayed ripening of fresh vegetables and fruits, and improved safety and sterilisation of fresh and frozen meat products, eggs, and seafood (Miller 2005a). Ionising radiation has the ability to strengthen food safety for both fresh and processed foods by two fundamental mechanisms of action (Mahapatra et al. 2005). Ions, excited molecules, or molecular fragments are generated as a result of primary processes or events. The second step involves the reaction of the product obtained from the first step with the material. The preliminary stage of the event is independent, whereas the second step depends on the factors such as pressure, temperature, and gases. Products obtained through the reaction are called radiolytic products (Nair and Sharma 2016; Gómez-López et al. 2021).

Ionising radiation effectively eradicates several common pathogenic organisms that cause foodborne illness, including *L. monocytogenes, E. coli*, *Vibrio* spp., and *Salmonella* spp. among others. Irradiation offers the benefit of allowing the food to be treated after being sealed in its final packaging, minimising or eliminating the chance of recontamination after this treatment. Like hot dogs and many other deli items, ready-to-eat foods are irradiated after packaging, which helps to avoid the entry of foodborne pathogens like *L. monocytogenes* (Borsa and Ph 2012). Ionising radiation is being used to diminish certain carcinogenic anti-nutritional substances, including tannin inhibitors, gossypol, and trypsin, in addition to antimicrobial action in various food commodities. Furthermore, irradiation can be used to dry food and produce traditionally fermented meals (Arvanitoyannis and Stratakos 2010).

2.1.1 Gamma Radiation

Electromagnetic radiation or photons emerging from the nucleus of radioisotopes, such as ^{60}Co and ^{137}Cs, are known as gamma (γ) rays (Mahapatra et al. 2005; Pillai and Shayanfar 2017). The waves are targeted at the food that is being irradiated; however, the food itself is never in contact with the caesium or cobalt source (Loaharanu 2003). The energy of this gamma radiation is sufficient to remove electrons from food molecules, leading them to become ions. On the other hand, gamma rays do not have enough energy to excite neutrons in the nucleus of molecules so they cannot cause radioactivity in food. The moisture content, antioxidant levels, density, and characteristics of the foods all influence the efficacy of a given dose of radiation. External environmental factors, like temperature, oxygen, and storage conditions, significantly affect radiation efficiency (Mahapatra et al. 2005).

Gamma irradiation is used to preserve food quality and extend shelf life, resulting in less food degradation-related economic loss, which is particularly essential in developing nations (Markov et al. 2015). This technique is a simple and effective food decontamination technology that involves inducing chemical DNA damage in biological contaminants, such as insects, moulds, yeasts, bacteria, and other microorganisms (Markov et al. 2015). Robichaud et al. (2021) explored the role of γ-irradiation alone and in combination with food additives (0.5 % Na2CO3, 2.5 % $Na_3C_6H_5O_7$, and 0.75% C6H8O7) in powdered and frozen infant formula on the microbial inactivation of foodborne pathogens. In frozen infant formula, the study revealed that irradiation alone was more effective against microorganisms. This study also showed that combining food additives, mostly Na2CO3, with γ-radiation can be a valuable technique to shorten the irradiation treatment time while still ensuring the safety of the infant formula product. Mycotoxins can be effectively decontaminated via irradiation. In a study conducted by Maatouk et al. (2019), the ochratoxin A concentration was dramatically reduced by 48% and 62% after gamma irradiation of naturally contaminated millet flour at doses of 2 and 4 kGy, respectively. The ionising radiation significantly reduced phytic acid without affecting the nutritional property of Carioca beans. According to the findings, irradiation (^{60}Co) up to 10 kGy can be applied to extend the shelf life of ordinary beans without compromising nutritional quality (Lima et al. 2019). Table 1 represents the applications of gamma radiation in food sectors.

Another study looked at the impact of irradiation and drying methods on the kola nuts' proximate, phytochemical, and free scavenging properties (Darfour et al. 2021). The administered dose was assessed using an ethanol chlorobenzene dosimeter after milled samples were wrapped in polythene bags and irradiated with a ^{60}Co source at a dose rate of 2 kGy/h. Notably, γ -irradiation had little effect on the health and nutritional properties of the kola nut and was effective in microbial decontamination (Darfour et al. 2021). In recent years, a minimal investigation has been performed on the impact of irradiation on the dehydration kinetics of dried fruit. Mohammad-Razdari et al. (2021) analysed the performance of hot-air dryers and proposed a new technique for improving the quality of dried bio-products by using gamma irradiation as a drying pre-treatment. The results revealed that varying doses of 0.5, 2, and 5 kGy reduced moisture content by an average of 35%, 21%, and 11%. Moisture content dropped to 42%, 31%, and 25%, respectively, by lowering the dehydration temperature. Increasing the irradiation dosage raised the effective diffusivity values at different temperatures (control, 0.5, 2, and 5 kGy increased to 12%, 17%, 29%, and 43%, correspondingly) and vice versa for the activation energy. Furthermore, the study found that gamma radiation has the potential to reduce drying quality while also improving product quality during the drying process.

2.1.2 Electron-Beam Irradiation

The usage of electron beams (EB) to irradiate foods is increasing due to their ease of use and inexpensive cost compared to ^{60}Co irradiation. Several studies have shown that EB treatment can decontaminate foods, reduce the occurrence of agricultural pests, and increase the shelf life of various items without compromising quality (Freita-Silva et al. 2015). Table 2 represents applications of electron beam radiation in food sectors. Electron-beam irradiation (EBI) is a new food decontamination technique that uses low-dose ionising radiation generated from particle accelerators to eradicate microbial contamination in produce or food (Deng et al. 2020). Linear accelerators are the equipment that produces EB and X-rays. There are diverse kinds of accelerators,

Table 1. Application of Gamma Radiation in Food Sectors

Sample	Treatment Condition	Other Conditions/ Materials Used	Major Findings	Application	Reference
Strawberries and Raspberries	1–11 kGy	Murine norovirus type 1 human adenovirus type 5	2 log CFU/g reduction at 4 kGy dose	Virucidal potential and food safety	Pimenta et al. (2019)
-	2 kGy	Thyme-based nanoemulsion coated with chitosan, 0.5% polyvinylpyrrolidone	Higher antimicrobial activity against S. aureus, E. coli, and C. albicans	Antibiofilm	Deghiedy et al. (2021)
Frozen and powdered infant formula	1–10 kGy	Hurdle technology 0.5% sodium carbonate 2.5% sodium citrate, and 0.75% citric acid	Showed a higher radio sensitization effect against S. Typhimurium, B. cereus, and L. monocytogenes	Microbial inactivation of foodborne pathogens in infant formula	Robichaud et al. (2021)
Brown rice based weaning food	0–10 kGy	12 months, 32 ± 2°C	Sensory characteristics, oxalate, phytate contents, and total plate count, decreased with an increase in irradiation dose	To assess the effect of gamma irradiation	A. Jan et al. (2020)
Grape juice, wine, water, and wheat flour	2.4, 10.1, and 30.5 kGy		Doses ≥ 2.5 kGy degrade OTA above 90% in aqueous solution. γ- radiation in aqueous solution is more effective but OTA showed higher resistance in dried food	Ochratoxin A decontamination	Calado et al. (2018)

depending on the quantity of energy required, the processing line speeds that can be attained, the electrical efficiency, and so forth. Pulsed accelerators, CW accelerators, and DC (direct current) accelerators are the three types of commercial e-beam accelerators (Pillai and Shayanfar 2017). Electrons are typically created at a maximum energy of 10 MeV for purposes of food preservation. In comparison to gamma radiation, the penetration power of EB into food is limited. However, EB radiation is convenient since it can be turned on and off as needed and does not require an isotope source (Sommers 2012). Treatment with EB technology is gaining popularity in fresh produce and food to improve overall quality, such as shelf life and microbiological quality. The increasing expense of procuring cobalt-60, the problems connected with gamma facility approvals, and the general drop in the cost of purchasing EB technology contribute to increased interest in EB technology (Pillai and Shayanfar 2018).

Based on the kinetic energy of the electrons, EB can be categorised as low-energy (< 300 keV) or high-energy (> 300 keV) electron beams (Hertwig et al. 2018). Electron energy and, to a considerable extent, the density of the treated material govern and define the depth of penetration. When electrons are bombarded into food, their kinetic energy is lost due to collisions with product particles (Hertwig et al. 2018). Food products can be penetrated to a depth of several centimetres with high-energy electrons but only a few micrometres with a low-energy electron beam (Hertwig et al. 2018). Low-energy electrons, often known as "soft-electrons" have an energy of less than 300 keV. Treatment with a low-energy electron beam (LEEB) can disinfect low-moisture foods without compromising their quality. LEEB treatment has been proven to inactivate spores using methods similar to conventional electron beams, and its efficacy is comparable to that of other ionising radiation technologies. Low-energy electrons are used in "Laatu", a commercial device invented

Table 2. Application of Electron Beam Radiation in Food Sectors

Sample	Treatment Condition	Major Findings	Application	Reference
Beef fajitas	15 kGy	10 log reduction *E. coli* 5 log reduction *C. sporogenes*	Sterilization of Space Foods	Bhatia et al. (2018)
Packed egg	Accelerating voltage of 20, 25, and 30 kV	Doses ≥ 5 kGy or more showed complete disinfection of the surface of packed eggs	Surface disinfection	Sokovnin (2021)
Ready-to-eat food	1 and 8 kGy	EB treatment showed increased concentration of 25-OH and 7-keto COP above the accepted limit in food	Safety and formation of cholesterol oxidation products	Lozada-Castro et al. (2011)
Idli	2.5, 5, and 7.5 kGy	7.5 kGy and subjected to combination processing at 2.5 kGy and thermal treatment were shelf-stable for 60 days.	Enhance the shelf-life	Mulmule et al. (2017)
Spices	300 keV and 10 MeV	The effectiveness of microbial inactivation was comparable for spices, where microorganisms exist in the surface layer	Microbial decontamination	Gryczka et al. (2020)

by Buhler for microbial inactivation in foods. The new technique is currently being developed for different uses and is available for spices (Gómez-López et al. 2021).

Electrons with an energy of 300 keV or less can disinfect the surfaces of various foods while causing minimum quality impairment. The EBI technique's success in decontaminating specific food products, such as vegetables, fruits, and meat products, was demonstrated by a significant reduction in the number of microorganisms (Khaneghah et al. 2020). Gryczka et al. (2020) proved the efficiency of microbial decontamination by both low (300 keV) and high intensity (10 MeV) EB was comparable for black pepper and white pepper. The mechanism of microbial inactivation is similar to those associated with other types of ionising radiation. Ionising radiation targets the DNA of the microorganism. EB can inactivate the pathogenic organism through direct and indirect interaction (Akhila et al. 2021). The energy transmission of electrons to target molecules, such as DNA, membrane proteins, RNA, and enzymes, is caused by direct interactions. The principal mechanism involved in microbial inactivation by direct effects is induced DNA degradation. The ionisation of water or oxygen molecules causes indirect effects of EB, resulting in several reactive species known as radiolytic products (Khaneghah et al. 2020). Microorganisms contain up to 70% water, and the produced electrons are absorbed resulting in highly reactive radicals, including hydroxyl and superoxide radicals that can damage DNA and cell membranes during exposure to EB radiation (Hertwig et al. 2018). Electrons can be focused directly at the object or pathogen, whereas γ-rays are discharged in all directions, thus making the e-beam approach more efficient than the γ-ray application (Freita-Silva et al. 2015).

Shawrang et al. (2011) proved that EB treatment assures the reduction of anti-nutritional factors without compromising the physicochemical property of the produce. Shawrang et al. (2011) investigated the impact of electron beam irradiation at dosages of 10, 15, 20, 25, and 30 kGy on the chemical composition, *in vivo* digestibility, and phytate and tannin levels of sorghum grain. Irradiation did not influence chemical compositions ($p > 0.05$), but phytate content was significantly lowered ($p > 0.05$) by 39%, 49%, 66%, 79%, and 90%; tannin content was lowered ($p > 0.05$) by 28%, 30 %, 42 %, 83 %, and 86%, respectively. Since it is efficient against most insects and has no effect on the quality of the products, electron irradiation is an ideal method for pest control in agriculture (Yun et al. 2014). The application of EB to stored products can be a solution to decrease the problems posed by fungi and their mycotoxins (Freita-Silva et al. 2015). Implementing electron beam-based methods for flue gas treatment and wastewater treatment on an industrial level is expected to provide unique ecological rehabilitation options in the future years. To expand electron accelerators in these future applications, the demand for efficient, very high-power electron beam machines will be critical (Sabharwal 2013).

2.1.3 X-Rays

X-rays, also known as Roentgen rays, are electromagnetic waves ranging from 10^{16} to 10^{19} Hz. They occur adjacent to gamma rays in the electromagnetic spectrum. X-rays emit lower-energy photons due to a charged particle's interaction with matter or by replacing displaced electrons from a low-lying orbit or by bremsstrahlung, also referred to as braking radiation (Moosekian et al. 2012). Foodstuffs can be treated using machine-generated x-rays with a maximum energy of 5 MeV (7 MeV in the US). The beam of an electron is bounced off a tantalum or gold target to generate x-rays. The same as γ- rays, X-rays can penetrate through thick foods, necessitating strong shielding for protection. Unlike radionuclide sources, which emit essentially monoenergetic photons, X-ray sources emit a broad spectrum of photons ranging from the electron's maximal energy to zero energy (Riganakos 2010). X-ray exhibits high penetration capacity (30–40 cm), and treatment does not produce radioactive waste (Valentina 2017).

The Compton scattering, photoelectric effect, and pair production are the three fundamental X-ray interaction methods that deposit energy as a product moves through the exposure zone. The most common x-ray interaction mechanism discovered in foods is relativistic Compton scattering (low atomic number elements). The important feature of each process, irrespective of the mechanism, is the production of energetically charged particles (positrons and electrons) that are then slowed by Coulombic interactions; thus, the energy deposition mechanism for X-ray is comparable to those considered in electron beam irradiation of food (Miller 2005b).

It can be used to achieve microbial decontamination more efficiently as a food preservation strategy. It successfully achieved a high efficacy of microbial inactivation in various food materials, such as strawberries, sweet potatoes, smoked catfish, ready-to-eat shrimp, milk, tomatoes, and spinach leaves (Wang et al. 2015). Begum et al. (2020) studied the radiation effect of foodborne microorganisms (*S. typhimurium, Escherichia coli*, and *L. monocytogenes*) using gamma (^{60}Co) and low-energy (125 keV) X-ray. Compared to gamma-irradiated pathogens, low-energy X-ray irradiated pathogens showed considerably greater D10 values (lower microbicidal efficiency) for all microbes. Oner and Wall (2013) looked into the influence of X-ray irradiation on the quality of fresh-cut, chilled purple-fleshed sweet potato (PFSP) cubes. Packaged sweet potato cubes were irradiated with X-rays (0, 250, 500, 750, or 1,000 Gy) and kept for 14 days at 4 ± 1°C. The total bacterial and mould-yeat count was 3.2 $\log_{10}$ CFU g^{-1} and 3.0 $\log_{10}$ CFU g-^{1} in 1 kGy irradiated sample after the 14-day storage. The findings demonstrate that irradiating PFSP cubes with X-rays at levels up to 1 kGy can suppress microbial populations while retaining the physical quality and anthocyanin content for up to 14 days. The initial microbiota on RTE shrimp samples was lowered from 3.8 ± 0.2 log CFU/g to below the detectable limit after treatment with 0.75 kGy X-ray (Mahmoud 2009). X-ray generation is inefficient; but for high-capacity plants, it can be competitive against gamma radiation. It is appealing to use electron beams and X-rays in the same irradiation facility. Many X-ray irradiators have been created by turning electrons into X-rays for achieving the deep penetration required for food pallets (Riganakos 2010). Nevertheless, because the majority of the energy required to generate X-rays is wasted as heat, the method is more expensive than gamma or electron-beam irradiation (Sommers 2012).

Through imaging, diffraction, and scattering modes, X-rays can be utilised to assess the quality. All of the approaches are non-destructive, durable, and have specific applications. The X-ray imaging mode system is simple, fast, and widely utilised in food science and industry for quality and safety evaluation (Nielsen et al. 2013). It can be used to do quality checks, such as screening for foreign objects and microparticles as well as assessing the density of the food. The structure of complex proteins and biomolecules with a noncrystalline characteristic can be determined via X-ray scattering (Purohit et al. 2019).

2.2 Non-Ionising Radiations Basic Principle and Application

Non-ionising radiation is electromagnetic radiation that travels at the speed of light and has frequencies ranging from 0 Hz to 1.1 THz. The spectrum of UV, VL, IR, MW, and RW is considered non-ionising radiation. Compared to ionising radiation, non-ionising radiation has a lesser energy level and the least concern with human safety and regulation. They offer the advantage of better retaining the sensory and nutritional properties of food than conventional heat processing.

2.2.1 Ultraviolet Radiation

Ultraviolet radiation with a wavelength of 100 nm to 400 nm is produced naturally by the Sun or artificially by mercury vapour lamps, incandescent sources, fluorescent, some types of lasers, and black lights. The four wavelength classes that make up the UV light spectrum include UV-A, UV-B, UV-C, and UV-V are 315–400 nm, 280–315 nm, 200–280 nm, and 100–200 nm, respectively. UV-C light, with a wavelength of 254 nm, seems to be the fatal and effective wavelength on microorganisms (Falguera et al. 2011). Because of their high penetration power into food, UV sources are critical in microbial inactivation efficiency. Its germicidal properties are used in liquid sterilisation, air disinfection, and the inactivation of microorganisms on the surface (Falguera et al. 2011). UV light's destruction and detoxifying effects on microorganisms and toxins have been studied extensively in food. It is used in the food sector as an air disinfectant in meat and vegetables, on water, and inactivates germs on the surface of fresh items, such as fish, eggs, poultry, liquid food, milk, fruit juices, and cider. Table 3 represents the applications of UV radiation in food sectors. However, due to its non-penetrating nature, its use is confined to surface disinfection and detoxification of food products (Koutchma and Popović 2019).

UV light technology to manage microbiological contaminants in the air, non-food and food contact surfaces, materials, packaging, and raw and processed foods can considerably enhance the reliability of food manufacturing and finished products (Koutchma and Popović 2019). UV light has been recognised by the US Food and Drug Administration (FDA) as a substitute for thermal pasteurisation for fresh juice products (US Food and Drug Administration 2001). The US FDA does not set a maximum or minimum dose by regulation. The UV dose levels given to the juice will be determined by the changes in the organoleptic characteristics following UV irradiation. Surface microbe control and sterilisation of drinkable water, shelf life extension of pasteurised milk, and vitamin D production in bread are permitted UV radiation applications for food processing (Koutchma and Popović 2019). In January 2016, the European Food Safety Authority's (EFSA) Panel on Dietetic Products, Nutrition, and Allergies (NDA) issued an opinion on UV-treated

Table 3. Application of UV Radiation in Food Sectors.

Sample	Treatment Condition	Major Findings	Application	Reference
Fish	0.30 to 0.79 J/cm^2	Shelf life of refrigerated fish improved to 2 to 6 days	Increase the quality and safety of fish	Monteiro et al. (2021)
Tomato	254 nm 0 to 223.1 mJ/cm^2	Significantly ($p < 0.05$) reduced Salmonella populations	Reduce Salmonella contamination	Lim and Harrison (2016)
Raspberries	7.8×10^2 mJ/cm^2	Surface of the fruit showed a reduction of *L. monocytogenes* to 1.5 log CFU/g	Inactivation of *L. monocytogenes*	Liao et al. (2017)
Mango	2.5, 5, 7.5 and 10 kJ/m^2	C. sakazakii showed 2.4–2.6 log cfu/g	Inhibition of E. coli and *C. sakazakii*	Santo et al. (2018)
Apple Cherry Strawberry Raspberry	1.2, 2.1, 3.3, and 3.3 kJ/m^2	1.8, 2.4,2.6,2.8 Log reduction *P. expansum* on apple, cherry, strawberry, and raspberry respectively	Inactivation of *P. expansum* on fruit surfaces	Syamaladevi et al. (2015)

milk (novel food) as mandated by Regulation (EC) No 258/97. EPSA stated that treatment of UV radiation after the pasteurisation of cow milk (whole, semi-skimmed, or skimmed) could be used as a novel technology to improve the shelf life and increase vitamin D3 by the conversion of 7-dehydrocholesterol and added that it is safe under the intended conditions of use as specified by the applicant (EFSA 2016).

UV-C radiation is thought to be the most effective germicidal part of the UV spectrum for destroying microorganisms, like bacteria, fungi, viruses, yeasts, protozoa, and algae. It has advantages, such as low installation and maintenance costs and minimal energy use to preserve food without the deleterious effects of heat treatment (Shin et al. 2016; Chawla et al. 2021). Absorbed UV light inactivates pathogenic microorganisms by photophysical, photothermal, or photochemical effects of absorbed radiation (Krishnamurthy et al. 2008). Because of their aromatic ring structure, DNA base pairs absorb UV light more quickly than other components in the cell of microorganisms. In particular, pyrimidines [cytosine (DNA and RNA), thymine (DNA), and uracil (RNA)] are potent absorbers of UV rays, resulting in structural alterations that cause microbial inactivation and death (Krishnamurthy et al. 2008; Shin et al. 2016). Santo et al. (2018) studied the efficacy of UV-C radiation against *E. coli* and *Cronobacter sakazakii* on fresh-cut mangos. UV-C treatments of 2.5 to 10 kJ/ m^2 were shown to reduce C. sakazakii (2.4-2.6 log CFU/g) more effectively than *E. coli*. Another analysis utilised a pilot-scale real-life aquaponic system to illustrate the efficiency of UV-radiation for coliform reduction. The treatment significantly ($p < 0.05$) reduced bacterial counts by 1.5 and 3.0 logs on 3-M Petri film and m-Endo agar. Lettuce was free from coliforms or *E. coli* after the treatment (Moriarty et al. 2018). Liao et al. (2017) used UV-C light to reduce the population of L. monocytogenes on frozen raspberries stored at –35°C for nine months.

The use of UVC as a germicidal method for the inactivation of hazardous microorganisms in foods is fast gaining acceptance. Syamaladevi et al. (2015) investigated the UV-C inactivation of *Penicillium expansum* inoculated into the surface of organic fruits in a study. Maximum reductions of 1.8, 2.4, 2.6, and 2.8 log CFU/ g were recorded after UV-C doses of 1.2, 2.1, 3.3, and 3.3 kJ/ m^2, respectively. According to this study, UV-C efficiently reduces *P. expansum* populations on fresh fruit surfaces, and the success of radiation relies on the morphology of the fruit surface. The sensitivity of different types of microorganisms, species, and strains to UV-C radiation varies substantially. The composition, size, and conformation of the genetic material, microbial cell size, pigment production, irradiation-generated photoproducts, and DNA repairability are all crucial in determining the effectiveness of the treatment. According to research, when compared to Gram-positive bacteria, UV radiation is more effective at inactivating Gram-negative bacteria, which is followed by yeast, bacterial spores, fungus, viruses, and protozoa (Delorme et al. 2020).

Furthermore, the hurdle technology (a combination of UVC and various non-thermal treatments like high hydrostatic pressure, ultrasound, pulsed electric fields, and antimicrobials) has shown excellent synergistic activity against microorganisms (H. Singh et al. 2021). The disinfection of packing materials and the air and water used in dairy plants are other possible UV light applications in the dairy business (Chawla et al. 2021). In terms of packaging, it has been discovered that UV radiation can be utilised on packaged products as well. According to a study on the effects of UV-C on packaged cheese surfaces, transparent polymers like polypropylene (PP) and polyethylene (PE) films combined with UV-C radiation can ensure the prevention of contamination after processing (Ha et al. 2016).

2.2.2 Pulsed Light (PL)

The use of inert gas flash lamps to convert high-speed electronic pulses into short-duration and high-power pulses of radiation with a broad spectrum similar to that of the Sun (including IR, VL, and UV), rich in UV-C light is referred to as pulsed light (PL) (Mahendran et al. 2019; Yousefi et al. 2021). In most cases, the number of light pulses required in a treatment process is less than 10, and each pulse can produce power ranging from 0.01 to 50 J/cm^2 (Yousefi et al. 2021). This technology is faster than traditional procedures for degradation, sterilisation, and decontamination because many pulses of light are supplied in a second. PL is a valuable method for reducing microbial counts in foods and food packaging materials in the industrial world. It can also be used to reduce microbial contamination of food contact surfaces, equipment, and media (such as air and water) utilised in the manufacturing process (Mahendran et al. 2019). Furthermore, PL treatment has been shown to be an appropriate procedure that does not alter the quality of the food. This claim is supported by better preservation of vitamin C and physicochemical properties (colour and TSS) of the fruit with PL treatments over the storage period (Denoya et al. 2020). The FDA has approved PL for the treatment in the production, processing, and handling of foods from 1996 onward, with a total cumulative treatment of fewer than 12.0 J/cm^2 (Oms-Oliu et al. 2010).

Purepulse Technologies Inc. (San Diego, California), a Xenon Corp company, was the first to market the PureBright™ system for use in water purification systems and virus eradication systems for biopharmaceutical firms. According to accessible literature reports, Xenon Corporation of the United States, SteriBeam Systems of Germany, and Claranor of France are the three largest commercial businesses producing disinfection systems based on PL (Bhavya and Umesh Hebbar 2017). The fact that the energy is given in a short time is one of the significant advantages of PL versus static UV treatment. PL systems have low operating costs, produce less solid waste, reduce public health risks from foodborne diseases, extend product shelf life, and are more cost-effective. It has potential applications in food processing, such as fresh whole fruit and vegetable, meat slices or hard cheeses, where surface contamination occur by microorganism (Bhavya and Umesh Hebbar 2017). The photochemical and photothermal action of PL generating structural changes in the DNA of bacteria, viruses, and other pathogens, stopping the cell from replicating and cell death, has been proposed as the mechanism of microbial inactivation by PL (Go'mez-Lo'pez et al. 2007). Agglomeration of cytoplasmic material leads to cell rupturing also happen during the PL treatment against microorganism (Aguiló-Aguayo et al. 2013; Yousefi et al. 2021). Physical elements—such as pulse fluence or light intensity, number of flashes, pulsed energy level, fluence rate, the voltage applied, the quantity of UV content, and distance between lamp and sample, as well as microbiological strain, sample type, and packaging—are all variables to consider the effectiveness of PL (Aaliya et al. 2021).

PL has been employed in various research over the last decade to understand better its germicidal action on the surface of food and food products, such as strawberries, blueberries, carrots, lettuce, spinach, fish fillets, cornmeal, milk, honey, and alfalfa seeds (Aaliya et al. 2021). Bernal et al. (2019) investigated the effect of PL (fluence = 1.2 to 47.8 J/cm^2) on *Botrytis cinerea* suspended in peptone water and artificially contaminated strawberries throughout storage. Strawberries intentionally contaminated with *B. cinerea* showed a 2-day delay at the beginning of infection, and a reduced incidence in PL-treated strawberries (11.9 and 23.9 J/cm^2) compared to control (16–20%) approximately 10 days of cold storage. In peptone water and infected strawberries, the results showed that PL inhibits *B. cinerea* growth considerably. The work promotes the synergetic approach of different preservation techniques that would boost PL activity, hence better control of this fungus. Ferrario and Guerrero (2018) found that PL combined with ultrasound exposure reduced the *Alicyclobacillus acidoterrestris* from 5.9 log CFU/g to 3.0 log CFU/g in apple juice when compared to individual PL administration. Another research utilised kinetic modelling to explore the impact of PL processing variables on the quality of mixed fruit beverages. At the extreme condition of 17.42 W/cm^2 fluence-rate for 3 minutes (2,400 V/2.4 cm/3 minutes), both aerobic mesophiles, yeast and mould were reduced by more than 5-logs or completely inactivated. The secondary kinetic model study help to find out the best condition of PL treatment to create a microbial-safe fruit beverage with minimal loss of bioactive components (Dhar and Chakraborty 2020). Preetha et al. (2021) examined *E. coli* elimination in orange juice, pineapple juice, and tender coconut water, using PL fluence rates (0.18, 2 and 5.6 W/cm2) and exposure times (0 and 15 s). Treatment with PL dosages of 95.2 J/cm^2 showed log reductions of up to 4.0, 4.5, and 5.33 of *E. coli* in each sample, respectively. The researchers concluded that PL could be a viable non-thermal alternative for pasteurising liquid foods.

2.2.3 Visible Light

Visible radiation is drawn to clinical and food-related pathogen decontamination because of its safety, convenience of use, and efficacy. Visible spectrum light has become an emerging method for inactivating microorganisms (Akhila et al. 2021). Not all visible spectral (400–700 nm) colours are microbial destructive. Greenlight has a modest antimicrobial impact, while red light has none. Microorganisms can be rendered inactive by a violet-blue light area with a wavelength of around 405 nm (Akhila et al. 2021). According to Roh et al. (2016), 465 nm and 405 nm LEDs offer antibacterial properties against bacterial infections that affect shellfish and fish. After delivering 137–262 J/cm^2 antimicrobial blue light at 405 nm to bacterial suspensions, about 3.2–3.8 $\log_{10}$ CFU inactivation was achieved. Effectiveness is pronounced in wavelengths of 405 nm than 465 nm.

Antimicrobial blue light inactivation has been studied in a broad range of microbial taxa during the last five years, comprising bacteria, yeasts, dermatophytes, moulds, and mycobacteria (Wang et al. 2017). One of the hypotheses connected with the inactivation of microbes by VL is photodynamic inactivation (PDI). When a microorganism is exposed to violet-blue light, its endogenous photosensitising chromophores (iron-free porphyrins or/and flavins) become stimulated (Akhila et al. 2021). These activated molecules produce deadly reactive oxygen species, such as singlet oxygen and hydroxyl radicals, which induce oxidative damage to cell components (proteins, nucleic acids, and lipids) and ultimately death. Biener et al. (2017) hypothesised that blue light kills bacteria by inducing significant changes in the transmembrane potential, like changes in membrane integrity or the disabling of ion pumps, making the membrane more susceptible to ions. Kim et al. (2017) evaluated the antibacterial activity of a 405 ± 5 nm LED on the surface of fresh-cut mango against *L. monocytogenes*, *E. coli*, and *Salmonella* spp. Treatment of LED at a total dose of 1.7 kJ/cm^2 lowered 1.2 log of *Salmonella* and prevented the growth of *E. coli* at 20°C for 24 hours. These findings suggest that 405 ± 5 nm LEDs coupled with temperatures of 20°C can be used to store fresh-cut fruits in food outlets without compromising their physicochemical quality, lowering the risk of foodborne illness. Salmonella is a bacterium that is frequently linked to foodborne diseases.

Bumah et al. (2015) investigated the ability of 470 nm light to inhibit the growth of two *Salmonella* strains. Blue light exposure resulted in a significant dose-dependent drop in the number of colonies generated by the *Salmonella* strain when compared to controls ($P < 0.001$). Irradiating *S. typhimurium* with 55 and 110 J/cm^2 decreased bacterial growth by 31% and 93%, respectively, inhibiting *S. heidelberg* by 11% and 84%. The application of 165 or 220 J/cm^2 was shown to inactivate each Salmonella strain in the investigation. Blue light is effective in eliminating bacteria, making it a feasible intervention approach for decontaminating food and environments that contain pathogenic microorganisms (Bumah et al. 2015). Hui et al. (2017) evaluated the effectiveness of different wavelengths on Botrytis cinerea mycelium using LED. Blue light and PL strongly suppressed *B. cinerea* mycelium growth with inhibition rates of 15.16 and 22.3%, respectively, and purple light had a better inhibitory impact than blue light. Uekert et al. (2020) conducted a photo-reforming study using visible light, a simple, sunlight-driven approach for converting food waste into valuable chemicals and pure H_2 fuel.

UV-VIS spectroscopy is one of the most often used laboratory techniques for food analysis. To establish the quantitative measurement of inorganic and organic compounds and functional groups on molecules, this technology exploits the absorption of UV-VIS radiation. The availability of reliable instrumentation accuracy, simplicity of use, accuracy, and rapidity of the technique make UV-VIS spectroscopy appealing. Colour information is directly provided by visible wavelength spectroscopy, and these wavelengths are utilised to detect pigments in samples. Food adulteration and authenticity of various food products, such as grapes for wine, olive oil, meats, fish, seafood etc., can be quickly confirmed by this technology (Martelo-Vidal and Vázquez 2016). Aleixandre-Tudo et al. (2015) assessed two types of tannins in South African red wines using UV-VIS spectral data and partial least-squares models.

2.2.4 Infrared Radiation

Infrared (IR) is a source of radiant electromagnetic energy with wavelengths shorter than microwaves but longer than visible light (Gómez-López et al. 2021). Based on their application, IR radiation is divided into three zones by the International Commission on Illumination; viz, near-infrared (NIR), mid-infrared (MIR), and far-infrared (FIR) with wavelengths of 0.76–2 μm, 2–4 μm, and 4–1,000 μm, respectively (Gómez-López et al. 2021). Commercially, FIR radiation operates from 2.5–30 μm and is extensively utilised in IR radiation in food applications. IR is used in a variety of food processing operations, including drying, baking, frying, peeling, roasting, blanching, heating, pasteurisation, and inactivating microbes. Despite the fact that this technology has been in use for over a century (mainly for drying purposes), it has just gotten increasing attention in the last two decades for its potential use in food safety applications. Typically, the food absorbs MIR and FIR more effectively via stretching vibration, resulting in heat generation. After absorbing IR and based on the emissivity, food material releases thermal radiation, which influences the heating process of various food processing applications (Gómez-López et al. 2021).

IR drying is seen as a viable drying technology in agriculture produce. Several food crops, including carrots, potatoes, mushrooms, eggplant, and others, have been effectively dried using infrared technology (Huang et al. 2021). When IR is used to dry high-moisture food commodities, the energy is absorbed to a small depth in the materials before being transformed into heat. IR drying technique has advantages over traditional drying technology—i.e., high energy efficiency, uniform heating of materials, superior end-product quality, easy control of material temperature, short drying time, environment-friendly, and cheap energy costs. Equipment simplicity, modifiability, adaptability, easy combination with other heating technologies, such as MW heating, vacuum, and convection, and inexpensive, are a few of the characteristics of IR heating (Huang et al. 2021). The heat flux produced by an IR drier is 6–10 times that of a typical convection hot air dryer, enabling substantial drying time efficiency (Gómez-López et al. 2021). Wu et al. (2019) investigated the freeze-drying and infrared-freeze drying of *Cordyceps militaries* in an experimental study. When compared to freeze-drying, the analysis indicates that infrared-freeze drying may save up to 18.37% of energy usage and 17.78% drying time at the same drying temperature. The final product quality was comparable to that of freeze drying when using infrared-freeze drying. However, IR drying alone is not ideal for heat-sensitive food products, so hybrid drying is commonly used to improve the efficiency and quality of the drying process.

IR heating can inactivate microorganisms through mechanisms comparable to MW heating (induction heating) and ultraviolet light, in addition to thermal effects (DNA damage). All cell components of microorganisms, including genetic material and protein, can be destroyed by heat inactivation. IR heating has been used to improve food safety by inactivating pathogens, decontaminating fruit surfaces, disinfesting grains, sterilising milk, and pasteurising nuts (Akhila et al. 2021). Bowie et al. (2019) have applied IR wavelengths with varying IR intensities and heating times, accompanied by a tempering phase, to inactivate microorganisms (mould and bacteria) on the grain. Mould and bacterial load were reduced the most in samples treated at wavelength 3.2 μm for the 30s; about 3.11 and 1.09 log reductions in the colony-forming unit of mould and bacteria, respectively. Moulds were shown to be more sensitive to IR decontamination than bacteria. After one-pass IR heating and subsequent tempering at 50°C for 4 hours, Wilson et al. (2017) reduced the fungal load in high moisture content (MC) corn [24% wet basis (w.b.)] by 4.03 log (CFU/g).

IR spectrometry's adaptability has led to its application in various fields, including medicine, biology, soil chemistry, agro-food, and cereals among others. IR spectroscopy is a valuable instrument for assessing raw materials, monitoring manufacturing processes, and verifying finished goods because it investigates the interactions between IR radiation and matter (Levasseur-Garcia 2018). In terms of instrumentation, accessories, application, and available chemometric software, NIR spectroscopy was deemed the most sophisticated technology among non-destructive quality assessment methods (Porep et al. 2015). Furthermore, the NIRS technique opens up a plethora of possibilities for other food research and identification and constituent analysis of oil, fruit, meat, wine, and corn (Cen and He 2007). NIR spectroscopy uses a spectral range of 780 to 2,500 nm to provide far more sophisticated structural information about the combination of bond and their vibration behaviour (Cen and He 2007). By utilising the vibrational energy change caused by NIR radiation, the molecular bond between OH, CH, CO, and NH may be easily recorded (Alizadeh et al. 2021). The NIRS technique is employed in common index food quality testing and the monitoring of constituent fluctuation in food processing and the study of dynamical behaviours. NIRS is utilised in the cheese industry to optimise sampling time and investigate the chemical and physical dynamics of cheese from various sources. To ensure food safety, NIRS is used to analyse the freshness and maturity of food based on colour change. Sivertsen et al. (2011) examined VIS/NIR spectroscopy for distinguishing between fresh and frozen-thawed cod fillets, as well as measuring freshness as days on the ice. The implementation of the NIRS technology was successful in identifying and, to some extent, measuring harmful chemical residues, such as pesticides, other agrochemicals, and mycotoxins in foods, according to the results of extensive research conducted earlier. However, NIR spectroscopy was shown to have low sensitivity to quantifying chemical residues in food materials (Teye et al. 2013). Sánchez et al. (2010) employed NIR reflectance spectroscopy to measure pesticide levels in peppers, and his findings proved that the NIRS method could be used for quick, non-destructive initial evaluation for pesticide residues in suspected food samples.

2.2.5 Dielectric Heating

Advanced thermal-processing methods in dielectric heating include microwave (MW) (300–300,000 MHz) and radio frequency (RF) (0.003–300 MHz) (Gómez-López et al. 2021). Dielectric heating has been used to successfully dry, bake, and pre-cook bacon, meat patties, or thaw frozen products (Awuah et al. 2014). This technology has the potential for rapid heating, volumetric heating, faster heat penetration, and reduced processing time. The dielectric and ionic mechanisms are used in MW and RF heating. Water in food is typically the primary source of dielectric heating. The dipolar nature of water molecules oscillates in the presence of EMR and generates heat. The second mechanism involves through the oscillatory movement of ions in the food that produces heat under the influence of the oscillating electric filed (Datta and Davidson 2000). It also has a fundamental flaw, namely, non-uniform heating, which leads to the establishment of a cold spot where microorganisms can begin to proliferate under favourable conditions.

The use of MW technology in the food industry has gained acceptability for a variety of food processing applications, including drying, pasteurisation, sterilisation, thawing, tempering, and baking of food products (Salazar-González et al. 2012; Chandrasekaran et al. 2013). MW ovens are now common household appliances (Gómez-López et al. 2021). Pre-cooking meat patties or bacon and tempering fish blocks or meat are popular industrial implementations of MW heating in food processing, whereas RF heating is often utilised in completing the drying of freshly baked items (Gómez-López et al. 2021). MW cooking has grown in popularity in the food processing industry because of its capacity to reach high heating rates, which significantly reduce cooking time, provide more uniform heating, and remain safe to handle, operate, and maintain. Furthermore, when compared to conventional heating during the cooking or reheating process, MW heating potentially alters the flavour and nutritional characteristics of food to a lower amount (Chandrasekaran et al. 2013). MW radiation has been restricted to a few specific bandwidths in scientific, industrial, and medical operations to avoid interfering with radar and telecommunications radio frequencies. MWs with a wavelength of 2,450 MHz are utilised in domestic and industrial MW ovens, while MW with a wavelength of 915 MHz is only employed in large-scale industrial food processing equipment (Tang 2015). The penetration depths ranging from 3 to 8 cm at 2,450 MHz and 8 to 22 cm at 915 MHz vary with food composition, state (fresh vs. frozen), resulting dielectric properties. Other specific frequencies, including 433.92, 896, and 2375 MHz, are also approved for MW heating outside of the United States (Gómez-López et al. 2021).

Commercial MW pasteurisation operations, on the other hand, have been established in the United States, Europe, and Japan (Tang 2015). Pasteurised foods, such as pasta, bread, refrigerated ready-to-serve meals, and cakes run at a frequency of 2,450 MHz with a total power of 10 to 40 kW. Micvac (Sweden) unveiled a novel approach in commercial MW pasteurised food preparation. A re-closable valve is introduced into each package, which helps to MW heating of sealed food by transferring thermal energy by internally produced steam in the headspace, thereby creating better uniform heating. The valve maintains a vacuum in the container, reducing the thermal impact of edge heating while also improving food quality (Tang 2015). The reduction of microbial populations in foods is a new approach to MW heating. Heating occurs when EMR at specific frequencies interacts with dielectric materials; this is also known as volumetric heating because, unlike conduction or convection mechanisms, MW radiation directly penetrates the material, causing volumetric heat generation, which leads to high energy efficiency and shorter heating times (Salazar-González et al. 2012). The oscillatory movement of ions in the food produces energy at a higher frequency, and an oscillating electric field could also cause MW heating. The dielectric characteristics and penetration depth are the two most critical parameters that influence MW heating and heat dispersion (Chandrasekaran et al. 2013).

MW treatment reduces anti-nutritional compounds in food, which increases protein digestibility, quality, and safety in food grains (Guo et al. 2017). In agricultural produce, including horse chestnut, velvet bean, black soybean, buckwheat, Kabuli chickpea, and others, MW treatment was already reported to diminish anti-nutritional components, such as trypsin inhibitors, saponins, tannins, oxalate, and phytic acid (Suhag et al. 2021). The dissociation of peptide bonds, deamidation (covalent bond splitting), and destruction or interchange of disulfide bonds are all involved in the heat deterioration of ANFs (Suhag et al. 2021). The impact of MW and traditional pasteurisation on the enzymatic activity and the principal bioactive components of grape juice was studied by Igual et al. (2010). The findings revealed that conventional treatment reduced citric acid (1,538 to 1,478 mg/100 g) and ascorbic acid (36 to 34.3 mg/100 g), but MW pasteurisation retained these components. Frozen non-treated and conventional pasteurised samples kept about 20% and 75% of antioxidant properties and total phenolic content. Still, frozen MW pasteurised juices preserved 33% and 82% of antioxidant capacity and total phenols. Based on these findings, MW energy could be recommended as a substitute for traditional heat pasteurisation for preserving grapefruit juice's inherent organoleptic qualities and key thermolabile components.

3. Sustainable Technology

There is an increasing demand for alternative and sustainable technology that can meet the food safety and security of consumers. This emerging interest in different processing and preservation technologies allied with the green approach paves the way to applying various EMRs to improve the usage of sustainable sources assuring food security. Among various EMRs, gamma radiation, MW, PL, and UV is gaining increased attention in the food industry because of their capacity to protect food quality, lowest nutrient loss during processing, inactivation of enzyme and microorganisms, and efficient decontamination of toxins and pesticides (Arshad et al. 2021). The application of EMR toward sustainability includes food safety, environmental sustainability, economic sustainability, and food security.

3.1 Food Safety

The concept of nutritious and safe food involves a wide range of factors. The term "safe food" refers to food devoid of harmful toxins, pesticides, and chemical, physical, and microbiological hazards, including bacteria, fungi, and viruses (Bonciu 2018). Contamination of microorganisms can occur at any stage of food processing from the initial production, processing, and marketing to the final consumption stage through the soil, irrigation water, unhygienic practices of workers, cross-contamination, and improper packaging storage and transportation (Gautam and Tripathi

2016). Bacteria, fungi, and their toxins cause major foodborne outbreaks throughout the world, which affect the human health and socioeconomics of the country (Akhila et al. 2021). Food safety must be assured at both the retail and consumer levels (Gautam and Tripathi 2016). Food irradiation has a proven safety record when used in conjunction with acceptable manufacturing methods. This provides a solid scientific foundation for using food radiation processing as an effective way to improve food safety (Gautam and Tripathi 2016). Pesticide residue in agricultural produce is a significant concern to the consumer because of its high toxicity and widespread use. Experiments with several EMR treatments revealed considerable pesticide decontamination. Basfar et al. (2012) examined the effect of gamma irradiation on pesticide elimination in aqueous solutions and plants and fruits. Ionising radiation reduced pirimiphos-methyl residues to below maximum residue limits.

3.2 Environmental Sustainability

Environmental sustainability defines a line that must be crossed in order to meet present needs without jeopardising the environment's ability to support the needs of future generations. EMR is a non-thermal, energy-efficient, green technology with renewable resources, which showed a very low impact on the environment by releasing greenhouse gases. Some radiation processes have a lot of potential in terms of environmental protection. The lack of progress in translating R&D results into commercialisation is owing to the industry terms of minimal involvement in the process (Chmielewski and Haji-Saeid 2004). The usage of EMR with the allowed limit is an environment-friendly and safe method. In wastewater treatment, ionising radiation, such as gamma and Eb, disinfects the water and lowers the bio-refractory nature of various persistent organic pollutants (Abdel Rahman and Hung 2020). Chemical pesticides, herbicides, and fungicides are used in accordance with national agricultural recommendations to increase the efficiency of agricultural production. Agricultural wastewater contains residues of these persistent pollutants and their harmful metabolites, which could spread to surface and ground waterways. Using a tertiary treatment method, ionising radiation therapy technology confirmed its efficacy in degrading a spectrum of these contaminants (Abdel Rahman and Hung 2020). The most effective treatment of wastewater in eliminating pathogenic microorganisms and organic decontaminants has been discovered to be EBs (S. Jan et al. 2015). A 1.5 kGy electron beam dose was shown to be an effective dose for the destruction of all coliforms (Maruthi et al. 2011). Depending on the source of energy, dose rate, and absorbed radiation dose, an electron beam causes the simultaneous destruction of several chemical molecules and the inactivation of microorganisms (S. Jan et al. 2015). The presence of an unpaired electron makes them a potent oxidising agent that degrades heavy metals and numerous organic and inorganic contaminants and disinfests pathogenic bacteria at the same time.

Despite the fact that there is no perfect way for extracting bioactive chemicals, recent studies have revealed that ionising radiation approaches can improve extraction efficiency and bioactive characteristics of agro-industrial residues. MW irradiation has acquired substantial recognition in recent years as an excellent technology for sludge treatment, owing to quick and selective heating, efficiency, capacity to enhance product quality and yield, and reduced toxic product generation and emissions (thus, rendering the technique environmentally friendly) (Tyagi and Lo 2013). Through the use of MW, Ahn et al. (2009) noticed a significant rise in soluble protein, carbohydrate, and fat concentrations. They used MW irradiation for 15 minutes (700 W; boiling point) to disintegrate the WAS [26 g/L total solids(TS)] and found that the concentrations of lipids, proteins, and carbohydrates increased significantly from 0.07 to 0.85 g/L, 0.15 to 0.9 g/L, and 0.07 to 0.9 g/L, respectively.

3.3 Economic Sustainability and Food Security

Through highly competitive food processing, economic sustainability promotes the effective use of energy, increased throughput, and high profitability (Arshad et al. 2021). The economic scale of radiation technology use is enormous when considering the overall value of products in which manufacturing or modification radiation is utilised (Chmielewski and Haji-Saeid 2004). Radiation technologies provide a wide range of tools that aid in the advancement of sustainable development. Ionising radiation produces changes in the matter without chemicals or heat, hence it has a lower carbon footprint than other methods. It is thus more cost-effective and environmentally friendly than traditional methods as it consumes less energy and produces less trash (Madureira et al. 2020). The United Nations recognises global food security as a part of the Sustainable Development Goals (SDGs; SDG2 in particular) through increasing sustainable agriculture production, reducing food losses and waste, improving nutrition, and ensuring zero hunger (Karthikeyan et al. 2020). Food irradiation has a vital role in maintaining food safety and lowering food losses as the wholesomeness of irradiated food has been proved by various scientific investigations.

4. Conclusion

The adoption of various EMRs continues to progress in the food industry. EMR processing has been widely studied and is currently used for a variety of food processing and preservation and quantitative and qualitative analysis of food components throughout the world. It's a green technology that follows the principles of sustainable development. Irradiation has been used to reduce harmful microorganisms, parasites, post-harvest sprouting, and the extension of shelf life of perishable foods. The United States has made the most progress in the implementation of ionising radiation, but steps have been taken in other regions of the world as well. Food safety and disinfestation in various disciplines are assured by ionising radiation. In comparison, non-ionising radiation is already exploited in conventional unit processing and continues research in different food processing sectors. MW ovens in the home have a high level of acceptance with a market penetration of 70%. Among various EMRs, gamma radiation, MW, PL, and UV is gaining increased attention in the food industry because of their capacity to protect food quality, lowest nutrient loss during processing, inactivation of enzyme and microorganism, and efficient decontamination of toxins and pesticides. It would be ecologically friendly and cost-effective than traditional processes as it consumes less energy and generates less waste.

Acknowledgements

The authors are thankful to Dr Shijin A, veterinary surgeon, Government of Kerala.

Reference

Aaliya, B., Sunooj, K.V., Navaf, M., Akhila, P.P., Sudheesh, C., Mir, S.A. et al. 2021. Recent Trends in Bacterial Decontamination of Food Products by Hurdle Technology: A Synergistic Approach Using Thermal and Non-Thermal Processing Techniques. Food Res. Int. 147. https://doi.org/10.1016/j.foodres.2021.110514

Abdel Rahman, R.O. and Hung, Y.T. 2020. Application of Ionizing Radiation in Wastewater Treatment: An Overview. Water 12.

Aguiló-Aguayo, I., Charles, F., Renard, C.M.G.C., Page, D. and Carlin, F. 2013. Pulsed Light Effects on Surface Decontamination, Physical Qualities and Nutritional Composition of Tomato Fruit. Postharvest Biol. Technol. 86: 29–36. http://dx.doi.org/10.1016/j.postharvbio.2013.06.011.

Ahn, J.H., Shin, S.G. and Hwang, S. 2009. Effect of Microwave Irradiation on the Disintegration and Acidogenesis of Municipal Secondary Sludge. Chem. Eng. J. 153: 145–150.

Akhila, P.P., Sunooj, K.V., Aaliya, B., Navaf, M., Sudheesh, C., Sabu, S. et al. 2021. Application of Electromagnetic Radiations for Decontamination of Fungi and Mycotoxins in Food Products: A Comprehensive Review. Trends Food Sci. Technol. 114: 399–409. https://doi.org/10.1016/j.tifs.2021.06.013.

Aleixandre-Tudo, J.L., Nieuwoudt, H., Aleixandre, J.L. and Du Toit, W.J. 2015. Robust Ultraviolet-Visible (UV-Vis) Partial Least-Squares (PLS) Models for Tannin Quantification in Red Wine. J. Agric. Food Chem. 63: 1088–1098.

Alizadeh, A.M., Hashempour-Baltork, F., Khaneghah, A.M. and Hosseini, H. 2021. New Perspective Approaches in Controlling Fungi and Mycotoxins in Food Using Emerging and Green Technologies. Curr. Opin. Food Sci. 39: 7–15. https://doi.org/10.1016/j.cofs.2020.12.006.

Arshad, R.N., Abdul-Malek, Z., Roobab, U., Munir, M.A., Naderipour, A., Qureshi, M.I. et al. 2021. Pulsed Electric Field: A Potential Alternative towards a Sustainable Food Processing. Trends Food Sci. Technol. 111: 43–54. https://doi.org/10.1016/j.tifs.2021.02.041

Arvanitoyannis, I.S. and Stratakos, A.C. 2010. Potential Uses of Irradiation. In Irradiation of Food Commodities, 635–669. First Edit. Academic Press. http://dx.doi.org/10.1016/B978-0-12-374718-1.10016-1.

Awuah, G.B., Ramaswamy, H.S. and Tang, J. 2014. Radio-Frequency Heating in Food Processing:Principles and Applications. CRC Press. CRC Press.

Barba, F.J., Ahrné, L., Xanthakis, E., Landerslev, M.G. and Orlien, V. 2018. Innovative Technologies for Food Preservation. In Innov. Technol. Food Preserv. 25–51. Academic Press.

Basfar, A.A., Mohamed, K.A. and Al-Saqer, O.A. 2012. De-Contamination of Pesticide Residues in Food by Ionizing Radiation. Radiat. Phys. Chem. 81: 473–478. http://dx.doi.org/10.1016/j.radphyschem.2011.12.040.

Begum, T., Follett, P.A., Hossain, F., Christopher, L., Salmieri, S. and Lacroix, M. 2020. Microbicidal Effectiveness of Irradiation from Gamma and X-Ray Sources at Different Dose Rates against the Foodborne Illness Pathogens Escherichia Coli, Salmonella Typhimurium and Listeria Monocytogenes in Rice. LWT - Food Sci. Technol. 132: 109841. https://doi.org/10.1016/j.lwt.2020.109841.

Bernal, A.R.R., Contigiani, E.V., González, H.H.L., Alzamora, S.M., Gómez, P.L. and Raffellini, S. 2019. Botrytis Cinerea Response to Pulsed Light: Cultivability, Physiological State, Ultrastructure and Growth Ability on Strawberry Fruit. Int. J. Food Microbiol. 309.

Bhatia, S.S., Wall, K.R., Kerth, C.R. and Pillai, S.D. 2018. Benchmarking the Minimum Electron Beam (EBeam) Dose Required for the Sterilization of Space Foods. Radiat. Phys. Chem. 143: 72–78. http://dx.doi.org/10.1016/j.radphyschem.2017.08.007.

Bhavya, M.L. and Umesh Hebbar, H. 2017. Pulsed Light Processing of Foods for Microbial Safety. Food Qual. Saf. 1: 187–201.

Biener, G., Masson-Meyers, D.S., Bumah, V.V., Hussey, G., Stoneman, M.R., Enwemeka, C.S. et al. 2017. Blue/Violet Laser Inactivates Methicillin-Resistant Staphylococcus Aureus by Altering Its Transmembrane Potential. J. Photochem. Photobiol. B Biol. 170: 118–124. http://dx.doi.org/10.1016/j.jphotobiol.2017.04.002.

Bonciu, E. 2018. Food Processing, a Necessity for the Modern World in the Context of Food Safety: A Review. Ann. Univ. Craiova-Agriculture. 391–398. http://anale.agro-craiova.ro/index.php/aamc/article/view/610.

Borsa, J. and Ph, D. 2012. Introduction: Food Irradiation Moving On. In Food Irradiation: Research and Technology, 1–8.

Bowie, R., Oduola, A.A., Shad, Z.M., Wilson, S.A. and Atungulu, G.G. 2019. Impact of Selected Infrared Wavelengths on Inactivation of Microbes on Rough Rice. Discov. Student J. Dale Bump. Coll. Agric. Food Life Sci. 20: 13–20.

Bumah, V.V., Masson-Meyers, D.S. and Enwemeka, C.S. 2015. Blue 470 Nm Light Suppresses the Growth of Salmonella Enterica and Methicillin-Resistant Staphylococcus Aureus (MRSA) *in Vitro*. Lasers Surg. Med. 47: 595–601.

Calado, T., Fernández-Cruz, M.L., Cabo Verde, S., Venâncio, A. and Abrunhosa, L. 2018. Gamma Irradiation Effects on Ochratoxin A: Degradation, Cytotoxicity and Application in Food. Food Chem. 240: 463–471.

Calado, T., Veñancio, A. and Abrunhosa, L. 2014. Irradiation for Mold and Mycotoxin Control : A Review. Compr. Rev. Food Sci. Food Saf. 13(5): 1049–1061.

Cen, H. and He, Y. 2007. Theory and Application of near Infrared Reflectance Spectroscopy in Determination of Food Quality. Trends Food Sci. Technol. 18: 72–83.

Chandrasekaran, S., Ramanathan, S. and Basak, T. 2013. Microwave Food Processing-A Review. Food Res. Int. 52: 243–261. http://dx.doi.org/10.1016/j.foodres.2013.02.033.

Chawla, A., Lobacz, A., Tarapata, J. and Zulewska, J. 2021. Uv Light Application as a Mean for Disinfection Applied in the Dairy Industry. Appl. Sci. 11: 7285.

Chmielewski, A.G. and Haji-Saeid, M. 2004. Radiation Technologies: Past, Present and Future. Radiat. Phys. Chem. 71: 17–21.

Darfour, B., Ofori, H. and Asare, I.K. 2021. Gamma Irradiation and Drying Method: The Effects on Kola Nut Powder. Radiat. Phys. Chem. 185: 109489. https://doi.org/10.1016/j.radphyschem.2021.109489.

Datta, A.K. and Davidson, P.M. 2000. Microwave and Radio Frequency Processing. J. Food Sci. 32–41.

Deghiedy, N.M., Elkenawy, N.M. and Abd El-Rehim, H.A. 2021. Gamma Radiation-Assisted Fabrication of Bioactive-Coated Thyme Nanoemulsion: A Novel Approach to Improve Stability, Antimicrobial and Antibiofilm Efficacy. J. Food Eng. 304: 110600. https://doi.org/10.1016/j.jfoodeng.2021.110600.

Delorme, M.M., Guimarães, J.T., Coutinho, N.M., Balthazar, C.F., Rocha, R.S., Silva, R. et al. 2020. Ultraviolet Radiation: An Interesting Technology to Preserve Quality and Safety of Milk and Dairy Foods. Trends Food Sci. Technol. 102: 146–154.

Deng, L.Z., Tao, Y., Mujumdar, A.S., Pan, Z., Chen, C., Yang, X.H. et al. 2020. Recent Advances in Non-Thermal Decontamination Technologies for Microorganisms and Mycotoxins in Low-Moisture Foods. Trends Food Sci. Technol. 106: 104–112.

Denoya, G.I., Pataro, G. and Ferrari, G. 2020. Effects of Postharvest Pulsed Light Treatments on the Quality and Antioxidant Properties of Persimmons during Storage. Postharvest Biol. Technol. 160: 111055. https://doi.org/10.1016/j.postharvbio.2019.111055.

Dhar, R. and Chakraborty, S. 2020. Influence of Voltage and Distance on Quality Attributes of Mixed Fruit Beverage during Pulsed Light Treatment and Kinetic Modeling. J. Food Process Eng. 43: 1–14.

EFSA. 2016. Safety of UV-treated Milk as a Novel Food Pursuant to Regulation (EC) No 258/97. EFSA Journal 14: 4370.

Falguera, V., Pagán, J., Garza, S., Garvín, A. and Ibarz, A. 2011. Ultraviolet processing of liquid food: A review. Part 2: Effects on microorganisms and on food components and properties. Food Research International 44(6): 1580–1588. https://doi.org/10.1016/j.foodres.2011.03.025.

Ferrario, M.I. and Guerrero, S.N. 2018. Inactivation of Alicyclobacillus Acidoterrestris ATCC 49025 Spores in Apple Juice by Pulsed Light. Influence of Initial Contamination and Required Reduction Levels. Rev. Argent. Microbiol. 50: 3–11. https://doi.org/10.1016/j.ram.2017.04.002.

Freita-Silva, O., de Oliveira, P.S. and Freire Júnior, M. 2015. Potential of Electron Beams to Control Mycotoxigenic Fungi in Food. Food Eng. Rev. 7: 160–170.

Gautam, S. and Tripathi, J. 2016. Food Processing by Irradiation—An Effective Technology for Food Safety and Security. Indian J. Exp. Biol. 54: 700–707.

Go´mez-Lo´pez, V.M., Ragaert, P., Debevere, J. and Devlieghere, F. 2007. Pulsed Light for Food Decontamination: A Review. Trends Food Sci. Technol. 18: 464–473.

Gómez-López, V.M., G. Pataro, B. Tiwari, M. Gozzi, M.Á.A. Meireles, S. Wang et al. 2021. Guidelines on Reporting Treatment Conditions for Emerging Technologies in Food Processing. Crit. Rev. Food Sci. Nutr. 1–25. https://doi.org/10.1080/10408398.2021.1895058.

Gryczka, U., Kameya, H., Kimura, K., Todoriki, S., Migdał, W. and Bułka, S. 2020. Efficacy of Low Energy Electron. Beam. on Microbial Decontamination of Spices. Radiat. Phys. Chem. 170.

Guo, Q., Sun, D.W., Cheng, J.H. and Han, Z. 2017. Microwave Processing Techniques and Their Recent Applications in the Food Industry. Trends Food Sci. Technol. 67: 236–247. http://dx.doi.org/10.1016/j.tifs.2017.07.007.

Ha, J.W., Back, K.H., Kim, Y.H. and Kang, D.H. 2016. Efficacy of UV-C Irradiation for Inactivation of Food-Borne Pathogens on Sliced Cheese Packaged with Different Types and Thicknesses of Plastic Films. Food Microbiol. 57: 172–177. http://dx.doi.org/10.1016/j.fm.2016.02.007.

Hertwig, C., Meneses, N. and Mathys, A. 2018. Cold Atmospheric Pressure Plasma and Low Energy Electron Beam as Alternative Nonthermal Decontamination Technologies for Dry Food Surfaces: A Review. Trends Food Sci. Technol. 77: 131–142. https://doi.org/10.1016/j.tifs.2018.05.011.

Huang, D., Yang, P., Tang, X., Luo, L. and Sunden, B. 2021. Application of Infrared Radiation in the Drying of Food Products. Trends Food Sci. Technol. 110: 765–777. https://doi.org/10.1016/j.tifs.2021.02.039.

Hui, X., Y. nan FU, T. lai L.I. and Wang, R. 2017. Effects of Different LED Light Wavelengths on the Resistance of Tomato against Botrytis Cinerea and the Corresponding Physiological Mechanisms. J. Integr. Agric.16: 106–114. http://dx.doi.org/10.1016/S2095-3119(16)61435-1.

Igual, M., García-Martínez, E., Camacho, M.M. and Martínez-Navarrete, N. 2010. Effect of Thermal Treatment and Storage on the Stability of Organic Acids and the Functional Value of Grapefruit Juice. Food Chem. 118: 291–299. http://dx.doi.org/10.1016/j.foodchem.2009.04.118.

Jan, A., Sood, M., Younis, K. and Islam, R.U. 2020. Brown Rice Based Weaning Food Treated with Gamma Irradiation Evaluated during Storage. Radiat. Phys. Chem. 177, no. April: 109158. https://doi.org/10.1016/j.radphyschem.2020.109158.

Jan, S., Kamili, A.N., Parween, T., Hamid, R., Parray, J.A., Siddiqi, T.O. et al. 2015. Feasibility of Radiation Technology for Wastewater Treatment. Desalin. Water Treat. 55: 2053–2068.

Karthikeyan, L., Chawla, I. and Mishra, A.K. 2020. A Review of Remote Sensing Applications in Agriculture for Food Security: Crop Growth and Yield, Irrigation, and Crop Losses. J. Hydrol. 586: 124905. https://doi.org/10.1016/j.jhydrol.2020.124905.

Khaneghah, A.M., Hashemi Moosavi, M., Oliveira, C.A.F., Vanin, F. and Sant'Ana, A.S. 2020. Electron Beam Irradiation to Reduce the Mycotoxin and Microbial Contaminations of Cereal-Based Products: An Overview. Food Chem. Toxicol. 143: 111557.

Kim, M.J., Tang, C.H., Bang, W.S. and Yuk, H.G. 2017. Antibacterial Effect of 405 ± 5 Nm Light Emitting Diode Illumination against Escherichia Coli O157:H7, Listeria Monocytogenes, and Salmonella on the Surface of Fresh-Cut Mango and Its Influence on Fruit Quality. Int. J. Food Microbiol. 244: 82–89. http://dx.doi.org/10.1016/j.ijfoodmicro.2016.12.023.

Koutchma, T. and Popović, V. 2019. UV Light-Emitting Diodes (LEDs) and Food Safety. In Ultraviolet LED Technology for Food Applications 91–117.

Krishnamurthy, K., Irudayaraj, J., Demirci, A. and Yan, W. 2008. 11 UV Pasteurization of Food Materials. In Food Processing Operations Modeling Design and Analysis. pp. 281–302. *In*: Jun, S. and Irudayaraj, J.M. (eds.). CRC Press.

Levasseur-Garcia, C. 2018. Updated Overview of Infrared Spectroscopy Methods for Detecting Mycotoxins on Cereals (Corn, Wheat, and Barley). Toxins 10: 38.

Liao, Y. Te, Syamaladevi, R.M., Zhang, H., Killinger, K. and Sablani, S. 2017. Inactivation of Listeria Monocytogenes on Frozen Red Raspberries by Using Uv-c Light. J. Food Prot. 80: 545–550.

Lim, W. and Harrison, M.A. 2016. Effectiveness of UV Light as a Means to Reduce Salmonella Contamination on Tomatoes and Food Contact Surfaces. Food Control. 66: 166–173. http://dx.doi.org/10.1016/j.foodcont.2016.01.043.

Lima, D.C., Miano, A.C., Augusto, P.E.D. and Arthur, V. 2019. Gamma Irradiation of Common Beans: Effect on Nutritional and Technological Properties. LWT - Food Sci. Technol. 116: 0–4.

Lima, F., Vieira, K., Santos, M. and Santos, M. 2018. Effects of Radiation Technologies Technologies on on Food Food Nutritional Nutritional Quality Quality. *In*: Descriptive Food Science 1–17.

Loaharanu, P. 2003. Irradiated Food [5]. New York: American Council of Science and Health.

Lozada-Castro, J.J., Gil-Díaz, M., Santos-Delgado, M.J., Rubio-Barroso, S. and Polo-Díez, L.M. 2011. Effect of Electron-Beam Irradiation on Cholesterol Oxide Formation in Different Ready-to-Eat Foods. Innov. Food Sci. Emerg. Technol. 12(4): 519–525. http://dx.doi.org/10.1016/j.ifset.2011.07.005.

Maatouk, I., Mehrez, A., Ben Amara, A., Chayma, R., Abid, S., Jerbi, T. et al. 2019. Effects of Gamma Irradiation on Ochratoxin a Stability and Cytotoxicity in Methanolic Solutions and Potential Application in Tunisian Millet Samples. J. Food Prot. 82: 1433–1439.

Madureira, J., Barros, L., Verde, S.C., Margaça, F.M.A., Santos-Buelga, C. and Ferreira, I.C.F.R. 2020. Ionizing Radiation Technologies to Increase the Extraction of Bioactive Compounds from Agro-Industrial Residues: A Review. J. Agric. Food Chem. 68: 11054–11067.

Mahapatra, A.K., Muthukumarappan, K. and Julson, J.L. 2005. Applications of Ozone, Bacteriocins and Irradiation in Food Processing: A Review. Crit. Rev. Food Sci. Nutr. 45: 447–461.

Mahendran, R., Ramanan, K.R., Barba, F.J., Lorenzo, J.M., López-Fernández, O., Munekata, P.E.S. et al. 2019. Recent Advances in the Application of Pulsed Light Processing for Improving Food Safety and Increasing Shelf Life. Trends Food Sci. Technol. 88: 67–79. https://doi.org/10.1016/j.tifs.2019.03.010.

Mahmoud, B.S.M. 2009. Effect of X-Ray Treatments on Inoculated Escherichia Coli O157: H7, Salmonella Enterica, Shigella Flexneri and Vibrio Parahaemolyticus in Ready-to-Eat Shrimp. Food Microbiol. 26: 860–864. http://dx.doi.org/10.1016/j.fm.2009.05.013.

Markov, K., Mihaljević, B., Domijan, A.M., Pleadin, J., Delaš, F. and Frece, J. 2015. Inactivation of Aflatoxigenic Fungi and the Reduction of Aflatoxin B1invitro and *in Situ* Using Gamma Irradiation. Food Control. 54: 79–85. http://dx.doi.org/10.1016/j.foodcont.2015.01.036.

Martelo-Vidal, M.J. and Vázquez, M. 2016. Advances in Ultraviolet and Visible Light Spectroscopy for Food Authenticity Testing. In Advances in Food Authenticity Testing 35–70. Woodhead Publishing.

Maruthi, Y.A., Das, N.L., Hossain, K., Sarma, K.S.S., Rawat, K.P. and Sabarwa, S. 2011. Advance Oxidation of Sewage Water, Reclamation and Hygienization by Radiation Technology: A Novel Approach. Hydrol Current Res 2.

Miller, R.B. 2005a. Introduction to Food Irradiation. *In*: Electronic Irradiation of Foods 1–15.

Miller, R.B. 2005b. Food Irradiation Using X-Rays. *In*: Electronic Irradiation of Foods 75–121.

Mohammad-Razdari, A., Ebrahimi, R., Zarein, M., Mahfeli, M. and Yoosefian, S.H. 2021. Investigation of 60Co Gamma Irradiation on Dehydration Kinetics: A Case Study of Apple Slices. J. Food Process Eng. 44: 1–6.

Monteiro, M.L.G., Rosário, D.K.A., de Carvalho, A.P.A. and Conte-Junior, C.A. 2021. Application of UV-C Light to Improve Safety and Overall Quality of Fish: A Systematic Review and Meta-Analysis. Trends Food Sci. Technol. 116: 279–289.

Moosekian, S.R., Jeong, S., Marks, B.P. and Ryser, E.T. 2012. X-Ray Irradiation as a Microbial Intervention Strategy for Food. Annu. Rev. Food Sci. Technol. 3: 493–510.

Moriarty, M.J., Semmens, K., Bissonnette, G.K. and Jaczynski, J. 2018. Inactivation with UV-Radiation and Internalization Assessment of Coliforms and Escherichia Coli in Aquaponically Grown Lettuce. LWT - Food Sci. Technol. 89: 624–630. https://doi.org/10.1016/j.lwt.2017.11.038.

Mudgett, R. 1988. Electromagnetic Energy and Food Processing. J. Microw. Power Electromagn. Energy 23: 225–230.

Mulmule, M.D., Shimmy, S.M., Bambole, V., Jamdar, S.N., Rawat, K.P. and Sarma, K.S.S. 2017. Combination of Electron Beam Irradiation and Thermal Treatment to Enhance the Shelf-Life of Traditional Indian Fermented Food (Idli). Radiat. Phys. Chem. 131: 95–99. http://dx.doi.org/10.1016/j.radphyschem.2016.10.014.

Nair, P. and Sharma, A. 2016. Food Irradiation. Innov. Food Process. Technol. Elsevier. http://dx.doi.org/10.1016/B978-0-08-100596-5.02950-4.

Nielsen, M.S., Lauridsen, T., Christensen, L.B. and Feidenhans'l, R. 2013. X-Ray Dark-Field Imaging for Detection of Foreign Bodies in Food. Food Control. 30: 531–535. http://dx.doi.org/10.1016/j.foodcont.2012.08.007.

Oms-Oliu, G., Martín-Belloso, O. and Soliva-Fortuny, R. 2010. Pulsed Light Treatments for Food Preservation. A Review. Food Bioprocess Technol. 3: 13–23.

Oner, M.E. and Wall, M.M. 2013. Quality of Fresh-Cut Purple-Fleshed Sweet Potatoes after X-Ray Irradiation Treatment and Refrigerated Storage. Int. J. Food Sci. Technol. 48: 2064–2070.

Pan, Y., Sun, D.W. and Han, Z. 2017. Applications of Electromagnetic Fields for Nonthermal Inactivation of Microorganisms in Foods: An Overview. Trends Food Sci. Technol. 64: 13–22. http://dx.doi.org/10.1016/j.tifs.2017.02.014.

Pillai, S.D. and Shayanfar, S. 2017. Electron Beam Technology and Other Irradiation Technology Applications in the Food Industry. *In*: Applications of Radiation Chemistry in the Fields of Industry, Biotechnology and Environment 249–268. Springer, Cham.

Pillai, S.D. and Shayanfar, S. 2018. Electron Beam Processing of Fresh Produce—A Critical Review. Radiat. Phys. Chem. 143: 85–88. http://dx.doi.org/10.1016/j.radphyschem.2017.09.008.

Pimenta, A.I., Margaça, F.M.A. and Cabo Verde, S. 2019. Virucidal Activity of Gamma Radiation on Strawberries and Raspberries. Int. J. Food Microbiol. 304: 89–96. https://doi.org/10.1016/j.ijfoodmicro.2019.05.011.

Porep, J.U., Kammerer, D.R. and Carle, R. 2015. On-Line Application of near Infrared (NIR) Spectroscopy in Food Production. Trends Food Sci. Technol. 46: 211–230. http://dx.doi.org/10.1016/j.tifs.2015.10.002.

Preetha, P., Pandiselvam, R., Varadharaju, N., Kennedy, Z.J., Balakrishnan, M. and Kothakota, A. 2021. Effect of Pulsed Light Treatment on Inactivation Kinetics of Escherichia Coli (MTCC 433) in Fruit Juices. Food Control. 121: 107547. https://doi.org/10.1016/j.foodcont.2020.107547.

Purohit, S.R., Jayachandran, L.E., Raj, A.S., Nayak, D. and Rao, P.S. 2019. X-Ray Diffraction for Food Quality Evaluation. In Evaluation Technologies for Food Quality 579–594. Elsevier Inc. http://dx.doi.org/10.1016/B978-0-12-814217-2.00022-6.

Ramaswamy, R., Krishnamurthy, K. and Jun, S. 2012. Microbial Decontamination of Food by Infrared (IR) Heating. Microbial Decontamination in the Food Industry. Woodhead Publishing Limited. http://dx.doi.org/10.1533/9780857095756.2.450.

Riganakos, K.A. 2010. Food Irradiation Techniques. In Irradiation of Food Commodities, 23–42. First Edit. Academic Press. http://dx.doi.org/10.1016/B978-0-12-374718-1.10002-1.

Robichaud, V., Bagheri, L., Salmieri, S., Aguilar-Uscanga, B.R., Millette, M. and Lacroix, M. 2021. Effect of γ-Irradiation and Food Additives on the Microbial Inactivation of Foodborne Pathogens in Infant Formula. LWT - Food Sci. Technol. 139: 110547. https://doi.org/10.1016/j.lwt.2020.110547.

Roh, H.J., Kim, A., Kang, G.S. and Kim, D.H. 2016. Photoinactivation of Major Bacterial Pathogens in Aquaculture. Fish. Aquat. Sci. 19(1): 1–7. http://dx.doi.org/10.1186/s41240-016-0029-5.

Sabharwal, S. 2013. Electron Beam Irradiation Applications. *In*: Proceedings of the PAC 745–748.

Salazar-González, C., Martín-González, M.F.S., López-Malo, A. and Sosa-Morales, M.E. 2012. Recent Studies Related to Microwave Processing of Fluid Foods. Food Bioprocess Technol. 5: 31–46.

Sánchez, M.T., Flores-Rojas, K., Guerrero, J.E., Garrido-Varo, A. and Pérez-Marín, D. 2010. Measurement of Pesticide Residues in Peppers by Near-Infrared Reflectance Spectroscopy. Pest Manag. Sci. 66: 580–586.

Santo, D., Graça, A., Nunes, C. and Quintas, C. 2018. Escherichia Coli and Cronobacter Sakazakii in "Tommy Atkins" Minimally Processed Mangos: Survival, Growth and Effect of UV-C and Electrolyzed Water. Food Microbiol. 70: 49–54.

Shawrang, P., Sadeghi, A.A., Behgar, M., Zareshahi, H. and Shahhoseini, G. 2011. Study of Chemical Compositions, Anti-Nutritional Contents and Digestibility of Electron Beam Irradiated Sorghum Grains. Food Chem. 125: 376–379. http://dx.doi.org/10.1016/j.foodchem.2010.09.010.

Shin, J., Kim, S., Kim, D. and Kang, D. 2016. Fundamental Characteristics of Deep-UV Light-Emitting Diodes and Their Application To Control Foodborne Pathogens Joo-Yeon. Appl. Environ. Microbiol. 82: 2–10.

Singh, H., Bhardwaj, S.K., Khatri, M., Kim, K.H. and Bhardwaj, N. 2021. UVC Radiation for Food Safety: An Emerging Technology for the Microbial Disinfection of Food Products. Chem. Eng. J. 417: 128084. https://doi.org/10.1016/j.cej.2020.128084.

Singh, R. and A. Singh. 2019. Food Irradiation: An Established Food Processing Technology for Food Safety and Security. Def. Life Sci. J. 4(4): 206–213.

Sivertsen, A.H., Kimiya, T. and Heia, K. 2011. Automatic Freshness Assessment of Cod (Gadus Morhua) Fillets by Vis/Nir Spectroscopy. J. Food Eng. 103: 317–323. http://dx.doi.org/10.1016/j.jfoodeng.2010.10.030.

Sokovnin, S. 2021. An Electron Beam Technology of Surface Disinfection of the Packed Egg. Food Bioprod. Process. 127: 276–281. https://doi.org/10.1016/j.fbp.2021.03.009.

Sommers, C.H. 2012. Microbial Decontamination of Food by Irradiation. In Microbial Decontamination in the Food Industry, 322–343. Woodhead Publishing Limited. http://dx.doi.org/10.1533/9780857095756.2.322.

Suhag, R., Dhiman, A., Deswal, G., Thakur, D., Sharanagat, V.S., K. Kumar et al. 2021. Microwave Processing: A Way to Reduce the Anti-Nutritional Factors (ANFs) in Food Grains. LWT - Food Sci. Technol. 150: 111960. https://doi.org/10.1016/j.lwt.2021.111960.

Supriya, P., Sridhar, K.R. and Ganesh, S. 2014. Fungal Decontamination and Enhancement of Shelf Life of Edible Split Beans of Wild Legume Canavalia Maritima by the Electron Beam Irradiation. Radiat. Phys. Chem. 96: 5–11. http://dx.doi.org/10.1016/j.radphyschem.2013.08.007.

Syamaladevi, R.M., Adhikari, A., Lupien, S.L., Dugan, F., Bhunia, K., Dhingra, A. and Sablani, S.S. 2015. Ultraviolet-C Light Inactivation of Penicillium Expansum on Fruit Surfaces. Food Control. 50: 297–303. http://dx.doi.org/10.1016/j.foodcont.2014.09.006.

Tang, J. 2015. Unlocking Potentials of Microwaves for Food Safety and Quality. J. Food Sci. 80: E1776–E1793.

Teye, E., Huang, X. and Afoakwa, N. 2013. Review on the Potential Use of Near Infrared Spectroscopy (NIRS) for the Measurement of Chemical Residues in Food. Am. J. Food Sci. Technol. 1: 1–8. http://pubs.sciepub.com/ajfst/1/1/1/index.html.

Tyagi, V.K. and Lo, S.L. 2013. Microwave Irradiation: A Sustainable Way for Sludge Treatment and Resource Recovery. Renew. Sustain. Energy Rev. 18(71): 288–305. http://dx.doi.org/10.1016/j.rser.2012.10.032.

U. S. Food and Drug Administration. 2001. 21 CFR Part 179. Irradiation in the production, processing and handling of food. Federal Register. 65: 71056–71058. https://www.govinfo.gov/app/details/CFR-2012-title21-vol3/CFR-2012-title21-vol3-part179.

Uekert, T., Dorchies, F., Pichler, C.M. and Reisner, E. 2020. Photoreforming of Food Waste into Value-Added Products over Visible-Light-Absorbing Catalysts. Green Chem. 22: 3262–3271.

Valentina, L. 2017. New Technologies for Sanitization of Fresh Dairy Products. In.

Wang, J.J., Xu, Z.C., Fan, J.L., Wang, Y., Tian, Z.J. and Y.T. Chen. 2015. Effects of X-Ray Irradiation on the Microbial Growth and Quality of Flue-Cured Tobacco during Aging. Radiat. Phys. Chem. 111: 9–13.

Wang, Y., Wang, Y., Wang, Y., Murray, C.K., Hamblin, M.R., Hooper, D.C. et al. 2017. Antimicrobial Blue Light Inactivation of Pathogenic Microbes: State of the Art. Drug Resistance Updates 33–35: 1–22.

Wilson, S.A., Okeyo, A.A., Olatunde, G.A. and Atungulu, G.G. 2017. Radiant Heat Treatments for Corn Drying and Decontamination. J. Food Process. Preserv. 41: 1–9.

Wu, X.F., Zhang, M. and Bhandari, B. 2019. A Novel Infrared Freeze Drying (IRFD) Technology to Lower the Energy Consumption and Keep the Quality of Cordyceps Militaris. Innov. Food Sci. Emerg. Technol. 54: 34–42. https://doi.org/10.1016/j.ifset.2019.03.003.

Yousefi, M., Mohammadi, M.A., Khajavi, M.Z., Ehsani, A. and Scholtz, V. 2021. Application of Novel Non-Thermal Physical Technologies to Degrade Mycotoxins. J. Fungi. 7: 395.

Yun, S.H., Lee, S.W., Koo, H.N. and Kim, G.H. 2014. Assessment of Electron Beam-Induced Abnormal Development and DNA Damage in Spodoptera Litura (F.) (Lepidoptera: Noctuidae). Radiat. Phys. Chem. 96: 44–49. http://dx.doi.org/10.1016/j.radphyschem.2013.08.008.

8

Cooling

Karen Hariantty Estévez-Sánchez,[1] *Carlos Enrique Ochoa-Velasco,*[2] *Hector Ruiz-Espinosa*[3] and *Irving Israel Ruiz-López*[3,*]

1. Introduction

The cooling of food products involves the permanent heat removal from their point of origin all the way to the end-user, which is a network known as a cold chain. For sustainability, cooling processes require an effective design of cold chain logistics to minimize food waste, which is associated with greenhouse gas emissions (carbon footprint), and potential consumer health issues if the chain is disrupted. Besides, the increasing global food demand imposes additional challenges for developing sustainable cooling processes as more eco-friendly, regulatory-conforming cooling devices, and refrigerants are required for keeping the temperature of perishable products within the desired range. This chapter discusses the critical steps for setting up optimized cold chains, from the precooling stage, cold storage in warehouses, and during transit up to domestic refrigerators, including hardware and software requirements, automation, monitoring, traceability, integrity checks, and logistics management; these are all through the use of current smart technologies that aim at making more efficient, environmentally responsible cooling operations, which include internet of things (IoT), big data, cloud computing, blockchain, and artificial intelligence methods.

2. Precooling

Precooling is the first operation in the cold chain logistic. It is used to prepare foods for transportation, and it either avoids or slows down their deterioration before reaching commercialization. Precooling techniques depend on the food type as each one has its requirements in terms of final temperature, processing speed, and allowable water loss among others. The following sections introduce the main characteristics of precooling for fruits and vegetables, meat and poultry, and fishery products. A summary of common precooling methods and their characteristics is presented in Table 1.

2.1 Fruits and Vegetables

Precooling is used to remove the field heat of fruits and vegetables immediately after harvesting (Wang and Zhang 2021), minimizing the deterioration and senescence of fruits and vegetables while

[1] Facultad de Ciencias de la Electrónica, Benemérita Universidad Autónoma de Puebla, Av. San Claudio y 18 Sur S/N, Puebla, Pue.72570, México.

[2] Facultad de Ciencias Química, Benemérita Universidad Autónoma de Puebla, Av. San Claudio y 18 Sur S/N, Puebla, Pue. 72570, México

[3] Facultad de Ingeniería Química, Benemérita Universidad Autónoma de Puebla, Av. San Claudio y 18 Sur S/N, Puebla, Pue.72570, México.

* Corresponding author: irving.ruiz@correo.buap.mx

Table 1. Common Precooling Methods and Their Characteristics.[I]

Characteristics	Room Cooling	Forced air Cooling	Hydrocooling	Vacuum Cooling	Package Icing
Heat removal speed	Low	Medium	High	Very high	Low
Relative cost	Low	Low	Low	Medium	High
Energy efficiency	Low	Low	High	High	Low
Relative water loss	Medium	Medium	Low	High	Low
Products	**Temperatures:**				
Beef carcasses	7°C				
Pork carcasses	5°C				
Poultry carcasses		4°C			
Fruits and vegetables	-1 to 5°C[II]	-1 to 5°C[III]	-1 to 5°C[IV]		0 to 1°C[V]
Fish and seafood products		1°C			-1 to 0°C
Vegetables				-1 to 5°C[VI]	
Advantages	Clean and simple, recommended for small-scale operations		High energy efficiency		High heat transfer performance
		Rapid			Avoid moisture losses
	Low installation and maintenance costs		Sanitization with chemicals	Uniform cooling	Can be used during transit and short-term storage
	Precooled and stored in the same place	High heat transfer performance			
Disadvantages	Dehydration of the product		Drying is required	Risk of wilting	Added weight and additional space required
	Very slow heat removal	Slower than vacuum cooling	Monitoring water quality	High cost	Water quality monitoring is a must
	Not suitable for leafy vegetables		Water may affect food commodities	Packages need holes for the evaporation process	Possible chilling injury over long-term storage
	Uneven cooling at the beginning		Possible microbial growth	limited capacity due to space and vacuum requirements	High initial cost for the slurry ice generator
	Not recommended for fast-deteriorating items				Water-resistant packaging
					Physical injury and rotting are likely

(I) From Abbas et al. (2009), Delele et al. (2019), Huff-Lonergan (2001), Carroll and Alvarado (2008), Elansari (2009), Rao (2016), Valtýsdóttir et al. (2010). (II) Apples, apricots, pears, berries, citrus fruit, zucchini, squash, pumpkin, beans, beets, cabbage, eggplant, potato, peppers, and tomato. (III) Berries, grapes, apples, nectarines, peaches, pears, melons, plums, citrus fruit, celery, potatoes, onions, sweet potatoes, beans, broccoli, cabbage, eggplant, peas, peppers, squash, and tomato. (IV) Melons, sweet cherry, apples, plums, apricots, nectarines, peaches, pears, peaches, peas, spinach, turnip, sweet corn, celery, onions, asparagus, beans, broccoli, brussels sprouts, cauliflower, Chinese cabbage, cucumber, greens, and lettuce. (V) Melon, turnip, spinach, radish, onions, lettuce, sweet corn, carrots, asparagus, cabbages, collards, kale, Brussels sprouts, broccoli, etc. (VI) Greens, mushrooms, and sweet corn.

increasing their shelf life and quality (Duan et al. 2020). Precooling of fruits and vegetables is needed for diminishing the required refrigeration capacity of the cold storage room, thus avoiding temperature fluctuation during storage. With certain exceptions, harvesting is generally done very early in the morning at the coldest day temperature possible to reduce the field heat, which is critical as fruit and vegetable metabolism increases with temperature. In this sense, some results indicate that an hour of exposition to field temperature reduces fruit and vegetable shelf life by one day. Moreover, precooling also reduces the spread of decay, moisture loss, and ethylene production (Ravikumar et al. 2018). Among existing precooling methods, those where fruits and vegetables are handled are the most diverse, which include room cooling, air precooling, forced air cooling, icing, vacuum cooling, and hydro cooling. The precooling method selection depends on factors, such as those product-related (nature, value, and quantity of the product) or process-related (cost, type of equipment, and materials required) (Han et al. 2021).

2.2 Meat and Poultry

Precooling of meats and poultry carcasses is primordially needed to rapidly reduce their temperature and maintain the microbial load as low as possible after slaughter (Gürbüz et al. 2018). Precooling also reduces meat weight loss and discoloration caused by hemoglobin oxidation. In beef carcasses, precooling is used to quickly reduce surface temperature, avoiding muscle shortening by cold during its pH drop, and is generally carried out through air fans in short periods (Delele et al. 2019). The European Union legislation requires a maximum final meat temperature of 7°C before transport or cutting. Although the cooling rate is not stipulated, some studies have indicated that quickly chilling may reduce the evaporation rate and consequently reduce weight loss and aging time required to produce an acceptable product sensory-wise, cold shrink, and marbling perception; however, lower temperatures (< –25°C) can darken meat color and increase drip losses at retail (Aalhus et al. 2002). On the other hand, pork industries generally use either blast chilling or conventional chilling (room cooling) to reduce carcass temperature after slaughter (Chang et al. 2003).

Poultry carcasses are commonly precooled by water immersion or air blast chilling before freezing; the former is the most used in the United States, while the latter is more common in Europe, Brazil, Canada, and other countries (Carroll and Alvarado 2008). The temperature of carcasses must be reduced from 40°C to 4°C at the chicken breast center (Lorenzetti et al. 2018) after 4 hours postmortem for slowing down microbial growth responsible for flavor, texture, and appearance defects (Carciofi and Laurindo 2007). One of the typical features of chilling tanks is the constant water exchange; if two tanks are used, the water of the first tank needs to be changed at 1.5 L per poultry carcass while in the second, the exchange rate is 1 L per carcass. Although microbial growth could be avoided with such a treatment, its cost is higher than that of air chilling. Moreover, poultry carcasses chilled with water may gain weight during the process, thus regulations need to be accomplished to avoid excess water in them (Buhr et al. 2014). Air-chilling may reduce average carcass weight by about 0.7% after slaughter; according to some processors, they may exhibit higher microbial quality (Carroll and Alvarado 2008).

2.3 Fish and Seafood Products

The quality of fresh seafood products depends on several factors, including bacterial growth, enzyme activity, physical damage, dehydration, chemical reactions, and contamination (Chun-hua et al. 2014). In this sense, precooling before packaging is desirable to diminish the quality losses of these products (Valtýsdóttir et al. 2010). The temperature during precooling or storage should not decrease below –0.9°C because crystallization can occur, causing protein denaturation and cell damage. Slurry ice, crushed plate ice, and air blast cooling are the most used methods for precooling fish and fillets; the former being the most efficient (Abbas et al. 2009; Margeirsson et al. 2010; Valtýsdóttir et al. 2010). Onboard refrigerated seawater systems are used for chilling caught fish, extending their shelf life beyond 20 days (Chun-hua et al. 2014). Moreover, a combination of air blast and contact cooling surface at a temperature of 1°C to 3°C lowers the initial freezing point of the food product; namely, a super chiller is a better option for precooling fish and fillets (Chun-hua et al. 2014; Margeirsson et al. 2010). On the other hand, precooling of crustaceans, which are transported alive, is beneficial for their recovery by reducing their metabolic stress (Lorenzo et al. 2020)

2.4 Common Precooling Methods

A brief description of the most important precooling methods is presented below.

i. Room cooling: This low-cost method only requires basic equipment. Products are either placed in wooden, fiberboard, or plastic boxes, which are simply stored in a room and exposed to cold air. Due to its limited cooling potential, this method is only appropriate for small amounts of products or for those that do not deteriorate rapidly (Han et al. 2021). One advantage of room cooling is that several kinds of products can be stored simultaneously in the same place (Kongwong et al. 2019).
ii. Forced-air cooling: In this simple and low-cost system, packaged foodstuffs are exposed to airflow with high air pressure. Different arrangements have been developed for the application of this precooling method; however, room cooling adapted with high-velocity fans and forced-air cooling tunnels is the most widespread. In comparison with room cooling, forced-air cooling may achieve a ten-fold or higher reduction in the cooling time, and the condensation phenomenon is not observed due to air convection (Kongwong et al. 2019).
iii. Hydrocooling: This method is an effective cooling technique since water possess a high heat transfer coefficient. Fruits and vegetables are sprinkled with cold water or immersed in cold water or an ice water mixture bath. This method permits fast cooling, and the water used may be added with chemical sanitizers to prevent food spoilage. The main problem with hydrocooling relates to a possible damage of water or the added chemical to the product (Wang and Zhang 2021).
iv. Vacuum cooling: Vacuum cooling is a rapid evaporative cooling technique used in food products with high moisture content, porous structure, and adequate surface area. The method is based on decreasing the temperature of the product through water evaporation from food at reduced pressure, avoiding moisture loss in the product by spraying water on them (Zhu et al. 2021).
v. Package icing: Package icing is done by applying crushed ice or ice slurry on the top of food containers or packages (Gillies and Toivonen 1995). Although this is a cooling method commonly used for fish and seafood products, some vegetables, such as broccoli, are precooled in this way (Gillies and Toivonen 1995). It is used for maintaining a high moisture content and avoiding weight loss. The main problems with this method are the added weight and its inefficiency both energy and cost-wise; approximately 14 kg of ice is required for 9 kg of broccoli (Reddy et al. 2010).

3. Transport

Refrigerated transportation represents an important element of the cold chain as it allows the movement of foodstuffs from their production or origin location to the end sale spot. Refrigerated transport can be achieved by air, land (road or railway), or marine routes; each one has different uses and strengths. Maintaining both constant and uniform environmental temperature and humidity throughout the shipment period is paramount for keeping the quality, safety, and shelf-life of food products during refrigerated transport, thus helping reduce food losses. This section aims to provide a general landscape of food transportation systems and their impact on the sustainability of the food cold chain.

3.1 Road Transportation

Transport refrigeration systems may be exposed to harsh conditions and experience multiple scenarios for cooling demands and constraints imposed by the available space and weight, thus leading to lower coefficients of performance (COP; ratio of useful cooling provided to work/energy required) than stationary refrigeration systems. Typically, the COP of transport refrigeration systems ranges from about 0.5 at –20°C space temperature to 1.5–1.75 at 3°C space temperature and 30°C ambient temperature (Tassau et al. 2009). Besides having a low COP, the increasing demand for transported goods and their related high-quality expectations has prompted an important rise in the carbon footprint of the refrigerated transport industry (Bagheri et al. 2017), which may have both direct (greenhouse gas emissions and leakages) and indirect (energy consumptions) components (Wu et al. 2013). The rise of online grocery shopping has produced an increase in the use of refrigerated light commercial vehicles or vans (around 3.5 tons gross weight) for home deliveries. A study performed in the UK demonstrated that average CO_2 and NO_X emissions from transport refrigerated units (TRUs) equipped vans are about 282 g/km and 0.43 g/km, respectively, which are about 15% and 18% higher than standard vehicles (Yang et al. 2021).

TRU increases the workload of the vehicle engine by (1) its additional weight (including insulation material), (2) increased aerodynamic drag caused by the frontal condenser unit, and (3) the refrigeration load. According to some studies, typical refrigerated trailers used in food transportation consume about 12 MWh per year (COP of 0.58), and it is estimated that a reduction of about 8 tons of CO_2 per year per trailer is possible by replacing current diesel engine-driven vapor compression refrigeration systems with battery-powered types (Bagheri et al. 2017). In this case, the use of onboard photovoltaic panels installed on the vehicle roof as support systems looks promissory to further reduce greenhouse gas emissions. Nevertheless, photovoltaic energy integration may not apply to all scenarios as their current profitability depends on the required cooling temperature; for example, moving from chilled (≈ 0°C) to frozen products (≈ –20°C) would require additional photovoltaic panels because of higher refrigeration load, thus impacting the vehicle weight and consequently the fuel costs (Meneghetti et al. 2021). Moreover, simple practices, such as keeping a clean white external surface in refrigerated trucks may lead to lower cooling demands (about 10 to 15%) in comparison with both dirty and black external surfaces (Artuso et al. 2019). Cryogenic technologies based on liquid CO_2 or liquid N_2 have also been explored as alternatives to vapor compression systems in refrigerated transports; however, their total greenhouse gases emissions (due to associated fuel consumption, refrigerant leakage, and refrigerant production) and running costs are similar, while infrastructure costs are higher and thus reduce their appeal. Besides, the mass intensity of liquid CO_2 or liquid N_2 (defined as the mass of cryogen per mass of product transported per km) is nearly 10 times higher than that of diesel (20 to 60 kg/h versus 2 to 4 L/h), limiting the food distribution range (Rai and Tassou 2017).

The carbon footprint of food transportation systems has also been modeled to evaluate their global warming impact. Selected studies have concluded that CO_2 emissions caused by energy consumption to power the refrigerator and carry the refrigeration unit weight contribute the most to the total CO_2 emissions; moreover, refrigerators driven by auxiliary engines have higher CO_2 emissions than refrigeration systems driven by the main vehicle engine or electricity. As a result, an increase in both the COP of the refrigerator and the efficiencies of other pieces of equipment could lead to a significant reduction in the greenhouse effects of road transportation systems (Wu et al. 2013). According to some studies, insulation of refrigerated transport reduces its efficiency between 3% to 5% per year; as a result, the thermal conductivity of a vehicle may increase from 0.4 $W/m^2/°C$ to 0.62 $W/m^2/°C$ after nine years, resulting in a 50% increase in energy consumption and CO_2 emissions (Tassau et al. 2009). Thus, the deterioration of insulating materials should be also considered when appraising the carbon footprint of the food transportation industry.

3.2 Marine Transportation

This system has a relatively low cost, and it is adequate for long-distance transportation of large volumes of long-life perishable and nonperishable foods. It is estimated that about 90% of the world trade is transported in international marine vessels (Hafner et al. 2018). Refrigerated marine transport may be categorized into refrigerated (reefer) and container ships. The reefer ship is well-suited to palletized or break-bulk cargo, which is loaded into top opening hatches below deck with a typical total cargo capacity of about 7,000 m^3 (Tanner 2016). On the other hand, container vessels are designed to transport containerized unit loads having their own refrigeration unit. The containers are standard with a 20 ft or 40 ft length, where the first one is often referred to as a 20-foot equivalent unit (TEU). The capacity of the container ship goes from a few hundred TEUs in feeder ships to several thousand TEUs in ultra-large container vessels. Besides providing cold storage for foods, reefer vessels, in some instances, are also able to perform cold sterilization protocols against some fruit plagues, such as fruit flies and false codling moths (Goedhals-Gerber and Khumalo 2020).

3.3 Air Transport

This system is typically used for high-value, short-shelf-life products as its cost may be up to 100 times higher than marine shipping (Tanner 2016). The goods are typically loaded in standard aircraft unit load devices (ULD) in the form of pallets or containers, which may have built-in refrigeration units. The ULDs are manufactured to several specs (with typical capacities ranging from 3 m^3 to 16 m^3) and may be compatible with different aircraft models (Boeing 2012).

4. Warehouses

Refrigerated warehouses (RWs) are used to store foods under controlled temperature and relative humidity conditions. RWs are an essential component of the food supply chain as they serve for inventory management, regulating the trade-off between supply and demand (Han et al. 2021). Moreover, adequate cold storage may help to assure the availability of fresh foods all year. Besides providing the temperature and humidity regulation, the atmosphere of RWs can be preset at the desired composition at the start of the storage (modified atmosphere storage) or continuously adjusted (controlled atmosphere storage) during the storage period, especially for fruits, vegetables, and some meat products (Han et al. 2021). Oxygen and carbon dioxide levels are mainly adjusted in fresh agricultural products, leading to a reduced metabolism (for example, it can delay ripening and lower both respiration rate and production of ethylene and other metabolites), enhanced storage life, lowered development of undesirable physical and chemical changes (texture, color, odor, and flavor), and reduced microbial growth (Dumont et al. 2016).

When mixed for storage or transport, fruits and vegetables should be compatible in terms of environmental temperature and relative humidity requirements, ethylene production and sensitivity, and odors production and absorption (McGregor 1987). RWs can be categorized in terms of their food-temperature requirements as high-temperature (5°C to 15°C), medium-temperature (–5°C to 5°C), low-temperature (–18°C to –25°C), quick freezing (–35°C to 40°C), and ultralow temperature (–45°C to –60°C) facilities for cold air storage. For modified atmosphere storage, the temperature categories of RWs are slightly modified to 8°C to 15°C, 0°C to 8°C, and –2°C to 0°C for high-temperature, medium-temperature, and low-temperature, respectively (Han et al. 2021). Refrigerated storage capacity is on the rise because of population growth and the need for minimizing food waste. It is estimated that the RWs capacity of the top 20 countries (including India, the USA, China, Japan, and the UK, which are in the top 5 in that order) was close to 5.3×10^8 m^3 between 2012 and 2014 (Zhao et al. 2018). The overall refrigerated storage capacity in the USA in 2018 (1.3×10^8 m^3), was split into 20% for cooling and the remaining 80% for freezer and convertible storage (Shanmugam and Mital 2019). As a result, the energy consumption of RWs is very large, representing a considerable portion of the worldwide energy demand (with each RW consuming from hundreds of kilowatts to dozens of megawatts) and operating with COPs ranging from 1.5 to 5 as a function of machine design, setpoint, and working (climatic) conditions. Therefore, the integration of renewable energy sources (such as solar sorption systems, solar thermo-mechanical systems, solar photo-voltaic systems, and wind turbines) into RWs is of high priority, offering several advantages as highlighted in selected studies (Fikiin et al. 2017). Besides, additional strategies to increase the efficiency of RWs include the adoption of intelligent systems (automatic control of refrigeration systems, automatic, safety monitor and protection systems, automatic information retrieval systems, and automatic transportation systems) (Zhao et al. 2018)

5. Retail and Domestic Systems

Food retail and domestic refrigeration stages happen at the end of the cold chain; as they precede consumption, they are critical for maintaining food quality while preventing food losses. However, temperature control tends to be laxer as the cold chain moves from production to the final user, and this is more evident as food products enter supermarkets and convenience stores; it is especially true in store-to-home transit and home (Foster and Evans 2015). Thus, a clear understanding of the process dynamics, including the refrigeration devices used, the type of retail store and its business practices, transportation conditions from store to home, and consumer practices for keeping refrigerated and frozen goods at the household refrigeration level, are paramount for detecting hazards to food shelf life and possible sources of foodborne illnesses. Besides, both retail and domestic food cold chains could have adverse effects sustainability-wise as mounting evidence indicates that retail cold chain plays a pivotal role in greenhouse gases production via both direct (refrigerant emissions) and indirect (energy utilization) contributions, while domestic refrigeration only produces large direct emissions because of the vast number of household appliances (Evans et al. 2018). According to Salehi et al. (2020), refrigeration systems account for about 17% of the electric power globally used, 8% of greenhouse gases emitted, and up to 50% of the energy consumption in supermarkets, although several factors could influence this expenditure including in-store relative humidity, refrigeration system configurations, compressor control, defrost control, condensing pressure, insulation efficiency as well as consumer behavior—all of which may be related to energy losses.

5.1 Retail Refrigeration

The most common commercial cold chain systems are refrigerated display cabinets, which could either be closed or open and have widespread use for storing, exhibiting, and selling chilled and frozen foods. Found in several configurations (e.g., vertical or horizontal; with or without internal divisions) and built according to the construction, features, and performance requirements specified in the ISO23953-2:2015 standard, no universal consensus exists on whether closed or open cabinets are more efficient. The former is considered to generate energy savings as they use doors to help maintain more constant product temperatures inside the cabinet and are not as exposed to wet, warm frontal air infiltration, as it has been reported with the latter, which tries to prevent this by solely using air curtains (Chaomuang et al. 2020). However, as foodstuffs are not readily available, closed cabinets could limit product exposure and buying intention while frequent door openings, fog formation, and the presence of retail refrigerator lights could require a constant readjustment of temperature conditions (Han et al. 2021; Laguerre and Chaomuang 2021). Additional retail cold chain units include vending machines, water coolers, and other devices.

Several types of refrigeration equipment could be used to maintain the food cold chain depending on the nature and size of the trading outlet. From large hypermarkets (with an average size of more than 2,500 m^2 of commercial area), medium supermarkets (400–2,499 m^2), and small convenience stores (< 399 m^2) to wholesale and farmer's markets, food supplies are distributed in short and long chains where their energy expenses and pollutant generation is markedly different; generally, the smaller the outlet, the lower the energy utilization efficiency with higher electricity and refrigerant consumption per square meter, although the refrigerant leakage values increased along with the size of the retail store (Majewksi et al. 2020). Refrigeration equipment in hypermarkets and supermarkets are classified as low-temperature systems used for frozen products (kept at –18°C) and high-temperature systems for non-frozen foodstuffs (0–8°C) (Brownbridge et al. 2021). These could be found as small, integral units with relatively low electrical consumption levels (< 3 kW) and multiple-cabinet devices with central refrigeration systems. Large variations in performance have been reported for the same type of cabinet in different locations influenced by variations in usage (e.g., business practices, types of products stored, and shopping activity) and environmental conditions, such as ambient temperature and closeness to hot spots. In the UK, the reported energy consumption for refrigerated and frozen cabinets in supermarkets is 2,920 kW/h and 5,475 kW/h per year, respectively (Evans et al. 2018).

The improper temperature of goods displayed at the retail level is a major concern for assessing food safety. Food that requires refrigeration and is not kept at the adequate temperature at retail can represent a diminishing factor in product quality (Talbot et al. 2020); it has been identified as the top contributing factor for foodborne disease outbreaks in foodservice (King 2020) and could affect pathogen growth rate in selected products (de Frias et al. 2018), which may reduce the shelf life of stored products (Giarratana et al. 2020; Xie et al. 2021). It is of key importance to operate the devices within the operational recommendations issued by the manufacturer as overloading, products blocking air vents, can thus prevent an adequate airflow and could cause important breaches in the cold chain; besides, maintenance of cabinets (e.g., defrost cycles) is essential to avoid fluctuations in their performance. Cabinet temperature monitoring is automatically carried out at large retail stores; sensors are linked to alarm systems, which are triggered once the temperature reaches the setpoint, thus preventing undesired changes in the food products. However, this feature could not be available in convenience stores and manual temperature verification should be a well-established procedure, coupled with appropriate corrective actions to prevent further food safety problems (Kyriakides 2014). It is also important to consider that products placed atop could exhibit significantly higher temperatures than those stored beneath, depending on the type of food, seasonal temperature variations throughout the year, and consumer habits (e.g., door opening frequency in closed cabinets) (Zubeldia et al. 2016).

5.2 Domestic Refrigeration

The consumer phase of the cold chain involves food transportation from retail to home and its ensuing storage until consumed. Both stages present challenges, as with the rest of the cold chain, where suitable temperature control should be observed. During transportation from retail to home, the food safety and shelf life of refrigerated goods can be compromised, especially if travel distances are long and temperature fluctuations are more likely to occur. The possible proliferation of pathogenic organisms while transporting certain foodstuffs from retail to home has been analyzed (Mishra et al. 2017; Gurman et al. 2018; Golden and Mishra 2021). Although traditionally consumers are responsible for choosing their groceries and bringing them back home, several online ordering/pick-up methods and delivery services have appeared in the last few years, prompting the need to verify how the cold chain is maintained during these types of buying and shipping alternatives (Stanton et al. 2018; Ndraha et al. 2019).

Domestic refrigeration devices are usually small, closed, integral systems with electrical spending in the range of 20 to 150 kW (Evans et al. 2018), and their energy consumption and efficiency depend on the technical characteristics of the components, consumer usage habits (with emphasis on thermal load), and the environmental conditions where they are located, including room temperature and relative humidity (Belman-Flores et al. 2019). Moreover, instigated by international regulations aiming at making more eco-friendly devices refrigerator performance, which could be enhanced with advanced insulation, high-efficiency compressors, or improved evaporators and condensers like those using phase change materials; besides, as with retail display cabinets, household refrigerators operation is affected by its configuration, geometry, positioning, and setting (Omara and Mohammedali 2020). Age-related efficiency loss in household appliances has been associated with the degradation of insulation material over time and could negatively affect the desired refrigerator temperature (Hueppe et al. 2020). Although it is well known that several foodborne disease outbreaks could originate in prepared-at-home products (or in prepared foods pick up and delivered through a third-party service), both raw materials and meals refrigerated at suboptimal temperature conditions pose an additional risk factor to food safety (Vithanage et al. 2017), a large proportion of household refrigerators display average temperatures higher than the standard (4°C) with some appliances reaching up to 14°C (Evans et al. 2018; Ovca et al. 2021).

6. Sustainable Alternatives to Conventional Cold Storage

Refrigeration systems are beneficial for food preservation, but their operation is costly and may hurt the environment. The continuous increase in demand for fresh and frozen products and the need for food waste reduction have motivated the development of sustainable low-energy consumption and eco-friendly storage technologies that are able to keep a low-temperature environment. For example, the evaporative cooling phenomenon can be used to produce a low temperature and high relative humidity environment, mainly in warm and arid regions (Durmont et al. 2016). These systems can achieve cooling temperatures between 16°C and 21°C and relative humidity higher than 80% for initial ambient temperatures ranging from 29°C to 32°C with a relative humidity of 55% to 70% (Sibanda and Workneh 2020). The efficiency (η) of evaporative cooling systems is expressed as the ratio:

$$\eta = 100\frac{T_{d,\infty} - T_{d,c}}{T_{d,\infty} - T_{w,\infty}} \tag{1}$$

In the above equation T, is temperature and subscripts d, w, ∞, and c indicate dry bulb, wet bulb, ambient (inlet), and achieved cold conditions, respectively. According to Equation (1), low air temperatures and high relative humidity at the cooler entrance led to a lowered potential for temperature drop and significatively reduced the evaporative efficiency. Evaporative cooling systems can achieve efficiencies higher than the 80% when the temperature difference between the inlet and outlet conditions is higher than 8°C (Ronoh et al. 2020; Sibanda and Workneh 2020). Refrigeration chambers and air-conditioning systems based on evaporative cooling have been explored as storage alternatives for agricultural products as well as to mitigate heat stress in poultry and livestock farms (Raza et al. 2021; Shahzad et al. 2021). Evaporative cooling chambers can be prefabricated (Nkolisa et al. 2018) or built from low-cost materials, such as bricks and charcoal (Rehman et al. 2020, Ronoh et al. 2020). Besides, the evaporative cooling principle can be also used in small-scale domestic systems; for example, in clay coolers, where the cooling system is formed by two concentrical clay pots with a layer of wet sand between them. These systems can achieve average cooling efficiencies ranging from 35% to 54% with water consumption between 2 L to 3 L per day and cooling chambers of about 6 L to 32 L (Rehman et al. 2020).

An additional strategy for sustainable cooling systems is using underground storage, that is, sites located below the surface, such as pits or caves destined for food storage purposes. The most basic form of underground storage is pits; these small-size structures are used in arid regions of Africa for long-term grain storage (up to 20 years; about 2 tons per pit), providing an atmosphere poor in O_2 and rich in CO_2 which prevents insect proliferation and enhances shelf-life of grains (Sunano 2020). Higher-scale underground cellars have been also used for centuries in the wine industry, providing adequate temperature ranges for wine aging (8°C to 15°C) without the need for air-conditioning systems. The construction of wine cellars has been analyzed in terms of their passive thermal design and has been demonstrated to attenuate the maximum external temperature difference by 65% to 95%, depending on their size (Mazarrón et al. 2012). Underground storage can be also equipped with refrigeration units with energy consumption for refrigerated and frozen foods of about 25% and 75% in similar surface-level facilities, respectively (Goel et al. 2012). The energy-saving arises from the lower initial underground temperature as well as the improved isolation. Recently, underground spaces resulting from mining activities have been prepared and evaluated for apple storage (about 30,000 tons), where the underground facility is equipped with refrigeration units to lower its temperature up to 1°C before the apples are loaded into the cells and atmosphere is replaced by a gas mixture, mainly composed of N_2 (98%). Humidity control is helped by the geological characteristics of the site and the wall covering (Galgaro et al. 2020).

7. Sustainable Logistics

High-quality logistics have been highlighted as an essential component for the successful implementation and operation of cold chains. For example, prolonged waiting times for loading and unloading of goods during transportation caused by inefficient distribution logistics are undesirable as they may originate important losses in highly perishable food products. The executive decisions in the supply chain directed to boost customer value, including the temperature control operations for perishable foods are known as cold chain management (Esmizadeh et al. 2021); it is responsible for route planning to the location of distribution centers. An efficient route planning must consider factors intrinsic to the cold chain, such as the number and location of required stops, route length, allowable time windows for deliveries, loading/unloading times, multi-compartment capacity constraints, restrictions for starting and arrival times as well as external factors, such as traffic uncertainty and demand and supply peaks. Poorly optimized routes do not only contribute to food waste but may also cause increased both operating costs and carbon footprints because of higher fuel consumption. Therefore, several studies have been devoted to vehicle routing problems for transit within and between cities to minimize operational costs as well as greenhouse gas emissions (Chen et al. 2019; Meneghetti et al. 2018; Stellingwerf et al. 2018; Zhang et al. 2019). Regarding the optimal location of distribution centers, selected studies have demonstrated that the hub location problem can be seen as a bi-level planning model with an upper level that can change the scale and location of a distribution center and set reasonable carbon emissions trading policies, while the lower level depends on logistic users aiming to minimize their costs by choosing a reasonable cold chain logistics service (Zhang et al. 2021). As a result, an effective location of distribution centers should consider characteristics such as constructions costs and budget restrictions, transportation costs, operating costs, carbon emissions, the capacity of the facilities, consumer habits, distance between origin-destination demand nodes, reliability of the distribution network to handle unexpected events, and refreshing and freezing operations among others (Esmizadeh et al. 2021; Zhang et al. 2021). Route planning and hub location problems for cold chain logistics have been expressed as multi-objective mixed-integer linear programming models and solved by diverse algorithms, such as adaptive large neighborhood search, ribonucleic acid-ant colony optimization, cloud particle swarm optimization, and genetic algorithms (Chen et al. 2019; Esmizadeh et al. 2021; Zhang et al. 2019; 2021).

8. Cold Chain Integrity: Monitoring and Traceability

As much as 9% to 23% of the total food production is lost by the lack of refrigeration systems in both developed and developing countries (Murrant and Radcliffe 2018). These losses occur in different stages of the cold chain from precooling to the retail and domestic refrigeration systems. A correct understanding of these elements is essential to minimize both the occurrence and severity of cold chain breaks. Refrigeration system malfunctions, incorrect temperature settings, human errors, packaging failures, irregular temperature distributions, or exposure to ambient air during delivery loading and unloading can cause failures in the cold chain (Loisel et al. 2021; Vivaldi et al. 2020). Monitoring of the cold chain refers to the process of collecting, analyzing, and using the information to quantify its progress, identify breaches causing a deviation from the required thermal objectives as well as appraise their severity. On the other hand, traceability of the cold chain refers to the ability to verify the information of foods employing recorded identifications. Traceability technologies include paper records, barcodes, radiofrequency identification (RFID), and Wireless Sensor Networks (WSNs) (Óskarsdóttir and Oddsson 2019).

Cold chain breaks can be identified by proper temperature monitoring using thermistors, resistance temperature detectors, thermocouples, or time-temperature indicators (lightweight and inexpensive colorimetric tags undergoing mechanical, microbiological, chemical, or enzymatic changes when temperature exceeds a predefined threshold) (Vivaldi et al. 2020). Wireless data transfer by bluetooth and RFID technologies have allowed temperature monitoring without direct access to packaging and the tracking of the refrigeration status in real-time from the production until the sale of the products (Vivaldi et al. 2020).

Besides, recent studies have demonstrated that the use of novel technologies such as Thermochron® iButton® (a small-sized device about 18 mm in diameter and 6 mm in thickness, allowing the reading and storage of more than 2,000 temperature values at logging rates from 1 second to several hours) to identify temperature breaks during export cold chains (Goedhals-Gerber and Khumalo 2020). Despite the temperature measurement method used, an effective temperature recording can be challenging in some cases; for example, for fruits, the core temperature history is often monitored at easy-to-access locations (at the side of pallets and cartons) and only a limited number of specimens are monitored, which do not necessarily represent the overall fruit quality and temperature. Therefore, computational fluid dynamics models have been recently proposed to predict the thermal history and quality of individual fruits during virtual cold chains (Wu et al. 2018). The severity of the cold chain break can be identified by measuring specific food quality indicators, including those of microbiological, chemical, and sensory nature; for example, biogenic amines have been determined in tuna via HPLC (Reguera et al. 2019). For fruits and vegetables, gas sensors provide fast measurements and may interact with atmospheric (O_2, CO_2, H_2O, and N_2), metabolic (O_2, CO_2, ethylene, aromatic hydrocarbons, and phenolic), and exogenous (1-methyl cyclopropane, O_3, SO_2, ClO_2, and H_2S) gases (Wang et al. 2021).

The IoT involving network transmission technologies, such as the 5G network, has also the potential to simplify the real-time tracking of cold chain items with the aid of GPS systems as demonstrated in recent investigations (Li 2021). The IoT may contribute to solving common manufacturer challenges, such as forecasting the right demand versus supply and reducing the manufacturing lead-time, while adopting the design principles of Industry 4.0, that is virtualization, interoperability, decentralization, real-time capability, service orientation, and modularity (Manavalan and Jayakrishna 2019). This paves the way for Industry 5.0, and unlike Industry 4.0, "Industry 5.0 is not a technology-driven revolution but a value-driven initiative that drives technological transformation with a particular purpose" (Xu et al. 2021). Industry 5.0 centers around humans, sustainability, and resilience and relies upon technologies, such as recyclable smart materials, digital twins, and artificial intelligence among others. Recently, machine learning algorithms have been suggested as a novel method to analyze different aspects of the cold chain breaks, such as occurrence, temperature abuse level, and duration, contributing to the optimization of cold chain management, reduction of food waste and operating costs, and consumer safety assessment (Loisel et al. 2021).

Blockchain is a database that is shared among the nodes of a computer network where the information is stored in interconnected blocks. The adoption of this technology by the food industry is considered a top priority (Ali et al. 2021). Blockchain traceability of the cold chain cannot be only used to minimize food waste but also to prevent frauds, such as counterfeiting, dilution, or adulteration (Garaus and Treiblmaier 2021). Besides blockchain technology, cold chain monitoring and traceability can also benefit from other novel technologies, such as big data and cloud computing. For example, big data analysis services can be used for competitiveness advantage maximization through transparency, while cloud computing could enable real-time monitoring of foods, thus providing transparency of the product movement in the supply chain (Ali et al. 2021). The application of technologies such as blockchain, big data, IoT, wireless sensor networks, cloud computing, and machine learning can improve agri-food supply chain efficiency and quality management (Saurabh and Dey 2021).

9. Refrigerants for the Food Industry

The 2016 Kigali Amendment established the basis for reducing the production and consumption of hydrofluorocarbons (HFCs) in all countries involved in the Montreal Protocol. The controlled use of HFC adds to the previous phasing-out of ozone-depleting substances (ODS), including chlorofluorocarbons (CFCs) and hydrochlorofluorocarbons (HCFCs), which are greenhouse gases used worldwide in refrigeration, air-conditioning, and foam applications (Park et al. 2021). HFCs originally covered the need for non-ODS refrigerants; however, they have a very high global warming potential (GWP). The GWP indicates the amount of heat trapped by 1 ton of a gas relative to the amount of heat trapped by 1 ton of CO_2 over a specific time window (very often a century and is standardized by the Kigali Amendment). Most HCFCs and HFCs have GWPs that are much higher than that of CO_2. A list of common refrigerants for food applications with their status and uses is shown in Table 2, including ozone-depleting substances that are currently being phased out. Here, the ozone-depleting potential (ODP) is the relative amount of degradation to the ozone layer it can cause, where a value of 1 is assigned to the trichlorofluoromethane (R-11 or freon-11) (Roy and Halder 2020).

Regarding their safety, refrigerants are designated by an ASHRAE 34-based label: an initial A indicates relatively low toxicity with an occupational exposure limit value ≥ 400 ppm, while B indicates relatively high toxicity with an occupational exposure limit value ≤ 400 ppm. The second alphanumeric letter refers to the flammability of the refrigerant: class 1 (non-flammable), class 2L

Table 2. Common Refrigerants for Food Applications.

Refrigerant	Status(I)	ODP(II)	GWP(III)	Safety(IV)	Main Uses(V)	References
CFC-12	PO-ODS	1	8,500	A1	S	Roy and Halder (2020)
R-502	PO-ODS	0.328	4,590	A1	S	Kim (2017)
HCFC-22	PO-ODS	0.055	1,780	A1	C1, C2, T, I	Kudoma and Tekere (2021)
HFC-134a	PO-HFC	0	1,430	A1	S, C1, C2, T, I	Pardo-Martínez et al. (2021)
R-404A	PO-HFC	0	3,922	A1	S, C1, C2, T, I	Hart et al. (2020)
HFO-1234yf	EF-S	0	<1	A2L	S	Pabon et al. (2020)
R-455A	EF-SB	0	146	A2L	S	Llopis et al. (2019)
R-448A	EF-SB	0	1,273	A1	C1, C2	Jacob et al. (2020)
R-449A	EF-SB	0	1,282	A1	C1, C1	Makhnatch et al. (2017)
R-454A	EF-SB	0	238	A2L	C1, C2	Oruç and Devecioglu (2021)
R-452A	EF-SB	0	1,945	A1	T	Devecioğlu and Oruç (2021)
HFO-1234ze	EF-S	0	< 1	A2L	I	Jacob et al. (2020)
HFO-1233zd	EF-S	0	7	A1	I	Nair (2021)
HC-600A (isobutane)	EF-N	0	3	A3	S	Calleja-Anta et al. (2021)
HC-290 (Propane)	EF-N	0	3	A3	S, C1	Calleja-Anta et al. (2021)
R-744 (CO_2)	EF-N	0	1	A1	S, C1, C2, T, I	Gullo (2019)
R-717 (NH_3)	EF-N	0	0	B2L	I	Nawaz and Ally (2019)

(I) PO: phasing out; ODS: ozone-depleting substance; HFC: hydrofluorocarbon; EF: environmentally friendly; S: synthetic; SB: synthetic blend; N: natural. (II) ODP: ozone-depleting potential. (III) GWP: global warming potential. (IV) Safety: ASHRAE 34-based label. (V) S: domestic and commercial (deli counters, bottle coolers, and ice cream displays) small factory-built systems (charge 0.1–1 kg); C1: commercial multi-compressor central systems (displays and storage rooms for chilled or frozen foods) (charge 50–200 kg); C2: commercial condensing units (displays for chilled or frozen foods) (charge 5–10 kg); T: road transport and container refrigeration units (charge 5–10 kg); I: industrial, liquid chillers and direct expansion, flooded or pumped systems (charge 50–1,000 kg) (OzonAction 2017).

(low flammability), class 2 (flammable), class 3 (high flammability) (Roy and Halder 2020; Devotta et al. 2021). Most of the widespread HCFCs and HFCs have high GWP; thus, all applications should change to refrigerants such as ammonia, CO_2, hydrocarbons, and several recently introduced fluorocarbons (hydrofluoro-olefins or HFOs) that have ultra-low GWP (Nair 2021). Nevertheless, ultra-low GWP fluids are not adequate for all applications; for example, the flammability of HCs limits them to small, sealed refrigeration products, such as domestic refrigerators. Alternative refrigerants for HFCs must meet the criteria of high energy efficiency, safe operation, competitive investment and operating costs, and low environmental impact.

The specific charge (kg/kW), defined as the ratio between the refrigerant charge (kg) to the cooling capacity (kW), can be used to compare the magnitude order of several refrigeration systems according to their applicability (Poggi et al. 2008). Domestic systems (fridges and freezers) have specific charges ranging from 0.7 to 1 kg/kW when using HFCs, HCFCs, and CFCs and from 0.3 to 0.5 kg/kW for HCs. On the other hand, commercial systems (water chillers, direct chilling with direct expansion evaporators, and chilling by secondary refrigeration with direct expansion evaporator) have specific charges ranging from 0.2 to 3.4 kg/kW for HFCs, HCFCs, and CFCs and about 0.15 kg/kW for ammonia. Finally, the specific charge of industrial systems (direct freezing/chilling with direct expansion evaporators or with flooded systems, freezing/chilling by secondary refrigeration with direct expansion evaporator, and direct chilling with flooded evaporator) ranges from 0.5 to 5 kg/kW for HFCs, HCFCs, and CFCs and from 1 to 5.8 kg/kW for ammonia. The refrigerant charge of common refrigerants for food applications is also provided in Table 2.

10. Final Remarks

The increased consumer demand for both local and foreign high-quality foods has driven up the need for effective cooling systems along the food cold chain; however, refrigeration systems have a great carbon footprint because their operation is energy-intensive and the adoption of environmentally friendly refrigerants by the food industry still has a long way to go. Recent studies have focused on developing sustainable refrigeration technologies, including the use of low GWP refrigerants, renewable energy sources, optimization of route planning, and hub location for refrigerated transport and warehouses among others. Future research needs to center on the integration and adoption of intelligent systems, including IoT, blockchain, and other information technologies to improve the monitoring and traceability of the cold chain to minimize not only food waste and potential risks to consumers but also to better handle supply and demand tradeoffs and lower the greenhouse effects of the food cold chain.

References

Aalhus, J.L., Robertson, W.M., Dugan, M.E.R. and Best, D.R. 2002. Very fast chilling of beef carcasses. Can. J. Anim. Sci. 82: 59–67.

Abbas, K.A., Abdulamir, A.S., Ebrahimian, M. and Jamilah, B. 2009. Thermal properties variation of Malaysian yellowtail catfish during precooling process and numerical verification. J. Food Agric. Environ. 7: 196–201.

Ali, M.H., Chung, L., Kumar, A., Zailani, S. and Tan, K.H. 2021. A sustainable Blockchain framework for the halal food supply chain: lessons from Malaysia. Technol. Forecast. Soc. Change 170: 120870.

Artuso, P., Rossetti, A., Minetto, S., Marinetti, S., Moro, L. and Del Col, D. 2019. Dynamic modeling and thermal performance analysis of a refrigerated truck body during operation. Int. J. Refrig. 99: 288–299.

Boeing, 2012. Pallets and containers. Boeing, USA.

Bagheri, F., Fayazbakhsh, M.A. and Bahrami, M. 2017. Real-time performance evaluation and potential GHG reduction in refrigerated trailers. Int. J. Refrig. 73: 24–38.

Basediya, A.I., Samuel, D.V.K. and Beera, V. 2011. Evaporative cooling system for storage of fruits and vegetables—a review. J. Food Sci. Technol. 59: 429–442.

Belman-Flores, J.M., Pardo-Cely, D., Gómez-Martínez, M.A., Hernández-Pérez, I., Rodríguez-Valderrama, D.A. and Heredia-Aricapa, Y. 2019. Thermal and energy evaluation of a domestic refrigerator under the influence of the thermal load. Energies 12: 400.

Brownbridge, O., Sully, M., Noons, J. and Albayati, I.M. 2021. Modeling and simulation of a retail commercial refrigeration system. J. Therm. Sci. Eng. Appl. 13: 061028.

Buhr, R.J., Walter, J.M., Bourassa, D.V., Caudill, A.B., Kiepper, B.H. and Zhuang, H. 2014. Impact of broiler processing scalding and chilling profiles on carcass and breast meat yield. Poult. Sci. 93: 1534–1541.
Calleja-Anta, D., Nebot-Andres, L., Cabello, R., Sánchez, D. and R. Llopis. 2021. A3 and A2 refrigerants: border determination and hunt for A2 low-GWP blends. Int. J. Refrig. DOI: 10.1016/j.ijrefrig.2021.11.012.
Carciofi, B.A.M. and J.B. Laurindo. 2007. Water uptake by poultry carcasses during cooling by water immersion. Chem. Eng. Process. 46: 444–450.
Carroll, C.D. and Alvarado, C.Z. 2008. Comparison of air and immersion chilling on meat quality and shelf life of marinated broiler breast fillets. Poult. Sci. 87: 368–372.
Chaomuang, N., Laguerre, O. and Flick, D. 2020. A simplified heat transfer model of a closed refrigerated display cabinet. Therm. Sci. Eng. Prog. 17: 100494.
Chang, V.P., Mills, E.W. and Cutter, C.N. 2003. Reduction of bacteria on pork carcasses associated with chilling method. J. Food Prot. 66: 1019–1024.
Chen, L., Y. Liu and A. Langevin. 2019. A multi-compartment vehicle routing problem in cold-chain distribution. Comput. Oper. Res. 111: 58–66.
Chun-hua, W., Y. Chun-hong, Y. Xing-Qian, H. Ya-qin, C. Shi-Guo and L. Dong-hong. 2014. A critical review on superchilling preservation technology in aquatic product. J. Integr. Agric. 13: 2788–2806.
de Frias, J.A., Y. Luo, B. Zhou, E.R. Turner, P.D. Millner and X. Nou. 2018. Minimizing pathogen growth and quality deterioration of packaged leafy greens by maintaining optimum temperature in refrigerated display cases with doors. Food Control 92: 488–495.
Delele, M.A., Kuffi, K.D., Geeraerd, A., De Smet, S., Nicolai, B.M. and Verboven, P. 2019. Optimizing precooling of large beef carcasses using a comprehensive computational fluid dynamics model. J. Food Process Eng. 42: e13053.
Devecioğlu, A.G. and Oruç, V. 2021. Experimental comparison of R404A and R452A in refrigeration systems. Sci. Technol. Built Environ. 27: 61–69.
Devotta, S., Chelani, A. and Vonsild, A. 2021. Prediction of flammability classifications of refrigerants by artificial neural network and random forest model. Int. J. Refrig. 131: 947–955.
Duan, Y., Wang, G.-B., Fawole, O.A., Verboven, P., Zhang, X.-R., Wu, D. et al. 2020. Postharvest precooling of fruit and vegetables: a review. Trends Food Sci. Technol. 100: 278–291.
Dumont, M.J., Orsat, V. and Raghavan, V. 2016. Reducing Postharvest Losses. pp. 135–156. *In*: Madramootoo, C. (ed.). Emerging Technologies for Promoting Food Security. Woodhead Publishing, Massachusetts, USA.
Elansari, A.M. 2009. Design aspects in the precooling process of fresh produce. Fresh Prod. 3 (Special Issue 1): 49–57.
Esmizadeh, Y., Bashiri, M., Jahani, H. and Almada-Lobo, B. 2021. Cold chain management in hierarchical operational hub networks. Transport. Res. E: 147: 102202.
Evans, J., Curlin, J.S. and Clark, E. 2018. Cold Chain Technology Brief: Commercial, Professional and Domestic Refrigeration. International Institute of Refrigeration.
Fikiin, K., Stankov, B., Evans, J., Maidment, G., Foster, A., Brown, T. et al. 2017. Refrigerated warehouses as intelligent hubs to integrate renewable energy in industrial food refrigeration and to enhance power grid sustainability. Trends Food Sci. Technol. 60: 96–103.
Foster, A.M. and Evans, J.A. 2015. Overview of retail display in food retailing. pp. 1–16. *In*: Evans, J.A and Foster, A.M. (eds.). Sustainable Retail Refrigeration. Wiley Blackwell, New Jersey, USA.
Foster, A., Hammond, E., Brown, T., Maidment, G. and Evans, J. 2018. Technological options for retail refrigeration. International Institute of Refrigeration/London South Bank University.
Galgaro, A., Santa, G.D., Cola, S., Cultrera, M., De Carli, M., Conforti, F. et al. 2020. Underground warehouses for food storage in the Dolomites (Eastern alps - Italy) and energy efficiency. Tunn. Undergr. Space Technol. 102: 103411.
Gao, E., Cui, Q., Jing, H., Zhang, Z. and Zhang, X. 2021. A review of application status and replacement progress of refrigerants in the Chinese cold chain industry. Int. J. Refrig. 128: 104–117.
Garaus, M. and Treiblmaier, H. 2021. The influence of blockchain-based food traceability on retailer choice: the mediating role of trust. Food Control. 129: 108082.
Giarratana, F., Nalbone, L., Ziino, G., Giuffrida, A. and Panebianco, F. 2020. Characterization of the temperature fluctuation effect on shelf life of an octopus semi-preserved product. Ital. J. Food Saf. 9: 60–64.
Goedhals, L.L. and Khumalo, G. 2020. Identifying temperature breaks in the export cold chain of navel oranges: a Western Cape case. Food Control 110: 107013.
Golden, C.E. and Mishra, A. 2021. Assessing the risk of salmonellosis from consumption of conventionally and alternatively produced broiler meat prepared in-home in the United States. Microb. Risk Anal.: 100160.
Goel, R.K., Singh, B. and Zhao, J. 2012. Underground Infrastructures Planning, Design, and Construction. Butterworth-Heinemann, Massachusetts, USA.
Gullo, P. 2019. Innovative fully integrated transcritical R744 refrigeration systems for a HFC-free future of supermarkets in warm and hot climates. Int. J. Refrig. 108: 283–310.
Gürbüz, Ü., Telli, A.E., Kahraman, H.A., Balpetekkülcü, D. and Yalçın, S. 2018. Determination of microbial contamination, pH and temperature changes in sheep and cattle carcasses during the slaughter and pre-cooling processes in Konya, Turkey. Ital. J. Food Sci. 30: 828–839.

Gurman, P.M., Ross, T. and Kiermeier, A. 2018. Quantitative microbial risk assessment of salmonellosis from the consumption of Australian pork: minced meat from retail to burgers prepared and consumed at home. Risk Analysis 38: 2625–2645.

Hafner, A., Gabrielii, C.H. and Widell, K. 2018. Refrigeration units in marine vessels. Alternative to HCFCs and high GWP HFCs. Nordic Council of Ministers, Denmark.

Han, J.-W., Zuo, M., Zhu, W.-Y., Zuo, J.-H., Lü, E.-L. and Yang, X.-T. 2021. A comprehensive review of cold chain logistics for fresh agricultural products: current status, challenges, and future trends. Trends Food Sci. Technol. 109: 536–551.

Hart M., Austin, W., Acha, S., Brun, N.L., Markides, C.N. and Shah, N. 2020. A roadmap investment strategy to reduce carbon-intensive refrigerants in the food retail industry. J. Clean. Prod. 275: 123039.

Hueppe, C., Geppert, J., Stamminger, R., Wagner, H., Hoelscher, H., Vrabec, J. et al. 2020. Age-related efficiency loss of household refrigeration appliances: development of an approach to measure the degradation of insulation properties. Appl. Therm. Eng. 173: 115113.

Huff-Lonergan, E. 2001. The role of carcass chilling in the development of pork quality. National Pork Producers Council, Des Moines, IA, USA.

Jacob, T.A., E.P. Matty and B.M. Fronk. 2020. Comparison of R404A condensation heat transfer and pressure drop with low global warming potential replacement candidates R448A and R452A. Int. J. Refrig. 116: 9–22.

Kim, N.H. 2017. Optimization of the water spray nozzle, refrigerant charge amount and expansion valve opening for a unitary ice maker using R-404A. Int. J. Air-Cond. Refrig. 25: 1750025.

King, H. 2020. Food Safety Management Systems. Achieving Active Managerial Control of Foodborne Illness Risk Factors in a Retail Food Service Business. Springer, Cham, Switzerland.

Kongwong, P., Boonyakiat, D. and Poonlarp, P. 2021. Extending the shelf life and qualities of baby cos lettuce using commercial precooling systems. Postharvest Biol. Tec. 150: 60–70.

Kudoma, B. and Tekere, M. 2021. Exploration of the environmental and socioeconomic implications of HCFC-22 phase-out for Botswana, Adv. Clim. Chang. Res. 12: 108–118.

Kyriakides, A. 2014. An Overview of Retail Food Hygiene in Europe. pp. 81–131. *In*: Farber, J., Crichton, J. and Snyder, P., Jr. (eds.). Retail Food Safety, Springer, New York, USA.

Laguerre, O. and Chaomuang, N. 2021. Closed refrigerated display cabinets: Is it worth it for food quality? pp. 99–121. *In*: M. Khosrow-Pour (ed.). Research Anthology on Food Waste Reduction and Alternative Diets for Food and Nutrition Security, IGI Global, Pennsylvania, USA.

Li, G. 2017. Comprehensive investigation of transport refrigeration life cycle climate performance. Sustain. Energy Technol. 21: 33–9.

Li, G. 2021. Development of cold chain logistics transportation system based on 5G network and Internet of things system. Microprocess. Microsyst. 80: 103565.

Llopis, R., D. Calleja-Anta, D. Sánchez, L. Nebot-Andrés, J. Catalán-Gil and R. Cabello. 2019. R-454C, R-459B, R-457A and R-455A as low-GWP replacements of R-404A: experimental evaluation and optimization. Int. J. Refrig. 106: 133–143.

Loisel, J., Duret, S., Cornuéjols, A., Cagnon, D., Tardet, M., Derens-Bertheau, E. et al. 2021. Cold chain break detection and analysis: can machine learning help? Trends Food Sci. Technol. 112: 391–399.

Lorenzetti, E., Puton, B.M.S., Fernandes, I.A., do Prado, N.V., Frigotto, R., Román, S.S. et al. 2018. Water absorption and dripping of chicken breast and carcasses during pre-cooling in an industrial system. Poult. Sci. 97: 4462–4469.

Lorenzo, R.A., Tapella, F. and Romero, M.C. 2020. Pre-cooling effect on live transport of the Southern king crab, Lithodes santolla. Fish. Res. 227: 105552.

Majewski, E., Komerska, A., Kwiatkowski, J., Malak-Rawlikowska, A., Wąs, A., Sulewski, P. et al. 2020. Are Short Food Supply Chains More Environmentally Sustainable than Long Chains? A Life Cycle Assessment (LCA) of the Eco-Efficiency of Food Chains in Selected EU Countries. Energies 13: 4853.

Makhnatch, P., Mota-Babiloni, A., Rogstam, J. and Khodabandeh, R. 2017. Retrofit of lower GWP alternative R449A into an existing R404A indirect supermarket refrigeration system. Int. J. Refrig. 76: 184–192.

Manavalan, E. and Jayakrishna, K. 2019. A review of Internet of Things (IoT) embedded sustainable supply chain for industry 4.0 requirements. Comput. Ind. Eng. 127: 925–953.

Margeirsson, B., Magnússon, H., Sveinsdóttir, K., Valtýsdóttir, K.L., Reynisson, E. and Arason, S. 2010. The effect of different precooling media during processing and cooling techniques during packaging of cod (Gadus morhua) fillets. Skýrsla Matís 15–10: 1–27.

Mazarrón, F.R., Cid-Falceto, J. and Cañas, I. 2012. An assessment of using ground thermal inertia as passive thermal technique in the wine industry around the world. Appl. Therm. Eng. 33–34: 54–61.

McGregor, B.M. 1987. Tropical Products Transport Handbook. U.S. Department of Agriculture, Agriculture Handbook No. 668. p. 148.

Meneghetti, A., Pagnin, C. and Simeoni, P. 2021. Decarbonizing the cold chain: long-haul refrigerated deliveries with on-board photovoltaic energy integration. Sustainability 13: 8506.

Meneghetti, A., Da Rold, G. and Cortella, G. 2018. Sustainable refrigerated food transport: searching energy efficient routes. IFAC-PapersOnLine 51–11: 618–623.

Mishra, A., Guo, M., Buchanan, R.L., Schaffner, D.W. and Pradhan, A.K. 2017. Prediction of Escherichia coli O157: H7, Salmonella, and Listeria monocytogenes growth in leafy greens without temperature control. J. Food Prot. 80: 68–73.

Murrant, D. and Radcliffe, J. 2018. Analysis of when and where the integration of LAES with refrigerated warehouses could provide the greatest value to Europe. Energy Procedia 151: 144–149.
Nair, V. 2021. HFO refrigerants: a review of present status and future prospects. Int. J. Refrig. 122: 156–170.
Nawaz, K. and Ally, M.R. 2019. Options for low-global-warming-potential and natural refrigerants Part 2: Performance of refrigerants and systemic irreversibilities. Int. J. Refrig. 106: 213–224.
Ndraha, N., Sung, W.-C. and Hsiao, H.-I. 2019. Evaluation of the cold chain management options to preserve the shelf life of frozen shrimps: a case study in the home delivery services in Taiwan. J. Food Eng. 242: 21–30.
Nkolisa, N., Magwasa, L.S., Workneh, T.S. and Chimpango, A. 2018. Evaluating evaporative cooling system as an energy-free and cost-effective method for postharvest storage of tomatoes (Solanum lycopersicum L.) for smallholder farmers. Sci. Hortic. 241: 131–143.
Omara, A.A. and Mohammedali, A.A. 2020. Thermal management and performance enhancement of domestic refrigerators and freezers via phase change materials: a review. Innov. Food Sci. Emerg. Technol. 66, 102522.
Oruç V. and Devecioglu, A.G. 2021. Experimental investigation on the low-GWP HFC/HFO blends R454A and R454C in a R404A refrigeration system. Int. J. Refrig. 128: 242–251.
Óskarsdóttir, K. and Oddsson, G.V. 2019. Towards a decision support framework for technologies used in cold supply chain traceability. J. Food Eng. 2040: 153–159.
Ovca, A., Škufca, T. and Jevšnik, M. 2021. Temperatures and storage conditions in domestic refrigerators-Slovenian scenario. Food control 123: 107715.
OzonAction. 2017. OzonAction Kigali Fact Sheet No. 4 - Low GWP Fluids and Technology. UN Environment Programme.
Pabon, J.J.G., Khosravi, A., Belman-Flores, J.M., Machado, L. and Revellin, R. 2020. Applications of refrigerant R1234yf in heating, air conditioning and refrigeration systems: a decade of researches. Int. J. Refrig. 118: 104–113.
Pardo-Martínez, C.I., Cotte-Poveda, A. and Torralba-Barreto, D.R. 2021. Diagnosis and baseline refrigerant use in the Colombian meat, dairy, and fruit and vegetable industries. Int. J. Refrig. 131: 448–458.
Park, W.G., Shah, N., Vine, E., Blake, P., Holuj, B., Kim, J.H. et al. 2021. Ensuring the climate benefits of the Montreal Protocol: Global governance architecture for cooling efficiency and alternative refrigerants. Energy Res. Soc. Sci. 76: 102068.
Poggi, F., Macchi-Tejeda, H., Leducq, D. and Bontemps, A. 2008. Refrigerant charge in refrigerating systems and strategies of charge reduction. Int. J. Refrig. 31: 353–370.
Rai, A. and Tassou, S.A. 2017. Environmental impacts of vapour compression and cryogenic transport refrigeration technologies for temperature-controlled food distribution. Energy Convers. Manag. 99: 288–299.
Rao, C.G. 2016. Engineering for Storage of Fruits and Vegetables. Academic Press. USA.
Ravikumar, M., Desai, C.S., Raghavendra, H.R. and Pooja, N. 2018. Effect of pre-cooling in extending the shelf life of banana cv. Grand naine stored under different storage conditions. Int. J. Chem. Stud. 6: 872–878.
Raza, H.M.U., Sultan, M., Bahrami, M. and Khan, A.A. 2021. Experimental investigation of evaporative cooling systems for agricultural storage and livestock air-conditioning in Pakistan. Build. Simul. 14: 617–631.
Reguera, C., Sanllorente, S., Herrero, A., Sarabia, L.A. and Cruz Ortiz, M. 2019. Detection of cold chain breaks using partial least squares-class modelling based on biogenic amine profiles in tuna. Talanta 202: 443–451.
Rehman, D., McGarrigle, E., Glicksman, L. and Verploegen, E. 2020. A heat and mass transport model of clay pot evaporative coolers for vegetable storage. Int. J. Heat Mass Transf. 162: 120270.
Ronoh, E.K., Kanali, C.L. and Ndirangu, S.N. 2020. Effectiveness of an evaporative charcoal cooler for the postharvest preservation of tomatoes and kales. Res. Agric. Eng. 66: 66–71.
Roy, Z. and Halder, G. 2020. Replacement of halogenated refrigerants towards sustainable cooling system: a review. Adv. Chem. Eng. 3: 100027.
Salehy, Y., Hoang, H.M., Cluzel, F., Leroy, Y., Delahaye, A., Fournaison, L. et al. 2020. Energy performances assessment for sustainable design recommendations: case study of a supermarket's refrigeration system. Procedia CIRP 90: 328–333.
Saurabh, S. and Dey, K. Blockchain technology adoption, architecture, and sustainable agrifood supply chains. J. Clean. Prod. 284: 124731.
Shahzad, K., Sultan, M., Bilal, M., Ashraf, H., Farooq, M., Miyazaki, T. et al. 2021. Experiments on energy-efficient evaporative cooling systems for poultry farm application in Multan (Pakistan). Sustainability 13: 2836.
Shanmugam, S.K.G. and Mital, M. 2019. An ultra-low ammonia charge system for industrial refrigeration. Int. J. Refrig. 107: 344–354.
Sibanda, S. and Workneh, T.S. 2020. Performance evaluation of an indirect air cooling system combined with evaporative cooling. Heliyon 6: e03286.
Stanton, J.L. 2018. A brief history of food retail. Br. Food J. 120: 172–180.
Stellingwerf, H.M., Kanellopoulos, A., van der Vost, J.G.A.J. and Bloemhof, J.M. 2018. Reducing CO_2 emissions in temperature-controlled road transportation using the LDVRP model Transport. Res. D: 58: 80–93.
Sunano, Y. 2020. Traditional storage pit, polota, for storing sorghum as a long-term survival strategy in Dirashe special Worenda, Southern Ethiopia. Agroecol. Sustain. Food Syst. 44: 536–559.
Talbot, L., Purnell, G., James, S.J. and James, C. 2020. Operating temperatures of supermarket frozen retail display cabinets. Int. J. Refrig. 117: 81–93.
Tanner, D. 2016. Refrigerated transport. Reference Module in Food Science 1–7.

Tassou, S.A., De-Lille, G. and Ge, Y.T. 2009. Food transport refrigeration approaches to reduce energy consumption and environmental impacts of road transport. Appl. Therm. Eng. 29: 1467–1477.

Valtýsdóttir, K.L., Margeirsson, B., Arason, S., Lauzon, H.L. and Martinsdóttir, E. 2010. Guidelines for precooling of fresh fish during processing and choice of packaging with respect to temperature control in cold chains. Skýrsla Matís 40–10: 1–37.

Vithanage, N.R., Dissanayake, M., Bolge, G., Palombo, E.A., Yeager, T.R. and Datta, N. 2017. Microbiological quality of raw milk attributable to prolonged refrigeration conditions. J. Dairy Res. 84: 92–101.

Vivaldi, F., B. Melai, A. Bonini, N. Poma, P. Salvo, A. Kirchhain et al. 2020. A temperature-sensitive RFID tag for the identification of cold chain failures. Sensors and Actuators A: Physical 313: 112182.

Wang, G.-B. and X.-R. Zhang. 2021. Thermodynamic evaluation of heat transfer heterogeneity: comparative case studies on air cooling methods for postharvest apples. Case Stud. Therm. Eng. 24: 100832.

Wang, X., Feng, H., Chen, T., Zhao, S., Zhang, J. and Zhang, X. 2021. Gas sensor technologies and mathematical modelling for quality sensing in fruit and vegetable cold chains: a review. Trends Food Sci. Technol. 110: 483–492.

Wu, X., Hu, S. and Mo, S. 2013. Carbon footprint model for evaluating the global warming impact of food transport refrigeration systems. J. Clean. Prod. 115–124.

Wu, W., Cronjé, P., Nicolai, B., Verboven, P., Opara, U.L. and Defraeye, T. 2018. Virtual cold chain method to model the postharvest temperature history and quality evolution of fresh fruit—A case study for citrus fruit packed in a single carton. Comput. Electron. Agric. 144: 199–208.

Xie, Y., Brecht, J.K., Abrahan, C.E., Bornhorst, E.R., Luo, Y., Monge, A.L. et al. 2021. Improving temperature management and retaining quality of fresh-cut leafy greens by retrofitting open refrigerated retail display cases with doors. J. Food Eng. 292: 110271.

Xu, X., Lu, Y., Vogel-Heuser, B. and Wang, L. 2021. Industry 4.0 and Industry 5.0-Inception, conception and perception. J. Manuf. Syst. 61: 530–535.

Yang, Z., Tate, J.E., Morganti, E. and Shepherd, S.P. 2021. Real-world CO_2 and NO_X emissions from refrigerated vans. Sci. Total Environ. 763: 142974.

Zhang, L.-Y., Tseng, M.-L., Wang, C.-H., Xiao, C. and Fei, T. 2019. Low-carbon cold chain logistics using ribonucleic acid-ant colony optimization algorithm. J. Clean. Prod. 233: 169–180.

Zhang, S., Chen, N., She, N. and Li, K. 2021. Location optimization of a competitive distribution center for urban cold chain logistics in terms of low-carbon emissions. Comput. Ind. Eng. 154: 107120.

Zhao, H., liu, S., Tian, C., Yan, G. and Wang, D. 2018. An overview of current status of cold chain in China. Int. J. Refrig. 88: 483–495.

Zubeldia, B.B., Jiménez, M.N., Claros, M.T.V., Andrés, J.L.M. and Martin-Olmedo, P. 2016. Effectiveness of the cold chain control procedure in the retail sector in Southern Spain. Food Control. 59: 614–618.

9

Food Freezing

Transformation of Conventional Technology to Smart Processing

Kyuya Nakagawa[1,*] and *Shinji Kono*[2]

1. Food Preservation as a Sustainable Action

1.1 SDGs and Food Tech

Freezing and drying are the most classic and commonly accepted methods for food preservation that avoid spoilage. Both these methods improve shelf life by controlling the mobility of water in foods. This was achieved by solidification during freezing and dehydration during drying. Freeze-drying (lyophilization) is a technique in which water in the food system is frozen once and then dehydrated via sublimation. This process avoids the structural damage caused by the evaporation of water and preserves the original structure of the food, unlike other dehydration methods. These methods are practices that have been passed down since ancient times, but in terms of industrialization technology, they have only been around for approximately a century. In the context of addressing recent global problems, freezing is an essential technique in the future food system to maximize food functionality while minimizing food losses because it is the most powerful and widely accepted preservation technology.

The Intergovernmental Panel on Climate Change (IPCC) estimates that the global food system is responsible for 21–37% of greenhouse gas (GHG) emissions. This amount is equivalent to one-third of the total GHG emissions (Crippa et al. 2021). Here, the food system covers land use, production (e.g., livestock, fishery, and agriculture), post-production (e.g., secondary processing and post-harvest processing), and consumption (e.g., household cooking and retail). Food loss and waste are generated during each stage. Food loss refers to food that is thrown away even though it could be eaten. Notably, over 10% of the world's total energy consumption is used to provide food that is lost or wasted (da Silva 2017). Therefore, reducing food loss can significantly contribute to reducing GHG emissions.

[1] Kyoto University, Japan.
[2] Mayekawa MFG. Co., Ltd., Japan.
* Corresponding author: nakagawa.kyuya.8e@kyoto-u.ac.jp

The U.S. Department of Agriculture (USDA) estimates that food loss and waste from the food supply at the retail and consumer levels account for 31% of the food supply in the United States. This amount corresponds to approximately 60 million tons. According to statistics from the Japanese Ministry of Agriculture, Forestry and Fisheries (MAFF), the food loss rate in households and restaurants is approximately 4%, and the loss rate in banquets is as high as approximately 13%. The total amounts of household and retail food losses are approximately 6 and 3 million tons per year, respectively. The Japanese government has set a policy to reduce the overall amount of food loss to 5.5 million tons or less by 2030 and has passed legislation to reduce food loss (Food Loss Reduction Promotion Act 2019) is in progress. These activities are in response to the second Sustainable Development Goals (SDGs), which aim to reduce food loss through agriculture and food systems. SDG 2 has eight targets (end hunger, food production and poverty, food and health, food security and climate change, food and gender equality, agriculture and labor, etc.), which call for "profound changes in the global food and agriculture system" and "a complete 'rethinking' of how we grow, share, and consume food" (Rickards and Shortis 2019). How can freezing technology be improved to meet these demands?

First, a correct assessment is required to determine whether freezing contributes to a reduction in carbon emissions. This is largely dependent on the freezing scenario. Second, we must consider how innovations in freezing technology can rewrite these scenarios. It is conceivable that innovative freezing technologies can alter food systems. However, in the authors' view, innovation in the way freezing technology (even conventional freezing) is used in food systems is more significant. Recent smart technologies will play an important role in building innovative food systems that generate low amounts of food loss and waste.

1.2 Life Cycle Assessment on Food Refrigeration

At first glance, transporting frozen food from its place of production to other parts of the world may seem to have negative environmental impacts. Pelletier and Tyedmers et al. reported a life cycle assessment (LCA) of frozen tilapia products farmed in Indonesian lakes and exported to the other countries (Pelletier and Tyedmers 2010). According to their results, the most important leverage point is the method of feed procurement, and the energy consumption to maintain water quality is also a major factor in reducing environmental performance. The study concluded that, despite long-distance transportation, the environmental impact of freeze-packed tilapia exported to Europe and North America is negligible when it is transported by sea in containers.

According to the LCA results on frozen octopuses from Mauritania by Vazquez-Rowe et al. (2012), the energy intensity of fisheries has a significant environmental impact. Post-harvest operations are reported to have a low impact on frozen octopuses, regardless of the export route when sea transport is chosen. The ability of the EU countries and Japan to expand octopus catches is limited and imports from developing countries, such as Mauritania, continue as a result. Even if intensive transportation by air cargo is advantageous in the market from the perspective of environmental impact, the shelf life needs to be extended through innovation in manufacturing technology, such as better frozen products, which will lead to a reduction in the use of resources during transportation. Refrigeration devices that consume excessive energy to achieve over-specified quality are not desirable, and devices or systems that can achieve the minimum energy input to achieve the required quality will be required for reducing the environmental impact of food production. This is supported by several LCA studies (Ingwersen 2012).

1.3 Expectations for Freezing Technology

This chapter focuses on freezing technologies that have the potential to contribute to an emerging smart society. According to the current reports of LCA results, energy-efficient freezers can contribute to a reduction in GHG emissions through the food chain. However, further reduction is required.

As reported, approximately 14% of the global agricultural production is lost during the post-harvest process and distribution (FAO). Approximately 8% of the world's annual food supply is wasted by households and restaurants in their daily meals (Lovelock et al. 2019). As freezing is a means of withholding time-of-use, improving the flexibility of food use may contribute to the reduction of these food losses. For instance, after harvesting, some fruits and vegetables cannot be shipped when their maturity exceeds a certain level, considering the risk of spoilage during the distribution process. According to interviews with farmers in Thailand, the amount of edible agricultural waste was not small, and several farmers wanted to convert it to processed products. However, ordinary farmers have few links with the intensive processing industry, and securing the necessary human resources for processing, which is conducted irregularly in combination with harvesting, is a barrier to processing. Thus, it would be more practical to be able to withhold work for a certain period, e.g. by freezing.

For this purpose, the conditions for proper freezing of each food should be systematized so that several variations of the food can be handled. Is there a sufficient basic science involved in freezing to realize this? Is there sufficient engineering to build a system to achieve this? Is such a system feasible for large-scale manufacturers and home cooks to help themselves? The following chapters introduce the fundamental aspects of freezing, which are important even though they are not yet fully understood. The engineering technologies that the authors believe are necessary for future freezing engineering are also introduced.

2. Recent Progress in the Freezing Process

2.1 Heat Transfer During Freezing

Freezing is a heat transfer process that involves a transition between the liquid and solid phases. The freezing progress can be expressed using a simplified model, as shown in Fig. 1. The process of ice crystal growth and freezing front progress in response to the heat flow in the system is presented in three zones: unfrozen zone, mashy zone (i.e., ice dispersed in the liquid phase), and frozen zone. Each zone has a different heat capacity (i.e., C_{p-uf}, C_{p-m}, C_{p-f}, and C_{p-f}, respectively) and thermal conductivity (i.e., k_{uf}, k_m, and k_f respectively). We assume that a linear temperature gradient ΔT and a thermal conductivity gradient are formed inside the mashy zone, and the fraction of suspended ice in contact with the unfrozen zone is zero and that in contact with the frozen zone has a maximum value. Therefore, the apparent heat capacity and thermal conductivity can be applied to each zone. In the mashy zone, heat is generated by the freezing enthalpy corresponding to the number of ice crystals produced (X_{ice}). The heat transfer between the external surface of the product and the contact air can be described by the heat transfer coefficient in the gas film layer. By applying these parameters and equations to the heat conduction equation, we can simulate a freezing process that proceeds dynamically. Some examples of the simulation results can be found in previous publications (Nakagawa et al. 2007; Nakagawa et al. 2010).

Ice nucleation and ice crystal growth during freezing are among the main causes of damage to food that is consumed after thawing. The formation of ice microstructures in food depends on the freezing rates and temperature gradients formed in the product. It is empirically known that the mean ice crystal size () formed in the freezing layer has the following relationship with respect to the progress speed of the freezing interface () and the temperature gradient of the freezing layer () (Kurz and Fisher 1986): , where , , and are empirical coefficients. This correlation has been studied in metal alloy solidification systems, but it can also be applied to food-freezing systems (van der Sman et al. 2013). For example, correlations for apples (Bomben and King, 1982), for starch gels (Reid 1982), and or for the unidirectional solidification of gelatin gels (Woinet et al. 1998) have been proposed. Miyawaki et al. reported a relationship of for the freezing of soy protein curd gels (Fig. 2) in the range of freezing front velocity about 3–15 cm/h corresponding to the mean ice crystal size around 50–350 μm (Miyawaki et al. 1992). Even when a product is cooled at a constant rate,

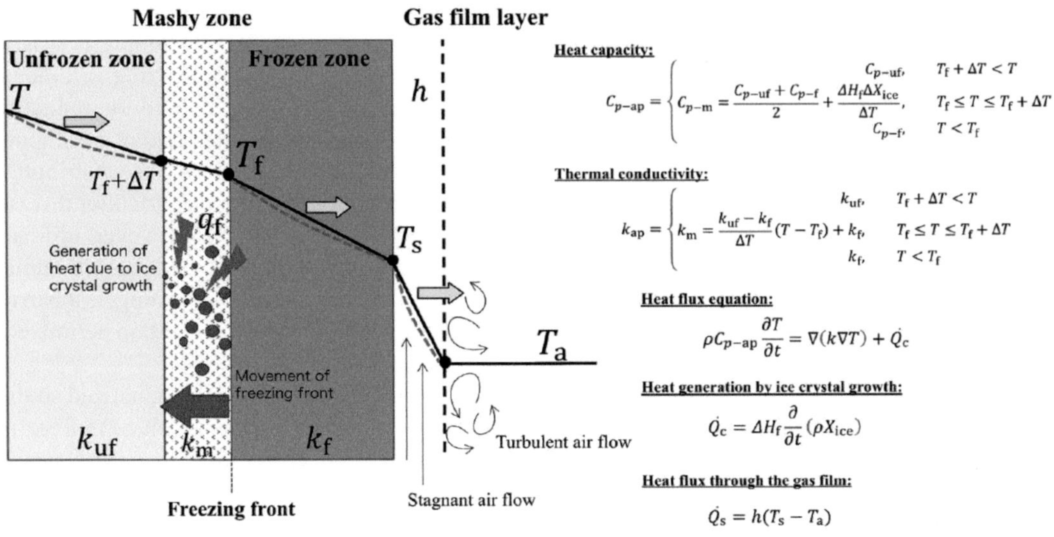

Figure 1. Schematic of the Progress of Freezing.

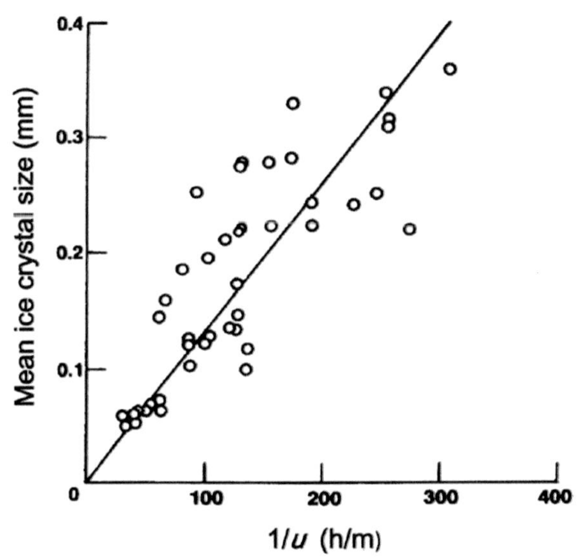

Figure 2. Correlation Between the Freezing Front Velocity (u) and Mean Ice Crystal Size Formed in Soy Protein Curd (Miyawaki et al. 1992).

the freezing rate and internal temperature gradients change over time, mainly because of changes in the frozen zone thickness. Consequently, this inhomogeneity influences the size distribution of ice crystals in frozen products. In addition, the formation of ice crystals proceeds quickly during the breakage of supercooling; therefore, small ice crystals are easily formed in the frozen region formed from the supercooled zones. The physics governing the formation of ice crystals is not yet well understood. Recent progress has been summarized in a review by van der Sman et al. (2013).

2.2 Innovations of Freezing Process by Heat Transfer Enhancement

As mentioned earlier, freezing is a heat transfer process, and the external surface area of the product affects the heat transfer through the external boundary film contacting the cooling medium, whereas

Table 1. Features of Industrial Freezers.

	Technical Feature	Specification
Air-blast freezing	Conventional technique by using cold air	Air velocity: 2–5 m/s Heat transfer coefficient: 20–55 $Wm^{-2}K^{-1}$
Impingement jet freezing	Improvement of heat transfer coefficient at the product surface by using cryogenic fluid jet	Air velocity: 15–45 m/s Heat transfer coefficient: 100–200 $Wm^{-2}K^{-1}$
Immersion freezing	Conventional technique by using liquid as cryogenic fluid	Heat transfer coefficient: 400–500 $Wm^{-2}K^{-1}$
Hydrofluidization	Improvement of heat transfer coefficient by strong agitation in the liquid cryogenic fluid. Commonly used to produce IQF foods.	

the geometry of the product affects the conductive heat flow inside it. Therefore, innovations in freezing technologies are mainly directed toward improving heat transfer properties. Given that rapid freezing is advantageous for the formation of small ice crystals and for maintaining the quality of the final product, improving the heat transfer velocity is a major challenge. However, as one can do little to alter the heat conduction inside the product directly, most techniques are used to enhance the heat transfer at the external boundary layer. Table 1 summarizes the features of freezers that are related to heat transfer enhancement, all of which have been commercially used.

Air-blast freezers are the most commonly used continuous freezers with conveyors installed inside the cold chambers (Fig. 3). The air velocity is in the range of 2–5 m/s, and the heat transfer coefficients at the product surface are approximately 20–55 $W/(m^2K)$. Impingement jet freezing is the process of applying cooling fluid jets to the surface of a product to enhance its convective heat transfer characteristics. The fluid velocity was set to approximately 15 m/s, and the heat transfer coefficients were more than 100 $W/(m^2K)$. Compared with air-blast freezers, impingement freezers can be operated at higher air temperatures while still achieving the same performance, resulting in shorter freezing times, lower weight loss, and lower operating costs (James et al. 2015; Salvadori and Mascheroni 2002).

2.3 Control of Supercooling, Nucleation, and Ice Crystal Growth

Ice crystal nucleation in a supercooled solution is a stochastic phenomenon, and it cannot be easily controlled with external devices. Undesirable continuation and random breakage of the supercooling state seriously affect the heat transfer (i.e., the reduction of cooling rate owing to the generation of enthalpy of solidification) in the freezing operation. As mentioned in the previous section, one of the reasons for the improvement in heat transfer characteristics in immersion freezing is that the effect of supercooling can be avoided by the suspended ice crystals acting as seed crystals. Therefore, a technique to control the breaking temperature of supercooling would provide the desired control of the heat transfer process during freezing. The breaking temperature of supercooling also affects subsequent ice crystal nucleation and growth processes. As is widely known, in most cases, the formation of small ice crystals in frozen products is beneficial for maintaining the quality of the product after thawing. Studies that have attempted to control supercooling to control the formation of ice crystals in frozen products have provided several insights. Nakagawa et al. employed ultrasound to control ice nucleation in a supercooled solution in vials and reported the relationship between the breaking temperature of supercooling and the microstructure formation of ice crystals in the frozen solution (Hottot et al. 2008; Nakagawa et al. 2006). They showed that the mean ice crystal size formed from the region that was supercooled was small and that the ice crystal size in the region that was subsequently frozen was large (Fig. 4). This can be explained by the freezing front velocity during freezing, which was verified through simulations based on the freezing progress model (Nakagawa et al. 2007). These studies also predicted that the variations in microstructure formation

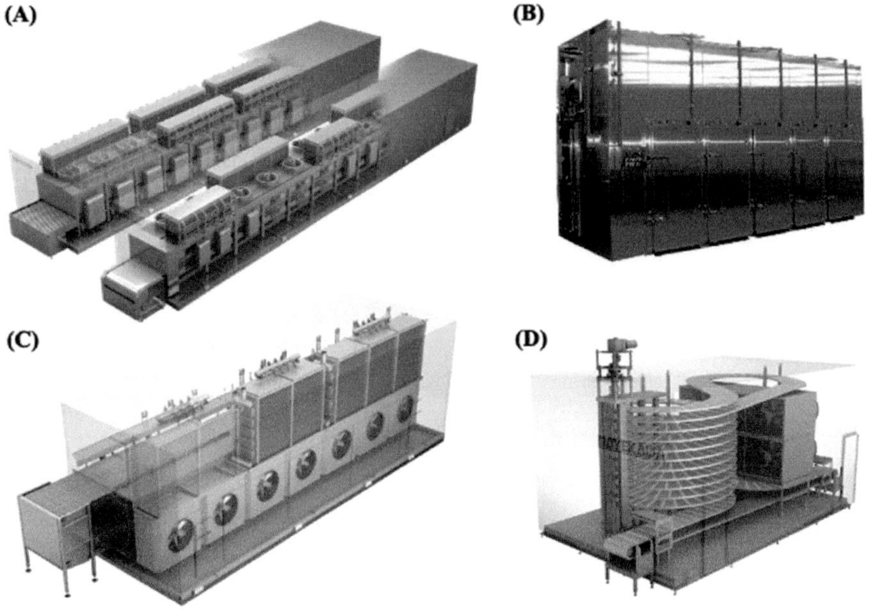

Figure 3. Illustrations of Industrial Freezers Reproduced With Kind Permission From Mayekawa MFG. Co., Ltd. (https://www.mayekawa.com), straight continuous tunnel freezer (A); batch freezer (B); fluidized bed freezer (C); spiral freezer (D).

hydrofluidization freezing is a technique in which the product is cooled in a fluidized medium to enhance the heat transfer (Dirita et al. 2007; Góral and Kluza 2006, 2012; Marazani et al. 2017; Sarkar et al. 2004; Sarkar and Singh 2004; Soto and Borquez 2001).

Air fluidization has been used as a conventional method, whereas fluidization with a liquid cooling medium is recognized as a recent innovation. Hydrofluidization freezing is a kind of immersion freezing technique in which a strong turbulent flow created by the circulating cooling medium that passes through an orifice or nozzle enhances the heat transfer coefficient. For example, the heat transfer coefficient achieved in immersion freezing operated under free convection is in the range of approximately 200–300 W/(m2K) (Lucas and Raoult-Wack 1998). However, this value can be improved to more than 900 W/(m2K) by applying forced convection (Fikiin 1992; Fikiin 2008). An aqueous NaCl solution is the most common cooling medium used for immersion freezing. Other solutions, such as CaCl2, MgCl2, ethanol, and ethylene glycol, can also be used. The use of ternary systems, such as water/ethanol/NaCl, is also advantageous for maintaining the low viscosity of the cooling medium and improving its heat transfer characteristics (Lucas and Raoult-Wack 1998). The use of ice slurry (i.e., a suspension of ice crystals in a NaCl solution) in hydrofluidization freezing can realize extremely high heat transfer coefficients on the order of 1,000–2,000 W/(m2K) (Fikiin and Fikiin 1999). Notably, this improvement is a contribution of the seeding action of the ice in the solution, which decreases the supercooling phenomena in the product to be frozen and consequently accelerates heat conduction (Lucas et al. 1998; Lucas and Raoult-Wack 1998). Hydrofluidization freezing is commercially accepted to produce individually quick-frozen (IQF) foods, which are foods frozen in pieces rather than in bulk, not only to secure rapid freezing but also to improve usability. IQF is widely used in the production of frozen fruits and vegetables and has been extensively studied, most notably by Fikiin et al. (Fikiin and Fikiin 1999; 2000), and validated in terms of quality attributes (Choi et al. 2017; Palacz et al. 2019; Palacz et al. 2021; Stebel et al. 2020; Stebel et al. 2021; Verboven et al. 2003).

owing to differences in the breaking temperature of supercooling are smaller when fast-freezing conditions are applied. Ice crystal microstructures formed by the breakage of the supercooling state have been reported to have a positive effect on food quality, contributing to the suppression of drip loss after freezing and thawing (Kobayashi et al. 2014).

Pressure-assisted freezing (pressure-shift freezing) is a freezing method in which products are rapidly and adiabatically released from a high-pressure state at subzero temperatures. The equilibrium freezing point temperature increases accordingly, causing the water in the ingredients to become supercooled instantaneously, and supercooling is immediately broken to produce ice crystals (Cheftel et al. 2002; LeBail et al. 2002; Petzold and Aguilera 2009). Several reports on

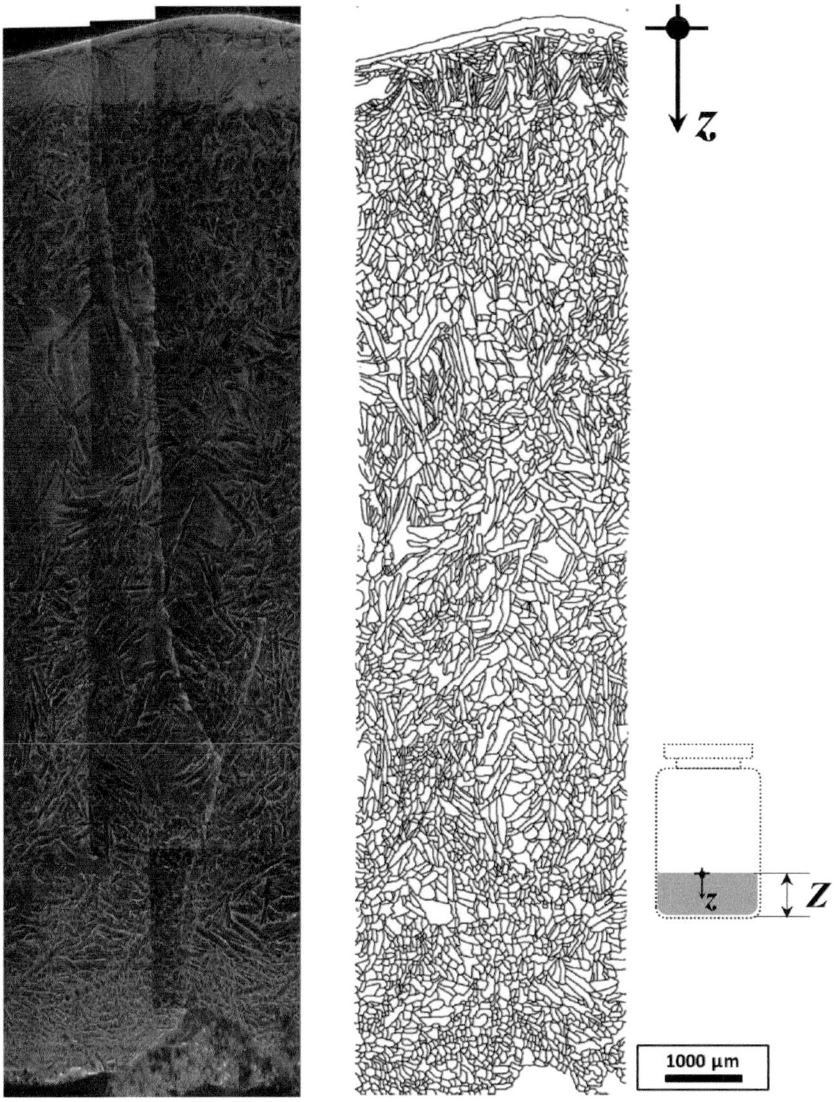

Figure 4. Formation of Ice Crystals Via the Directional Freezing of a Solution in a Vial (Goshima et al. 2016).

pressure-assisted freezing have been published, and their positive impact on food quality has been verified. The details have been summarized in review papers (Cheng et al. 2017; James et al. 2015).

2.4 Freezing Process Intensification in Combination with External Device

Emerging freezing techniques assisted by ultrasound, electrostatic fields, pulsed electric fields, microwaves, radio frequencies, etc., have been reported to control the formation of ice crystals during freezing. The details have been well summarized in the review by James et al. published in 2015, and as commented here, these technologies have not been established yet but are in the development stage. Freezers that claim to improve the quality of frozen products by applying a magnetic field have been commercialized in Japan, but there is strong skepticism about the idea that magnetic fields affect freezing and frozen products (Dalvi-Isfahan et al. 2017; Han et al. 2018; James et al. 2015; Otero et al. 2016; Suzuki et al. 2010). Notably, scientific evidence on the effect of magnetic fields on freezing is not available even in 2022.

Numerous scientific studies have been published on the application of ultrasound, electrostatic fields, pulsed electric fields, microwaves, and radio waves to a freezing process to alter the ice crystal growth rate and/or the location of ice crystal formation, which would improve the product quality. Water has a dipole moment such that the motion of water molecules is affected by the electric field. The alignment of water changes with the application of electrostatic fields. When this is the case during freezing, the polarization of water molecules is expected to affect the free energy of the system involved in the nucleation process (Orlowska et al. 2009). Orlowska et al. reported that the ice nucleation temperature increased (i.e., the reduction of the supercooling degree) with increasing applied electrostatic field in the range of 0–6.0 × 10^6 V/m. Xanthakis et al. applied an electrostatic field to the freezing of pork tenderloin and observed an improvement in the quality of the final frozen product by reducing the damage to the meat microstructure, apart from controlling the nucleation temperature (Xanthakis et al. 2013). In this study, the effect of the electrostatic field on the reduction of the ice crystal size was investigated. Interestingly, the authors also suggested that the shape irregularity in the formed ice crystals increased with the increasing voltage of the electric field, which would largely contribute to maintaining the final product quality. The positive effect of the electrostatic field and high-voltage electric field on food freezing has further been confirmed to be advantageous in terms of energy consumption and quality retention as reviewed by Dalvi-Isfahan et al. (Dalvi-Isfahan et al. 2016a; Dalvi-Isfahan et al. 2016b).

In this context, it is reasonable to attempt to apply other electric field sources, such as pulsed electric fields, radio frequencies, and microwaves. The application of a pulsed electric field breaks the original tissue structures in food products prior to the subsequent freezing and drying processes (Ben Ammar et al. 2010; Jalte et al. 2009; Rastogi et al. 1999; Shayanfar et al. 2014). Therefore, the electric field does not directly alter the formation of ice crystals but indirectly modifies frozen food microstructures (Ben Ammar et al. 2010; Shayanfar et al. 2014).

Microwave-assisted freezing has been studied to avoid damage to cellular tissues owing to ice formation (Hanyu et al. 1992; Jackson et al. 1997). Hanyu et al. applied a 2,450 MHz microwave for 50 ms to freeze cell tissues on a cooling plate and confirmed a significant reduction in cell damage. They considered that microwave irradiation improved the rapidity of freezing because it restricted ice nucleation at the localized zone, which improved the freezing rate owing to the reduction of latent heat generation (Hanyu et al. 1992). Xanthakis et al. also applied 2,450 MHz microwaves to the freezing of pork tenderloin and obtained positive results, less damage to the frozen meat microstructure, and a reduction in the supercooling degree (Xanthakis et al. 2014). A recent modeling and experimental approach by Sadot et al. on microwave-assisted freezing elucidated the influence of the irradiation time and duty cycle on the formation of ice crystals (Sadot et al. 2020a; Sadot et al. 2020b). Sadot et al. suggested that an increase in irradiation time tends to decrease the ice crystal size at a certain microwave power. A significant reduction in the ice crystal size (approximately 25%) was achieved in all cases where microwaves of different duty cycles were applied. The mechanism of reduction of the ice crystal size is still not well understood. Temperature oscillations are likely to alter the formation of ice crystals; however, considering the results of this study, other reasons must be considered. Sadot et al. hypothesized that the H-bond network in water is disrupted by microwaves and consequently becomes ice crystal nuclei (in the literature, it was expressed as "a precursor of the crystalline structure"). However, further research is required.

Freezing assisted by radio frequency is a competing technique with microwave-assisted freezing (Anese et al. 2012; Hafezparast-Moadab et al. 2018). Hafezparast-Moadab et al. reported that a radio frequency of 27.12 MHz was applied to the freezing of rainbow trout fish, which could reduce the ice crystal size by approximately 75% through the selection of appropriate duty cycles (Hafezparast-Moadab et al. 2018). The mechanism of reduction of the ice crystal size using this technique can be explained by the investigation of microwave-assisted freezing.

Ultrasound affects the formation of ice crystals through a mechanism different from that of microwaves—i.e., by transmitting longitudinal waves through a cooling medium. Ultrasound not only triggers nucleation but also affects the growth of ice crystals (Dalvi-Isfahan et al. 2017; Kiani et al. 2013; Petzold and Aguilera 2009; Sun et al. 2019; Xin et al. 2014; Zheng and Sun 2006). Ultrasound-assisted freezing is commonly used in combination with immersion freezing (James et al. 2015; Sun et al. 2019; Xin et al. 2014). As reported, ultrasound assistance contributes to a reduction in ice crystal size and operating time and an improvement in final quality.

2.5 Formation of Ice Crystals in Biological Tissues

The current technology does not allow us to control the formation of ice microstructures as we wish. For products of biological origin (particularly livestock and fishery products), numerous reports have shown that ice crystals formed during freezing have a significant impact on tissue structure, suggesting that this results in food quality loss and deterioration (Li and Sun 2002; Zhu et al. 2019). In products with biological tissue structures, such as meat, fish, and plants, ice nucleation proceeds from outside the cells because of the lower concentration of extracellular fluid than intracellular fluid (Sahagian and Goff 1994). After ice is formed extracellularly and is placed in a condition that allows ice crystals to grow slowly, structures that look like damaged tissue are observed. In addition, significant tissue damage during the phase transition of water in the tissue causes drip loss during thawing (Kong et al. 2017). However, when cells are rapidly frozen, a relatively dispersed formation of micro-ice inside and outside the cell is observed, resulting in less damage to the tissue structure (Chassagne-Berces et al. 2009; Liang et al. 2015; Su et al. 2014). These differences are recognized as extracellular and intracellular freezing and are classified based on the difference in the location of ice formation (Bevilacqua et al. 1979). The differences in the location of ice formation are mainly attributed to the freezing rate; rapid freezing is widely recognized as one of the easiest and most effective methods for maintaining the quality of the resultant product.

The formation of ice crystals in food products depends on the complex interactions of various factors, such as the cooling rate, phase transition, locational differences, and components. However, as most of these factors are reflected in the movement of the freezing front, the freezing front needs to be tracked to understand the formation of ice crystals (Miyawaki 2018; Miyawaki et al. 1992). However, the mechanisms of ice nucleation and formation in tissues and the factors that determine the location of ice formation remain unresolved. For example, in fish muscle tissue, postmortem rigor and its associated phenomena have been reported to have a significant effect on the properties of frozen tissue and the formation of ice crystals (Hashimoto et al. 2015; Love 1966; Love and Haraldsson 1961). Furthermore, it has been suggested that changes in tissue properties, such as a reduction in water binding ability and damage to the endomysium, associated with postmortem changes in fish muscle tissue can affect the location and growth of ice crystals during freezing (Kominami et al. 2014). The authors observed that the shape of ice crystals formed in tuna meat at different levels of freshness differed significantly when the meat was frozen under identical conditions (Fig. 5). Freezing fresh fish meat can minimize connective tissue damage. These results support the longstanding argument that freshness is related to the formation of ice crystals and indicate the need to reconsider the conventional view that slow freezing causes extracellular freezing and tissue damage (Nakazawa and Okazaki 2020; Nakazawa et al. 2020). Even large ice crystals formed in fish meat by slow freezing have a small effect on tissue damage after thawing, depending on the degree of denaturation of the fish protein under freezing (Fukuda 1996; Fukuda et al. 2011; Nakazawa and Okazaki 2020). According to these reports, the cellular damage caused by freezing is not simply due to the cell deformation by volumetric expansion but to the protein denaturation in the frozen state. These results suggested that the quality of various frozen foods is affected not only by the size of ice crystals but also by freshness and storage conditions. Rapid freezing has been widely used industrially based on the strong conventional view that the formation of small ice crystals achieved by rapid freezing leads to high-quality frozen foods. However, the factor that determines product quality is not necessarily the ice crystal size, and the quality-determining factors must be properly controlled.

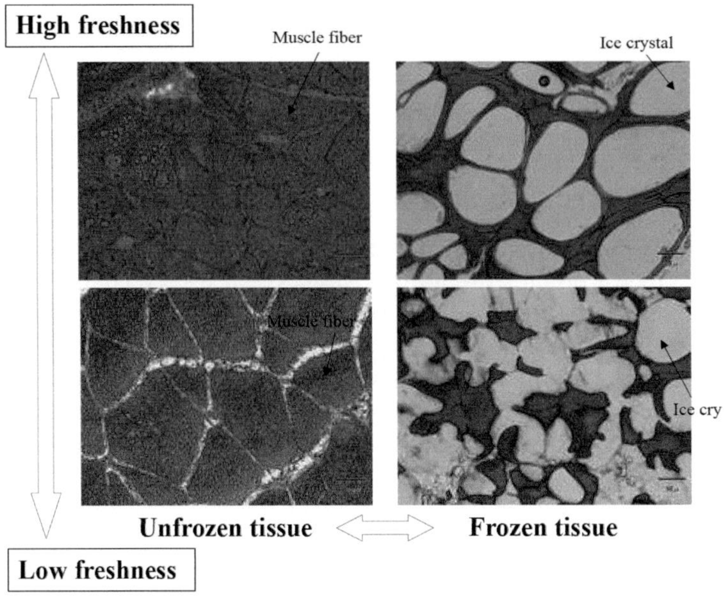

Figure 5. Micrograph of Tissue Structure (Thunnus Orientalis) Affected by Ice Crystals.

To date, several studies related to food freezing have focused on the evaluation of the formation of ice crystals in products, the restorability of food structure after thawing, compositional changes, and drip loss. The location and size of ice crystal formation would affect food quality in different ways depending on the type of food. However, there are few reports on the effects of the formation of ice crystals on eating quality (Choi et al. 2015; Fuchigami et al. 2002; Kono et al. 2015; Russell et al. 1999), and there is an overwhelming lack of data on practical frozen food production. This is a major deficiency in the field of food freezing research and should be the focus of future research. The use of large amounts of energy to maintain product quality, such as extremely rapid freezing and cryopreservation, is inconsistent with recent sustainable actions through energy conservation. A new perspective in the future may be a way to balance quality and energy load.

2.6 Application of Freezing to Food Structurization

As mentioned above, freezing is always associated with the formation of ice crystals and the resulting microstructures inside the product. This principle can be used to construct food products. For example, "kori-tofu; koya-tohu; shimi-tofu," a traditional food in Japan, is a porous, spongy product made from soybean curd that is produced via freezing. First, fresh soybean curd gel is frozen in cold air at night (with partial drying in progress). During the daytime, the frozen portions thaw, and the water is removed by drip-off or evaporation; this cycle is repeated for several days to remove the water from the tofu. Today, the entire process has been industrialized, and homogeneous spongy kori-tofu is produced by applying the appropriate freezing conditions.

Further application of this freeze-structuring process has been reported for the production of meat analogs (Chantanuson et al. 2022; Consolacion and Jelen 1986; Yuliarti et al. 2021). The production of plant-based meat analogs is regarded as a recent sustainable action (van der Goot et al. 2016). To mimic the unique fibrous structure of livestock meat, the formation of porous and fibrous microstructures consisting of multiple parallel and densely connected protein sheets has been intensively studied in recent decades (Dekkers et al. 2018). The freeze-structuring process is advantageous in producing block-like products of arbitrary shapes and sizes, unlike the extrusion technique, as shown in Fig. 6. However, the mechanism of formation of ice crystals in gel products is complex and is highly dependent on the plant protein source used. This is because different

Product appearance **Internal structure (XCT image)**

Figure 6. Plant-Based Meat Analogs Texturized by Freezing.

protein sources have different structure-function relationships in terms of hydration and solubility, gelation, and gel network formation, all of which affect frozen microstructures (Yuliarti et al. 2021). Chantanuson et al. prepared meat analogs using soybean powder in combination with sodium alginate as the main component. They reported that for generating porous and anisotropic layered structures by freezing, soybean powder with a less soluble fraction is more advantageous, and the application of secondary freezing is more effective in terms of the formation of frozen microstructures. They also reported that there was an optimal solid concentration to achieve this result and that structuring affected the rheological properties of the final product (Chantanuson et al. 2022).

In the context of structurization by freezing, a new technique that exploits the changes in the freeze-concentrated phase can be described. Biopolymer composites made from polysaccharides and proteins have shown promise in the design of food particulate systems for the protection and delivery of nutraceutical substances (Scholten et al. 2014). A freezing step is a potential processing pathway for an encapsulated system (Nakagawa 2013; Nakagawa and Nagao 2012; Nakagawa et al. 2013; Sowasod et al. 2013; Sowasod et al. 2012). During the freezing process, a freeze-concentrated phase is formed with the ice crystal phase. Freezing creates a highly concentrated microregion in which the concentration of solutes is controlled by the phase equilibrium. For example, at certain proportions of this polysaccharide mixture, a gel is formed upon freezing because of the freeze concentration, even though no gel is formed under ambient conditions (Nakagawa et al. 2013; Sowasod et al. 2013; Sowasod et al. 2012). Different freezing conditions result in different gel properties, which serve as the coating matrix for the encapsulation system. Gelatin and gum acacia form a complex coacervate under appropriate pH conditions. Nakagawa et al. applied the freezing process to a solution of gelatin and gum acacia, but at a pH insufficient to induce complexation; here, the complexes were formed by freeze concentration (Nakagawa 2013; Nakagawa and Nagao 2012). Freezing rate and aging time could be important controllable parameters in this process (Nakagawa and Fujii 2015). Casein salts are known to form aggregated clusters under low-pH conditions. The degree of aggregation depends on the pH. Nakagawa et al. (2016) and Nakagawa and Kagemoto (2013) reported that the formation of aggregates is induced by freezing (Nakagawa et al. 2016; Nakagawa and Kagemoto 2013). They produced casein nanoclusters encapsulating lipophilic substances by freezing and observed that aging under freezing conditions changed the microstructure of the clusters, which was related to the digestive properties. These results may provide an interesting research opportunity to control the bioavailability of foods through structural modifications owing to freezing.

3. Monitoring of the Freezing Process

3.1 Concept of Smart Processing

It is commonly accepted in the global food industry and distributors to store frozen foods at temperatures below −18°C. Therefore, the freezing process lowers the temperature of food to approximately −18°C, and the cryopreservation process maintains the temperature at −18°C or below. As mentioned in the previous sections, the freezing process requires energy that corresponds to the solidification enthalpy released from the products. The total amount of heat that must be removed from the products during the freezing process can be estimated from the amount of frozen fraction in the product, which is a function of temperature (e.g., as expressed by Hess's formula). However, when maintaining the target temperature during the cryopreservation process, the required energy is determined by the operating efficiency of the refrigeration unit and the heat transfer between the freezer and the external environment. However, current industrial food freezing processes and freezer systems do not allow the direct measurement of the temperature of food products (internal product temperature); therefore, product temperature is evaluated indirectly. This leads to excessive equipment installation and/or energy loads to ensure that the target temperatures are achieved. Therefore, the in-line measurement of food temperature is an important aspect of the next-generation freezer systems. If the internal temperature of food products can be measured continuously and directly in a non-destructive manner, efficient operating conditions for freezer systems can be derived and fed back to the operation.

Freezing level sensors using millimeter waves with frequency bands in the gigahertz region have already been commercialized (Al-Empowered Microwave Technology; AQUANTIS). This sensor not only measures the freezing level in a nondestructive and noncontact manner but also uses the measured data to intensify product traceability and automate the optimal operation control of freezers and refrigeration systems (CALLIFREEZE®; GEA). Efficient operation and energy conservation using sensing technology will be an important direction for the food freezing industry in the future.

3.2 Monitoring Techniques for Freezing Process

Conventional monitoring methods for measuring the freezing rate and internal temperature of food products involve the insertion of sensors directly into the product during or after freezing. However, this method is not applicable to continuous processes or total inspections. In addition, there is a risk of contamination associated with the sensor insertion. Therefore, in conventional food freezing processes, temperature measurements are performed indirectly based on experience with preliminary testing, sampling measurements, and freezer temperature control. This requires a large safety factor, which increases freezing time and energy consumption. Thus, a monitoring technique that can measure the degree of freezing and the internal temperature of food instantaneously and nondestructively needs to be developed.

The dielectric properties of ice differ significantly from those of water. Impedance monitoring is a known analytical tool for freeze-drying. It measures the dielectric and conductive properties of a freeze-drying material with electrodes in contact with the material (Rey and May 2004). The through-vial impedance spectroscopy technique has been proposed as an in-line analytical tool for pharmaceutical freeze-drying (Smith et al. 2017; Smith et al. 2018; Smith et al. 2013). This technique provides information on a frozen layer, such as temperature, ice fraction, and glass transition, which affect the dielectric properties.

An open-ended microwave coaxial cavity resonator that can detect the moisture content of wet materials is useful for nondestructive monitoring (Bianchi et al. 1999; Nakayama 1987a, b, 1995). Microwave resonators can be employed to monitor the moisture content of a material by detecting its electrical permittivity and permeability. This method can be applied to the measurement of moisture content in food and biological systems (McKeown et al. 2012; Trabelsi et al. 2016; Xu et al. 1987).

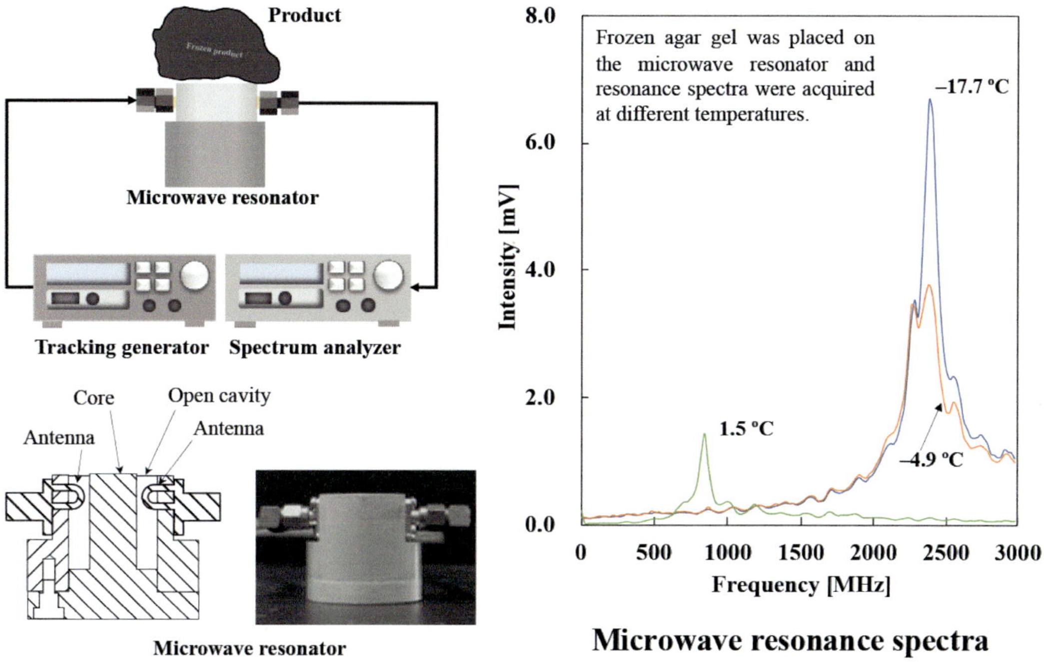

Figure 7. Microwave Resonance Spectroscopy Measurement.

Microwaves emitted from an antenna attached to the resonator pass through the material and are received by another antenna as shown in Fig. 7. The received microwave intensity is affected by the dielectric properties of the material. A resonance spectrum that reflects the dielectric properties of the material can be obtained by applying microwaves with different frequencies (Fig. 7). When a food product is subjected to freezing, an ice crystal phase is formed with the freeze-concentrated phase. Owing to the difference in the dielectric properties of these phases, the spectral patterns change to reflect the volume ratio of the ice and freeze-concentrated phases, and this ratio is significantly correlated with the product temperature, particularly the internal temperature during freezing (Kono et al. 2021). Kono et al. applied an open-ended microwave coaxial cavity resonator to monitor the freezing process using a microwave network system placed between a tracking generator and a spectrum analyzer. The microwave resonance intensity at a particular frequency strongly depends on the water state of the frozen concentrate phase in the sample and dominantly represents the state of the core portion of the product. The trend of the intensity change with temperature (Fig. 8) is different above and below the glass transition temperature owing to the difference in dielectric properties (Kono et al. 2021).

This microwave resonator was used to measure the internal temperature of ice cream in a water-resistant paper cup as shown in Fig. 9. The voltage intensity obtained from the microwave sensing system was a strong linear relationship with the internal sample temperature, which was simultaneously measured by a fiber optic temperature probe. This system has the advantage of nondestructively measuring product temperature through its package, and is very practical, as it can detect the internal condition of a product within one second simply by placing the object to be measured on the resonator.

3.2.1 Temperature Measurement of Ice Cream Using Microwave Resonator

The use of millimeter waves has been proposed to monitor freezing and drying processes (Pandey et al. 2018). Pandey et al. developed a dedicated sensor operating at 60 GHz to monitor the freezing process of potato slices. Millimeter waves are less affected by the composition of the food and

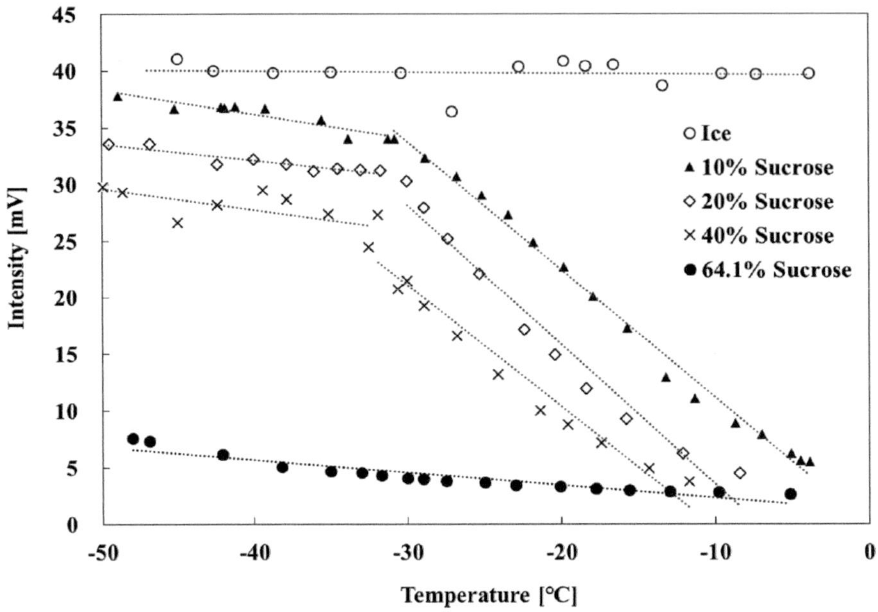

Figure 8. Microwave Resonance Intensity of Sucrose Solutions as a Function of Temperature (Kono et al. 2021).

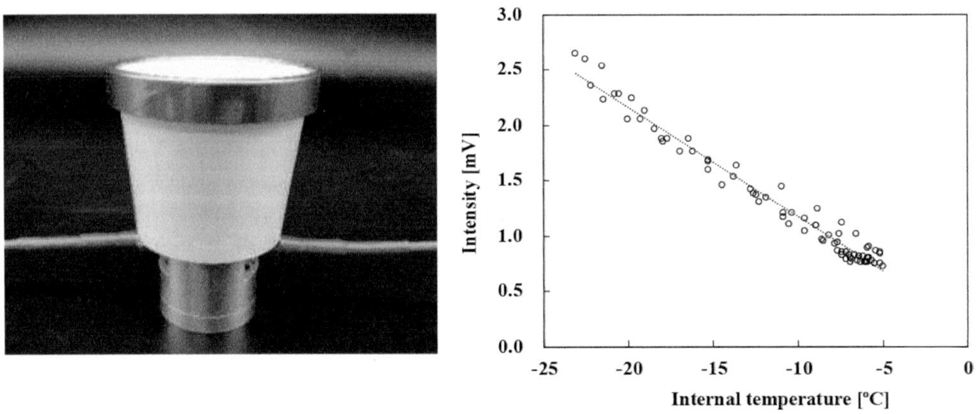

Figure 9. Application performance of a microwave resonator to measure the internal temperature of ice cream in a water-resistant paper cup.

are less susceptible to scattering owing to particle roughness and inhomogeneity when transmitted through the food, compared with wavelengths in the visible and infrared regions. These spectral measurements using electromagnetic waves require only a few seconds for each measurement and are powerful nondestructive monitoring techniques that can practically be applied to the freezing process in the industry. This method is expected to be a core technology in the smart industry.

4. Conclusion

Recent progress in freezing technology is reported in this chapter. The reduction of GHG emissions through the reduction of food losses is required for the future food industry, including the freezing industry. To develop innovative freezing technologies, it is necessary to determine whether they are compatible with the sustainable society implemented in food value chains. Given that several innovative freezing technologies are most likely to be used as additions to existing refrigeration systems, research should focus not only on the new technology but also on its role in the system or hybridization. As mentioned in the review by James et al., rapid freezing technologies, such as impingement, hydrofluidization, and high-pressure-assisted and ultrasound-assisted freezing, are "potentially complementary, and thus combined or sequential applications could offer advantages not seen when applied on their own" (James et al. 2015). It is also recognized that rapid freezing is not always essential in terms of product quality. Rapid freezing, which is more over-specified than necessary, is inappropriate for both the environmental impact and product quality. The key to future research is to develop a system that optimizes processing capacity, energy efficiency, and quality. For this purpose, fundamental issues, such as the formation mechanism and kinetics of ice crystals in food and the effects of the external environment (electric field, temperature fluctuation, etc.), need to be tackled.

References

Anese, M., Manzocco, L., Panozzo, A., Beraldo, P., Foschia, M. and Nicoli, M.C. 2012. Effect of radiofrequency assisted freezing on meat microstructure and quality. Food Research International 46: 50–54.

Ben Ammar, J., Lanoisellé, J.-L., Lebovka, N.I., Van Hecke, E. and Vorobiev, E. 2010. Effect of a pulsed electric field and osmotic treatment on freezing of potato tissue. Food Biophysics 5: 247–254.

Bevilacqua, A., Zaritzky, N. and Calvelo, A. 1979. Histological measurements of ice in frozen beef. International Journal of Food Science & Technology 14: 237–251.

Bianchi, G., Dionigi, M., Fioretto, D. and Sorrentino, R. 1999. A microwave system for moisture monitoring in wet powders for industrial applications, 1999 IEEE MTT-S International Microwave Symposium Digest (Cat. No.99CH36282), pp. 1603–1606 vol. 1604.

Bomben, J. and King, C. 1982. Heat and mass transport in the freezing of apple tissue. International Journal of Food Science & Technology 17: 615–632.

Chantanuson, R., Nagamine, S., Kobayashi, T. and Nakagawa, K. 2022. Preparation of soy protein-based food gels and control of fibrous structure and rheological property by freezing. Food Structure, 100258.

Chassagne-Berces, S., Poirier, C., Devaux, M.-F., Fonseca, F., Lahaye, M., Pigorini, G., Girault, C., Marin, M. and Guillon, F. 2009. Changes in texture, cellular structure and cell wall composition in apple tissue as a result of freezing. Food Research International 42: 788–797.

Cheftel, J., Thiebaud, M. and Dumay, E. 2002. Pressure-assisted freezing and thawing of foods: a review of recent studies. International Journal of High Pressure Research 22: 601–611.

Cheng, L., Sun, D.-W., Zhu, Z. and Zhang, Z. 2017. Emerging techniques for assisting and accelerating food freezing processes: A review of recent research progresses. Critical Reviews in Food Science and Nutrition 57: 769–781.

Choi, E.J., Park, H.W., Chung, Y.B., Park, S.H., Kim, J.S. and Chun, H.H. 2017. Effect of tempering methods on quality changes of pork loin frozen by cryogenic immersion. Meat Science 124: 69–76.

Choi, Y.-S., Ku, S.-K., Jeong, J.-Y., Jeon, K.-H. and Kim, Y.-B. 2015. Changes in ultrastructure and sensory characteristics on electro-magnetic and air blast freezing of beef during frozen storage. Korean Journal for Food Science of Animal Resources 35: 27.

Consolacion, F.I. and Jelen, P. 1986. Freeze Texturation of Proteins: Effect of the Alkali, Acid and Freezing Treatments on Texture Formation. Food Structure 5: 5.

Crippa, M., Solazzo, E., Guizzardi, D., Monforti-Ferrario, F., Tubiello, F. and Leip, A. 2021. Food systems are responsible for a third of global anthropogenic GHG emissions. Nature Food 2: 198–209.

da Silva, J. 2017. FAO-The future of food and agriculture: Trends and challenges. Report of the Food and Agriculture Organization of the United Nations.

Dalvi-Isfahan, M., Hamdami, N. and Le-Bail, A. 2016a. Effect of freezing under electrostatic field on the quality of lamb meat. Innovative Food Science & Emerging Technologies 37: 68–73.

Dalvi-Isfahan, M., Hamdami, N., Le-Bail, A. and Xanthakis, E. 2016b. The principles of high voltage electric field and its application in food processing: A review. Food Research International 89: 48–62.

Dalvi-Isfahan, M., Hamdami, N., Xanthakis, E. and Le-Bail, A. 2017. Review on the control of ice nucleation by ultrasound waves, electric and magnetic fields. Journal of Food Engineering 195: 222–234.

Dekkers, B.L., Boom, R.M. and van der Goot, A.J. 2018. Structuring processes for meat analogues. Trends in Food Science & Technology 81: 25–36.

Dirita, C., De Bonis, M.V. and Ruocco, G. 2007. Analysis of food cooling by jet impingement, including inherent conduction. Journal of Food Engineering 81: 12–20.

FAO, Sustainable Development Goals: Indicator 12.3. 1—Global Food Loss and Waste.

Fikiin, A. 1992. New method and fluidized water system for intensive chilling and freezing of fish. Food Control 3: 153–160.

Fikiin, K. 2008. Emerging and novel freezing processes, Frozen food science and technology. Blackwell Publishing Ltd., Oxford, UK, pp. 101–123.

Fikiin, K. and Fikiin, A. 1999. Novel cost-effective ice-slurry based technology for individual quick freezing of foods by hydrofluidisation, Proceedings of the 20th international congress on refrigeration, Sydney, Australia.

Fikiin, K. and Fikiin, A. 2000. Individual quick freezing of foods by hydrofluidisation and pumpable ice slurries. Science et technique du froid 319–326.

Fuchigami, M., Ogawa, N. and Teramoto, A. 2002. Trehalose and hydrostatic pressure effects on the structure and sensory properties of frozen tofu (soybean curd). Innovative Food Science & Emerging Technologies 3: 139–147.

Fukuda, Y., 1996. Denaturation by freezing of fish muscle proteins. Bulletin of the National Research Institute of Fisheries Science (Japan).

Fukuda, Y., Okazaki, E. and Wada, R. 2011. Effect of fluctuations of temperature during frozen storage on denaturation of fish myofibrillar protein. Transactions of the Japan Society of Refrigerating and Air Conditioning Engineers 23: 335–340.

Góral, D., Kluza, F. 2006. Physical changes of vegetables during freezing by conventional and impingement methods. Acta Agrophysica 7.

Góral, D. and Kluza, F. 2012. Heat transfer coefficient in impingement fluidization freezing of vegetables and its prediction. International Journal of Refrigeration 35: 871–879.

Goshima, H., Do, G. and Nakagawa, K. 2016. Impact of Ice Morphology on Design Space of Pharmaceutical Freeze-Drying. J. Pharm. Sci. 105: 1920–1933.

Hafezparast-Moadab, N., Hamdami, N., Dalvi-Isfahan, M. and Farahnaky, A. 2018. Effects of radiofrequency-assisted freezing on microstructure and quality of rainbow trout (Oncorhynchus mykiss) fillet. Innovative Food Science & Emerging Technologies 47: 81–87.

Han, Z., Cai, M.-j., Cheng, J.-H. and Sun, D.-W. 2018. Effects of electric fields and electromagnetic wave on food protein structure and functionality: A review. Trends in Food Science & Technology 75: 1–9.

Hanyu, Y., Ichikawa, M. and Matsumoto, G. 1992. An improved cryofixation method: cryoquenching of small tissue blocks during microwave irradiation. Journal of microscopy 165: 255–271.

Hashimoto, K., Kawashima, T., Yoshino, N., Shirai, T. and Takiguchi, A. 2015. Effects of freshness on thawing drip and ice crystal formation in frozen spotted mackerel Scomber australasicus. Nippon Suisan Gakkaishi 81: 124–129.

Hottot, A., Nakagawa, K. and Andrieu, J. 2008. Effect of ultrasound-controlled nucleation on structural and morphological properties of freeze-dried mannitol solutions. Chemical Engineering Research and Design 86: 193–200.

Ingwersen, W.W. 2012. Life cycle assessment of fresh pineapple from Costa Rica. Journal of cleaner production 35: 152–163.

Jackson, T.H., Ungan, A., Critser, J.K. and Gao, D. 1997. Novel microwave technology for cryopreservation of biomaterials by suppression of apparent ice formation. Cryobiology 34: 363–372.

Jalte, M., Lanoisellé, J.-L., Lebovka, N.I. and Vorobiev, E. 2009. Freezing of potato tissue pre-treated by pulsed electric fields. LWT-Food Science and Technology 42: 576–580.

James, C., Purnell, G. and James, S.J. 2015. A review of novel and innovative food freezing technologies. Food and Bioprocess Technology 8: 1616–1634.

Kiani, H., Zhang, Z. and Sun, D.-W. 2013. Effect of ultrasound irradiation on ice crystal size distribution in frozen agar gel samples. Innovative Food Science & Emerging Technologies 18: 126–131.

Kobayashi, R., Kanesaka, N., Watanabe, M. and Suzuki, T. 2014. Effect of the Breaking Temperature of Supercooling on Ice Characteristics and Drip Loss of Foods in Supercooled Freezing Method. Transactions of the Japan Society of Refrigerating and Air Conditioning Engineers advpub.

Kominami, Y., Watanabe, M. and Suzuki, T. 2014. Effects of postmortem physiological changes on ice crystal morphology in the fish muscle tissue. Trans JSRAE 31: 47–56.

Kong, C.H., Hamid, N., Ma, Q., Lu, J., Wang, B.-G. and Sarojini, V. 2017. Antifreeze peptide pretreatment minimizes freeze-thaw damage to cherries: An in-depth investigation. LWT 84: 441–448.

Kono, S., Imamura, H. and Nakagawa, K. 2021. Non-destructive monitoring of food freezing process by microwave resonance spectroscopy with an open-ended coaxial resonator. Journal of Food Engineering 292: 110293.

Kono, S., Kawamura, I., Yamagami, S., Araki, T. and Sagara, Y. 2015. Optimum storage temperature of frozen cooked rice predicted by ice crystal measurement, sensory evaluation and artificial neural network. International Journal of Refrigeration 56: 165–172.

Kurz, W. and Fisher, D.J. 1986. Fundamentals of solidification. trans tech publications Aedermannsdorf, Switzerland.

LeBail, A., Chevalier, D., Mussa, D. and Ghoul, M. 2002. High pressure freezing and thawing of foods: a review. International Journal of Refrigeration 25: 504–513.

Li, B., Sun and D.-W. 2002. Novel methods for rapid freezing and thawing of foods—a review. Journal of Food Engineering 54: 175–182.

Liang, D., Lin, F., Yang, G., Yue, X., Zhang, Q., Zhang, Z. and Chen, H. 2015. Advantages of immersion freezing for quality preservation of litchi fruit during frozen storage. LWT-Food Science and Technology 60: 948–956.

Love, R. 1966. Protein denaturation in frozen fish. XI.—The proportion of tissue water converted to ice. Journal of the Science of Food and Agriculture 17: 465–471.

Love, R. and Haraldsson, S. 1961. The expressible fluid of fish fillets. XI.—Ice crystal formation and cell damage in cod muscle frozen before rigor mortis. Journal of the Science of Food and Agriculture 12: 442–449.

Lovelock, C., Evans, C., Barros, N., Prairie, Y., Alm, J., Bastviken, D., Beaulieu, J., Garneau, M., Harby, A. and Harrison, J. 2019. Refinement to the 2006 IPCC Guidelines for National Greenhouse Gas Inventories, Chap. 7. Wetlands, IPCC Kyoto.
Lucas, T., François, J. and Raoult-Wack, A.L. 1998. Transport phenomena in immersion-cooled apples. International Journal of Food Science & Technology 33: 489–499.
Lucas, T. and Raoult-Wack, A. 1998. Immersion chilling and freezing in aqueous refrigerating media: review and future trends: Réfrigération et congélation par immersion dans des milieux réfrigérants: revue et tendances futures. International Journal of Refrigeration 21: 419–429.
Marazani, T., Madyira, D.M. and Akinlabi, E.T. 2017. Investigation of the parameters governing the performance of jet impingement quick food freezing and cooling systems—a review. Procedia Manufacturing 8: 754–760.
McKeown, M.S., Trabelsi, S., Tollner, E.W. and Nelson, S.O. 2012. Dielectric spectroscopy measurements for moisture prediction in Vidalia onions. Journal of Food Engineering 111: 505–510.
Miyawaki, O. 2018. Water and freezing in food. Food Science and Technology Research 24: 1–21.
Miyawaki, O., Abe, T. and Yano, T. 1992. Freezing and ice structure formed in protein gels. Bioscience, biotechnology, and biochemistry 56: 953–957.
Nakagawa, K. 2013. Characterization of Freeze-Dried Core-Shell Nanoparticles Prepared via Gelatin-Acacia Complex Coacervation: A Study on Particle Formation upon Freezing. Drying Technology 31: 1466–1476.
Nakagawa, K. and Fujii, Y. 2015. Protein-Based Microencapsulation with Freeze Pretreatment: Spray-Dried Oil in Water Emulsion Stabilized by the Soy Protein Isolate–Gum Acacia Complex. Drying Technology 33: 1541–1549.
Nakagawa, K., Hottot, A., Vessot, S. and Andrieu, J. 2006. Influence of controlled nucleation by ultrasounds on ice morphology of frozen formulations for pharmaceutical proteins freeze-drying. Chemical Engineering and Processing: Process Intensification 45: 783–791.
Nakagawa, K., Hottot, A., Vessot, S. and Andrieu, J. 2007. Modeling of freezing step during freeze-drying of drugs in vials. AIChE Journal 53: 1362–1372.
Nakagawa, K., Jarunglumlert, T. and Adachi, S. 2016. Structural changes in casein aggregates under frozen conditions affect the entrapment of hydrophobic materials and the digestibility of aggregates. Chemical Engineering Science 143: 287–296.
Nakagawa, K. and Kagemoto, M. 2013. Characterization of casein-based nanoparticles formed upon freezing by in situ SAXS measurement. Colloids Surf B Biointerfaces 103: 366–374.
Nakagawa, K. and Nagao, H. 2012. Microencapsulation of oil droplets using freezing-induced gelatin–acacia complex coacervation. Colloids and Surfaces A: Physicochemical and Engineering Aspects 411: 129–139.
Nakagawa, K., Sowasod, N., Tanthapanichakoon, W. and Charinpanitkul, T. 2013. Hydrogel based oil encapsulation for controlled release of curcumin by using a ternary system of chitosan, kappa-carrageenan, and carboxymethylcellulose sodium salt. LWT - Food Science and Technology 54: 600–605.
Nakagawa, K., Thongprachan, N., Charinpanitkul, T. and Tanthapanichakoon, W. 2010. Ice crystal formation in the carbon nanotube suspension: A modelling approach. Chemical Engineering Science 65: 1438–1451.
Nakayama, S. 1987a. Microwave sensor of basis weight and moisture content of sheet materials by re-entrant cavity. Japanese journal of applied physics 26: 1935.
Nakayama, S. 1987b. Simultaneous measurements of basis weight and moisture content of sheet materials by microwave cavity. Japanese Journal of Applied Physics 26: 1198.
Nakayama, S. 1995. Development of a microwave moisture sensor for aggregates. Measurement Science and Technology 6: 429.
Nakazawa, N. and Okazaki, E. 2020. Recent research on factors influencing the quality of frozen seafood. Fisheries Science 86: 231–244.
Nakazawa, N., Wada, R., Fukushima, H., Tanaka, R., Kono, S. and Okazaki, E. 2020. Effect of long-term storage, ultra-low temperature, and freshness on the quality characteristics of frozen tuna meat. International Journal of Refrigeration 112: 270–280.
Orlowska, M., Havet, M. and Le-Bail, A. 2009. Controlled ice nucleation under high voltage DC electrostatic field conditions. Food Research International 42: 879–884.
Otero, L., Rodríguez, A.C., Pérez-Mateos, M. and Sanz, P.D. 2016. Effects of magnetic fields on freezing: Application to biological products. Comprehensive Reviews in Food Science and Food Safety 15: 646–667.
Palacz, M., Adamczyk, W., Piechnik, E., Stebel, M. and Smolka, J. 2019. Experimental investigation of the fluid flow inside a hydrofluidisation freezing chamber. International Journal of Refrigeration 107: 52–62.
Palacz, M., Piechnik, E., Halski, M., Stebel, M., Adamczyk, W., Eikevik, T.M. and Smolka, J. 2021. Experimental analysis of freezing process of stationary food samples inside a hydrofluidisation freezing chamber. International Journal of Refrigeration 131: 68–77.
Pandey, G., Vandermeiren, W., Dimiccoli, L. and Stiens, J. 2018. Contactless monitoring of food drying and freezing processes with millimeter waves. Journal of Food Engineering 226: 1–8.
Pelletier, N. and Tyedmers, P. 2010. Life cycle assessment of frozen tilapia fillets from Indonesian lake-based and pond-based intensive aquaculture systems. Journal of Industrial Ecology 14: 467–481.
Petzold, G. and Aguilera, J.M. 2009. Ice Morphology: Fundamentals and Technological Applications in Foods 4: 378–396.

Rastogi, N., Eshtiaghi, M. and Knorr, D. 1999. Accelerated mass transfer during osmotic dehydration of high intensity electrical field pulse pretreated carrots. J. Food Sci. 64: 1020–1023.
Reid, D. 1982. Cryomlcroscope studies of the freezing of model solutions of cryobiological interest. Cryobiology 19: 652.
Rey, L. and May, J.C. 2004. Freeze-Drying/Lyophilization Of Pharmaceutical & Biological Products, Revised and Expanded. CRC Press.
Rickards, L. and Shortis, E., 2019. SDG 2: End hunger, achieve food security and improved nutrition and promote sustainable agriculture. RMIT University.
Russell, A., Cheney, P. and Wantling, S. 1999. Influence of freezing conditions on ice crystallisation in ice cream. Journal of Food Engineering 39: 179–191.
Sadot, M., Curet, S., Chevallier, S., Le-Bail, A., Rouaud, O. and Havet, M. 2020a. Microwave assisted freezing part 2: Impact of microwave energy and duty cycle on ice crystal size distribution. Innovative Food Science & Emerging Technologies 62: 102359.
Sadot, M., Curet, S., Le-Bail, A., Rouaud, O. and Havet, M. 2020b. Microwave assisted freezing part 1: Experimental investigation and numerical modeling. Innovative Food Science & Emerging Technologies 62: 102360.
Sahagian, M.E. and Goff, H.D. 1994. Effect of freezing rate on the thermal, mechanical and physical aging properties of the glassy state in frozen sucrose solutions. Thermochimica Acta 246: 271–283.
Salvadori, V.O. and Mascheroni, R.H. 2002. Analysis of impingement freezers performance. Journal of Food Engineering 54: 133–140.
Sarkar, A., Nitin, N., Karwe, M. and Singh, R.P. 2004. Fluid flow and heat transfer in air jet impingement in food processing. J. Food Sci. 69: CRH113–CRH122.
Sarkar, A. and Singh, R. 2004. Modeling flow and heat transfer during freezing of foods in forced airstreams. J. Food Sci. 69: E488–E496.
Scholten, E., Moschakis, T. and Biliaderis, C.G. 2014. Biopolymer composites for engineering food structures to control product functionality. Food Structure 1: 39–54.
Shayanfar, S., Chauhan, O., Toepfl, S. and Heinz, V. 2014. Pulsed electric field treatment prior to freezing carrot discs significantly maintains their initial quality parameters after thawing. International Journal of Food Science & Technology 49: 1224–1230.
Smith, G., Arshad, M.S., Polygalov, E., Ermolina, I., McCoy, T.R. and Matejtschuk, P. 2017. Process understanding in freeze-drying cycle development: applications for through-vial impedance spectroscopy (TVIS) in mini-pilot studies. Journal of Pharmaceutical Innovation 12: 26–40.
Smith, G., Jeeraruangrattana, Y. and Ermolina, I. 2018. The application of dual-electrode through vial impedance spectroscopy for the determination of ice interface temperatures, primary drying rate and vial heat transfer coefficient in lyophilization process development. European Journal of Pharmaceutics and Biopharmaceutics 130: 224–235.
Smith, G., Polygalov, E., Arshad, M., Page, T., Taylor, J. and Ermolina, I. 2013. An impedance-based process analytical technology for monitoring the lyophilisation process. Int. J. Pharm. 449: 72–83.
Soto, V. and Borquez, R. 2001. Impingement jet freezing of biomaterials. Food Control. 12: 515–522.
Sowasod, N., Nakagawa, K., Charinpanitkul, T. and Tanthapanichakoon, W. 2013. Encapsulation of curcumin loaded oil droplets with chitosan based cryogel: Influence of freezing condition on nanocapsule properties. Food Science and Technology Research 19: 633–640.
Sowasod, N., Nakagawa, K., Tanthapanichakoon, W. and Charinpanitkul, T. 2012. Development of encapsulation technique for curcumin loaded O/W emulsion using chitosan based cryotropic gelation. Materials Science and Engineering: C 32: 790–798.
Stebel, M., Smolka, J., Palacz, M., Adamczyk, W. and Piechnik, E. 2020. Numerical investigation of the fluid flow distribution for the hydrofluidisation food freezing method. International Journal of Thermal Sciences 151: 106284.
Stebel, M., Smolka, J., Palacz, M., Piechnik, E., Halski, M., Knap, M., Felis, E., Eikevik, T.M., Tolstorebrov, I., Peralta, J.M. and Zorrilla, S.E. 2021. Numerical modelling of the food freezing process in a quasi-hydrofluidisation system. Innovative Food Science & Emerging Technologies 74: 102834.
Su, G., Ramaswamy, H.S., Zhu, S., Yu, Y., Hu, F. and Xu, M. 2014. Thermal characterization and ice crystal analysis in pressure shift freezing of different muscle (shrimp and porcine liver) versus conventional freezing method. Innovative Food Science & Emerging Technologies 26: 40–50.
Sun, Q., Zhao, X., Zhang, C., Xia, X., Sun, F. and Kong, B. 2019. Ultrasound-assisted immersion freezing accelerates the freezing process and improves the quality of common carp (Cyprinus carpio) at different power levels. LWT 108: 106–112.
Suzuki, T., Takeuchi, Y., Masuda, K., Watanabe, M., Shirakashi, R., Fukuda, Y., Tsuruta, T., Yamamoto, K., Koga, N., Hiruma, N., Ichioka, J. and Takai, K. 2010. Experimental Investigation of Effectiveness of Magnetic Field on Food Freezing Process. Reitou 85: 470–471.
Trabelsi, S., Lewis, M.A. and Nelson, S.O. 2016. Microwave moisture meter for in-shell peanut kernels. Food Control. 66: 283–290.
van der Goot, A.J., Pelgrom, P.J.M., Berghout, J.A.M., Geerts, M.E.J., Jankowiak, L., Hardt, N.A., Keijer, J., Schutyser, M.A.I., Nikiforidis, C.V. and Boom, R.M. 2016. Concepts for further sustainable production of foods. Journal of Food Engineering 168: 42–51.

van der Sman, R.G.M., Voda, A., van Dalen, G. and Duijster, A. 2013. Ice crystal interspacing in frozen foods. Journal of Food Engineering 116: 622–626.

Vazquez-Rowe, I., Moreira, M.T. and Feijoo, G. 2012. Environmental assessment of frozen common octopus (Octopus vulgaris) captured by Spanish fishing vessels in the Mauritanian EEZ. Marine Policy 36: 180–188.

Verboven, P., Scheerlinck, N. and Nicolaï, B.M. 2003. Surface heat transfer coefficients to stationary spherical particles in an experimental unit for hydrofluidisation freezing of individual foods. International Journal of Refrigeration 26: 328–336.

Woinet, B., Andrieu, J., Laurent, M. and Min, S. 1998. Experimental and theoretical study of model food freezing. Part II. Characterization and modelling of the ice crystal size. Journal of Food Engineering 35: 395–407.

Xanthakis, E., Havet, M., Chevallier, S., Abadie, J. and Le-Bail, A. 2013. Effect of static electric field on ice crystal size reduction during freezing of pork meat. Innovative Food Science & Emerging Technologies 20: 115–120.

Xanthakis, E., Le-Bail, A. and Ramaswamy, H. 2014. Development of an innovative microwave assisted food freezing process. Innovative Food Science & Emerging Technologies 26: 176–181.

Xin, Y., Zhang, M. and Adhikari, B. 2014. Ultrasound assisted immersion freezing of broccoli (Brassica oleracea L. var. botrytis L.). Ultrasonics Sonochemistry 21: 1728–1735.

Xu, D., Liu, L. and Jiang, Z. 1987. Measurement of the dielectric properties of biological substances using an improved open-ended coaxial line resonator method. IEEE transactions on microwave theory and techniques 35: 1424–1428.

Yuliarti, O., Kovis, T.J.K. and Yi, N.J. 2021. Structuring the meat analogue by using plant-based derived composites. Journal of Food Engineering 288: 110138.

Zheng, L. and Sun, D.-W. 2006. Innovative applications of power ultrasound during food freezing processes—a review. Trends in Food Science & Technology 17: 16–23.

Zhu, Z., Zhou, Q. and Sun, D.-W. 2019. Measuring and controlling ice crystallization in frozen foods: A review of recent developments. Trends in Food Science & Technology 90: 13–25.

10

Food Drying

João Borges Laurindo, Bruno Augusto Mattar Carciofi, Aline Iamin Gomide* and *Ricardo Lemos Monteiro*

1. Introduction

Drying is an ancient and valuable technology to preserve foods for extended periods without refrigeration. Besides dropping the product volume and weight, the water removal reduces the availability of the remaining water for chemical reactions and microbial growth. The drying processes also impact foods' characteristics. However, in some cases, they promote undesirable nutrient degradation, color changes, and matrix shrinkage.

Drying is a complex operation with simultaneous heat and mass transfer that must be done carefully, considering the desired properties of the dried product. For example, the dehydration of thermolabile foods must consider a maximum temperature to avoid undesirable changes in their nutritional and sensory properties. Furthermore, food materials of different consistency demand different dryers and operational conditions. For example, fragile solids must be manipulated with care to avoid or reduce excessive breaks, while deformable solids and semisolids can change their structures when submitted to mechanical shocks and stresses. Furthermore, selecting the equipment and drying conditions for a liquid suspension depends on its composition and thermolability. The drying of sticking solutions and suspensions (e.g., fruit pulps) is a challenge that is frequently solved by adding carrying agents, but the surface energy of the dryer wall is another aspect to be considered.

The design of food dryers is very flexible. They can operate in continuous or batch mode, which is useful, and can adapt to different production scales. The heat for water evaporation is supplied by convection using hot air in most cases, which also carries out the water vapor. Alternatively, the heat can be provided by conduction, thermal radiation, or microwave, depending on the dryer configuration.

This chapter presents some fundamental aspects of the drying operation and practical examples of food drying as well as considering the processing aspects, driers design, and product properties.

2. Food Drying Principles

Drying is the operation of removing water by evaporation that results in dried-like materials. The drying operation depends on the supply of energy to the food matrix (heat transfer) and the ability of the surroundings to carry out the water evaporated from the food (Fig. 1).

Departament of Chemical and Food Engineering, Federal University of Santa Catarina, EQA/CTC/UFSC, 88040-900, Florianópolis, SC, Brazil.

Emails: bruno.carciofi@ufsc.br; aline.iamin.gomide@gmail.com; ricardolemosmonteiro@hotmail.com

* Corresponding author: jb.laurindo@ufsc.br

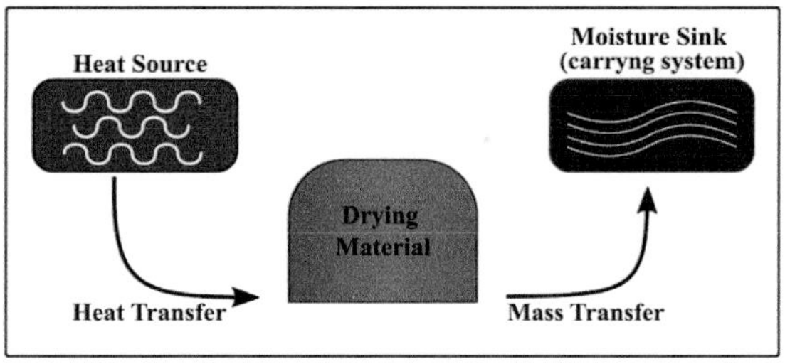

Figure 1. Schematic Representation of Food Drying.

When a moisty food is in contact with hot dry air, it causes mass transfer from the food surface to the air—triggered by the difference in the water chemical potential—which is the carrying agent that removes the evaporated water from the drying surroundings. Therefore, the direct impact of drying is to extend food shelf-life due to reducing food water activity and, in turn, reducing or stopping microbial growth and chemical reaction rates. Also, other consequences are inherent to this process: (i) changes in sensory aspects by creating different microstructure, crispness, and unique flavors—which may be adverse when desirable flavors and colors are damaged; (ii) changes in nutritional composition—enhancement of nutrients bioavailability and micro components degradation, depending on the pretreatments and time-temperature profile during drying; (iii) reduction of weight and volume, facilitating and saving energy in transport and storage; (iv) extended shelf-stable condition at ambient temperature (no need of refrigeration and cold chain); and (v) creation of new and convenient food products for industries and consumers.

The remaining moisture in a dried food is strongly attracted by hydrogen bonding to the food matrix macromolecules (e.g., protein, carbohydrates, cellulose/hemicellulose, and pectin) or present in the porous space as very concentrated solutions of salts, sugars, organic acids, and minerals. The firmly attached water gives the dry perception while keeping water activity below 0.3–0.4 in the dried product, which avoids microbial growth.

Water activity a_w is a thermodynamic concept intrinsically connected to the chemical potential of the water (μ_w) in a system (food or drying environment) at a given absolute temperature (T), as shown in Equation 1.

$$\mu_w = \mu^0{}_w + RT\, ln\; aw \qquad `(1)$$

In this, μ^0_w is the standard chemical potential of the water, and R is the ideal gas constant. The higher the $\alpha\ w\alpha$ the higher the $\mu\ w\mu\ w$ and the tendency/availability (or simply potential) for the water present in a medium to react, move, or be used in microbial metabolism. In practice, $\mu\ w\mu\ w$ is correlated to the potential for water mass transfer—essential in drying—and food stability after processing.

2.1 The Driving Forces of Drying

Transport phenomena always result from a non-equilibrium situation (driving force), such as the temperature difference (heat transfer) and difference in chemical potential or activity of a component (mass transfer) between two phases. Drying occurs while both potential differences exist—i.e., in a non-equilibrium situation with simultaneous heat and mass transfer. The temperature difference is directly measurable and quite noticeable. However, mass transfer potential is not directly obtained, and some assumptions and strategies are required to measure it.

In a direct relationship, the water activity of a substance i (α_i) is estimated by the ratio of the actual fugacity in a real medium (*fi*) by the fugacity of that pure component (*f io*) at the same temperature (Equation 2).

$$\alpha_i = f_i/f_i^0 = f_i/f_i^0 \qquad (2)$$

In this way, the driving force for water mass transfer in drying is given by the difference between the water fugacity of the food system (*f wL* - liquid phase) and the air (*f wV* - vapor phase) (Equation 3) since *f io* is the same for both phases.

$$\textit{Driving force} = f_w^L - f_w^V \quad (3)$$

For the liquid phase (Equation 4):

$$f_w^L = \alpha_w f_w^0 \quad (4)$$

In this, $f\ wo = \phi\ wsatp\ \ wsat$ when the global temperature and pressure are relatively low (far from the critical point). Here, $\phi\ wsat$ is the fugacity coefficient of saturated water and $p\ wsat$ is the water vapor pressure, both at the same temperature. Additionally, when the global pressure is close to the atmospheric condition, $\phi\ wsat$ is very close to 1. Thus, Equation 4 results in Equation 5.

$$f_w^L = \alpha_w p_w^{sat} \quad (5)$$

For the vapor phase, the definition of fugacity coefficient (Equation 6):

$$\phi_w^V = f_w^V / p_w \quad (6)$$

In this, $\phi\ wV$ is the fugacity coefficient of water vapor and p_w is the vapor partial pressure in the air. Again, as drying is commonly under relatively low pressure, is very close to 1. Thus, finally, the driving force can be expressed as in Equation 7.

$$\textit{Driving force} = \alpha_w p_w / {}^{sat}_{pw} - p_w \quad (7)$$

The equilibrium condition, when the driving force is zero, leaves to $\alpha_w p_w / {}^{sat}_{pw} = p_w$, which gives the well-known approximation (Equation 8):

$$a_w = p_w / p_w^{sat} \quad (8)$$

Air relative humidity (*RH*) is an indirect measurement of water chemical potential in the air that depends on temperature and air moisture content (*U*). It is defined in percentage by the ratio of the actual *U* by the maximum *U* value (saturation) at the same given temperature. Assuming an ideal gas behavior for the air results in Equation 9.

$$RH = 100\, p_w / {}^{sat}_{pw} \quad (9)$$

Note that, at the equilibrium (Equation 10):

$$\alpha_w = RH/100 \quad (10)$$

Thus, in a simplified way, water removal occurs when the relationship between food α_w and air *RH* is not satisfied by Equation 10, and a heat source (from the surroundings or the drying material itself) is present to supply vaporization enthalpy.

2.2 Water-Food Interaction

Food products are complex systems that may consist of numerous components and heterogeneous macro and microstructures. Thus, the water molecules interact differently with each piece of food, which strongly depends on the local temperature. Additionally, the moisture content changes those interactions, and water molecules' bonding intensities may differ inside the same piece. The literature presents thermodynamics-based equations for estimating the water activity of some food solutions. However, water interaction with the complex structure of solid foods and their soluble solids is still im α_w sible. However, there is some equipment available for the measurement of food-water activity, α_w.

On the other hand, a practical approach to obtaining a food α_w is correlating the food moisture with air *RH* when both are equilibrated at a given temperature. Moisture content is represented mainly on a dry basis (*X*)—i.e., the mass ratio of water and all other components except the water. This equilibrium relationship *X* vs. *RH* at a given temperature is called moisture sorption isotherm, as plotted in Fig. 2. A solid porous material reaches the equilibrium with the same *RH* at a lower *X* as the temperature increases. It means that, for the same moisture content, α_w increases with the temperature (see the isotherms for different temperatures in Fig. 2).

The Guggenheim-Anderson-de Boer (GAB) equation, Equation 11, is the most common mathematical model for representing food's water sorption isotherm, *X* vs a_w.

$$\mathrm{X} = \frac{X_0\, CKa_w}{(1 - Ka_w)[1 + (C - 1)Ka_w]} \tag{11}$$

X_0 is the monolayer moisture—an estimative of the water present at the first layer of sorption; represents the total sorption energy of this first layer—there is a strong interaction of water molecules with the primary sorption sites, leading to exclusively positive *C* values; and *K* is related to the moisture multilayer sorption energy is to keep the GAB equation rational, $0 < K \leq 1$. Despite the proposed physical meaning of X_0, *C*, and *K*, foods are multicomponent systems with complex microstructures, and those parameters are only found by fitting Equation 11 to the experimental data. Details on estimating GAB parameters can be found in the literature, such as in Blahovec and Yanniotis (2008).

The typical isotherm behavior of solid foods in Fig. 2 shows the highly nonlinear reduction of a_w because of removing water from a moisty food matrix. The water molecules slightly bonded to the matrix, with a higher a_w, evaporate more easily. As the drying progresses and the moisture gradually decreases, the water strongly bound to the solid also evaporates. At the limit, just a monolayer of water adsorbed on the surface of the solid porous structure remains.

Figure 3 represents the water-solid food interaction in four stages, starting from a high-moisture food before the drying process and ending with a food matrix with very low a_w. The image in Fig. 3a is typical of food parts with a_w very close to 1, indicating that most water behaves like pure water. In this situation, an aqueous phase solvates ions (e.g., salts and organic acids) and polar molecules (e.g., sugars), while macromolecules (e.g., starch, pectin, and proteins) are surrounded by water molecules, forming hydration layers depending on their hydrophilicity. One can observe in Fig. 2 that the sorption isotherm has a very high slope in the region of a_w close to 1, representing the water on the material surface and in large pores. In this region, moisture reductions do not cause

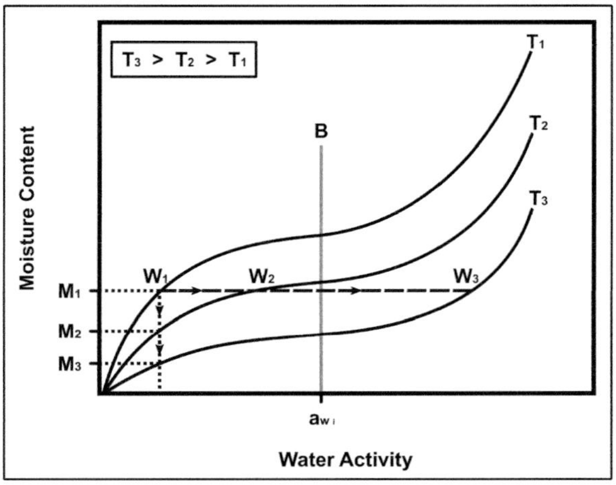

Figure 2. Moisture sorption isotherm in different temperatures. X is the moisture content at equilibrium in each *RH* and temperature, a_w is the water activity at equilibrium in each moisture content and temperature, and T is the temperature.

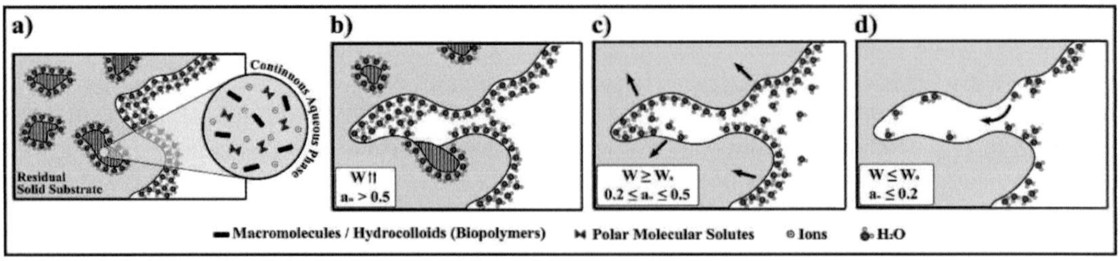

Figure 3. Representation of the water-food interaction during drying for different water activity values (a) a_w close to 1; (b) $a_w > 0.5$; (c) $0.2 \leq a_w \leq 0.5$ and (d) $a_w \leq 0.2$ (adapted from Martinez-Navarrete et al. 1998).

significant drops in water activity. However, this behavior changes when the weakly bonded water molecules are evaporated, moving toward the second stage.

Figure 3b represents food-water interaction in regions that reached intermediate a_w between 0.5–0.6 and 0.8. The aqueous phase became more concentrated due to water removal, mostly inside the porous spaces. Also, there are water molecules on the surface of the solid matrix, forming adsorbed multiple layers. As the drying process continues, part of the water in the pores evaporates, forming an even more concentrated and viscous solution, in Fig. 3c, reducing a_w below 0.5 (until approximately 0.2–0.3). The last image (Fig. 3d) represents very low a_w values since most water molecules are bonded to the solid matrix surface, requiring much more energy to remove them. In general, at this last stage, very little moisture is present in the food and X is very close to the monolayer value X_0.

These generic images in Fig. 3 help in understanding the physical-chemical aspects of water-food interactions at different moistures and water activities, which are crucial for predicting food shelf-stability. Figure 4 is a diagram demonstrating how the rate of the main food changes—microorganisms' growth, enzyme activities, oxidation, anthocyanin degradation, and browning reactions—are related to water-food matrix interactions.

Intermediate moisture foods ($0.5 < a_w \leq 0.8$) do not present microbial proliferation as a problem, but enzyme activity remains high as does non-enzymatic browning (at high temperature) and lipid oxidation. These food products are often stored under refrigeration. As much as the product goes toward a_w around 0.3, the stability increases since no microbial growth occur (although some microorganisms can remain latent), and chemical/biochemical reactions are minimized. Further drying, reaching the fourth stage in Fig. 3, leads to moisture monolayer disruption. Thus, oxidation increases drastically because of the exposure of the matrix wall to the air, frequently at a relatively high temperature when under drying.

It is important to note that all values and stages mentioned above are generic. The construction of the food sorption isotherm at the temperature expected for its storage is the correct way to estimate foods' stability based on their moisture and water activity.

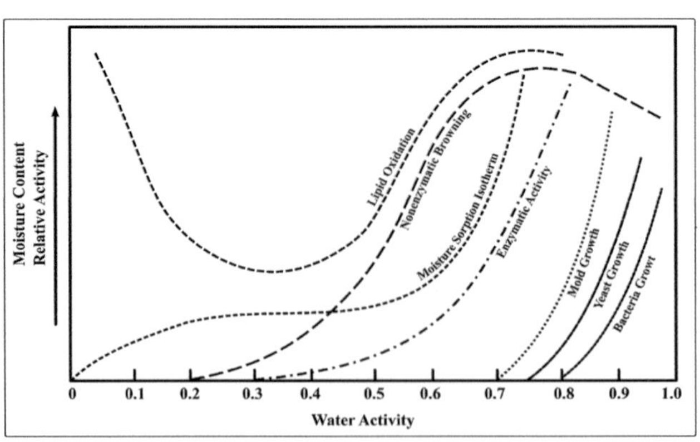

Figure 4. Relation between water activity and relative rate of the most important changes in a food material.

2.3 Food Drying: Phenomena and Processing

Drying is a unit operation in which several physical and chemical phenomena co-occur and interact, which depend on the initial moisture of the product and operating conditions. It occurs because of the difference in the water chemical potential between the food surface and the drying air, the most common drying method. However, the difference in water activity or partial water vapor pressure between the material surface and the air is often used to represent the chemical potential difference between the phases.

During drying, simultaneous and coupled heat and mass transfer are caused by heating the product surface with hot air (convective drying). Drying is an intrinsically transient process, with temperature and moisture gradients that are continually changing. These gradients, in turn, generate mechanical stresses in the material structure, which deforms and shrinks, modifying the pore space and its transport properties.

A schematic of the heat and mass transfer phenomena involved in drying processes is depicted in Fig. 5. Heat is transferred from the hot air to the material surface by convection and then to its interior by conduction. Therefore, drying occurs initially at the surface in contact with the hot air, creating moisture concentration gradients between the interior and the material surface. Thus, convective heat transfer occurs from the air to the food surface, and water transfer occurs from the surface to the drying air, which acts as the heating and water vapor-carrying media. Therefore, the interface food-environment governs the coupled heat and mass transfer phenomena, with convective heat and mass transfer, including diffusion through the boundary layers.

Depending on the characteristics of the solid matrix and drying period, the internal moisture gradients cause water to flow to the surface by capillarity or liquid diffusion (e.g., in gels). Evaporation can also occur inside the solid matrix, causing water vapor to diffuse through the porous structure to the food surface. Finally, viscous flow can appear in the pores if pressure gradients are produced by temperature gradients or rapid structure shrinkage. Generally, a combination of these transfer mechanisms is present during drying. The relative importance of each transport mechanism depends on the food moisture and frequently changes during the process. When the drying process occurs under a vacuum, the boundary layer and airflow are mostly absent, and the chamber air-vapor drainage is crucial for efficient drying. Several physical-mathematical models,

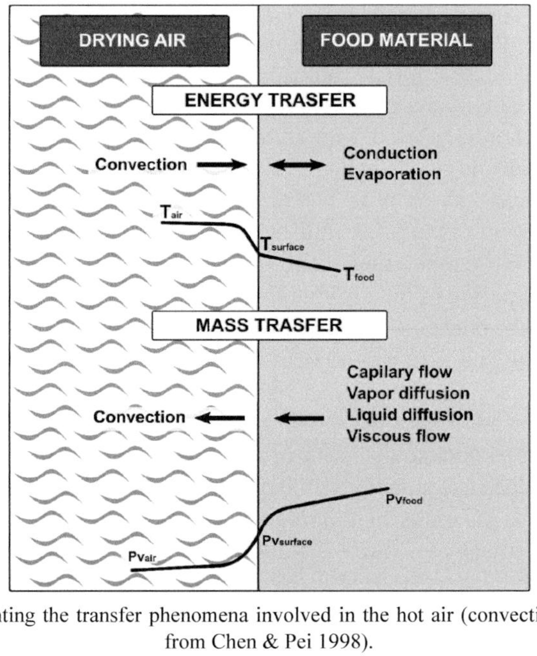

Figure 5. A diagram representing the transfer phenomena involved in the hot air (convective drying) operation (adapted from Chen & Pei 1998).

called phenomenological models, have been proposed to describe and predict drying processes. However, this chapter will not address the modeling of drying processes.

Another dimension of the operation is linked to the goal of producing dehydrated foods that meet consumer demands. Thus, it should be considered that chemical and biochemical reactions that occur during drying need to be known so that the process conditions reduce the transformations that can degrade the product's nutritional and sensory characteristics.

On the other hand, biochemical changes that occur during drying are already part of the identity of the resulting product, as is the case for intermediate moisture bananas. The process carried out at high temperatures causes caramelization of the sugars, darkens the fruit, and produces its characteristic flavor. Another typical example is dehydrated grapes (raisins), which undergo significant shrinkage, and it is a part of its identity but is not a real technological problem. Dehydrated grapes that maintain their initial volumes would be another (new) product.

The nature of the food to be dried should be considered when selecting the type of dryer to use. If the food is composed of solids that can withstand mechanical shocks with the equipment walls and other hard surfaces, there is a wide variety of dryers available for its dehydration (tunnel dryers and rotary dryers among others). On the other hand, if the food can be damaged during its movement, one should avoid rotating and fluidized bed dryers. If the food is a liquid solution or suspension (milk, coffee extract, and fruit pulp), suitable dryers are different from those used for solids, and fewer models are available. Drum dryers and spray dryers are the most common, as discussed later. However, before presenting the usual dryers found in the food industry, an overall analysis of convective drying is presented based on the experimental approach and theory developed over a century.

2.4 Food Drying: The Experimental Approach

The experimental drying curve represents the variation of the material mass (or its moisture) during the drying process, obtained by weighing and recording the sample mass during the drying process done at given air conditions. Since the initial sample mass and moisture are known, the masses of the dry solid (constant during drying) and water in the sample are easily calculated. This way, one can plot the drying curve obtained under experimental conditions close to an expected industrial condition.

The drying rate curve can be obtained from the mathematical derivative of the drying curve. If many records of the sample mass during drying are available, the derivative can be performed numerically, using the mass change at each time interval. However, it is usual to perform a smoothing of the drying curve data using sums of polynomials or filters that follow the trend of the drying curve. Empirical models should be used with great care, as the goodness of fit does not guarantee that the derivatives result in physically meaningful drying rates. In other words, the drying curve smoothed by the methods cited above must respect the physics of the drying process.

Figure 6 shows the drying curve—drying rate (dX/dt vs. X)—and related drying rate curve, and it also shows the sample temperature curve during convective drying. The drying rate curve shows an initial adaptation period of the drying process, when the drying rate increases rapidly until reaching a plateau, called the constant drying rate period. For this period to exist, the surface of the solid exposed to drying must be very wet and covered by a film of water. The physical situation is like the wet bulb thermometer. The evaporated water is weakly bound to the food solid matrix, and evaporation occurs as if the solid does not exist. In porous solids with high moisture, the capillary mechanism feeds liquid water back to the surface, keeping the surface wet. The constant drying rate period continues if the surface has sufficient moisture to maintain the water vapor pressure, like the saturation vapor pressure, at the material temperature.

The literature calls critical moisture (X_c) the value at which there is a transition from the constant to the falling drying rate period. This transition occurs when the rate of liquid reaching the surface from the matrix inside can no longer balance the evaporation rate. However, X_c depends on

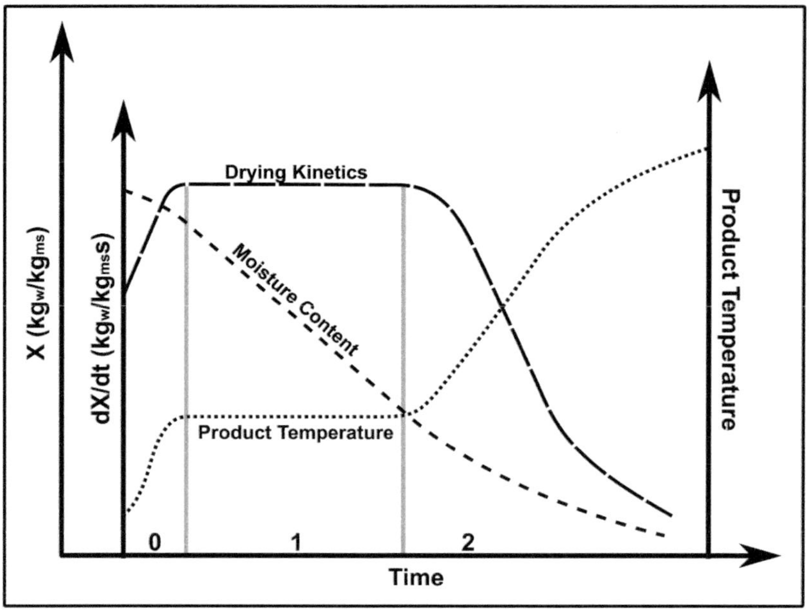

Figure 6. Generic drying curve, determined from the continuous weighing of a sample during the drying process, drying rate curve, and temperature curves of a sample during convective drying.

the solid dimensions, transport properties, and drying conditions. For a given solid and dryer, the critical moisture is expected to be higher for higher air temperatures and velocities, increasing both the driving forces and the convective coefficients for heat and mass transfer. On the contrary, for low and constant drying rates (e.g., drying air at a temperature just above the solid's temperature), the moisture transfer from the solid interior to its surface will ensure a wet surface for longer times, thus resulting in higher X_c.

From X_c onwards, the liquid film on the solid surface loses continuity, and filmless zones appear and grow continuously, leading to a second drying period, the falling rate period. Heat transfer from the air to the solid surface and then to its interior by heat conduction increases. The temperature of the solid surface increases, and the evaporation front moves inside the solid. At the same time, water vapor diffuses from the evaporation front to the solid surface and the drying air (Fig. 5).

The water removed during the falling rate period may be much less than in the constant rate period, but the time required is frequently more prolonged because the drying process is controlled by the relatively slow diffusion of water from the interior to the solid surface. Therefore, the shape of the drying curve in the falling rate period depends on the mechanisms controlling the moisture transfer from the interior of the solid to the drying surface.

The liquid theory has been used to describe what runs in non-porous solids, where an aqueous solution is formed in materials like pastes, gels, soaps, and glues. This theory has also been used to describe the final stages of food drying and the drying of leather, fabrics, and clay. Overall, the water diffusion coefficient in the solid decreases for lower moisture; in this way, decreases during the falling rate drying period. On the other hand, the capillarity theory has been applied to describe the water transfer in capillary porous solids. The typical shapes of the drying curves of both solid materials are depicted in Fig. 7.

The prediction of food drying time has been reported for many situations but is not the goal of this chapter. The lector can find fundamental and more advanced approaches to the drying operation in classical books of unit operations (Geankoplis 1993; Ibarz and Barbosa-Cánovas 2002; McCabe et al. 1987). A simple and reliable way to estimate the drying time is from the food drying curve obtained from drying experiments that mimic industrial conditions.

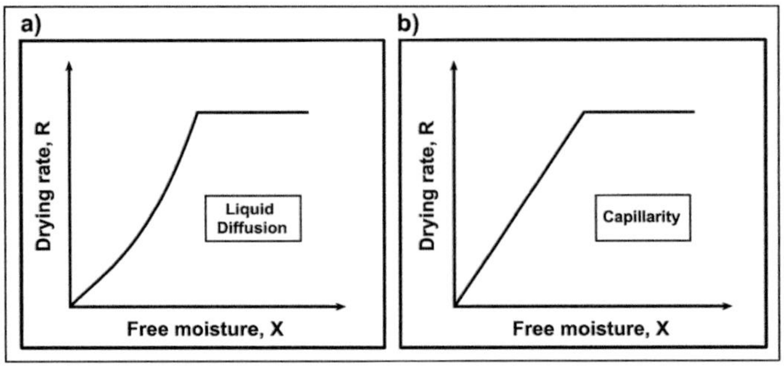

Figure 7. Mechanisms of transfer of liquids into solids and the shape of the drying curve in the period at decreasing rate.

2.5 Food Dryers

Foods to be dried can be liquid, solid, or semisolid; they can also be fragile, deformable, or rigid. They have different sizes and mechanical requirements for manipulation and are particularly impacted by heating. Just as important, one needs to ponder if the process respects environmental, ethical, and economic aspects and if the product obtained matches the nutritional and sensory requirements. In this sense, many approaches resulted in several dryers and operational modes developed over the years.

The first generation of dryers used trays, cabinets, and beds with direct air or solar heating. Then, atomizers and rotating drums represented better solutions for drying liquids, purees, and pastes. Next, freeze-drying and osmotic dehydration remarkably moved towards obtaining high-quality products. Finally, the development of new approaches led to continuous dryers (conveyor belts and fluidized bed), intermittent heating, intermittent vacuum drying, and the use of electromagnetic energy that results in volumetric heating (microwave and radio frequency), infrared radiation (superficial heating), and cast-tape drying. All these drying techniques and equipment aim to increase product quality, productivity, efficiency, and control. Figure 8 presents the sequential overcame drawbacks and developed features while the dryers and drying processes evolved over the years.

A classification considering the material heating and operational modes of dryers is presented in the following.

- *Direct or convective drying*: It is a class of drying processes in which the air is the heating medium, and the carrying agent removes the water evaporated. Mostly called convective drying since the airflow exchanges heat and mass with the food surface by the convective mechanism.
- *Indirect drying*: In these processes, a heating source other than the drying air supplies the evaporation heat. This source can be thermal radiation (including solar radiation), microwaves, radiofrequency, or conductive heating (through food and their support).
- *Batch drying*: This operating mode is most common on a small scale when all food load is placed inside the drier and, after processing, the dried product is removed.
- *Continuous drying*: In this operating mode, the drying equipment is designed to be fed and deliver the product continuously.
- *Semi-continuous drying*: In this case, the dryer operates continuously, but the drying material input and output are shifted at a specific time interval after the last one.

There are many food dryers available for different foods. The most common dryers used in the industry are presented with a short description and some comments about their use.

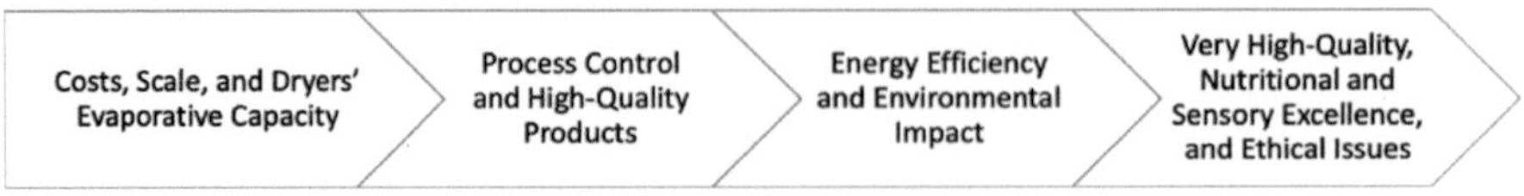

Figure 8. Sequential dryers and drying processes development features, reaching consumer-faced criterium.

2.5.1 Cabinet Tray Dryers

It is the most versatile dryer operating in batches with direct and indirect heating options, capable of drying most foods, which remain static while processing (Fig. 9a). The trays can be fed outside the drying chamber and inserted into the dryer after each batch is completed, favoring productivity. The food is spread evenly over perforated trays that allow the hot air to circulate after being heated by electric resistances or a heat exchanger operating with steam. The air temperature and *RH* can be controlled by a thermostat and recirculating part of the hot air, which can also save energy. Although these processing parameters can be controlled, drying heterogeneities may happen depending on the dryer design. It tends to be simpler and cheaper than other dryers, especially at atmospheric pressure. However, cabinet tray dryers can operate under a vacuum, resulting in lower material temperatures and high drying rates. Vacuum dryers do not have airflow; thus, they are coupled to a vapor condenser and a vacuum pump to remove incondensable gases (Fig. 9b). The lower drying temperature and lower chamber oxygen concentration are suitable for thermolabile or fatty foods, which are susceptible to oxidation.

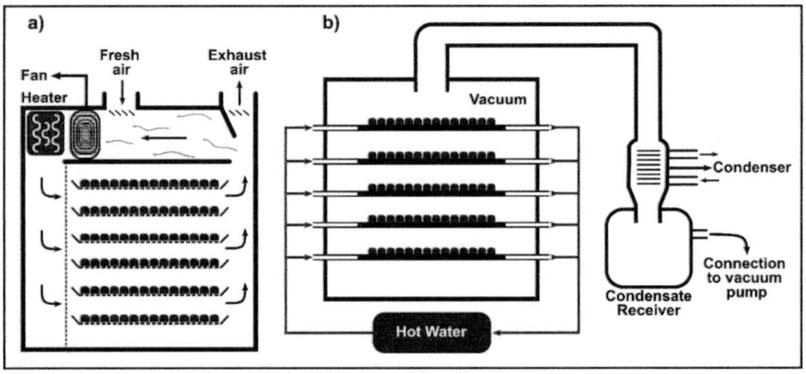

Figure 9. (a) Cabinet tray dryer operating at atmospheric pressure and (b) under vacuum.

2.5.2 Tunnel Dryers

It enhances the cabinet tray dryer, increasing productivity without adding complexity and operational costs. Here, the trays (individually or grouped into a troller), containing the food material, are fed at the tunnel entrance. Thus, they go through the drying chamber to be removed at the other end. The drying air frequently flows in countercurrent, but the concurrent flow can be chosen if it is convenient, aiming at a given product characteristic. It is classified as a semi-continuous dryer mainly using direct heating (hot air). Air blowers' arrangement along the tunnel can reduce heterogeneities and improve food quality (Fig. 10).

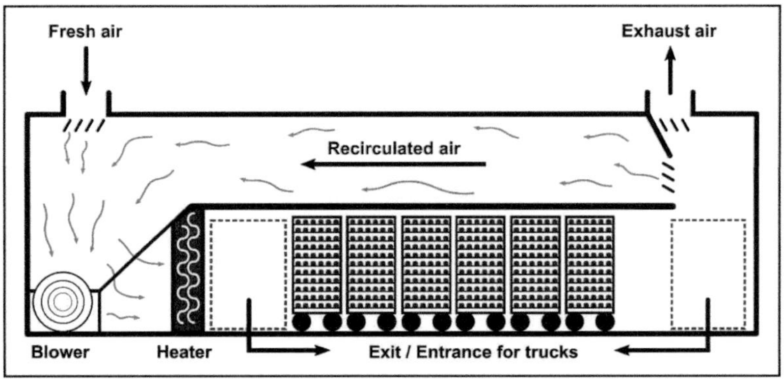

Figure 10. Tunnel Dryer

2.5.3 Belt Dryers

It is a continuous dryer where a conveyor belt carries the wet food through the drying chamber (Fig. 11). Direct drying is the most common operational mode with concurrent-countercurrent or cross-airflow arrangements (for perforated conveyor belts). Also, many of these driers use indirect heating—microwaves (MW), infrared radiation (IR), and heated belts (conductive heating)—as the only heat source. It is possible to combine directly with indirect heating in hybrid dryers. In any case, air flows through the dryer to carry out the water evaporated. MW and IR must be generated along the chamber length, and some controlling system may be required to avoid overheating the drying material.

Conductive heating can be done by heating an impervious belt with attached electric resistances or by hot water or vapor touching its bottom side (cast-tape drying and refractance window methods). Thermal radiation emitted by infrared lamps can also heat the bottom side belt or the food material if the lamps are placed above the conveyor belt. Belt dryers are suitable for solid, semisolid, and viscous foods directly applied to the belt, and low-viscous liquids can also be dried if put in containers (like baking trays). The belt material needs to be food grade, flexible, and resilient, ensuring safety and productivity. In this way, the correct belt for each application is a vital aspect to consider. The final products are dried pieces, sheets, films, flakes, or powders, and collecting them may require a cooling down section and a scraping system. Some details of this kind of pulp dryer are presented in this chapter.

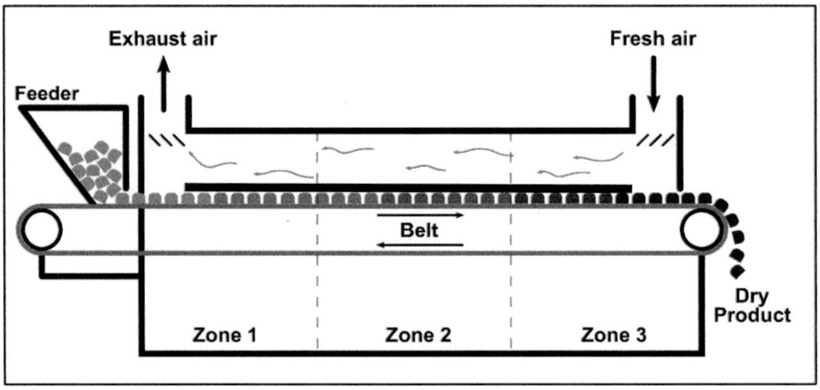

Figure 11. Belt Dryer.

2.5.4 Drum Dryers

It consists of a hot rotary drum coated with a thin layer of the product to be dried (Fig. 12), such as starches and cereal-based foods, fruit pulps, solutions, suspensions, pastes, and semisolids, milk and milk-based foods, and yeast. Drum dryers work by indirect heating with vapor spraying inside de dryer wall that is heated, promoting conduction heat transfer to the thin layer of liquid submitted to drying. Drum dryers operate continuously, similarly to belt dryers, with the removal of the final sheets, films, flakes, or powders enabled by cooling and a scrapper. Frequently, the drum dryer operates under atmospheric pressure, but it can also operate inside a vacuum chamber, thus reducing the drying temperature and increasing the drying rate. Some configurations use auxiliary applicator rolls and twin drums, changing the application mode aiming at the drying of many food liquids.

2.5.5 Spray Dryers

Spray dryers operate in continuous mode, where the liquid is fed under pressure through an infeed nozzle, producing micrometric droplets that have direct contact with the hot air flowing concurrently or counter currently inside the drying chamber (Fig. 13). Spray Dyers' take advantage of the

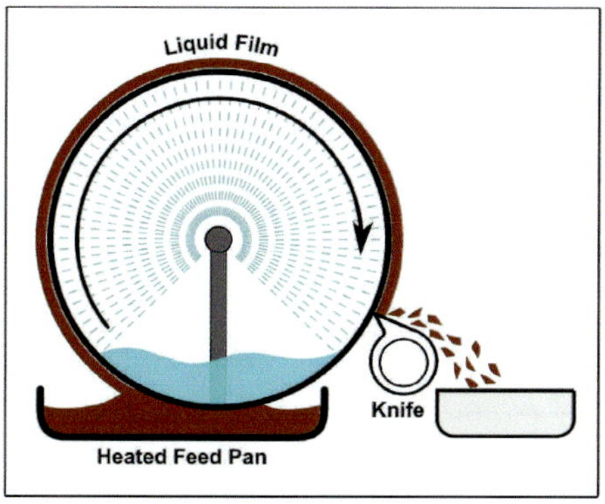

Figure 12. Drum Dryer.

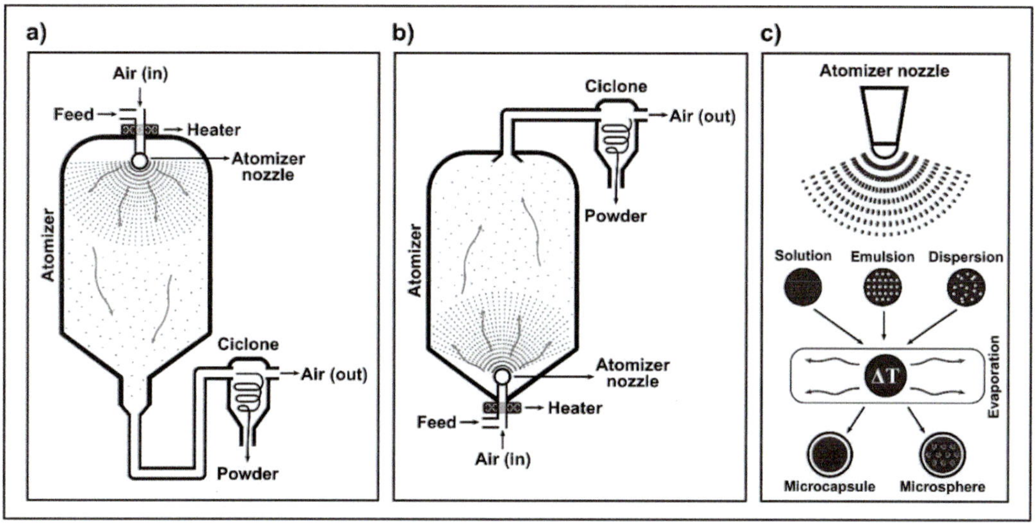

Figure 13. Spray Dryers and compounds encapsulation.

substantial interfacial area between the sprayed liquid food and the hot air to obtain higher heat and mass transfer rates, resulting in drying times of seconds. The inlet air is heated to temperatures between 120°C and 180°C, drastically lowering its *RH*. The operating conditions in spray dryers create high potential differences for heat and mass transfer between the droplets and the drying air, resulting in very high drying rates.

These dryers are cylindrical chambers vertically oriented with a radius big enough to reduce droplet agglutination and dried material deposition on the walls, ranging from a few to tens of meters high and the capacity to produce 30,000 kg of milk powder per hour. The selection of the atomizer mechanism (nozzle and pumping system) depends on the liquid rheology, composition, suspended particles, emulsion presence, and powder required size. Spray dryers have been successfully applied to produce powdered products from milk, coffee, tea, and specific fruit pulps.

Chemical compound encapsulation has also been carried out in this dryer by pumping into the chamber an emulsion or different liquids (one with the wall material and another containing the molecules to be enclosed) concomitantly. The final powder, commonly collected by cyclone and bag filter systems, has a highly effective surface area that facilitates further dissolution. However, at the same time, the dried material has high surface energy for the same reason and can become as sticky as the amount of sugar and organic acids on it. The latter problem is contoured by additives (such as carbohydrate oligomers) that increase the glass transition temperature of the dried product, reducing its stickiness and the caking phenomenon.

2.5.6 Fluidized Bed Dryers

A fluidized bed dryer uses airflow designed to suspend particles, increasing the particle-drying air contact. Particles inside a chamber form a fluidized bed when the fluid through the bed has the combined velocity, density, and viscosity that results in an equilibrium between the drag and weight forces acting on the suspended particles (they look like they are flowing). Thus, the heat and mass transfer coefficients are significant, and the drying rate increases in continuous or batch-fluidized bed dryers. It is advantageous for not too fragile food products since mechanical damage can occur due to particle movements. A schematic of a fluidized bed dryer is done in Fig. 14.

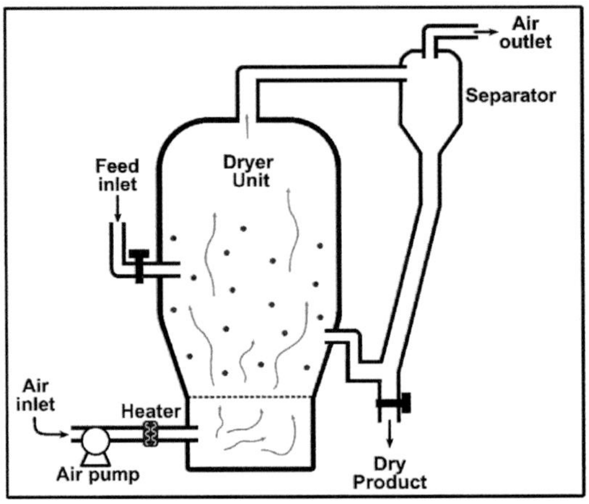

Figure 14. Fluidized Bed Dryer.

2.5.7 Freeze Dryers

The concept of freeze-drying (or lyophilization) is to remove water by ice sublimation. It requires operating below the water triple point (around 0°C and 611 Pa or 6.11 millibars), avoiding ice melting. The food product may be frozen before or inside the drying chamber, which operates under a vacuum in batch cycles. The low temperatures (–20°C; –30°C) preserve thermolabile components. Also, depending on how the ice was produced, this method helps to protect the solid food or microorganism microstructure. Fast freezing produces tiny ice crystals, thus avoiding cell wall rupture. Besides, the ice phase reduces the liquid water plasticizing effect, increasing the local glass transition temperature during the water removal and reducing polymeric chain mobility.

Consequently, in many cases, the original structure is almost perfectly reestablished after rehydration, such as pieces of meat and vegetables added to instantaneous soups or infant formula. Thus, the advantage here is to preserve food structure and nutritional compounds. However, the relatively long processing time and high energy consumption are drawbacks that increase the costs of lyophilized products. A diagram of the equipment is shown in Fig. 15.

The drying systems above are examples covering usual dryers. However, there are many other dryer systems, including combinations or variations of those described. In addition, product needs, such as short drying time, reducing oxidation, or thermal damage, may lead to specific adaptations of their geometry, operational mode, drying conditions, or even new system development.

3. Advanced Food Drying Technologies

Dried foods' quality and process efficiency are crucial in industrial drying operations. Therefore, efforts have been made to develop drying technologies that reduce product damage with improved efficiency, resulting in products with desirable quality characteristics at affordable costs. Some drying technologies have demonstrated their potential for these purposes.

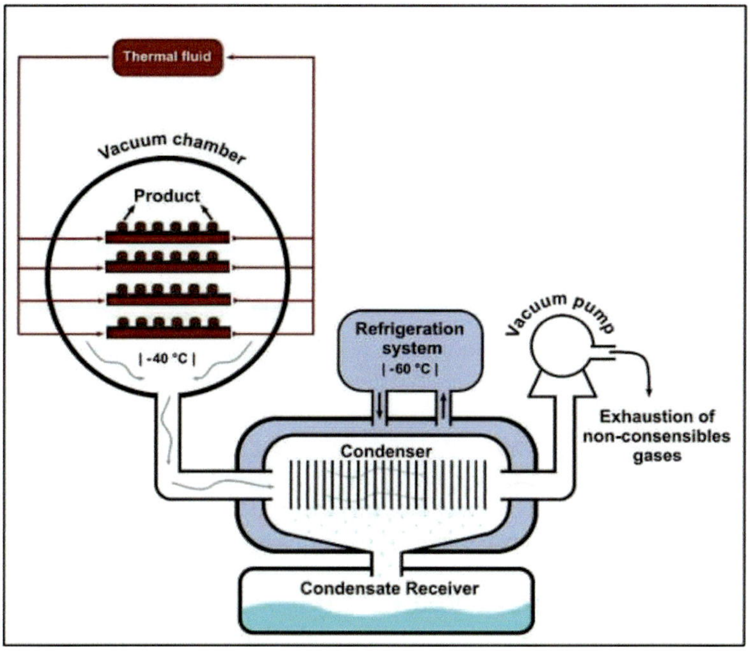

Figure 15. Freeze-dryer.

3.1 Infrared Drying

Infrared (IR) is an electromagnetic wave that is divided into three categories according to its wavelength: near-infrared (NIR, 0.75–3.0 μm), middle-infrared (MIR, 3.0–25 μm), and far-infrared (25–1,000 μm) (Niu et al. 2021; Ratti and Mujumdar 2020). IR radiation can be absorbed, reflected, or transmitted when it hits a foodstuff. The ability of food to absorb energy and convert it into heat depends on the source's wavelength and the food composition. For instance, amino acids, nucleic acids, and proteins have strong absorption bands centered at 3–4 μm and 6–9 μm, while sugars absorb IR with wavelengths of 3 μm and 7–10 μm, and lipids absorb IR with wavelengths of 3–4 μm, 6 μm, and 9–10 μm. In addition, water presents a broad IR absorption spectrum, making IR a suitable method for food drying (Sakare et al. 2020).

After penetration into the food surface, the IR changes the vibrational state of molecules, which vibrate at a frequency of 60,000–150,000 MHz, causing intermolecular friction and rapid internal heating. In this way, IR heating has been used for food drying since it can penetrate deep within moist materials and heat the trapped water (Sakare et al. 2020). IR penetration depth is calculated as the $1/e$ attenuation, in which e is the Euler number. For instance, IR can penetrate the human skin from 0.1 mm to 4 cm.

IR drying has become more used in recent years because of its high efficiency in converting electrical energy into heat, saving energy while resulting in uniform heating in flat and thin materials and short drying times. Also, the equipment is easy to install and operate with a low maintenance cost and clean operation. Furthermore, it is suitable for process control, which helps to produce good-quality dried foods. Therefore, IR has been combined with other drying methods (air-drying, vacuum drying, microwave drying, and freeze-drying) to improve food drying efficiency. However, the disadvantage of IR drying is the non-uniformity of heating in foods with complex shapes and sizes once the energy absorption is different from one place to another. In addition, as IR absorption differs for proteins, fats, carbohydrates, and water, the drying uniformity can be affected when a thick food material is dried (Ratti and Mujumdar 2020).

3.2 Microwave Drying

Microwaves and radiofrequency are electromagnetic waves with different frequencies. Microwaves are in the frequency range of 300 MHz to 300 GHz and have wavelengths from 1 m to 1 mm. Radiofrequency wave systems operate at 13.56, 27.12, and 40.68 MHz. Domestic microwave equipment operates at 2,450 MHz while in the industry, the equipment operates at 2,450 MHz and 915 MHz (Verma et al. 2020).

Microwave heating occurs by dipole rotation and ionic conduction. Dipole rotation mainly occurs with water molecules due to its dipolar nature. When exposed to an oscillating electric field, water molecules try to realign according to the electric field direction. The realignment occurs a billion times per second (2.45×10^9 in the case of 2,450 MHz) because of the high frequency of the electric field, causing molecules friction and food heating. Ionic conduction occurs due to the ions' oscillatory migration that accompanies the electric field, causing collisions between ions and molecules and frictional heat dissipation (Regier et al. 2016).

Therefore, the thermal effect resulting from food-microwaves interaction depends on the food composition that states its ability to absorb and convert microwaves into heat, expressed by its dielectric proprieties. The material dielectric properties are strongly influenced by the waves' frequency, temperature, moisture, specific mass, and food composition (salt content and fat among others).

Drying using microwaves heats the food volumetrically, producing temperature gradients in many directions (instead of only from the surface), and water vaporization may occur inside the food matrix. Like in convective drying, heat and mass transfer by diffusion and capillary liquid flow are present. Besides, if the energy supply is high enough, a water phase change occurs inside the food, producing a significant vapor volume. If the dryer operates under a vacuum, the vapor volume can be huge because of its high specific volume. In addition, internal water vaporization produces pressure gradients that cause liquid and vapor flow inside the food matrix. These complex and intricate phenomena that cause liquid and water vapor flow may accelerate drying and change the food temperature and moisture gradients.

Microwave heating has gained popularity in food processing due to the short heating time, but it has several advantages over conventional air-drying (Datta 2001):

- Selective heating as the energy goes directly to the solvent (wetter areas), reducing wasteful heating;
- Volumetric heating;
- High drying rates that result in short processing times;
- Energy efficiency;
- Can be environmentally friendly, if using renewable energy;
- Fast start-up and shut down times.

Even though microwave heating has the above advantages, some drawbacks limit its application, such as the inherent lack of uniformity of the electromagnetic field within the microwave cavity which can lead to non-uniform heating (Raaholt and Isaksson 2017). Foods with low moisture, low loss constant, and low attenuation factor can favor the presence of stationary waves inside the sample, which cause hot spots and product quality damage. Moreover, another disadvantage of microwave drying is that excessive heating of food corners and edges produces off-flavors, especially during the final drying stages (Zhang et al. 2006).

However, drying temperature and uniformity can be improved by a suitable microwave cavity design, numerical simulations to optimize the appropriate magnetron arrangement (position and numbers), installation of wave stirrers, use of a turntable, rotary drum, and temperature control from power modulation (Monteiro et al. 2020). Using microwave drying combined with air drying can reduce the problems presented above. The use of microwave drying operating under a vacuum is another excellent alternative to reduce the drawbacks of microwave drying significantly. Both alternatives are discussed in the following.

3.2.1 Microwave Heating Coupled With Hot Air

Microwave heating can be combined with hot air in different stages of the drying process to increase the drying rate, which has been applied to many food products. When microwaves are applied at the initial drying stages, the interior gets heat rapidly, forcing vapor outside that creates a porous structure that facilitates further mass transfer, increasing the drying rate. Additionally, applying only hot air at the end of drying helps control the final product temperature that will not exceed the hot air temperature. Microwave drying can also be applied in the final stage of air-drying to remove bound water to overcome the effects of product shrinkage, which causes drops in the drying rate over time (Chandrasekaran et al. 2013).

3.2.2 Microwave Vacuum Drying

Microwave vacuum drying is suitable for producing crispy products with high nutritional and sensory quality in a short drying time. Vacuum drying is particularly suitable for thermosensitive products since it lowers the water boiling point, reducing the drying temperature. Furthermore, as presented before, microwave heating under a vacuum generates water vapor inside the material, producing pressure gradients that drive vapor and liquid flow to the surface. These phenomena result in product expansion (puffing effect), which forms a new porous structure that results in crispy dried products (Gomide et al. 2022). Moreover, microwave vacuum drying prevents or reduces oxidation, preserving color and flavor, leading to high-quality products with affordable costs in shorter drying times than freeze-drying.

It is possible to use a rotating drum to move the product inside the dryer to improve the heating uniformity. For that, the material must support the mechanical shocks resulting from the slow drum rotation, which can be built with sections to reduce the amplitude of food moving inside the dryer. Using automatic temperature control by power modulation is another powerful strategy to reduce excessive product heating. For that, an infrared temperature sensor must measure the average temperature of the food ensemble during drying (Monteiro et al. 2022).

3.3 Explosion Puffing Drying

The explosion puffing drying was developed at the Eastern Regional Research Center in the 1960s to dehydrate fruits and vegetables (Sullivan et al. 1977). This process is based on the compression-decompression of the drying chamber, which causes self-vaporization of the material's moisture, resulting in crispy products with good rehydration capacity (Saca and Lozano 1992; Sullivan and Craig Jr 1984). Figure 16 presents a schematic of the puffing effect. First, the fruits and vegetables to be processed must be partially dehydrated, placed in a pressurized chamber, and heated by superheated steam. After that, the chamber is instantly depressurized to atmospheric pressure, and the water temperature in the food quickly decreases to 100°C, causing instantaneous evaporation that results in a porous and expanded product (puffing effect). After the explosion puffing, the foods need to be dried to 2% to 4% moisture by other conventional methods before packaging.

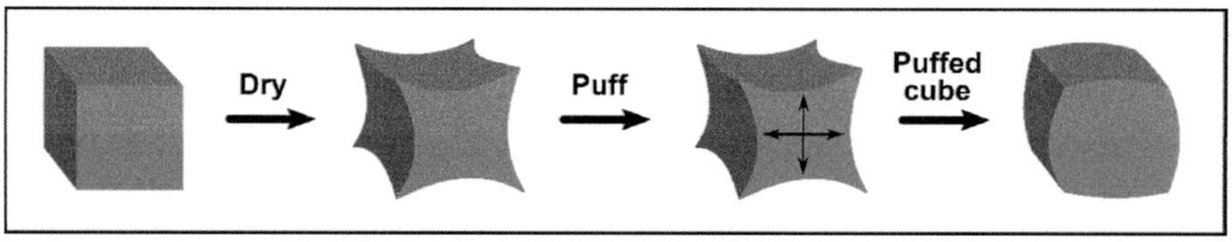

Figure 16. Schematic representation of the puffing mechanism (adapted from Varnalis, Brennan, & MacDougall (2001)).

3.4 Multiflash Drying

Multiflash drying (MFD) is a technology developed at the Federal University of Santa Catarina, Brazil, to quickly produce dried-and-crispy foods with high porosity (Patent No. PI 1107173-

1107177, 2011). The process is based on successive cycles of heating followed by vacuum pulses in the food to be dried. The heating step is done at atmospheric pressure by contacting the food with heated plates (conductive-multiflash drying–KMFD) or microwaves (microwave-multiflash drying–MWMFD). After reaching the desired temperature (e.g., 60°C), a sudden decompression (vacuum pulse) is applied (Monteiro et al. 2016; Porciuncula et al. 2016). The vacuum pulses cause flash evaporation and food texturization, creating a new porous structure.

The flash evaporation results in food cooling, which supplies the required phase change enthalpy. The amount of water evaporated depends on the difference between the sample temperature before and after the vacuum pulse and is represented by Equation 12 (Louka and Allaf 2002; Zotarelli et al. 2012).

$$\Delta m_w = \frac{c_p m_p \Delta T}{\Delta H_v} \tag{12}$$

Δm_w is the mass m_p of water evaporated, c_p is the specific heat of the product that is a function of the moisture content, is the instantaneous mass of the product (water plus dry solids), is the decrease in the sample temperature that results from the pressure drop, and Δm_v is the specific enthalpy of vaporization of water at the instantaneous pressure.

The puffing effect is like the one discussed before (explosion puffing drying section), but the food is submitted to lower temperatures and does not need any pre-drying before the MFD process. Moreover, the drying chamber is not pressurized, and many cycles can be applied to reach the desired product texture.

After the vacuum pulse, the atmospheric pressure is reestablished and additional heating-vacuum pulse cycles can be applied. The number of heating-vacuum pulse cycles can be varied to achieve the desired textural characteristics and moisture content. Figure 17 shows a schematic pressure and temperature evolution during three heating-vacuum pulse cycles. The MFD has the advantage of drying at mild temperatures, vital to preserving nutritional and sensory characteristics, unlike the explosion puffing performed at high temperatures. In addition, the MFD is simpler and less expensive since the heating step occurs at atmospheric pressure.

Figure 18 presents pumpkin pieces dehydrated by air-drying, freeze-drying, conductive-multiflash drying (KMFD), and microwave-multiflash drying (MWMFD). Air-drying caused pronounced fruit shrinkage that formed a compact structure. Freeze-drying retained the fruit's original structure, which is spongy with tiny pores of uniform size. In contrast, samples dehydrated by MFD presented a porous matrix with large pores, which contributes to their noticeable crispness (Link et al. 2017; Monteiro et al. 2018a; 2018b; Monteiro et al. 2021).

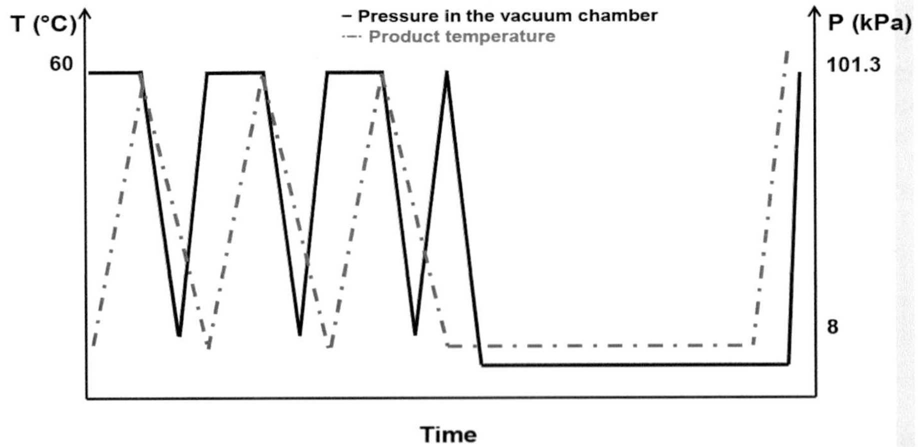

Figure 17. Schematic pressure and temperature profiles during the multiflash drying process (adapted from Monteiro et al. (2016)).

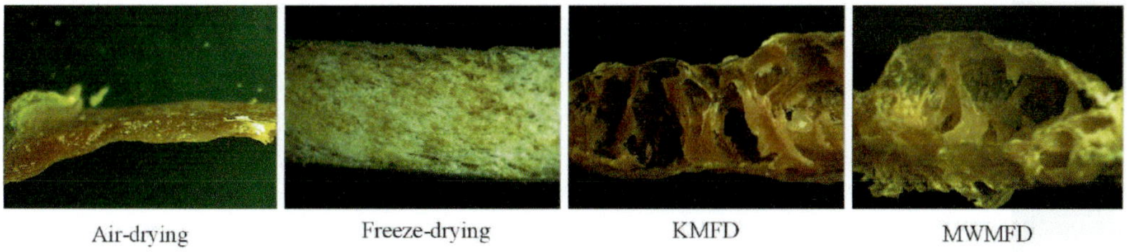

Figure 18. Pumpkins dehydrated by air-drying, freeze-drying, conductive-multiflash drying (KMFD), and microwave multiflash drying (MWMFD) (adapted from Monteiro et al. (2018a)).

The MFD process can be an alternative to freeze-drying for some food dehydration since it requires low capital and operational costs. Additionally, it can be performed in a shorter drying time (1–4 hours) compared to freeze-drying (12–16 hours).

3.5 Cast-tape Drying and Refractance Window

Cast-tape drying (CTD) and refractance window™ (RW) are new-and-similar drying techniques for drying liquid foods. A thin layer of viscous solutions or suspensions is spread with a controlled thickness on heated supports to be concentrated or dried. In the last case, the resulting dried materials are films or flakes, which can be ground to produce powders.

In the RW process, the material to be dried is spread over a flexible and transparent belt conveyor (Mylar®) heated by water in direct contact with the conveyor underside (Nindo and Tang 2007). The transparent belt conveyor is supposed to transmit the infrared radiation emitted by the hot water that heats the moist material (the radiation window is open). As the product loses its moisture, the pulp reflectivity increases, and the radiation is reflected into the heated water source, closing the drying window (Raghavi et al. 2018).

Zotarelli et al. (2015) showed that radiant heat transfer is less than 5% of the total heat transferred during the RW drying process. Ortiz-Jerez et al. (2015) reported results that corroborated the negligible effect of infrared radiation emitted by the hot water during RW drying. As most heat is transferred by conduction through the conveyor belt, using supports that are not transparent to infrared radiation results in similar drying rates. Consequently, Zotarelli et al. (2015) called this drying technique CTD, which includes RW and is controlled by heat conduction but not by infrared radiation. The main components of continuous CTD equipment are shown in Fig. 19.

CTD consists in drying a thin layer of a liquid (e.g., fruit pulp) on low surface energy and impervious conveyor belt, which is heated from its underside by vapor released by hot water (Durigon et al. 2018; Simão et al. 2020). Controlling the temperature of the hot water, which releases the vapor that touches the belt underside, directly controls the drying temperature. An exhaust tunnel with adjustable air velocity removes the water evaporated from the product. At the end of drying, the belt

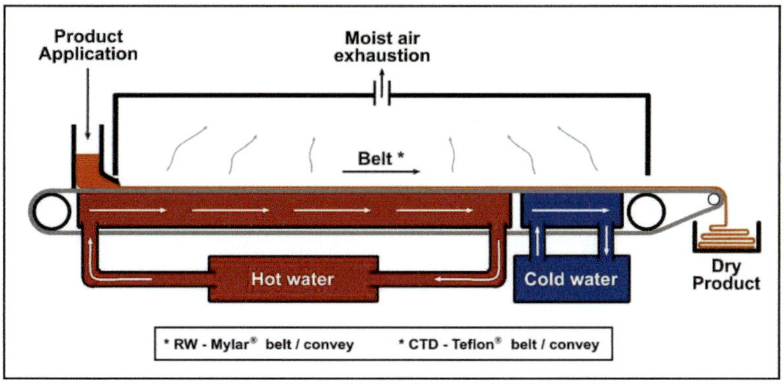

Figure 19. Scheme of CTD or RW process in a continuous regime.

passes over a cooling system that contains cold water to reduce the product temperature, facilitating its removal. Using low surface energy belts (e.g., polyesters or Teflon-coated belts) facilitates the removal of the dried product, requiring minimal use of carrier agents. CTD has been reported as a technique that can dry viscous and sticking solutions and suspensions, like some fruit pulps.

The CTD (RW) has gained much attention due to its short drying time, scalability, energy efficiency, cost, and product quality (Raghavi et al. 2018). This technique has a short drying time, retaining the product color, aroma, antioxidant compounds, and nutritional properties. The dying time can be some minutes to approximately 30 minutes, depending on the pulp thickness and drying conditions. The drying system costs 30–50% of freeze dryers and requires less than 50% energy to dry the same product mass (Nindo et al. 2003; Nindo and Tang 2007).

These advantages have attracted the attention of the food industry and researchers. As a result, several fruits and vegetable suspensions, such as açai, carrot, cherry, guava, mango, strawberry, tomato, and others have been dehydrated by this technique to obtain high-quality products with preserved bioactive compounds responsible for sensory and nutritional attributes (Abonyi et al. 2002; Castoldi et al. 2015; da Silva Simão et al. 2019; Durigon et al. 2016; Frabetti et al. 2021; Frabetti et al. 2018; Pavan et al. 2012; Tontul et al. 2018; Zotarelli et al. 2017).

4. Quality of Dried Foods

Water is an essential part of moist food structure, and as long their molecules leave the medium, significant changes happen and affect the product quality, requiring special attention. Shrinkage is one of the most significant problems when drying solid foods. It is a consequence of intermolecular forces into the dried structure balanced by water molecules' bonds. As a result, the dried food may have undesirable sensory aspects, including excessive hardness and wrinkled aspects. Some alternative drying methods, discussed in this chapter, minimize this drawback by controlling process parameters. Moreover, the shrinkage that causes size and shape modification also impacts heat and mass transfer since the porous structure, characteristic lengths, and superficial area are changed.

The quality attributes can be classified into physical, biological, and chemical properties that influence the shelf life, and the nutritional and sensory characteristics of the dried products, determining their success on the market. The quality changes include loss of aroma and flavor, structural and color modification, and a decrease in nutritional value (Ratti 2001; Sablani 2007). All these attributes are interrelated since microstructural changes affect the drying behavior (heat and mass transfer) that influences biochemical aspects, product texture, and sensory quality.

The damage during drying depends on the food's original composition and structural elements and the drying method and operating temperature since it affects the principles of water removal, drying rate, and drying time (Aguilera et al. 2003). In addition, consumers demand high-quality products with appealing sensory properties and high in healthy compounds. Therefore, understanding each drying process's physical and chemical changes is crucial to food scientists and engineers. Thus, some essential quality attributes of dried foods are discussed.

4.1 Microbial Safety

Fresh foods usually have a high moisture content and are susceptible to microbial deterioration. Dehydration reduces food water activity to avoid microorganisms (bacteria, molds, and yeasts) growth, resulting in products with extended shelf life. The water activity of dehydrated foods is usually lower than 0.6, inhibiting any microbial growth. Pathogenic bacteria cannot grow below water activities of 0.85–0.86, while yeasts and molds usually require values higher than 0.61, being the microorganisms of major relevance in dehydrated products (Barbosa-Cánovas et al. 2020).

The microbiota in dehydrated foods depends on the product's characteristics, such as composition, pH, pretreatments, types of endogenous and contaminating microorganisms, process conditions (drying time, temperature, and pressure), and the drying method (freeze-drying, solar-drying, vacuum-drying, and air-drying) (Bourdoux et al. 2016; Sagar and Suresh Kumar 2010).

However, fresh foods with a significant microbial load require great attention if the drying process is performed under mild temperature conditions to minimize sensory and nutritional changes. Thus, undesirable or harmful microorganisms can survive after food drying, threatening the consumer as vegetative cells and spores can remain viable for months, even at a low water activity (Bourdoux et al. 2016). Drying processes performed at high temperatures can reduce the microbial count. However, pretreatments, such as blanching or pasteurization, are recommended to reduce the microbial count before drying, depending on the food characteristics and its initial microbial load.

4.2 Physical Properties

4.2.1 Texture

Water removal during drying affects the product structure and composition and frequently causes state transitions with significant texture changes. Crispness and brittleness are appealing texture attributes of dried products and are related to the glass transition caused by water removal (Bhandari and Howes 1999; Link et al. 2018).

The glass transition occurs in amorphous materials with a disorderly molecular structure, involving the food transition from an amorphous glassy state to an amorphous rubbery state at a given temperature called glass transition temperature (T_g). Above T_g, the product is in a rubbery state in which the solid matrix has significant mobility with a "plastic" or "chewy" texture. Below T_g, the solid matrix has much lower mobility, and the amorphous material becomes vitreous, with a solid-like structure, and mechanically "hard", "brittle", and "crisp" (Ratti 2009; Roos 2010).

This transition depends on the temperature, food composition, and moisture content. Water acts as a strong plasticizer that increases the food matrix's free volume and molecular mobility, reducing T_g (Huang and Hsieh 2005; Simão et al. 2020). Thus, drying can increase food T_g to a value above the storage temperature, keeping the dried food in a glassy state with low molecular mobility (low biochemical changes rates) and its characteristic textural properties (Fellows 2017).

Monteiro et al. (2020) studied the texture evolution of sweet potatoes during microwave vacuum drying. The puncture test was applied to the samples submitted to different drying times, resulting in force-penetration curves for each sample (Fig. 20).

The force-deformation data were smoothed by a non-parametric method to show the global behavior. Then, the differences between each experimental and smoothed force were calculated to

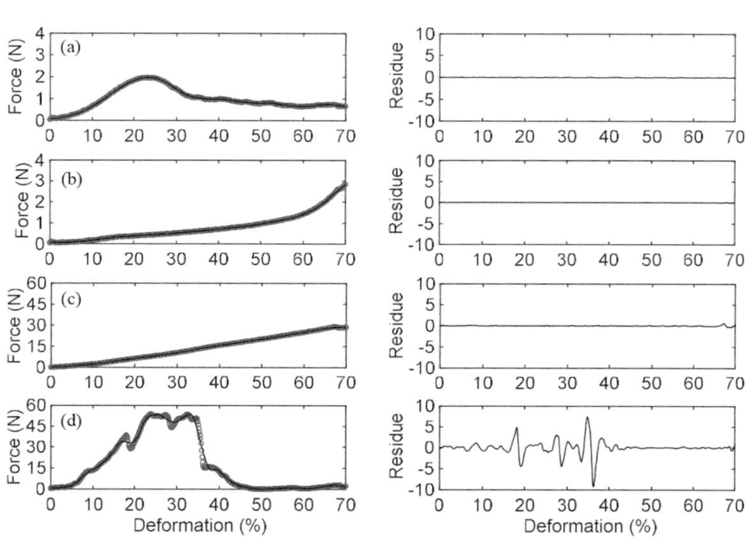

Figure 20. Instrumental texture analysis of sweet potato during drying. Left column: force-deformation data (○) and non-parametric adjustment (—); Right column: curves of residues. (a) blanched potatoes, (b) 4 min, (c) 10 min, (d) 27 min (adapted from Monteiro et al. (2020)).

have the force residues representing the local oscillations. The higher the irregularity of the residues curve, the higher the product crispness (Laurindo and Peleg 2008; 2007). Initially, the potato had a soft texture but changed with water removal, as shown by the residues' curves. A jagged residue curve, typical of crispy products, is observed after a given drying time and food moisture.

Some products appear wet or with a leathery or chewy texture, even at lower moisture content. This characteristic is observed in dehydrated sugar-rich (fructose, glucose, and sucrose) products such as fruit leathers (Simão et al. 2020). The low molecular weight sugars have a plasticizing effect that reduces T_g to below room temperature, preventing the transition of the dried product to the glassy state. In the dehydration of fruit pulps to produce leathers, the addition of maltodextrin before drying increases T_g, which reduces the material stickiness, easing the leather removal from the drying support and its storage (Valenzuela and Aguilera 2015).

The texture of dried foods is also affected by the physical changes the product undergoes during dehydration, such as shrinkage, crust formation (case hardening), crack formation, and changes in the material's porosity. Thus, these physical parameters are discussed below.

4.2.2 Shrinkage, Case Hardening, and Cracks Formation

Moist solid foods are multiphasic and heterogeneous systems formed by a solid matrix holding a large amount of a diluted aqueous solution, adsorbed water, and vapor. During hot air drying, the food surface is the first part that is heated, causing most of the evaporation in this region, which creates moisture gradients in the solid matrix. The moisture gradients result in stresses between dry and moist regions (contraction stresses), causing the food structure to collapse and shrink, changing its shape and size. These geometrical changes are tridimensional and complex but are frequently represented by just the relative reduction of the food volume (Equation 13).

$$\% \, Shrinkage = \frac{V_0 - V_0}{V_0} \times 100 \tag{13}$$

In this, V_o and V are the food volume before and after drying.

As a result of shrinkage, the dried food may have undesirable sensory qualities, including excessive hardness and wrinkled aspects. Thus, food shrinkage reduces consumers' acceptance, except for dried fruits that are well-known for their shrunken aspect, such as dried raisins, plums, peaches, or dates (Mayor and Sereno 2004). Therefore, shrinkage is one of the most critical food physical changes during drying.

As more water evaporates during drying, more contraction stresses appear, increasing food shrinkage (Mayor and Sereno 2004). However, several factors can affect shrinkage, such as material type (ability and resistance of the material to deform), glass transition temperature, drying operational conditions, and case hardening.

When food is dried above its glass transition temperature, its solid matrix is in a rubbery state with high mobility, favoring shrinkage. Thus, the moisture loss is almost compensated by shrinkage, and the material volume decreases linearly with its moisture loss. The higher the difference between the process temperature and T_g, the higher the collapse (Xiao and Mujumdar 2008). On the other hand, below T_g, the low mobility and rigidity of the solid matrix in the glassy state significantly decrease the rate and extension of the material shrinkage, resulting in a relatively more porous structure. As T_g varies according to water content, the extent of shrinkage can change during the drying process. For instance, during the air drying of potatoes, apples, and squid flesh, it was reported that shrinkage decreased at the final stages of drying because of the higher solid matrix rigidity. It is explained by the moisture reduction that causes the T_g to increase to a value closer to the process temperature (Krokida 1997; Mayor and Sereno 2004; Moreira et al. 2000; Rahman and Potluri 1990).

The drying process performed at low temperatures tends to prevent shrinkage. However, as most air-drying processes are performed at temperatures higher than T_g, substantial shrinkage is usual, and the dried food is shrunken and dense (Ratti 2001).

Processes that generate counterbalancing effects that minimize the capillary forces (increased during drying) are suitable for reducing shrinkage. These counterbalancing forces can be produced by (i) drastically reducing the environmental pressure, as observed in freeze-drying and vacuum drying, and (ii) increasing the internal food pressure, which is possible in microwave vacuum drying because of the significant vapor pressure produced by volumetric heating (Jagan et al. 2011).

The counterbalancing forces can also be created by strengthening the solid matrix during ice formation in the freeze-drying process (freezing step) or because of the case hardening. Case hardening is a common phenomenon during drying and consists of a rigid outer shell formation on the material's surface (commonly known as crust formation). It is facilitated at high drying rates when the surface moisture decreases fast enough to significantly reduce water diffusion from the inner to the food surface. The low moisture content of the external surface may induce a rubber-glass transition early in the drying and a rigid crust formation that reduces shrinkage of the still rubbery inner part of the food. On the other hand, the residual stress increases, and the dried material can undergo cracking and breakage during storage. Surface cracking happens when shrinkage is not uniform during the drying process, leading to unbalanced stress and material failure (Mayor and Sereno 2004).

Materials shrinking reveal a significant reduction in the number of pores and pore size (Lewicki 1998a). Thus, the extent of shrinkage influences essential physical properties of dried foods, such as porosity and bulk density described in the following.

4.2.3 Porosity and Bulk Density

Porosity (ε) is defined as the volumetric fraction of voids in a solid material. In other words, porosity is the ratio between the volume of voids (V_v) to the total volume of a solid (V_t) (Jagan et al. 2010), as described by Equation 14.

$$\varepsilon = \frac{V_v}{V_t} \tag{14}$$

Porosity directly affects the solid transport properties, such as thermal conductivity and mass and thermal diffusion coefficients. Mechanical and textural properties also correlate to porosity (Hussain et al. 2002). A porous structure has been associated with the crispness of dried products, a critical quality attribute that influences the sensory acceptance of dried foods (Gomide et al. 2022). The lower the porosity, the higher the agglomerate strength of dried products.

Porosity depends on the initial empty spaces existent before drying and the porous spaces formed by water removed during drying. Thus, the shrinkage of the initial solid structure during drying significantly influences the dried food porosity. Pores formation during drying depends on very complex phenomena and food-water content. In an idealized scenario, the water removed is replaced by new empty spaces (filled by air), and the initially existing porosity remains intact during the entire process (without shrinkage). Conversely, the removed water does not form empty spaces in the worst scenario, and the initial porosity disappears (total collapse with significant shrinkage). Intermediate cases are the most frequent (Khalloufi et al. 2009).

Drying conditions that lead to structural preservation or promote the expansion of the food matrix, balancing the shrinkage, are suitable to increase food porosity. For instance, freeze-drying and microwave vacuum-drying produce highly porous dried materials (Khalloufi et al. 2015).

Figure 21 shows the porous and porosity of sweet potato chips submitted to different drying methods. Air drying of sweet potatoes caused pronounced shrinkage that resulted in a compact structure with low porosity (18%). Freeze-dried samples retained the original structure, which is spongy with tiny pores of uniform size and high porosity (58%). Microwave vacuum drying and conductive multiflash drying resulted in dried sweet potatoes with the most expanded solid matrix and the largest pores and intermediate porosity of 44% (microwave vacuum drying) and 38% (multiflash drying) (Monteiro et al. 2022).

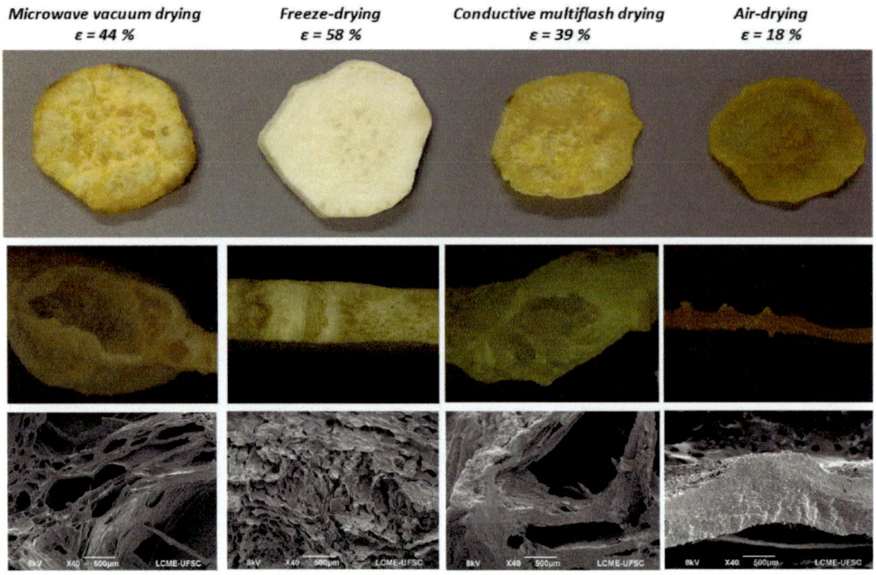

Figure 21. Images of the dried sweet potatoes obtained by different drying methods (adapted from Monteiro et al. (2022)).

The dried food porosity influences its bulk density, defined as the relation between the sample mass and its total (bulk) volume. This property depends on the solid matrix density (most food dry solids have a density near 1.4–1.6 g/cm^3), porosity, and water content. Then, the bulk density of dried foods depends on the drying method and conditions and can be higher or lower than the bulk density of fresh foods. Higher values are consequences of shrinkage during dehydration, while lower values indicate significant porosity creation. Thus, measuring the bulk density of the dried food is also an indirect way to understand how drying has changed the food structure (Krokida and Maroulis 2001).

The porosity and bulk density evolution of sweet potatoes during microwave vacuum drying was reported by Monteiro et al. (2020) and is presented in Fig. 22. At the beginning of the drying, the potatoes presented a porosity of $\approx$ 2.5% and a bulk density of $\approx$ 1,060 g cm^3. However, the porosity increased during drying, reaching $\approx$ 68% after a given elapsed time. This high porosity was created by vapor expansion inside the sample because of the volumetric heating caused by the microwaves during drying. As a result, the bulk density of the potatoes decreased and reached a low value of 0.46 g cm^3.

4.2.4 Rehydration Capacity of Dried Foods

Some dried foods are rehydrated before use to restore the raw product properties. Thus, rehydration has been used as a quality criterion (van der Sman et al. 2013). It offers consumers convenience in preparing meals and is commonly used for powders and vegetables in instantaneous foods.

Rehydration consists of moistening dry material where the water is initially taken up via capillary suction (Xiao and Mujumdar 2008). However, the vegetable cell wall material absorbs water from the pore space and swells. This swelling constricts the pore space, retarding capillary suction. Thus, the absorption rate decreases gradually as the moisture content approaches the equilibrium and is primarily determined by the microstructure (Datta et al. 2012). Therefore, the rehydration capacity highly depends on the changes the material undergoes during the drying process. As the changes are irreversible, dehydration cannot be considered simply a reverse dehydration process, and the product does not regain the initial quality characteristics (Lewicki 1998a).

The shrinkage decreases the rehydration capacity because of the dense and collapsed structure of the dried material with shrunken capillaries (Jayaraman and Das Gupta 1992; Mayor and Sereno

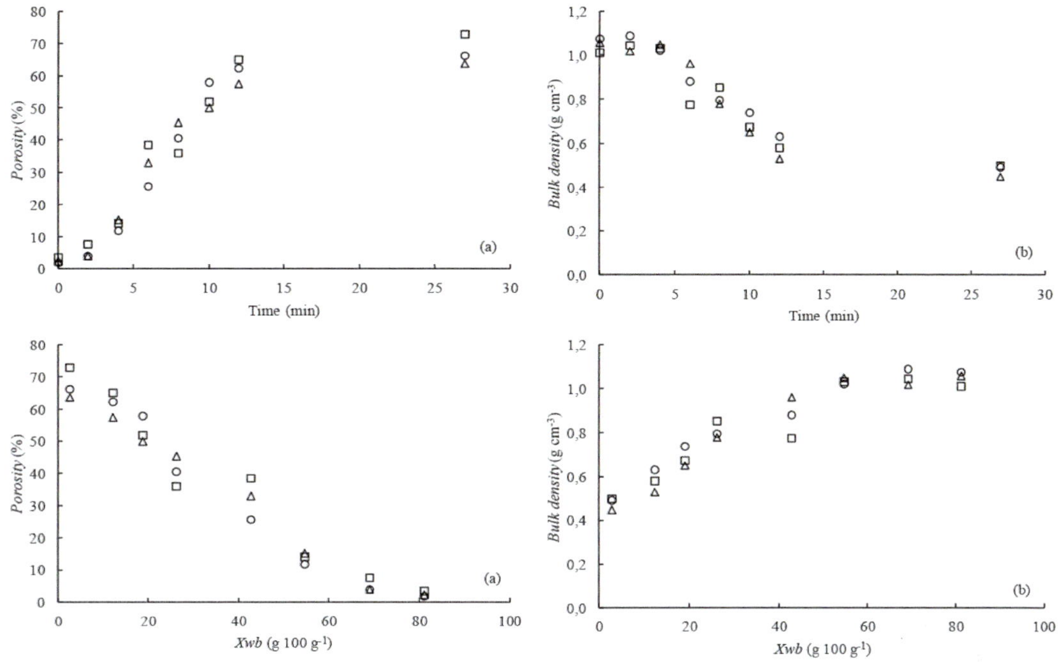

Figure 22. Evolution of porosity (a) and bulk density (b) with time and moisture content in wet bases (X_{wb}) of sweet potato slices during microwave vacuum drying (adapted from Monteiro et al. (2020)).

2004). Therefore, drying methods and conditions that produce products with high porosity and preserved structure are desirable to obtain high-quality rehydrated food with characteristics close to the fresh product. This behavior is illustrated by Tribuzi and Laurindo (2016) (Fig. 23) when studying the rehydration of mussels dried by different methods. Air-drying and vacuum drying cause more significant damage to the structure of the mussels, while the freeze-drying process causes little modifications, producing rehydrated mussels close to the fresh ones.

The rehydration capacity is calculated by the Rehydration Ratio (RR) (Equation 15). During water absorption, the leaching of soluble solids occurs concomitantly. Thus, two other indexes are used to quantify food rehydration, Water Absorption Capacity (WAC) (Equation 16) and Dry Matter Holding Capacity (DHC) (Equation 17) (Lewicki 1998b; Monteiro et al. 2018b).

$$RR = \frac{m_r}{m_d} \tag{15}$$

$$WAC = \frac{m_{wr} - m_{wd}}{m_{wo} - m_{wd}} \tag{16}$$

$$DHC = \frac{m_{sr}}{m_{sd}} \tag{17}$$

Figure 23. Effect of the drying and rehydration (20°C) processes on the external aspect of mussel (adapted from Tribuzi & Laurindo (2016)).

In this, m_r is the mass of the rehydrated sample, m_d is the mass of the dried sample, m_{wr} is the water mass in the rehydrated sample, m_{wd} is the water mass in the dried sample, m_{wo} is the water mass in the sample before drying, m_{sr} is the dry matter mass of rehydrated samples, and m_{sd} is the dry matter mass of dried samples.

Figure 24 presents results on the temporal evolution of the rehydration ratio of dried pumpkins produced by different drying processes, illustrating the influence of food structure on rehydration (Monteiro et al. 2018b). The preserved structure of freeze-dried pumpkins provides a shorter rehydration time and high RR (10.5) and WAC (70%), while the air-dried compact samples took a higher time to reach the final equilibrium of moisture content and presented lower RR (3.1) and WAC (60%). The microwave vacuum-dried samples showed the highest RR (15.3) and WAC (130%) because of their highly porous structure with large pores facilitating the rehydration. However, the microwave vacuum-dried pumpkins presented lower values of DHC due to the loss of solid mass by the fragile and expanded structure.

	Microwave vacuum drying	Conductive multiflash drying	Freeze-drying	Air-drying
Equilibrium rehydration ratio	15.3	5.8	10.5	3.1
Water holding capacity	1.3	0.7	0.7	0.6
Dry matter holding capacity	0.2	0.5	0.3	0.5

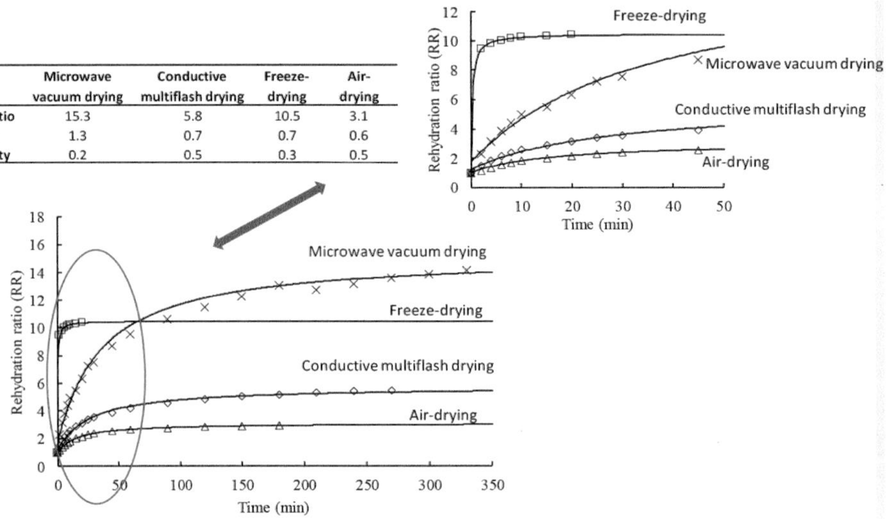

Figure 24. Temporal evolution of the rehydration ratio of dried pumpkins at 25°C and the indexes values of quality rehydration. The curve was magnified for better visualization (adapted from Monteiro et al. (2018b)).

4.3 Chemical Quality of Dried Foods

Chemical reactions occur during drying, causing color, flavor, aroma, and biochemical compound degradation that affect dried food's nutritional and sensory properties.

Color is an essential aspect of a product's quality since it is a visual attribute that the consumers have first contact with, influencing their opinions. Consumers tend to associate color with other properties, such as taste, safety, shelf life, nutritional value, and personal satisfaction (Pedreschi et al. 2006; Zielinska and Michalska 2016).

Drying changes the surface characteristics of a food, altering its reflectivity and color (Fellows 2017). However, color changes are mainly attributed to browning reactions (enzymatic and non-enzymatic browning) and chemical changes to color pigments (carotenoid, chlorophyll, anthocyanins, etc.) because of heat, oxidation, and pH changes among others (Medeni Maskan 2008).

Browning reaction lowers the product quality and is considered undesirable for most dried fruit and vegetables, except for some traditionally consumed darker than fresh ones, such as raisins and bananas. The browning can result from enzymatic reactions catalyzed by polyphenol oxidase (PPO) and peroxidase (POD), which involves the oxidation of phenolic compounds to brown pigment compounds when exposed to air. Brown compounds can also be produced by Maillard reaction, caramelization, and ascorbic acid oxidation. These chemical reactions involve multiple food constituents without enzymes (non-enzymatic browning) (Moon et al. 2020). Browning reactions also induce losses or changes in flavor, aroma, nutritional value, and textural properties (Ioannou 2013; Xiao and Mujumdar 2008).

The changes in aroma and flavor are highly associated with the oxidation of volatile components and lipids during the drying process or storage. Oxidation of unsaturated fatty acids produces hydroperoxides, aldehydes, ketones, and acids responsible for rancid and off-flavor. During storage, the open porous structure of dried food allows access to oxygen, facilitating oxidation reactions (Fellows 2000). The loss of flavor and aroma is also because of the moisture removal process that carries away the volatile compounds responsible for these properties. The release of these compounds during drying and storage is highly associated with the product's microstructure, created during dehydration (Jagan et al. 2010). For instance, when drying occurs above the glass transition temperature, the high mobility of food molecules (food in the rubbery state) increases the degradation and release of that compounds during processing. Conversely, drying foods in a glassy state (temperature below the glass transition) reduces the release and the perception of the flavor of dried products during consumption.

Simão et al. (2021) studied the profile of volatile compounds during the dehydration of pineapple by two different methods and expressed by its relative content. Compounds with high vapor pressure tend to be lost during drying, whereas compounds with low vapor pressure can be preserved. This behavior is mainly observed in the drying process performed under vacuum conditions.

Carrier agents used in pulp dehydration have been reported to improve the retention of some volatile compounds. For instance, Frabetti et al. (2021) reported that pectin helped to preserve the volatile compounds of fresh strawberry pulps during cast-tape drying. However, some new compounds appeared in the dried product, possibly due to the thermal degradation of sugars or formed by secondary Maillard reactions.

The loss of nutritional value of a dried product is associated with the sensitivity of phytochemicals to heat and light (Oliveira et al. 2016). Oxidation reactions affect lipids and other essential nutritional substances, such as vitamins, phenolic compounds, and proteins. For example, lipid oxidation causes the loss of fat-soluble vitamins and pigments in some foods (Fellows 2017). Protein can be denatured and release amino acids, which could react with other chemical compounds via the Maillard and enzymatic browning reactions with possible degradation of essential amino acids. Additionally, the sugar content can be reduced since it is a substrate of the Maillard reaction (Di Scala et al. 2011; Oliveira et al. 2016).

The extent of color, aroma, flavor, and nutrient changes usually increases as the temperature increases since the reaction constants are temperature-dependent. Thus, drying at lower temperatures minimizes chemical reactions and reduces valuable compound loss. For instance, freeze-dried products and those dried under a vacuum using hot air or microwaves are likely to have low degradation of bioactive compounds. On the other hand, drying at a high temperature for short times can inactivate enzymes and reduce some chemical reactions and product degradation (Chen and Martynenko 2018; Tontul and Topuz 2017). It is important to say that the extent of food thermal damage depends on the drying period. During the constant drying rate period, damages are not extreme due to the product's initially high moisture, which is at the wet-bulb temperature. However, the increasing temperature during the falling rate period can cause more product quality degradation. Thus, prolonged exposure to elevated temperature, especially during the falling rate period, may substantially degrade product quality.

The glass transition also affects chemical reaction rates and extent since materials above T_g are in the rubbery state and have higher molecular mobility and diffusion coefficients. Therefore, the quality of dehydrated foods could be improved by keeping the material close to the glass transition (Slade et al. 1991; Roos 2010).

Food pretreatments, such as blanching, acid solution immersion, and edible coating, can be applied before drying to prevent or reduce chemical changes. These treatments inactivate enzymes that catalyze undesirable reactions and help protect the foods from oxygen during drying. Also, antioxidants, storage temperature reduction, and light exclusion are essential to prevent oxidation reactions.

4.4 Shelf Life of Dried Foods

The shelf-life of dried foods is strongly related to the product composition, processing history, packaging, transport, and storage conditions. Proper packaging and storage conditions are crucial to reducing changes in food appearance, aroma, flavor, and texture. Vacuum packaging and food packaging using a modified atmosphere (e.g., nitrogen) and films with a high barrier to gases and light (e.g., laminated packages) help avoid oxidation. In addition, dried foods must be packaged with high water vapor barrier films to avoid moisture and water activity from increasing. Otherwise, the increase in water activity could lead to microbial growth and unwanted chemical reactions and decrease the glass transition temperature, promoting physical changes.

Figure 25 shows that the banana crispness decreased as the water activity increased during storage. The crispness was determined indirectly by the irregularities of jagged mechanical and acoustic curves. The higher the irregularities, the higher the crispness (Moraes et al. 2022). The texture change is related to the increase of the solid matrix mobility because of the water-plasticizing effect that reduces the glass transition temperature. Link et al. (2018) reported how the T_g varies as the relative humidity increases during the storage of dehydrated mangoes (Fig. 26). This stability diagram helps predict the stability and texture of dehydrated foods during storage.

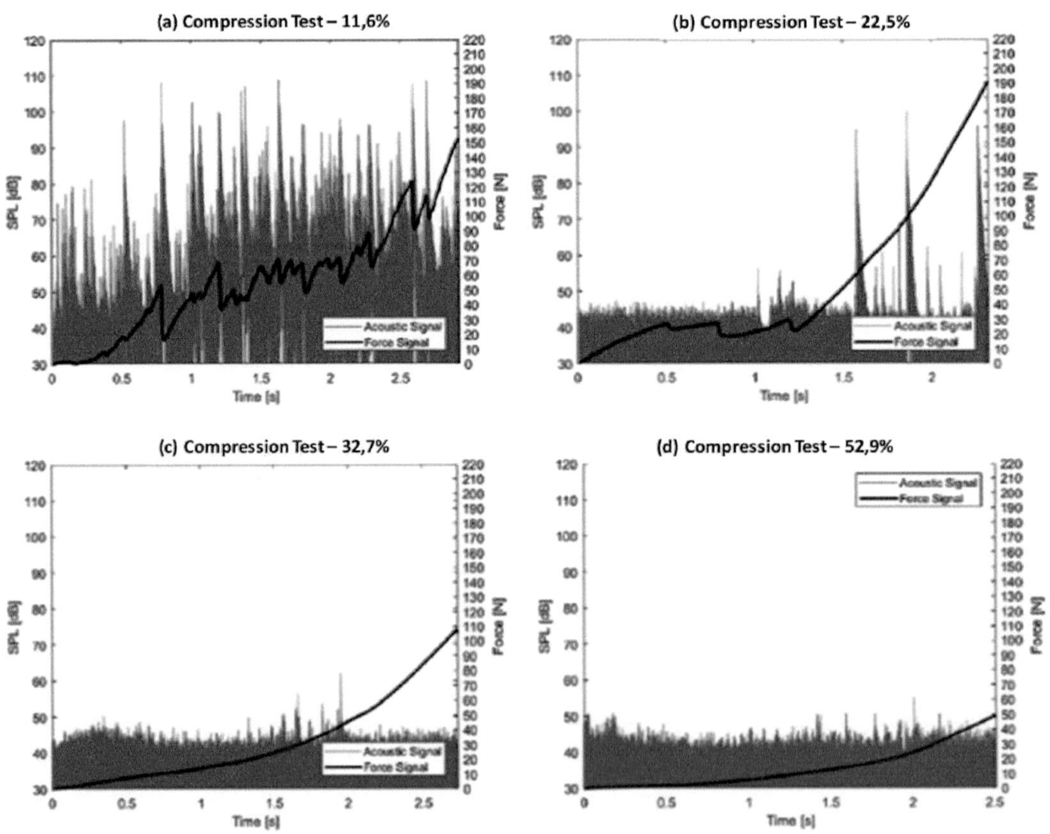

Figure 25. Data of force and Sound Pressure Level (SPL) over time during the compression of bananas stored at (a) 11.6%, (b) 22.5%, (c) 32.7%, (d) 52.9% (adapted from de Moraes et al. (2022)).

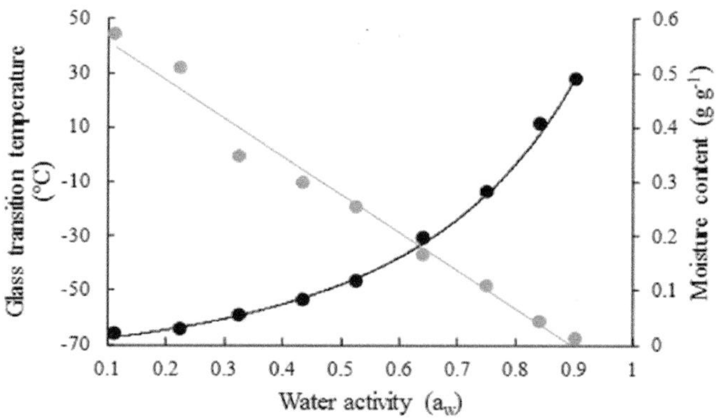

Figure 26. Glass transition temperature in function of water activity and moisture content (dry bases), at 25°C, of mango samples dried by air-drying drying (adapted from Link et al. (2018)).

Acknowledgments

The authors thank the Federal University of Santa Catarina, Brazil, for its support.

References

Abonyi, B.I., Feng, H., Tang, J., Edwards, C.G., Chew, B.P., Mattinson, D.S. and Fellman, J.K. 2002. Quality retention in strawberry and carrot purees dried with Refractance WindowTM system. Journal of Food Science 67(3): 1051–1056.

Aguilera, J.M., Chiralt, A. and Fito, P. 2003. Food dehydration and product structure. Food Science and Technology 14: 432–437. https://doi.org/10.1016/S0924-2244(03)00122-5

Barbosa-Cánovas, G.V, Fontana Jr, A.J., Schmidt, S.J. and Labuza, T.P. 2020. Water activity in foods: fundamentals and applications. John Wiley & Sons.

Bhandari, B.R. and Howes, T. 1999. Implication of glass transition for the drying and stability of dried foods. Journal of Food Engineering 40(1): 71–79. https://doi.org/10.1016/S0260-8774(99)00039-4

Blahovec, J. and Yanniotis, S. 2008. GAB generalized equation for sorption phenomena. Food and Bioprocess Technology 1(1): 82–90.

Bourdoux, S., Li, D., Rajkovic, A., Devlieghere, F. and Uyttendaele, M. 2016. Performance of drying technologies to ensure microbial safety of dried fruits and vegetables. Comprehensive Reviews in Food Science and Food Safety 15(6): 1056–1066.

Castoldi, M., Zotarelli, M.F., Durigon, A., Carciofi, B.A.M. and Laurindo, J.B. 2015. Production of tomato powder by refractance window drying. Drying Technology 33(12): 1463–1473.

Chandrasekaran, S., Ramanathan, S. and Basak, T. 2013. Microwave food processing—A review. Food Research International 52(1): 243–261.

Chen, P. and Pei, D.C. 1989. A mathematical model of drying processes. International Journal of heat and mass transfer 32(2): 297–310.

Chen, Y. and Martynenko, A. 2018. Combination of hydrothermodynamic (HTD) processing and different drying methods for natural blueberry leather. Lwt 87: 470–477. https://doi.org/10.1016/j.lwt.2017.09.030

Chirife, J. and Buera, M.P. 1995. A critical review of some non-equilibrium situations and glass transitions on water activity values of foods in the microbiological growth range. Journal of Food Engineering 25(4): 531–552.

da Silva Simão, R., de Moraes, J.O., de Souza, P.G., Carciofi, B.A.M. and Laurindo, J.B. 2019. Production of mango leathers by cast-tape drying: Product characteristics and sensory evaluation. LWT 99: 445–452.

da Silva Simão, R., de Moraes, J.O., Monteiro, R.L., Schaidt, A.L., Carciofi, B.A.M. and Laurindo, J.B. 2021. Conductive drying methods for producing high-quality restructured pineapple-starch snacks. Innovative Food Science & Emerging Technologies 70: 102701.

Datta, A.K. 2001. Handbook of microwave technology for food application. CRC Press.

Datta, A.K., van der Sman, R., Gulati, T. and Warning, A. 2012. Soft matter approaches as enablers for food macroscale simulation. Faraday Discussions 158: 435–459.

de Moraes, J.O., Andreani, P., Murta, B.H.P., Link, J.V, Tribuzi, G., Laurindo, J.B. and Carciofi, B.A.M. 2022. Mechanical-acoustical measurements to assess the crispness of dehydrated bananas at different water activities. LWT 154: 112822.

Di Scala, K., Vega-Gálvez, A., Uribe, E., Oyanadel, R., Miranda, M., Vergara, J. and Lemus-Mondaca R. 2011. Changes of quality characteristics of pepino fruit (Solanum muricatum Ait) during convective drying. International Journal of Food Science and Technology 46(4): 746–753. https://doi.org/10.1111/j.1365-2621.2011.02555.x

Durigon, A., de Souza, P.G., Carciofi, B. A.M. and Laurindo, J.B. 2016. Cast-tape drying of tomato juice for the production of powdered tomato. Food and Bioproducts Processing, 100: 145–155.

Durigon, A., Parisotto, E.I.B., Carciofi, B.A.M. and Laurindo, J.B. 2018. Heat transfer and drying kinetics of tomato pulp processed by cast-tape drying. Drying Technology 36(2): 160–168.

Fellows, P. 2000. (2nd ed.). Boca Raton: CRC Press.

Fellows, P.J. 2017. Food Processing and Technology (4th ed.). Woodhad Publishing.

Frabetti, A.C.C., de Moraes, J.O., Porto, A.S., da Silva Simão, R. and Laurindo, J.B. 2021. Strawberry-hydrocolloids dried by continuous cast-tape drying to produce leather and powder. Food Hydrocolloids 121: 107041.

Frabetti, A.C.C., Durigon, A. and Laurindo, J.B. 2018. Effect of process variables on the drying of guava pulp by cast-tape drying. LWT 96: 620–626.

Geankoplis, C.J. 1993. Transport Processes and Unit Operations (3rd ed.). New Jersey: Prentice-Hall.

Gomide, A.I., Monteiro, R.L. and Laurindo, J.B. 2022. Impact of the power density on the physical properties, starch structure, and acceptability of oil-free potato chips dehydrated by microwave vacuum drying. *LWT*, *155*, 112917. https://doi.org/https://doi.org/10.1016/j.lwt.2021.112917

Huang, X. and Hsieh, F. 2005. Physical properties, sensory attributes, and consumer preference of pear fruit leather. Journal of Food Science 70(3): E177–E186.

Hussain, M.A., Shafiur Rahman, M. and Ng, C.W. 2002. Prediction of pores formation (porosity) in foods during drying: Generic models by the use of hybrid neural network. Journal of Food Engineering 51(3): 239–248. https://doi.org/10.1016/S0260-8774(01)00063-2

Ibarz, A. and Barbosa-Cánovas, G.V. 2002. Unit operations in food engineering. CRC press.

Ioannou, I. 2013. Prevention of enzymatic browning in fruit and vegetables. European Scientific. Journal. 9(30).

Jagan, S.V., Law, C.L. and Mujumdar, A.S. 2010. Drying of Foods, Vegetables and Fruits (1st ed.). Singapore.

Jagan, S.V., Law, C.L. and Mujumdar, A.S. 2011. Drying of Foods, Vegetables and Fruits (2nd ed.).

Jayaraman, K.S. and Das Gupta, D.K. 1992. Dehydration of fruits and vegetables-recent developments in principles and techniques. Drying Technology 10(1): 1–50.

Khalloufi, S., Almeida-Rivera, C. and Bongers, P. 2009. A theoretical model and its experimental validation to predict the porosity as a function of shrinkage and collapse phenomena during drying. Food Research International 42(8): 1122–1130. https://doi.org/10.1016/j.foodres.2009.05.013

Khalloufi, S., Kharaghani, A., Almeida-Rivera, C., Nijsse, J., van Dalen, G. and Tsotsas, E. 2015. Monitoring of initial porosity and new pores formation during drying: A scientific debate and a technical challenge. Trends in Food Science and Technology 45(2): 179–186. https://doi.org/10.1016/j.tifs.2015.06.011

Krokida, M.K. and Maroulis, Z.B. 2001. Structural properties of dehydrated products during rehydration. *International* Journal of Food Science and Technology 36(5): 529–538. https://doi.org/10.1046/j.1365-2621.2001.00483.x

Krokida, P. 1997. Effect of drying method on shrinkage and porosity. Drying Technology 15(10): 2441–2458.

Laurindo, J.B., Porciuncula, B.D.A. and Zotarelli, M.F. 2011. Patent No. PI 1107173-1107177. Brazil.

Laurindo, J.B. and Peleg, M. 2008. Mechanical characterization of shredded wheat. Journal of Texture Studies 39(5): 444–459.

Laurindo, João B. and Peleg, M. 2007. Mechanical measurements in puffed rice cakes. Journal of Texture Studies 38(5): 619–634.

Lewicki, P.P. 1998a. Effect of pre-drying treatment, drying and rehydration on plant tissue properties: A review. International Journal of Food Properties 1(1): 1–22. https://doi.org/10.1080/10942919809524561

Lewicki, P.P. 1998b. Some remarks on rehydration of dried foods. Journal of Food Engineering, 36(1): 81–87.

Link, J.V., Tribuzi, G., de Moraes, J.O. and Laurindo, J.B. 2018. Assessment of texture and storage conditions of mangoes slices dried by a conductive multi-flash process. Journal of Food Engineering 239: 8–14.

Link, J.V., Tribuzi, G. and Laurindo, J.B. 2017. Improving quality of dried fruits: A comparison between conductive multi-flash and traditional drying methods. LWT 84: 717–725.

Link, J.V., Tribuzi, G. and Laurindo, J.B. 2018. Conductive multi-flash drying of mango slices: Vacuum pulse conditions on drying rate and product properties. Journal of Food Processing and Preservation 42(2): e13440.

Louka, N. and Allaf, K. 2002. New process for texturizing partially dehydrated biological products using controlled sudden decompression to the vacuum: application on potatoes. Journal of Food Science 67(8): 3033–3038.

Mayor, L. and Sereno, A.M. 2004. Modelling shrinkage during convective drying of food materials: A review. *Journal of* Food Engineering 61(3): 373–386. https://doi.org/10.1016/S0260-8774(03)00144-4

McCabe, W., Smith, J. and Harriot, P. 1987. Unit Operations Of Chemical Engineering, McCabe And. pdf (5th ed.). McGraw-Hill Chemical Engineering Series.

Medeni Maskan. 2008. Effect of thermal processing on tristimulus colour changes of fruits. Stewart Postharvest Review 2(5): 1–8. https://doi.org/10.2212/spr.2006.5.10

Monteiro, R.L., Gomide, A.I., Link, J.V., Carciofi, B.A.M. and Laurindo, J.B. 2020. Microwave vacuum drying of foods with temperature control by power modulation. Innovative Food Science and Emerging Technologies 65: 102473. https://doi.org/10.1016/j.ifset.2020.102473

Monteiro, R.L., Link, J.V., Tribuzi, G., Carciofi, B.A.M. and Laurindo, J.B. 2018a. Effect of multi-flash drying and microwave vacuum drying on the microstructure and texture of pumpkin slices. Lwt 96(February), 612–619. https://doi.org/10.1016/j.lwt.2018.06.023

Monteiro, R.L., Link, J.V, Tribuzi, G., Carciofi, B.A.M. and Laurindo, J.B. 2018b. Microwave vacuum drying and multi-flash drying of pumpkin slices. Journal of Food Engineering 232: 1–10. https://doi.org/10.1016/j.jfoodeng.2018.03.015

Monteiro, R.L., Carciofi, B.A.M. and Laurindo, J.B. 2016. A microwave multi-flash drying process for producing crispy bananas. Journal of Food Engineering 178: 1–11. https://doi.org/10.1016/j.jfoodeng.2015.12.024

Monteiro, R.L., de Moraes, J.O., Domingos, J.D., Carciofi, B.A.M. and Laurindo, J.B. 2020. Evolution of the physicochemical properties of oil-free sweet potato chips during microwave vacuum drying. Innovative Food Science and Emerging Technologies 63: 102317. https://doi.org/10.1016/j.ifset.2020.102317

Monteiro, Ricardo L., Domschke, N.N., Tribuzi, G., Teleken, J.T., Carciofi, B.A.M. and Laurindo, J.B. 2021. Producing crispy chickpea snacks by air, freeze, and microwave multi-flash drying. LWT 140: 110781.

Monteiro, Ricardo Lemos, de Moraes, J.O., Gomide, A.I., Carciofi, B.A.M. and Laurindo, J.B. 2022. Temperature control for high-quality oil-free sweet potato CHIPS produced by microwave rotary drying under vacuum. LWT 113047.

Moon, K.M., Kwon, E. Bin, Lee, B. and Kim, C.Y. 2020. Recent Trends in Controlling the Enzymatic Browning of Fruit and Vegetable Products. Molecules 25(12). https://doi.org/10.3390/molecules25122754

Moreira, R., Figueiredo, A. and Sereno, A. 2000. Shrinkage of apple disks during drying by warm air convection and freeze drying. Drying Technology 18: 279–294.

Martínez-Navarrete, Andrés, A., Chiralt, A. and Fito, P. 1998. Termodinámica y cinética de sistemas alimento entorno. Universidad Politécnica de Valencia. España.

Nindo, C.I., Feng, H., Shen, G.Q., Tang, J. and Kang, D.H. 2003. Energy utilization and microbial reduction in a new film drying system. Journal of Food Processing and Preservation 27(2): 117–136.

Nindo, C.I. and Tang, J. 2007. Refractance Window Dehydration Technology: A Novel Contact Drying Method. Drying Technology 25(1): 37–48. https://doi.org/10.1080/07373930601152673

Niu, J., Wang, Y., Zou, X., Tan, Y., Jia, C., Weng, X. and Deng, L. 2021. Infrared electrochromic materials, devices and applications. Applied Materials Today 24: 101073.

Oliveira, S.M., Brandão, T.R.S. and Silva, C.L.M. 2016. Influence of Drying Processes and Pretreatments on Nutritional and Bioactive Characteristics of Dried Vegetables: A Review. Food Engineering Reviews 8(2): 134–163. https://doi.org/10.1007/s12393-015-9124-0

Ortiz-Jerez, M.J., Gulati, T., Datta, A.K. and Ochoa-Martínez, C. I. 2015. Quantitative understanding of refractance window™ drying. Food and Bioproducts Processing 95: 237–253.

Pavan, M.A., Schmidt, S.J. and Feng, H. 2012. Water sorption behavior and thermal analysis of freeze-dried, Refractance Window-dried and hot-air dried açaí (Euterpe oleracea Martius) juice. LWT-Food Science and Technology 48(1): 75–81.

Pedreschi, F., León, J., Mery, D. and Moyano, P. 2006. Development of a computer vision system to measure the color of potato chips. Food Research International 39(10): 1092–1098.

Peleg, M. 1995. Description of mechanical changes in foods at their glass region. *In*: Barbosa-Canovas, G.V. and Welti-Chanes (eds.). Food preservation by moisture control: fundamentals and applications. Switzerland: Journal of Technomic Publishing.

Porciuncula, B.D.A., Segura, L.A. and Laurindo, J.B. 2016. Processes for controlling the structure and texture of dehydrated banana. Drying Technology 34(2): 167–176. https://doi.org/10.1080/07373937.2015.1014911

Raaholt, B.W. and Isaksson, S. 2017. Improving the heating uniformity in microwave processing. pp. 381–406. *In*: The microwave processing of foods. Elsevier.

Raghavi, L.M., Moses, J.A. and Anandharamakrishnan, C. 2018. Refractance window drying of foods: A review. Journal of Food Engineering 222: 267–275.

Rahman, M.S. and Potluri, P.L. 1990. Shrinkage and density of squid flesh during air drying. Journal of Food Engineering, 12: 133–143.

Ratti, C. 2001. Hot air and freeze-drying of high-value foods: a review. Journal of Food Engineering 49: 311–319.

Ratti, C. 2009. Advances in Food Dehydration. CRC Press/Taylor & Francis Group.

Ratti, C. and Mujumdar, A.S. 2020. Infrared drying. pp. 567–588. *In*: *H*andbook of Industrial Drying. CRC Press.

Regier, M., Knoerzer, K. and Schubert, H. 2016. The microwave processing of foods. Woodhead publishing.

Roos, Y.H. 2010. Glass transition temperature and its relevance in food processing. Annual Review of Food Science and Technology 1(1): 469–496. https://doi.org/10.1146/annurev.food.102308.124139

Sablani, S.S. 2007. Drying of Fruits and Vegetables: Retention of Nutritional/Functional Quality. Drying Technology, 24(2): 123–135. https://doi.org/10.1080/07373930600558904

Saca, S.A. and Lozano, J.E. 1992. Explosion puffing of bananas. International Journal of Food Science & Technology 27(4): 419–426.

Sagar, V.R. and Suresh Kumar, P. 2010. Recent advances in drying and dehydration of fruits and vegetables: a review. Journal of Food Science and Technology 47(1): 15–26.

Sakare, P., Prasad, N., Thombare, N., Singh, R. and Sharma, S.C. 2020. Infrared drying of food materials: Recent advances. Food Engineering Reviews 12(3): 381–398.

Simão, R. da S., de Moraes, J.O., Carciofi, B.A.M. and Laurindo, J.B. 2020. Recent advances in the production of fruit leathers. Food Engineering Reviews 12(1): 68–82.

Slade, L., Levine, H. and Reid, D.S. 1991. Beyond water activity: recent advances based on an alternative approach to the assessment of food quality and safety. Critical Reviews in Food Science & Nutrition 30(2–3): 115–360.

Sullivan, J.F. and Craig Jr, J.C. 1984. The development of explosion puffing. Food Technology (USA).

Sullivan, J.F., Konstance, R.P., Aceto, N.C., Heiland, W.K. and Craig Jr, J.C. 1977. Continuous explosion-puffing of potatoes. Journal of Food Science 42(6): 1462–1463.

Tontul, I., Eroğlu, E. and Topuz, A. 2018. Convective and refractance window drying of cornelian cherry pulp: Effect on physicochemical properties. Journal of Food Process Engineering 41(8): e12917.

Tontul, I. and Topuz, A. 2017. Effects of different drying methods on the physicochemical properties of pomegranate leather (pestil). LWT - Food Science and Technology 80: 294–303. https://doi.org/10.1016/j.lwt.2017.02.035

Tribuzi, G. and Laurindo, J.B. 2016. Dehydration and rehydration of cooked mussels. International Journal of Food Engineering 12(2): 173–180.

Valenzuela, C. and Aguilera, J.M. 2015. Effects of different factors on stickiness of apple leathers. Journal of Food Engineering 149: 51–60.

van der Sman, R.G.M., Paudel, E., Voda, A. and Khalloufi, S. 2013. Hydration properties of vegetable foods explained by Flory-Rehner theory. Food Research International 54(1): 804–811. https://doi.org/10.1016/j.foodres.2013.08.032

Varnalis, A.I., Brennan, J.G. and MacDougall, D.B. 2001. Proposed mechanism of high-temperature puffing of potato. Part I. The influence of blanching and drying conditions on the volume of puffed cubes. Journal of Food Engineering 48(4): 361–367. https://doi.org/10.1016/S0260-8774(00)00197-7

Verma, D.K., Mahanti, N.K., Thakur, M., Chakraborty, S.K. and Srivastav, P.P. 2020. Microwave heating: Alternative thermal process technology for food application. Emerging Thermal and Nonthermal Technologies in Food Processing 25–67.

Xiao, D.C. and Mujumdar, A.S. 2008. Drying Technologies in Food Processing (1st ed.). Blackwell.

Zhang, M., Tang, J., Mujumdar, A.S. and Wang, S. 2006. Trends in microwave-related drying of fruits and vegetables. Trends in Food Science and Technology 17(10): 524–534. https://doi.org/10.1016/j.tifs.2006.04.011

Zielinska, M. and Michalska, A. 2016. Microwave-assisted drying of blueberry (Vaccinium corymbosum L.) fruits: Drying kinetics, polyphenols, anthocyanins, antioxidant capacity, colour and texture. Food Chemistry 212: 671–680. https://doi.org/10.1016/j.foodchem.2016.06.003

Zotarelli, M.F., Carciofi, B.A.M. and Laurindo, J.B. 2015. Effect of process variables on the drying rate of mango pulp by Refractance Window. Food Research International 69: 410–417.

Zotarelli, M.F., da Silva, V.M., Durigon, A., Hubinger, M.D. and Laurindo, J.B. 2017. Production of mango powder by spray drying and cast-tape drying. Powder Technology, 305: 447–454.

Zotarelli, M.F., Porciuncula, B.D.A. and Laurindo, J.B. 2012. A convective multi-flash drying process for producing dehydrated crispy fruits. Journal of Food Engineering 108(4): 523–531. https://doi.org/10.1016/j.jfoodeng.2011.09.014

11

Extrusion for the Sustainable Development of Novel Foods

Basics, Principles, and Applications

Guido Rolandelli,*[1,2] *Abel Eduardo Farroni*[3] and *María del Pilar Buera*[1,2]

1. Basics and Principles of Extrusion Processing

Extrusion processing is a thermo-mechanical technology that was initially developed for the plastic industry, but then it was adapted for cereals' food development using extruders with different lengths and diameters. During this process, the raw ingredients are fed into a cylinder, where one or two co- or counter-rotating screws mix and shear the materials through the different sections of the barrel (Figs. 1 and 2a). Extrusion flexibility and versatility allow obtaining a wide variety of products using the same technology: breakfast cereals, snacks, cereal bars, pellets, different types of pasta, texturized proteins, and meat analogues among others.

Cold extrusion is referred to the single-screw extrusion process with temperatures slightly over room temperature that is used to produce different types of pasta (Bouvier and Campanella 2014). In such cases, water and semolina, which are the main ingredients, are mixed at low shear rates and then pushed through dies of different shapes to produce a wide variety of pasta (Fig. 2b). On the contrary, to produce ready-to-eat cereal products, high shear and high temperatures extrusion processing is required. This makes extrusion a high-temperature short-time (HTST) methodology for the generation of food products. Single screw extruders (Fig. 1a) are mainly used for the generation of simple products, such as expanded cereal-based snacks, breakfast cereals, pellets, and pet food among others, where the materials are not uniformly melted or mixed, and the obtained products are generally heterogeneous (Tiwari and Jha 2017). In contrast, twin screw extrusion (Fig. 1b) allows the obtention of more complex, homogeneous, and better-mixed materials, such as cereal-based filled products, snacks made by flours blends, crispy flatbreads, meat analogues, texturized proteins, snacks with encapsulated flavors, etc. (Bouvier and Campanella 2014).

[1] Universidad de Buenos Aires, Facultad de Ciencias Exactas y Naturales, Departamentos de Industrias y Química Orgánica. Intendente Güiraldes 2160, Ciudad Universitaria, C1428EGA, Buenos Aires, Argentina.

[2] CONICET – Universidad de Buenos Aires, Instituto de Tecnología de Alimentos y Procesos Químicos (ITAPROQ). Intendente Güiraldes 2160, Ciudad Universitaria, C1428EGA, Buenos Aires, Argentina.

[3] Instituto Nacional de Tecnología Agropecuaria, Estación Experimental Agropecuaria Pergamino (INTA – EEA Pergamino). Av. Frondizi km 4.5, 2700, Pergamino, Buenos Aires, Argentina.

* Corresponding author: rolandelliguido@gmail.com

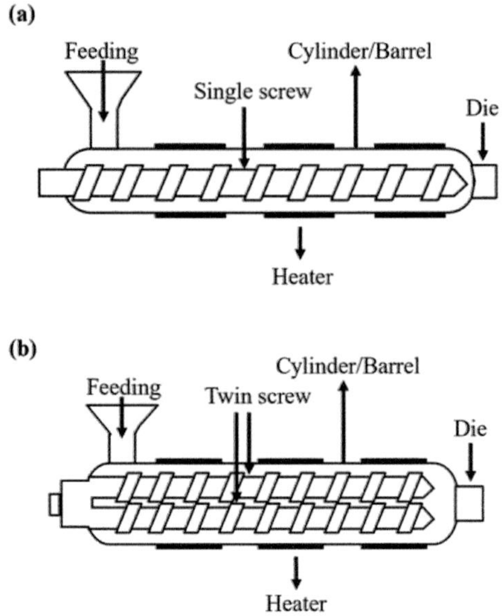

Figure 1. Extrusion Technology Used in Food Industry. (a) Single-Screw Extruder. (b) Twin-Screw Extruder.

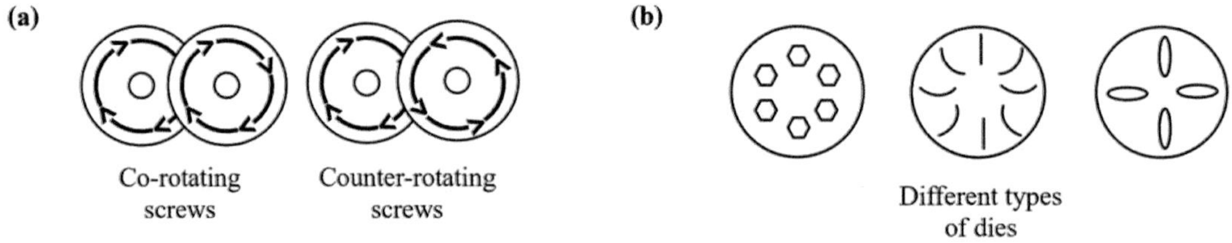

Figure 2. (a) Co- or Counter-Rotating Screws in Twin-Screw Extruders. (b) Different Designs of Dies.

During cereal extrudates generation, the materials are compressed, plasticized, and cooked along their way to the end of the barrel by the combination of the high temperature and pressure applied (Bouvier and Campanella 2014). The molten matrix fills the small spaces between the screws and the barrel, which limits its movement and increases the shear and inner pressure. As the plasticized and compacted mass is forced through the different sections of the extruder, the temperature and pressure rise due to the shearing friction (Tiwari and Jha 2017). In addition, the temperature is controlled by different heating elements placed along the barrel. Finally, the material comes out of the extruder through a die of reduced diameter and the product expands due to the difference between the inner and the atmospheric pressure. Additionally, the water steam rapidly flashes off the food product, favoring the expansion while generating a porous structure and reducing its moisture content (Prabha et al. 2021). According to the die design and diameter, the final product results in different sizes and shapes (Fig. 2b).

The main independent variables of the extrusion process are the water content of the raw ingredients (14–30% wet basis), the screws speed (rpm), the specific mechanical energy (SME) applied (J/g), and the temperature of each section of the extruder (60–160°C) (Bouvier and Campanella 2014; Tiwari and Jha 2017). SME is the ratio of total mechanical energy applied by the screw during the shearing of the mix to the mass flow rate during feeding (Dalbhagat et al. 2019). The combination of the different processing conditions defines the characteristics of the final products. Thus, it is possible to obtain foodstuffs with varied attributes by controlling and modifying the different processing variables as well as with the use of distinctive combinations of raw ingredients.

2. Thermo-Mechanical and Nutritional Modifications of Food Components During Extrusion

During extrusion, food components (carbohydrates, proteins, lipids, fiber, pigments, and antioxidants) are exposed to high temperatures, pressure, and mechanical energy in the presence of limited water content. This combination promotes the occurrence of a series of physicochemical transformations while destroying and creating structures, as well as changing the molecular associations and interactions of their components (Table 1). Nevertheless, given the short residence times of the process, food components are more preserved than in other common technologies used for food development (Reyes Moreno et al. 2018). Extrudates final quality, their organoleptic properties, and consumers' sensory perceptions depend on these transformations and the level of development of the reactions that take place during processing and storage, which are intimately related with samples composition (Arora et al. 2020; Reyes Moreno et al. 2018). Overall, nutritional losses will occur because of the destruction of these components of interest, but molecular modifications may also take place, thus enhancing the exposure and/or bioavailability of the nutrients. These transformations induced by the process can be analyzed through different and complementary assays (Table 2), such as scanning electron microscopy (SEM), confocal laser scanning microscopy (CLSM), Fourier-transformed infrared and Raman spectroscopies (FT-IR and FT-Raman, respectively), X-ray diffraction (XRD), differential scanning calorimetry (DSC), and time-resolved proton nuclear magnetic resonance (^{1}H-NMR) among others (Rolandelli et al. 2020, 2021b; Mosio et al. 2015; Jebalia et al. 2019; Cueto et al. 2018; Jafari et al. 2017; Masavang et al. 2019; Makowska et al. 2017).

2.1 Starch

Starch is the main component of cereal-based extrudates. During extrusion, starch gelatinization occurs because of the combination of heat and mechanical energy in the presence of water. As a result of this process, ordered starch granules melt and form a disordered and continuous matrix of gelatinized starch that constitutes the basic structure of the obtained products (Dalbhagat et al. 2019; Wang et al. 2021). Gelatinized starch leads to the expansion of the extrudates and enhances their

Table 1. Modifications and Transformations of Main Food Components During Extrusion.

Food Component	Modifications/Transformations During Extrusion	References
Starch	Gelatinization Dextrinization Crystallinity Amylose-lipid complexes	Wang et al. 2021; Reyes Moreno et al. 2018; Dalbhagat et al. 2019; Arora et al. 2020; Gulati et al. 2020
Proteins	Denaturation Conformational and structural changes Maillard reaction Protein aggregation	Shah et al. 2021; Reyes Moreno et al. 2020; Gulati et al. 2020; Arora et al. 2020; Jebalia et al. 2019; Moisio et al. 2015; Chanvrier et al. 2015
Lipids	Isomerization Oxidation Amylose-lipid complexes	Krishnan et al. 2021; Ilo et al. 2000; Reyes Moreno et al. 2018; Gulati et al. 2020; Lampi et al. 2015; Moisio et al. 2015
Fiber	Physicochemical transformations	Robin et al. 2012; Gulati et al. 2020; Reyes Moreno et al. 2018
Pigments and antioxidants	Oxidation	Debelo et al. 2020; Reyes Moreno et al. 2018; Cueto et al. 2017; Dalbhagat et al. 2019

Table 2. Physicochemical and Mechanical Properties of Interest and Instrumental Analysis of Cereal-Based Extruded Products.

Characterization		**Importance**	**References**
Physical Properties	Expansion rate (ER)	Evaluation of volume growth correlated with starch gelatinization	Dalbhagat et al. 2021; Kristiawan et al. 2018
	Bulk density (BD)	Inversely correlated with ER in association with porosity degree	Bisharat et al. 2013; Chanvrier et al. 2014
	Water solubility index (WSI)	Indicator of starch degradation degree	Logié et al. 2018; Chanvrier et al. 2015
	Water absorption index (WAI)	Starch status, degree of gelatinization, and association with water	Rolandelli et al. 2020; Kasprzak et al. 2013
	Water content	Stability, the definition of mechanical properties and texture	Tiwari and Jha 2017; Rolandelli et al. 2020
	Water activity (a_w)	Evaluation of susceptibility toward deteriorative reactions	Lampi et al. 2015; Saldanha do Carmo et al. 2019
	Color	Correlation with sensory acceptance	Masatcioglu et al. 2014; 2015; Cueto et al. 2017
Mechanical Properties	Hardness	Sensory acceptance	Bouvier and Campanella 2014; Chanvrier et al. 2005; 2014; 2015; Dalbhagat et al. 2019; Kristiawan et al. 2018; Jebalia et al. 2019
	Fragility		
	Adhesiveness		
	Crunchiness		
	Crispness		
	Porosity		
	Chewiness		
Chemical Properties, Development of Chemical Reactions	Nutritional profile	Nutritional value	Reyes Moreno et al. 2018; Shah et al. 2021; Gulati et al. 2020; Arora et al. 2020
	Maillard reaction development: quantification of Maillard reaction products (MRPs)	Definition of color and flavor, nutritional value, and food safety	Masatcioglu et al. 2014; 2015; Mulla et al. 2011; Rannou et al. 2016
	Lipid oxidation: quantification of lipids oxidation products, rancidity	Sensory acceptance, food safety, nutritional value, stability, and flavor	Yang et al. 2021; Lampi et al. 2015
	Sensory evaluation	Consumer's acceptance and preferences	Pęksa et al. 2016; Chalermchaiwat et al. 2015; Saldanha do Carmo et al. 2019
	Total polyphenols content and antioxidant activity	Functional activity	Reis et al. 2014; Zhang et al. 2018
	Pigments stability	Color and functionality	Cueto et al. 2017; Obradović et al. 2015

Table 2. contd. ...

... Table 2. contd.

Characterization		**Importance**	**References**
Instrumental Analysis	Microscopic assays: Scanning electron microscopy (SEM), confocal laser scanning microscopy (CLSM), atomic force microscopy (AFM)	Microstructural characterization, correlation with physicochemical, and mechanical properties	Moisio et al. 2015; Jebalia et al. 2019; Bisharat et al. 2013; Kasprzak et al. 2013; Cueto et al. 2015
	Fourier-transformed infrared spectroscopy (FT-IR), Fourier-transformed Raman spectroscopy (FT-Raman)	Components' modifications and associations, definition of mechanical properties	Jafari et al. 2017; Cueto et al. 2018; Rolandelli et al. 2021b
	X-ray diffraction (XRD)	Crystallinity, correlation with mechanical properties	Jafari et al. 2017; Cueto et al. 2018
Water-Related Properties	Water sorption isotherms	Analysis of water-solids associations, prediction of products stability during storage	Masavang et al. 2019; Rolandelli et al. 2020
	Differential scanning calorimetry (DSC): glass transition, enthalpic relaxations, starch gelatinization, and starch retrogradation	Analysis of thermal transitions, starch characterization and degree of cooking, correlation with mechanical properties	Jafari et al. 2017; Masavang et al. 2019
	Molecular mobility: time-resolved proton nuclear magnetic resonance (^{1}H-NMR)	Analysis of water solids associations, susceptibility towards physicochemical modifications, and definition of mechanical properties	Makowska et al. 2017; Zhang et al. 2015; Chen et al. 2010

water-absorbing capacity. The gelatinization temperature varies according to the starches' botanic source, differences in composition (amylose/amylopectin ratio), crystallinity, and structure among other factors, but it is usually in the range between 60–70°C (Singh et al. 2003). It is because of gelatinization that starch turns digestible (Gulati et al. 2020), which is also reflected in reductions of the crystalline portion (Cueto et al. 2018) and fiber contents after extrusion (Wang et al. 2021; Reyes Moreno et al. 2018). Starch conformational changes can impact the mechanical, textural, microstructural, and sensory properties of the products (Jebalia et al. 2019; Chanvrier et al. 2005).

Besides gelatinization, dextrinization may also occur, especially when thermal treatment is more favorable for melting than for gelatinization: high temperatures, screw speed, and low feed moisture conditions (Arora et al. 2020). Dextrinization is a depolymerization reaction particularly favored by a high shearing rate in which fragments of starch of low molecular weight—i.e., dextrins, are released (Gulati et al. 2020). These fragments are more available for digestive enzymes and can also participate in the Maillard reaction (Dalbhagat et al. 2019).

Nevertheless, starch digestibility may be reduced by the formation of complexes between amylose and lipids. This inclusion is facilitated by the hydrophobic character of the interior of the amylose helix, given by the spatial disposition of its glucose molecules (Krishnan et al. 2021; Arora et al. 2020). These complexes are type-V crystalline structures that constitute resistant starch (thus, account as fiber) and are also used to protect unsaturated fatty acids from oxidation (Gulati et al.

2020). Amylose-lipids complexes formation is favored by high extrusion temperatures and feeding moisture levels (Dalbhagat et al. 2019; Gulati et al. 2020).

2.2 Proteins

Denaturation is the most common modification of proteins during extrusion, given their heat-labile nature. Extrusion at high feeding moisture levels diminishes shear forces but facilitates protein orientation and alignment following the extrusion flow (Arora et al. 2020; Rolandelli et al. 2021b). The mechanical shear and high temperatures lead to the disruption of the molecular arrangements and breakage of hydrogen and disulfide bonds, improving protein digestibility but reducing their solubility (Dalbhagat et al. 2019; Arora et al. 2020; Reyes Moreno et al. 2018). This phenomenon produces the loss of their native structure, disruption of intramolecular bonds, and creation of new intermolecular associations that increase their random conformation with an exhibition of reactive groups (Shah et al. 2021; Jafari et al. 2017).

The exposure of thiol groups, for example, can facilitate the cross-linking among proteins through disulfide bonds, which conducts to the formation of agglomerates that negatively influence extrudates' mechanical properties (Arora et al. 2020; Jebalia et al. 2019; Chanvrier et al. 2015) and affect proteins functionality (Gulati et al. 2020). Protein aggregation is also favored by the associations of hydrophobic groups that are highly exposed after denaturation (Moisio et al. 2015). Simultaneously, denaturation may favor the exposure of hidden amino groups and promote their condensation with reducing sugars through the Maillard reaction (which will be discussed in Section 3.3) with consequent changes in organoleptic properties and nutritional value (Wang et al. 2022; Reyes Moreno et al. 2018). Considering lysine is the limiting essential amino acid in most cereals, its reduction upon extrusion via the Maillard reaction is of high importance (Gulati et al. 2020).

Protein denaturation may also cause the disruption of complexes between native proteins and other food components, such as carotenoids, components of cell walls, and polyphenols among others, improving their functionality (antioxidant properties) but also their susceptibility toward oxidation or other deteriorative reactions (Arora et al. 2020; Gulati et al. 2020).

2.3 Lipids

The main reaction suffered by lipids during extrusion is oxidation, which will be detailed later in Section 3.3. The cereal grains milling process for the obtention of flour breaks structures or bonds between lipids and other food components, releasing and/or exposing fatty acids (Lampi et al. 2015; Wang et al. 2022). This favors their isomerization and oxidation during extrusion by the combination of thermal and mechanical energies, which negatively affects their nutritional value and organoleptic properties (Dalbhagat et al. 2019).

Nevertheless, extrusion also triggers several reactions that protect lipids from oxidation (Moisio et al. 2015; Ilo et al. 2000; Makowska et al. 2017; Dalbhagat et al. 2019; Gulati et al. 2020): thermomechanical energy produces the inactivation of oxidative enzymes (such as lipases and lipoxygenases) and favors the formation of complexes of lipids with amylose that enhances lipids stability during thermomechanical processing. As previously mentioned, because of protein denaturation, polyphenols or other antioxidants are released or exposed and some products with antioxidant properties derived from the Maillard reaction are also formed, which can altogether protect lipids against oxidation processes (Reyes Moreno et al. 2018).

2.4 Fiber

Technological treatment via the extrusion process of cereal-based formulations does not usually affect fiber content and characteristics. Nevertheless, this depends on several factors, including the type of material, fiber content, associations with other components, disposition and structure of the matrix, processing conditions, etc. (Wang et al. 2022; Gulati et al. 2020; Robin et al. 2012).

Some of these transformations include starch gelatinization, which in its native conformation cannot be digested and thus is considered as fiber, or the formation of amylose-lipid complexes that by their structure and conformations cannot be degraded by digestive enzymes. It has also been reported the increment of dietary fiber by modification of insoluble fiber or by disruption of bonds between fiber molecules and other food components of the matrix by thermomechanical processing, which favors its extraction and quantification by conventional methods (Wang et al. 2022; Arora et al. 2020; Shah et al. 2021; Gulati et al. 2020). Higher fiber content is favorable for intestinal health and prevents the development of diseases, such as type II diabetes, obesity, and cardiovascular diseases among others, and reduces foods' glycemic index (Reyes Moreno et al. 2018). Thus, the use of whole flour for the development of food products is highly recommended.

2.5 Pigments and Antioxidants

Extrusion can have positive and/or negative effects on the structure and functionality of pigments and antioxidants (Reyes Moreno et al. 2018; Wang et al. 2022). On one hand, during extrusion decarboxylation, oxidation, isomerization, and polymerization reactions of these thermo-labile and susceptible compounds may occur, leading to a reduction of their respective levels and changes in products color (Cueto et al. 2017; Dalbhagat et al. 2019). Simultaneously, and as previously mentioned, extrusion generates disruption of complexes between phenolic compounds, carotenoids, and proteins/fiber components. These modifications enhance their exposition, availability, and functionality (Debelo et al. 2020). This is highly dependent on processing conditions, initial composition, and matrix characteristics (Reyes Moreno et al. 2018; Debelo et al. 2020). For example, high temperatures and pressures in systems with low water contents favor degradative reactions. Moreover, the generation of MRPs with antioxidant properties increases the functionality of the products (Cueto et al. 2017).

While, as discussed before, the use of whole flour is convenient for the development of healthy food products, certain processing conditions may impair some nutrients and impact their healthfulness. However, it has been shown that the extrusion of cereal grains leads to improved bioavailability of phenolic acids and mineral and protein digestibility. Moreover, increased content of soluble dietary fiber and phenolic acids and destruction of antinutritional factors, such as trypsin inhibitors and phytates in extruded products have been documented (Hole et al. 2013).

3. Extrudates Characterization and Modulation Through Processing and Formulation Strategies

Table 2 lists some of the physicochemical and mechanical properties of interest as well as several instrumental analyses that are necessary to perform an adequate characterization of cereal-based extrudates. When defining the proper combination of processing conditions, it is important to consider that, for example, the increment in feed moisture, barrel temperature, and screw speed reduce the required energy to push the molten material through the die (torque) because of an increment of the plasticity of the melt and the reduction of its viscosity (Abeykoon et al. 2021). Likewise, the addition of ingredients with high-fat contents reduces the melt viscosity since lipids act as lubricants (Cueto et al. 2015). On the contrary, the incorporation of sticky ingredients generates blends with higher viscosities, which yields higher required energy to force the material through the die (Dalbhagat et al. 2019).

3.1 Effect of Process Conditions and Composition on Physical Properties

Starch gelatinization depends on SME since as higher the applied energy, the higher is the disruption of starch granules (Kristiawan et al. 2018). Particularly, at low moisture values, high SME increases starch transformation. Similarly, either higher temperatures or screw speed favor starch melting and promote lower melt viscosity because of a faster heat transfer and higher shear rate, respectively (Logié et al. 2018; Bisharat et al. 2013; Kristiawan et al. 2018). Regarding this, larger starch gelatinization conducts to higher expansion rates (ER) and lower bulk densities (BD) of the extrudates. Besides, higher temperatures favor water steam accumulation inside the barrel and its greater release at the exit of the extruder, enhancing the expansion. On the contrary, higher moisture contents of the raw ingredients diminish the shear rate and thus SME, which negatively affects ER (Bisharat et al. 2013; Chanvrier et al. 2015).

Reduced ER and higher BD were also observed when starchy ingredients were replaced by others containing high levels of proteins and/or fiber (Bisharat et al. 2013; Chanvrier et al. 2014). These hydrophilic components compete with starch for the available water and reduce the gelatinization degree, which in turn affects the expansion rate (Kristiawan et al. 2018). In such cases, higher feed moisture is necessary to reach similar or higher ER. Likewise, the inter and intramolecular interactions generated by protein molecules negatively affect the generation of a porous structure (Jebalia et al. 2019; Chanvrier et al. 2005). Other authors reported that the incorporation of ingredients with high fiber contents weakens the microstructure of the extrudates, narrowing the cell walls, and generating fragile products with reduced porosity and lower ER (Tas and Shah 2021). Nevertheless, opposite results have been reported by different authors regarding ER and the addition of protein-rich and fiber-rich ingredients (Rolandelli et al. 2020; Cueto et al. 2015; Kasprzak et al. 2013).

The water absorption index (WAI) is an indicator of the capacity of starch to retain or hold water molecules. It measures the ability of starch to interact with water during further processing and/or consumption. On the contrary, the water solubility index (WSI) measures the degradation of starch during extrusion due to the mechanical energy applied, which generates its dextrinization (Dalbhagat et al. 2019). Due to this phenomenon, polysaccharides with lower molecular weight (dextrins) are formed. Thus, the greater the starch degradation, the lower the WAI and the higher the WSI. Thus, regarding processing conditions, increasing feeding moisture limits starch degradation, generating extrudates with higher WAI. On the contrary, when the moisture content is reduced, higher screw speed and/or temperature favor starch degradation and dextrinization due to a higher shear rate, promoting its solubility (Logié et al. 2018; Chanvrier et al. 2015). Likewise, the replacement of starchy materials for others rich in proteins and/or fiber causes a dilution effect, which reduces WAI (Dalbhagat et al. 2019). Nevertheless, some authors reported higher WAI because of the incorporation of protein-rich or fiber-rich ingredients, which is due to the increase of hydrophilic groups that bound free water molecules and promotes their retention after extrusion (Rolandelli et al. 2020; Kasprzak et al. 2013).

Lower values of final water content and water activity (a_w) will be obtained with increasing water steam evaporation at the end of the extrusion process. A direct relationship between final water content and feed moisture and an inverse association with high extrusion temperatures have been reported (Tiwari and Jha 2017). On the other side, the incorporation of hydrophilic ingredients may favor the retention of water molecules and the generation of extrudates with higher final water contents (Rolandelli et al. 2020). This parameter is important since it is related to products' stability, mechanical properties, and susceptibility toward deteriorative reactions (Tiwari and Jha 2017). Nevertheless, extrudates' final water contents are usually low, which grants their microbial stability and adequate mechanical and textural properties, but has the drawback of increased susceptibility towards lipid oxidation (Lampi et al. 2015; Saldanha do Carmo et al. 2019). The latter can be addressed by packing the products with materials with low permeability towards oxygen and light to prevent oxidative reactions (Section 3.3).

Since during extrusion the ingredients are exposed to high temperatures and mechanical energy, non-enzymatic browning reactions such as the Maillard reaction and the degradation of thermolabile pigments are expected to happen (Masatcioglu et al. 2014; 2015; Cueto et al. 2017). These aspects, which will be discussed below (Section 3.3), cause a reduction of essential amino acids like lysine and changes in the color and flavor of the products. The latter has an important role in the definition of consumers' acceptance. The generation of brown pigments because of the Maillard reaction reduces products' lightness (L^*) and increases redness (a^*) and yellowness (b^*) (Pepa et al. 2020; Félix-Medina et al. 2020). Processing conditions are crucial for the definition of browning degree since temperature favors the Maillard reaction rate while higher feeding moisture delays it since water is a product of the reaction (Rolandelli et al. 2021a). The incorporation of protein-rich ingredients (such as legumes, peas, or protein-rich crops like quinoa) favors browning due to the increase of reacting amino groups (Félix-Medina et al. 2020; Rolandelli et al. 2020). Similarly, higher starch dextrinization or depolymerization generates more reducing sugars, which also promotes browning due to caramelization and the Maillard reaction (Mulla et al. 2010). Nevertheless, the natural color of the ingredients also influences the products' coloration. For example, corn flour has quite large amounts of carotenoids (lutein and zeaxanthin) that cause the yellow-orange coloration, and their degradation after extrusion reduces b^* values (Cueto et al. 2017).

3.2 Mechanical Properties as Related to Texture

Extrudates' mechanical properties (hardness, crispness, and adhesiveness among others) are key factors in the definition of texture and affect consumers' acceptance (Dalbhagat et al. 2019). As mentioned above, they are directly related to products' physical properties (ER; final water content) but also depend on microstructural characteristics, such as porosity, cell wall thickness, and associations of the main components, including starch, proteins, lipids, and fiber (Cueto et al. 2015; Farroni et al. 2008; Bouvier and Campanella 2014). For example, the hardness increases with feed moisture while the opposite was observed with the rise in temperature in association with product porosity (Chanvrier et al. 2015).

The addition of protein-rich or fiber-rich ingredients negatively affects ER and porosity, generating products with higher BD and hardness values (Jebalia et al. 2019; Chanvrier et al. 2014). Regarding this, some authors reported cell wall awakening with the incorporation of fiber-rich ingredients, which generated lower hardness values than control systems (Rolandelli et al. 2020). These opposite results indicate that not only the amount, but the type of ingredients used as well as the processing conditions applied and the consequent components associations define the microstructural characteristics (Chanvrier et al. 2015; Kristiawan et al. 2018). On the other side, crispness, i.e. the brittleness of the product that generates a noisy structure when subjected to fracture as in mastication, was reduced with increased protein-rich or fiber-rich ingredients, as measured by acoustic and microscopic analysis (Chanvrier et al. 2014). In such cases, the cell wall thickness defines the crispness of the product; larger amounts of pores with thin cell walls favor crispness perception (Dalbhagat et al. 2019; Chanvrier et al. 2014; Cueto et al. 2015).

Adhesiveness and chewiness, measured by textural analysis, were related to starch status and the characteristics of the melted network. Regarding this, starch dextrinization favors the liberation of sticky polysaccharides of low molecular weight, reducing the strength of the starch network (Dalbhagat et al. 2019). On the other side, the formation of protein agglomerates was also related to obtaining products with undesired mechanical properties (Jebalia et al. 2019; Chanvrier et al. 2005). Moreover, Chanvrier et al. (2015) reported that the type of interactions among proteins defined the aggregation degree, extrudates porosity, and thus, their mechanical properties. Particularly, covalent interactions through disulfide bonds produced extrudates with large protein aggregates, thick cell walls, reduced porosity, and hard structures. Kristiawan et al. (2018) reported higher protein aggregation with the increment of temperature and SME. On the contrary, associations through hydrogen and hydrophobic bonds favored depolymerization and formation of smaller aggregates, thinner cell walls, and reduced hardness (Chanvrier et al. 2015).

3.3 Chemical Properties and Development of Chemical Reactions

Maillard reaction and lipid oxidation are the primary chemical reactions occurring during extrudates' processing and storage, which affect mainly products' color, appearance, and flavor (Yang et al. 2021; Rannou et al. 2016). Particularly, lipid oxidation is one of the key processes involved in the definition of extrudates' shelf life (Yang et al. 2021; Viscidi et al. 2004). The development of these reactions, up to a certain level, favors extrudates organoleptic properties. In association with these reactions, pigments, and polyphenols among other antioxidants may be oxidized, modified, or transformed to prevent or delay the occurrence of these deteriorative reactions (Viscidi et al. 2004; Utama-ang et al. 2017). Besides the organoleptic and functional modifications, the degradation of proteins and lipids conducts to a reduction of the nutritional value of the products (Lampi et al. 2015; Masatcioglu et al. 2014). In addition to the development of desired sensory characteristics, some harmful products may also be formed, such as furfurals, furans, heterocyclic amines, and acrylamide (Rannou et al. 2016; Mulla et al. 2011; Masatcioglu et al. 2014; 2015), as well as off-flavor compounds (Yang et al. 2021), making necessary to control the progress of these reactions.

Maillard reaction is a non-enzymatic browning reaction of which mechanisms are not still fully elucidated. It initiates with the condensation of an amino group with reducing sugar, generating an instable Schiff base that rapidly converts to substituted glucosamine (Rufián-Henares and Pastoriza 2016). Up to this point, the reaction is reversible but, as it progresses, an Amadori or Heyns product is formed, and the reaction turns irreversible (Rufián-Henares and Pastoriza 2016). From this step, multiple pathways may be followed according to the processing conditions applied (temperature, pH, water activity, and substrates), leading to a wide variety of products, including early products with antioxidant properties (Rannou et al. 2016). More advanced Maillard reaction products (MRPs), such as hydroxymethylfurfural (HMF), furosine, and acrylamide, are frequently used as heat-damage indicators as they can be quantified by spectroscopic and chromatographic techniques (Masatcioglu et al. 2014; 2015; Rannou et al. 2016). These are also precursors of the late reaction products, melanoidins, which are the pigments that generate the typical brown color of the reaction (Rufián-Henares and Pastoriza 2016). Moreover, some flavor-related compounds, mainly volatile aldehydes and ketones, are also formed, which are responsible for the aroma modifications (Rufián-Henares and Pastoriza 2016).

In extrusion cooking, the Maillard reaction is favored by high temperature and reduced feed moisture (Masatcioglu et al. 2014; 2015; Guerrero et al. 2012). On the other hand, the short residence time, typical of extrusion, limits the development of the reaction. Moreover, the addition of reducing sugars or protein-rich ingredients favors non-enzymatic browning and can enhance the formation of specific compounds (Guerrero et al. 2012; Rannou et al. 2016). On the contrary, higher water addition diminishes browning reactions during extrusion (Saldanha do Carmo et al. 2019). In line with this, the adequate degree of browning allows the formation of aroma-related compounds, which are highly accepted by consumers (Saldanha do Carmo et al. 2019) but, simultaneously, harmful products are formed (Masatcioglu et al. 2015; Rannou et al. 2016). Therefore, it is important to control the development of the reaction to achieve desired characteristics while minimizing the formation of harmful products (Rolandelli et al. 2021a).

Lipid oxidation is initiated by the free radical formation and proceeds as a chain reaction in the presence of oxygen. It is favored by light exposition, low water activities, high temperatures, and the presence of pro-oxidants, such as metals or photosensitizers (Schaich 2013). Unsaturated fatty acids are more susceptible to oxidation, and susceptibility increases with the number of unsaturation (Barden and Decker 2013). Lipid oxidation has a *lag* phase prior to the exponential accumulation of the products (Barden and Decker 2013). The presence of antioxidant compounds extends this *lag* phase and preserves lipids stability during longer times, but their stability also depends on extrusion conditions (Yang et al. 2021; Utama-ang et al. 2017; Viscidi et al. 2004). According to the progress of the reaction, different compounds are formed. During the first stage, hydroperoxides are the main products (Viscidi et al. 2004). In this step lipids degradation occur and the nutritional value of the product diminishes, but organoleptic properties are yet not affected (Schaich 2013; Barden and Decker 2013). During advanced stages, hydroperoxides suffer β-scission reactions and volatile compounds of low molecular weight, mainly aldehydes, ketones, and acids, are formed. These compounds are responsible for the perception of rancidity and are usually called off-flavor products (Yang et al. 2021). Moreover, some radicals formed during lipid oxidation can also cross-link with proteins and other components of the matrix, affecting their functionality and availability and generating changes in products' color (Schaich 2013). Lipid oxidation is a time-dependent reaction. Thus, it is responsible for the changes in the organoleptic properties of extrudates during storage, especially at high temperatures.

Regarding the effects of extrusion processing conditions, lipid oxidation is favored in extrudates obtained at high temperatures and low feeding moistures, since they are highly expanded (Lampi et al. 2015). Products with high ER present increased surface area, large pores, and thin cell walls, thus increasing oxygen exposure and susceptibility towards oxidation (Barden and Decker 2013). Under these conditions, antioxidants are also less stable and can be destroyed, which increase oxidation rates (Viscidi et al. 2004). Nevertheless, high temperatures may be useful for the inactivation of lipases and other enzymes that cause lipids oxidation (Moisio et al. 2015; Lampi et al. 2015). Once again, the proper combination of extrusion conditions needs to be studied and optimized in every case, according to the characteristics of the ingredients and the desired final properties.

4. Extrusion as a Sustainable Food Development Technique Towards Circular Economy

Current consumers are widely aware of the importance of including more plant-based food products in their diets to assure a healthy lifestyle (Kim et al. 2021; Grasso 2020). Regarding this, the demand for novel foodstuffs with minimum processing technology and with the replacement of additives by natural ingredients keeps raising every year (Zamuz et al. 2021; Lillford and Hermansson 2021; Prabha et al. 2021; Dalbhagat et al. 2019). The food industry also faces the challenge of developing these products in a sustainable way and minimizing the negative environmental impacts. Due to the continuous reduction of cultivable land, the impacts of climate change, the exploitation of natural resources, and the enormous generation of waste in the production chain, the United Nations have established the Sustainable Development Goals (SDG) in which food industry is involved in at least four of them, including the pursuit of zero hunger, responsible production and consumption, and assuring good health and well-being (Lillford and Hermansson 2021; Usmani et al. 2021).

To respond to this scenario, traditional methodologies need to be adapted or replaced by novel technologies with reduced energy consumption and waste generation that also minimize the impact on food ingredients with the aim of generating innovative and sustainable food products (Lillford and Hermansson 2021). The inclusion of edible food by-products, mostly from fruits and vegetables, and the revalorization of undervalued flours, such as those from millets, sorghums, and other non-traditional grains and crops, have been proposed as an alternative processing strategy. With this approach, it is possible to promote the circular economy through the complete exploitation of natural resources while improving the nutritional and functional properties of products (Wang et al. 2022; Usmani et al. 2021; Augustin et al. 2020; Grasso 2020). The utilization of these ingredients also improves food safety and expands the number of available materials for the generation of food products, allowing the social and economic development of the vulnerable regions where these raw materials are cultivated (Wang et al. 2022; Augustin et al. 2020; Lillford and Hermansson 2021). Nevertheless, since this is a recent approach, there are still several challenges that need to be addressed, such as scaling up the process to industrial size, collecting, characterizing, and assuring the safety of the food wastes, and facing the higher costs of this sustainable production system (Grasso 2020; Usmani et al. 2021; Lillford and Hermansson 2021; Augustin et al. 2020).

Extrusion technology responds to all these challenges since it is a continuous, versatile, simple, and economic technique that allows the development of a wide variety of plant-based food products with minimum waste generation (Prabha et al. 2021; Grasso 2020). Its versatility allows the inclusion of edible by-products (peels, bagasse, pomaces, and brans) and the utilization of non-traditional flours (from beans, peas, pulses, cereals, and pseudo cereals) for the generation of products with distinctive characteristics (Tas and Shah 2021; Grasso 2020; Prabha et al. 2021). When compared with other common food processing technologies, extrusion requires less amount of water and energy, showing its distinctive advantages for sustainable production (Abeykoon et al. 2021; Prabha et al. 2021). Besides, extruded ready-to-eat products with minimum nutritional loss are obtained in a short time in comparison with other traditional technologies (Reis et al. 2014).

4.1 Incorporation of Food By-Products

Currently, during each step of the food production chain, tons of materials are converted to food waste, discarded, or even never utilized. The efficiency of food production is estimated at 50%, with developed countries generating more residues than developing countries (Usmani et al. 2021). Fruits and vegetables comprise the major part of the total food loss worldwide (Augustin et al. 2020). With them, not only nutrients but also water, energy, land usage, man work, fertilizers, and other resources involved in their production are lost. This represents a waste of natural resources and does not help in resolving the hunger and poverty crisis.

The recovery of these materials, or their use for obtaining active ingredients, presents double advantages; it diminishes wastes generation while simultaneously improving the nutritional and functional values of the food products (higher levels of proteins, fiber, polyphenols, vitamins, and minerals) (Augustin et al. 2020; Grasso 2020). Extrusion versatility allows the easy incorporation of these materials in cereal-based formulations in a more green and sustainable way. Moreover, the process can improve the bioavailability and functionality of bioactive compounds, such as polyphenols and antioxidants (Hegazy et al. 2017).

Table 3 summarizes the application of several fruits and vegetable by-products in cereal-based formulations and the benefits of their inclusion. Apple pomace is the by-product obtained from the production of apple juice or other apple-based products. Its inclusion as a powder in corn and rice/wheat-based extrudates showed improvement in fiber and phenolic contents and antioxidant activity (Reis et al. 2014; Paraman et al. 2015). Similar results were obtained with the addition of grape pomace (from grape juice) in cassava-soy flour blends and corn grits, where increments in soluble and insoluble fiber contents, lower glycemic indexes, higher polyphenols content, and antioxidant activity were reported (Oladiran and Emmambux 2018; Oliveira et al. 2013). Corn and corn-wheat formulations enriched with tomato pomace also showed increased functionality for their higher antioxidant activity and phenolic contents, as well as a rise in fiber and lycopene levels (Yagci et al. 2022; Hegazy et al. 2017). Likewise, the addition of naranjita bagasse to corn-based extrudates improved products functionality for their higher carotenoids, fiber, and phenolics contents and antioxidant activity (Ruiz-Armenta et al. 2017; Delgado-Nieblas et al. 2019). The addition of orange bagasse to corn-based extrudates also improved their fiber and ashes levels (Navarro-Cortez et al. 2016).

Brewer's spent grain is the insoluble fraction of the malted barley and constitutes the main by-product of the brewing industry. This material is rich in fiber (lignin, cellulose, and arabinoxylans), proteins, and minerals (Grasso 2020). Its nutritional and functional benefits were confirmed when brewer's spent grain was added to cereal-based snacks (Makowska et al. 2013; Reis and Abu-Ghannam 2014; Ainsworth et al. 2007). Brans are by-products of refined cereal flours (rice, corn, wheat, rya, and oat) or are obtained when grains are polished and have high amounts of proteins, fiber, and minerals with values varying according to the botanic source (Grasso 2020). Their incorporation in cereal-based extrudates improved their final nutritional value and increased their phenolic contents and antioxidant activity (Wang et al. 2018; Makowska et al. 2015; Fleischman et al. 2016).

4.2 Revalorization of Undervalued Cereal and Pseudocereal Flours

Global trends in food science and technology include the utilization of crops that can grow in arid or semi-arid conditions, which have resistance to high temperatures and salinity and low water requirements (Wang et al. 2022; Lillford and Hermansson 2021). Millets, sorghums, quinoa, and canary seed are some examples of crops that fit with these characteristics; they have high protein and

Table 3. Fruits and Vegetables By-Products Incorporated in Cereal-Based Extrudates.

Fruit and Vegetable By-Products	Cereal-Based Formulation	Benefits	Reference
Apple pomace	Corn starch	Higher total phenolic content, antioxidant activity, and fiber content	Paraman et al. 2015
	Rice flour and wheat semolina	Higher fiber, total phenolic, total flavonoid, proanthocyanidins contents, and antioxidant activity	Reis et al. 2014
Grape pomace	Cassava and soy flours	Higher soluble and insoluble fiber contents and increase in phenolics and antioxidant activity. Lower glycemic index.	Oladiran and Emmambux 2018
	Corn grits	Higher fiber and phenolics contents	Oliveira et al. 2013
Tomato pomace	Corn starch and corn and wheat semolina	Higher phenolics and lycopene contents and increased antioxidant activity	Yagci et al. 2022
Chickpea flour and tomato pomace	Corn grits	Increase in protein, fiber, and polyphenols levels and antioxidant activity	Hegazy et al. 2017
Naranjita bagasse	Corn flour	Higher carotenoids contents	Ruiz-Armenta et al. 2017
	Corn grits, wheat bran, and oat bran	Increment of fiber and phenolics contents and antioxidant activity	Delgado-Nieblas et al. 2019
Orange bagasse	Corn flour	Higher fiber and ash levels	Navarro-Cortez et al. 2016
Brewer's spent grain	Corn grits	Increase in lignin, cellulose, and hemicellulose contents	Makowska et al. 2013
	Rice flour and wheat semolina	Higher fiber and phenolics contents and antioxidant activity	Reis and Abu-Ghannam 2014
	Corn, oat, and chickpea flour	Increment of protein, fiber, and resistant starch contents	Ainsworth et al. 2007
Rice bran	Rice flour	Increase in polyphenols contents, antioxidant activity, and resistant starch levels	Wang et al. 2018
Wheat bran	Wheat flour	Higher protein, ash, and fiber levels and antioxidant activity	Fleischman et al. 2016
Oat, wheat, and rye bran	Corn grits	Increment in protein, ash, and fiber contents	Makowska et al. 2015

fiber levels and are rich in bioactive compounds (Rolandelli et al. 2020; 2021b; Wang et al. 2022). Although having these agricultural and nutritional advantages, these cultivars are mostly grown locally in developing countries, where they are considered staple food and are only included in domestic recipes or used as animal feed. As a result, there is scarce industrial application, which can be seen as an opportunity for economic development in such areas. Nevertheless, some cereal-based extrudates have been elaborated including these undervalued ingredients (Table 4).

The addition of different varieties of millets flours to corn or rice-based formulations improved the protein, minerals, fiber, and phenolics contents of the products (Rolandelli et al. 2021b; McSweeney et al. 2017; Lakshmi Devi et al. 2012). Similarly, sorghum addition to corn flour improved its nutritional value and antioxidant activity (Buitimea-Cantúa 2017; 2018; Rolandelli et al. 2021b; Licata et al. 2013). Canary seed is mainly used for animal feed, but it was recently approved for human consumption. Thus, there are still scarce reports of its application in extrusion technology. Rolandelli et al. (2021b) reported an increment in fiber, ash, and up to 40% of protein levels with the replacement of 25% of corn flour by canary seed flour.

Table 4. Undervalued Cereals and Pseudocereals Incorporated in Cereal-Based Extrudates.

Undervalued Cereal/ Pseudocereal	Cereal-Based Formulation	Benefits	Reference
Millet	Corn flour	Higher protein, fiber, ash levels, and lower lipids levels	Rolandelli et al. 2021b
	Corn flour	Increment in protein, fiber, resistant starch, and phenolics contents. Lower glycemic index.	McSweeney et al. 2017
	Rice flour	Higher protein, fiber, calcium, phosphorus, and iron levels	Lakshmi Devi et al. 2012
Sorghum	Corn flour	Increment in phenolics contents and antioxidant activity	Buitimea-Cantúa et al. 2017; 2018
	Corn flour	Higher phenolics and resistant starch contents	Licata et al. 2013
	Corn flour	Higher protein, fiber, and ash levels	Rolandelli et al. 2021b
Canary seed	Corn flour	Increment of protein, ash, and fiber values	Rolandelli et al. 2021b
Quinoa	Corn flour	Increment in protein and fiber values	Ramos-Díaz et al. 2019
	Corn flour	Higher protein and ash levels, reduced lipids contents	Rolandelli et al. 2021b
	Corn flour	Increment of protein, fiber, zinc, iron, and essential fatty acids contents	Giménez et al. 2016
Amaranth	Corn flour	Increment in protein and fiber values	Ramos-Díaz et al. 2019
	Corn flour	Higher polyphenolics contents and antioxidant activity	Pęksa et al. 2016
	Corn flour	Increment in iron, zinc, and calcium levels	Galan et al. 2013
Kiwicha	Corn flour	Increment in iron, zinc, and calcium levels	Galan et al. 2013
Kañiwa	Corn flour	Increment in protein and fiber values	Ramos-Díaz et al. 2019

As for pseudocereals, quinoa is the most studied for its inclusion in cereal-based extrudates. It is a valuable Andean crop recognized for its high protein, lipids, and bioactive compounds. Its inclusion in corn-based extrudates improved their protein, fiber, lipids, and ash values (Ramos-Díaz et al. 2019; Rolandelli et al. 2021b; Giménez et al. 2016). Other pseudocereals that have been less studied for extrudates developments are amaranth and kañiwa (Ramos-Díaz et al. 2019; Pęksa et al. 2016; Galan et al. 2012). Their inclusions in corn flour-based extrudates increased their protein, fiber, iron, zinc, calcium, and phenolics contents and antioxidant activity. Another pseudocereal with high potential to be applied in extrusion technology is kiwicha (*Amaranthus caudatus*). It has a valuable nutritional composition, such as 13% proteins with high amounts of lysine, 16% of dietary fiber, as well as high phenolics contents and antioxidant activity, even after extrusion (Basilio-Atencio et al. 2020; Galan et al. 2012; Repo-Carrasco-Valencia et al. 2009). To date, there is only one study that elaborated corn-based extrudates with the addition of kiwicha (Galan et al. 2012). Authors reported increased levels of minerals, particularly, iron, zinc, and calcium with the inclusion of kiwicha flour. The use of these undervalued resources that can be grown with low soil and water requirements will offer the possibility of providing food security and economic means to economically vulnerable populations.

5. Conclusions

Extrusion represents a simple, versatile, and economic technique for the sustainable obtention of a wide variety of food products. The final characteristics of the extrudates will be defined by the combined effects of raw ingredients used and processing conditions applied and their influence on the structure and degree of associations of macro and micronutrients.

Extrusion may provide a feasible way of processing whole grain flour for the development of healthy food products with increased content of soluble dietary fiber and improved bioavailability of phenolic acids and minerals in extruded cereal grain. Also, it is possible to enhance protein digestibility with the simultaneous destruction of anti-nutritional factors.

This technique allows the control of thermo-mechanical transformations of food components that will directly influence the physicochemical, organoleptic, and mechanical properties of the extrudates, as well as their stability during storage.

The need to produce foodstuffs in a more sustainable way allows the wider utilization of extrusion for the development of distinctive foods in an eco-friendly way. The revalorization of undervalued flours and the incorporation of fruits and vegetable by-products can successfully be done through extrusion technology given its versatility. This combines several advantages: reduction of waste generation, nutritional and functional enhancement, and generation of innovative food products at low costs. Several benefits of the inclusion of novel ingredients in cereal-based extrudates were summarized in this chapter, which confirms the usefulness of extrusion technology for the sustainable generation of food products.

References

Abeykoon, C., McMillan, A. and Nguyen, B.K. 2021. Energy efficiency in extrusion-related polymer processing: A review of state of the art and potential efficiency improvements. Renewable and Sustainable Energy Reviews 147: 111219. https://doi.org/10.1016/j.rser.2021.111219

Ainsworth, P., İbanoğlu, Ş., Plunkett, A., İbanoğlu, E. and Stojceska, V. 2007. Effect of brewers spent grain addition and screw speed on the selected physical and nutritional properties of an extruded snack. Journal of Food Engineering 81: 702–709. https://doi.org/10.1016/j.jfoodeng.2007.01.004

Arora, B., Yoon, A., Sriram, M., Singha, P. and Rizvi, S.S.H. 2020. Reactive extrusion: A review of the physicochemical changes in food systems. Innovative Food Science and Emerging Technologies. 64: 102429. https://doi.org/10.1016/j.ifset.2020.102429

Augustin, M.A., Sanguansri, L., Fox, E.M., Cobiac, L. and Cole, M.B. 2020. Recovery of wasted fruit and vegetables for improving sustainable diets. Trends in Food Science and Technology 95: 75–85. https://doi.org/10.1016/j.tifs.2019.11.010

Barden, L. and Decker, E.A. 2013. Lipid oxidation in low-moisture food: a review. Critical Reviews in Food Science and Nutrition 56: 2467–2482. https://doi.org/10.1080/10408398.2013.848833

Basilio-Atencio, J., Condezo-Hoyos, L. and Repo-Carrasco-Valencia, R. 2020. Effect of extrusion cooking on the physical-chemical properties of whole kiwicha (*Amaranthus caudatus* L) flour variety *centenario*: Process optimizarion. LWT —Food Science and Technology 128: 109426. https://doi.org/10.1016/j.lwt.2020.109426

Bisharat, G.I., Oikonomopoulou, V.P., Panagiotou, N.M., Krokida, M.K and Maroulis, Z.B. 2013. Effect of extrusion conditions on the structural properties of corn extrudates enriched with dehydrated vegetables. Food Research International 53: 1–14. https://doi.org/10.1016/j.foodres.2013.03.043

Bouvier, J.-M. and Campanella, O.H. 2014. Extrusion Processing Technology. Food and Non-Food Biomaterials. Wiley Blackwell, West Sussex.

Buitimea-Cantúa, N.E., Torres-Chávez, P.I., Ramírez-Wong, B., Ledesma-Osuna, A.I., Gutiérrez-Uribe, J.A., Serna-Guerrero, D.A. et al. 2018. Ferulic, *p*-coumaric, diferulic and triferulic acids contents of corn tortillas prepared with extruded corn flour and enriched with sorghum (*Sorghum bicolor* (L.) Moench) bran. Journal of Food Measurement and Characterization 12: 1633–1640. https://doi.org/10.1007/s11694-018-9778-4

Buitimea-Cantúa, N.E., Torres-Chávez, P.I., Ramírez-Wong, B., Serna-Saldívar, S.O., Rouzaud-Sández, O., Rosas-Burgos, E.C. et al. 2017. Phenolics compounds and antioxidant activity of extruded nixtamalized corn flour and tortillas enriched with sorghum bran. Cereal Chemistry 94: 277–283. https://doi.org/10.1094/CCHEM-04-16-0115-R

Chalermchaiwat, P., Jangchud, K., Jangchud, A., Charunuch, C. and Prinyawiwatkul, W. 2015. Antioxidant activity, free gamma-aminobutyric acid content, selected physical properties and consumer acceptance of germinated brown rice extrudates as affected by extrusion process. LWT – Food Science and Technology 64: 490–496. https://doi.org/10.1016/j.lwt.2015.04.066

Chanvrier, H., Pillin, C.N., Vandeputte, G., Haiduc, A., Leloup, V. and Gumy, J.-C. 2015. Impact of extrusion parameters on the properties of rice products: A physicochemical and X-ray tomography study. Food Structure 6: 29–40. https://doi.org/10.1016/j.foostr.2015.06.004

Chanvrier, H., E. Jakubczyk, E. Gondek and J.-C. Gumy. 2014. Insights into the texture of extruded cereals: Structure and acoustic properties. Innovative Food Science and Emerging Technologies 24: 61–68. https://doi.org/10.1016/j.ifset.2013.11.013

Chanvrier, H., Colonna, P., Della Valle, G. and Lourdin, D. 2005. Structure and mechanical behaviour of corn flour and starch-zein based materials in the glassy state. Carbohydrate Polymers 59: 109–119. https://doi.org/10.1016/j.carbpol.2004.09.005

Chen, F.L., Wei, Y.M. and Zhang, B. 2010. Characterization of water state and distribution in textured soybean protein using DSC and NMR. Journal of Food Engineering 100: 522–526. https://doi.org/10.1016/j.jfoodeng.2010.04.040

Cueto, M., Farroni, A., Schoenlechner, R., Schleining, G. and Buera, P. 2017. Carotenoid and color changes in traditionally flaked and extruded products. Food Chemistry 229: 640–645. https://doi.org/10.1016/j.foodchem.2017.02.138

Cueto, M., Farroni, A., Rodríguez, S.D., Schoenlechner, R., Schleining, G. and Buera, M.P. 2018. Assessing changes in enriched maize flour formulations after extrusion by means of FTIR, XRD, and chemometric analysis. Food and Bioprocess Technology 11: 1586–1595. https://doi.org/10.1007/s11947-018-2113-6

Cueto, M., Porras-Saavedra, J., Farroni, A., Alamilla-Beltrán, L., Schöenlechner, R., Schleining, G. et al. 2015. Physical and mechanical properties of maize extrudates as affected by the addition of chia and quinoa seeds and antioxidants. Journal of Food Engineering 167: 139–146. https://doi.org/10.1016/j.jfoodeng.2015.07.027

Dalbhagat, C.D., Mahato, D.K. and Mishra, H.N. 2019. Effect of extrusion processing on physicochemical, functional and nutritional characteristics of rice and rice-based products: A review. Trends in Food Science and Technology 85: 226–240. https://doi.org/10.1016/j.tifs.2019.01.001

Debelo, H., Li, M. and Ferruzzi, M.G. 2020. Processing influences on food polyphenol profiles and biological activity. Current Opinion in Food Science 32: 90–102. https://doi.org/10.1016/j.cofs.2020.03.001

Delgado-Nieblas, C., Ruiz-Beltrán, K., Sánchez-Lizárraga, J., Zazueta-Morales, J.J., Aguilar-Palazuelos, E., Carrillo-López, A. et al. 2019. Effect of extrusion on physicochemical, nutritional and antioxidant properties of breakfast cereals produced from bran and dehydrated naranjita pomace. CyTA – Journal of Food. 17: 240–250. https://doi.org/10.1080/19476337.2019.1566276

Farroni, A.E., Matiacevich, S.B., Guerrero, S., Alzamora, S. and Buera, M.P. 2008. Multi-level approach for the analysis of water effects in corn flakes. Journal of Agricultural and Food Chemistry 56: 6447–6453. https://doi.org/10.1021/jf800541f

Félix-Medina, J.V., Montes-Ávila, J., Reyes-Moreno, C., Perales-Sánchez, J.X.K., Gómez-Favela, M.A., Aguilar-Palazuelos, E. et al. 2020. Second-generation snacks with high nutritional and antioxidant value produced by an optimized extrusion process from corn/common bean flours mixtures. LWT – Food Science and Technology 124: 109172. https://doi.org/10.1016/j.lwt.2020.109172

Fleischman, E.F., Kowalski, R.J., Morris, C.F., Nguyen, T., Li, C., Ganjyal, G. et al. 2016. Physical, textural, and antioxidant properties of extruded waxy wheat flour snack supplemented with several varieties of bran. Journal of Food Science 81: 2726–2733. https://doi.org/10.1111/1750-3841.13511

Galan, M.G., Drago, S.R., Armada, M. and González José, R. 2013. Iron, zinc and calcium dialyzability from extrude product based on whole grain amaranth (*Amaranthus caudatus* and *Amaranthus cruentus*) and amaranth/*Zea mays* blends. International Journal of Food Sciences and Nutrition 64: 502–507. https://doi.org/10.3109/09637486.2012.753038

Giménez, M.A., Drago, S.R., Bassett, M.N., Lobo, M.O. and Sammán, N.C. 2016. Nutritional improvement of corn pasta-like product with broad bean (*Vicia faba*) and quinoa (*Chenopodium quinoa*). Food Chemistry 199: 150–156. https://doi.org/10.1016/j.foodchem.2015.11.065

Grasso, S. 2020. Extruded snacks from industrial by-products: A review. Trends in Food Science and Technology 99: 284–294. https://doi.org/10.1016/j.tifs.2020.03.012

Guerrero, P., Beatty, E., Kerry, J.P. and de la Caba, K. 2012. Extrusion of soy protein with gelatin and sugars at low moisture content. Journal of Food Engineering 110: 53–59. https://doi.org/10.1016/j.jfoodeng.2011.12.009

Gulati, P., Brahma, S. and Rose, D.J. 2020. Impacts of extrusion processing on nutritional components in cereals and legumes: Carbohydrates, proteins, lipids, vitamins, and minerals. pp. 415–443. *In*: Ganjyal, G.M. (ed.). Extrusion cooking. Cereal grains processing. Woodhead Publ. & Cereals & Grains Assoc. Bookstore. https://doi.org/10.1016/B978-0-12-815360-4.00013-4

Hegazy, H.S., El-Bedawey, A.E.-F.A., Rahma, E.-S.H. and Gaafar, A.M. 2017. Effect of extrusion process on nutritional, functional properties and antioxidant activity of germinated chickpea incorporated corn extrudates. American Journal of Food Science and Nutrition Research 4: 59–66.

Hole, A.S., Kjos, N.P., Grimmer, S., Kohler, A., Lea, P., Rasmussen, B. et al. 2013. Extrusion of barley and oat improves the bioaccessibility of dietary phenolic acids in growing pigs. Journal of Agricultural and Food Chemistry 61: 2739–2747. https://doi.org/10.1021/jf3045236

Ilo, S., Schoenlechner, R. and Berghofe, E. 2000. Role of lipids in the extrusion cooking processes. Grasas y aceites 51: 97–110. https://doi.org/10.3989/gya.2000.v51.i1-2.410

Jafari, M., Koocheki, A. and Milani, E. 2017. Effect of extrusion cooking on chemical structure, morphology, crystallinity and thermal properties of sorghum flour extrudates. Journal of Cereal Science 75: 324–331. https://doi.org/10.1016/j.jcs.2017.05.005

Jebalia, I., Maigret, J.-E., Réguerre, A.-L., Novales, B., Guessasma, S., Lourdin, D. et al. 2019. Morphology and mechanical behaviour of pea-based starch-protein composites obtained by extrusion. Carbohydrate Polymers 223: 115086. https://doi.org/10.1016/j.carbpol.2019.115086

Kasprzak, M., Rzedzicki, Z., Wirkijowska, A., Zarzycki, P., Sobota, A., Sykut-Domańska, E. et al. 2013. Effect of fibre-protein additions and process parameters on microstructure of corn extrudates. Journal of Cereal Science 58: 488–494. https://doi.org/10.1016/j.jcs.2013.09.002

Kim, J., Kim, H. and Giovannucci, E.L. 2021. Plant-based diet quality and the risk of total and disease-specific mortality: A population-based prospective study. Clinical Nutrition 40: 5718–5725. https://doi.org/10.1016/j.clnu.2021.10.013

Krishnan, V., Mondal, D., Thomas, B., Singh, A. and Praveen, S. 2021. Starch-lipid interaction alters the molecular structure and ultimate starch bioavailability: A comprehensive review. International Journal of Biological Macromolecules 182: 626–638. https://doi.org/10.1016/j.ijbiomac.2021.04.030

Kristiawan, M., Micard, V., Maladira, P., Alchamieh, C., Maigret, J.-E., Réguerre, A.-L. et al. 2018. Multi-scale structural changes of starch and proteins during pea flour extrusion. Food Research International 108: 203–215. https://doi.org/10.1016/j.foodres.2018.03.027

Lakshmi Devi, N., Shobha, S., Alavi, S., Kalpana, K. and Soumya, M. 2012. Utilization of extrusion technology for the development of millet based complementary foods. Journal of Food Science and Technology 51: 2845–2850. https://doi.org/10.1007/s13197-012-0789-6

Lampi, A.-M., Damerau, A., Li, J., Moisio, T., Partanen, R., Forssell, P. et al. 2015. Changes in lipids and volatile compounds of oat flours and extrudates during processing and storage. Journal of Cereal Science 62: 102–109. https://doi.org/10.1016/j.jcs.2014.12.011

Licata, R., Chu, J., Wang, S., Coorey, R., James, A., Zhao, Y. et al. 2013. Determination of formulation and processing factors affecting slowly digestible starch, protein digestibility and antioxidant capacity of extruded sorghum-maize composite flour. International Journal of Food Science and Technology. 49: 1408-1419. https://doi.org/10.1111/ijfs.12444

Lillford, P. and A.-M Hermansson. 2021. Global missions and the critical needs of food science and technology. Trends in Food Science and Technology 111: 800–811. https://doi.org/10.1016/j.tifs.2020.04.009

Logié, N., Della Valle, G., Rolland-Sabaté, A., Descamps, N. and Soulestin, J. 2018. How does temperature govern mechanisms of starch changes during extrusion?. Carbohydrate Polymers 184: 57–65. https://doi.org/10.1016/j.carbpol.2017.12.040

Makowska, A., Polcyn, A., Chudy, S. and Michniewicz, J. 2015. Application of oat, wheat and rye bran to modify nutritional properties, physical and sensory characteristics of extruded corn snacks. Acta Scientiarum Polonorum Technologia Alimentaria 14: 375–386. https://doi.org/10.17306/J.AFS.2015.4.37

Makowska, A., Baranowska, H.M., Michniewicz, J., Chudy, S. and Kowalczewski, P.Ł. 2017. Triticale extrudates—Changes of macrostructure, mechanical properties and molecular water dynamics during hydration. Journal of Cereal Science 74: 250–255. https://doi.org/10.1016/j.jcs.2017.03.002

Makowska, A., Mildner-Szkudlarz, S. and Obuchowski, W. 2013. Effect of brewer's spent grain addition on properties of corn extrudates with an increased dietary fibre content. Polish Journal of Food and Nutrition Sciences 63: 19–24. https://doi.org/10.2478/v10222-012-0061-9

Masatcioglu, T.M., Ng, P.K.W. and Koksel, H. 2014. Effects of extrusion cooking conditions and chemical leaving agents on lysine loss as determined by furosine content in corn based extrudates. Journal of Cereal Science 60: 276–281. https://doi.org/10.1016/j.jcs.2014.06.008

Masatcioglu, T.M., Ng, P.K.W. and Koksel, H. 2015. Effects of formulation and extrusion cooking conditions on furfural and hydroxymethylfurfural content. Journal of Cereal Science 65: 31–38. https://doi.org/10.1016/j.jcs.2015.06.003

Masavang S., Roudaut, G. and Champion, D. 2019. Identification of complex glass transition phenomena by DSC in expanded cereal-based extrudates: Impact of plasticization by water and sucrose. Journal of Food Engineering 245: 43–52. https://doi.org/10.1016/j.jfoodeng.2018.10.008

McSweeney, M.B., Seetharaman, K., Dan Ramdath, D. and Duizer, L.M. 2017. Chemical and physical characteristics of proso millet (*Panicum miliaceum*)-based products. Cereal Chemistry 94: 357–362. https://doi.org/10.1094/CCHEM-07-16-0185-R

Moisio, T., Forssell, P., Partanen, R., Damerau, A. and Hill, S.E. 2015. Reorganisation of starch, proteins and lipids in extrusion of oats. Journal of Cereal Science 64: 48–55. https://doi.org/10.1016/j.jcs.2015.04.001

Mulla, M.Z., Bharadwaj, V.R., Annapure, U.S. and Singhal, R.S. 2010. Effect of damaged starch on acrylamide formation in whole wheat flour based Indian traditional staples, *chapattis* and *pooris*. Food Chemistry 120: 805–809. https://doi.org/10.1016/j.foodchem.2009.11.016

Mulla, M.Z., Bharadwaj, V.R., Annapure, U.S. and Singhal, R.S. 2011. Effect of formulation and processing parameters on acrylamide formation: A case study on extrusion of blends of potato flour and semolina. LWT – Food Science and Technology 44: 1643–1648. https://doi.org/10.1016/j.lwt.2010.11.019

Navarro-Cortez, R.O., Gómez-Aldapa, C.A., Aguilar-Palazuelos, E., Delgado-Licon, E., Castro Rosas, J., Hernández-Ávila, J. et al. 2016. Blue corn (*Zea mays* L.) with added orange (*Citrus sinensis*) fruit bagasse: novel ingredients for extruded snacks. CyTA – Journal of Food. 14: 349–358. https://doi.org/10.1080/19476337.2015.1114026

Obradović, V., Babić, J., Šubarić, D., Jozinović, A., Ačkar, Đ. and Klarić, I. 2015. Influence of dried Hokkaido pumpkin and ascorbic acid addition on chemical properties and colour of corn extrudates. Food Chemistry 183: 136–143. https://doi.org/10.1016/j.foodchem.2015.03.045

Oladiran, D.A. and Emmambux, N.M. 2018. Nutritional and functional properties of extruded cassava-soy composites with grape pomace. Starch. 70: 1700298. https://doi.org/10.1002/star.201700298

Oliveira, D.M., Rodrigues Marques, D., Kwiatkowski, A., Giriboni Monteiro, A.R. and Clemente, E. 2013. Sensory analysis and chemical characterization of cereal enriched with grape peel and seed flour. Acta Scientiarum. Technology 35: 427–431. https://doi.org/10.4025/actascitechnol.v35i3.13176

Paraman, I., Sharif, M.K., Supriyadi, S. and Rizvi, S.S.H. 2015. Agro-food industry byproducts into value-added extruded foods. Food and Bioproducts Processing 96: 78–85. https://doi.org/10.1016/j.fbp.2015.07.003

Pęksa, A., Kita, A., Carbonell-Barrachina, A.A., Miedzianka, J., Kolniak-Ostek, J., Tajner-Czopek, A. et al. 2016. Sensory attributes and physicochemical features of corn snacks as affected by different flour types and extrusion conditions. LWT – Food Science and Technology 72: 26–36. https://doi.org/10.1016/j.lwt.2016.04.034

Pepa, L., Rodríguez, S.D., dos Santos Ferreira, C. and Buera, M.P. 2020. Interpretation of the color due to the ubiquitous nonenzymatic browning phenomena in foods. Color Research and Application 46: 1–10. https://doi.org/10.1002/col.22574

Prabha, K., P. Ghosh, A.S, Joseph, R.M., Krishnan, R., Singh Rana, S. et al. 2021. Recent development, challenges, and prospects of extrusion technology. Future Foods 3: 100019. https://doi.org/10.1016/j.fufo.2021.100019

Ramos-Diaz, J.M., Rinnan, Å. and Jouppila, K. 2019. Application of NIR imaging to the study of expanded snacks containing amaranth, quinoa and kañiwa. LWT – Food Science and Technology 102: 8–14. https://doi.org/10.1016/j.lwt.2018.12.029

Rannou, C., Laroque, D., Renault, E., Prost, C. and Sérot, T. 2016. Mitigation strategies of acrylamide, furans, heterocyclic amines and browning during the Maillard reaction in foods. Food Research International 90: 154–176. https://doi.org/10.1016/j.foodres.2016.10.037

Reis, S.F. and Abu-Ghannam, N. 2014. Antioxidant capacity, arabinoxylans content and *in vitro* glycaemic index of cereal-based snacks incorporated with brewer's spent grain. LWT – Food Science and Technology. 55: 269–277. https://doi.org/10.1016/j.lwt.2013.09.004

Reis, S.F., Rai, D.P. and Abu-Ghannam, N. 2014. Apple pomace as a potential ingredient for the development of new functional foods. International Journal of Food Science and Technology 49: 1743–1750. https://doi.org/10.1111/ijfs.12477

Repo-Carrasco-Valencia, R., Peña, J., Kallio, H. and Salminen, S. 2009. Dietary fiber and other functional components in two varieties of crude and extruded kiwicha (*Amaranthus caudatus*). Journal of Cereal Science 49: 219–224. https://doi.org/10.1016/j.jcs.2008.10.003

Reyes Moreno, C., Reyes Fernández, P.C., Cuevas Rodríguez, E.O., Milán Carrillo, J. and Rochín, S.M. 2018. Changes in nutritional properties and bioactive compounds in cereals during extrusion cooking. pp. 103–124. *In*: Qamar, S.Z. (ed.). Extrusion of metals, polymers, and food products. IntechOpen. http://dx.doi.org/10.5772/intechopen.68753

Robin, F., Schuchmann, H.P. and Palzer, S. 2012. Dietary fiber in extruded cereals: Limitations and opportunities. Trends in Food Science and Technology 28: 23–32. https://doi.org/10.1016/j.tifs.2012.06.008

Rolandelli, G., Favre, L.C., Mshicileli, N., Vhangani, L.N., Farroni, A.E., van Wyk, J. et al. 2021a. The complex dependence of non-enzymatic browning development on processing conditions in maize snacks. LWT – Food Science and Technology 147: 111636. https://doi.org/10.1016/j.lwt.2021.111636

Rolandelli, G., Gallardo-Navarro, Y.T., García-Pinilla, S., Farroni, A.E., Gutiérrez-López, G.F. and Buera, M.P. 2021b. Components interactions and changes at molecular level in maize flour-based blends as affected by the extrusion process. A multi-analytical approach. Journal of Cereal Science 99: 103186. https://doi.org/10.1016/j.jcs.2021.103186

Rolandelli, G., García-Navarro, Y.T., García-Pinilla, S., Farroni, A.E., Gutiérrez-López, G.F. and Buera, M.P. 2020. Microstructural characteristics and physical properties of corn-based extrudates affected by the addition of millet, sorghum, quinoa and canary seed flour. Food Structure 25: 100140. https://doi.org/10.1016/j.foostr.2020.100140

Rufián-Henares, J.Á. and Pastoriza, S. 2016. Browning: Non-enzymatic browning. pp. 515–521. *In*: Caballero, B., Finglas, P.M. and Toldrá, F. (eds.). Encyclopedia of Food and Health. Academic Press. https://doi.org/10.1016/B978-0-12-384947-2.00089-1

Ruiz-Armenta, X.A., Zazueta-Morales, J.J., Aguilar-Palazuelos, E., Delgado-Nieblas, C.I., López-Díaz, A., Camacho-Hernández, I.L. et al. 2017. Effect of extrusion on the carotenoid content, physical and sensory properties of snacks added with bagasse of naranjita fruit: optimization process. CyTA – Journal of Food 16: 172–180. https://doi.org/10.1080/19476337.2017.1368717

Saldanha do Carmo, C., Varela, P., Poudroux, C., Dessev, T., Myhrer, K., Rieder, A. et al. 2019. The impact of extrusion parameters on physicochemical, nutritional and sensorial properties of expanded snacks from pea and oat fractions. LWT – Food Science and Technology 112: 108252. https://doi.org/10.1016/j.lwt.2019.108252

Schaich, K.M. 2013. Challenges in elucidating lipid oxidation mechanisms. When, where, and how do products arise? pp. 1–52. *In*: Logan, A., Nienaber, U. and Pan, X. (eds.). Lipid oxidation. Challenges in food systems. Academic Press and AOCS Press. https://doi.org/10.1016/B978-0-9830791-6-3.50004-7

Shah, M.A., Mir, S.A. and Dar, B.N. 2021. Advances in extrusion technologies. pp. 147–163. *In*: Pathania, S. and Tiwari, B.K. (eds.). Food Formulation: Novel Ingredients and Processing Techniques. John Wiley & Sons Ltd. https://doi.org/10.1002/9781119614760.ch8

Singh, N., Singh, J., Kaur, L., Sodhi, N.S. and Gill, B.S. 2003. Morphological, thermal and rheological properties of starches from different botanical sources. Food Chemistry 81: 219–231. https://doi.org/10.1016/S0308-8146(02)00416-8

Tas, A.A. and Shah, A.U. 2021. The replacement of cereals by legumes in extruded snack foods: Science, technology and challenges. Trends in Food Science and Technology 116: 701–711. https://doi.org/10.1016/j.tifs.2021.08.016

Tiwari, A. and Jha, S.K. 2017. Extrusion cooking technology: Principal mechanisms and effect on direct expanded snacks – an overview. International Journal of Food Studies 6: 113–128. 10.7455/ijfs/6.1.2017.a10

Usmani, Z., Sharma, M., Kumar Awasthi, A., Dutti Sharma, G., Cysneiros, D., Chandra Nayak, S. et al. 2021. Minimizing hazardous impact of food waste in a circular economy—Advances in resource recovery through green technologies. Journal of Hazardous Materials 416: 126154. https://doi.org/10.1016/j.jhazmat.2021.126154

Utama-ang, N., Phawatwiangnak, K., Naruenartwongsakul, S. and Samakradhamrongthai, R. 2017. Antioxidative effect of Assam Tea (*Camellia sinesis* var. *Assamica*) extract on rice bran oil and its application in breakfast cereal. Food Chemistry 221: 1733–1740. https://doi.org/10.1016/j.foodchem.2016.10.112

Viscidi, K.A., Dougherty, M.P., Briggs, J. and Camire, M.E. 2004. Complex phenolic compounds reduce lipid oxidation in extruded oat cereals. LWT – Food Science and Technology 37: 789–796. https://doi.org/10.1016/j.lwt.2004.03.005

Wang, P., Yang, Q., Zheng, D., Wang, Q., Wang, N., Saleh, A.S.M. et al. 2018. Physicochemical and antioxidant properties of rice flour based extrudates enriched with stabilized rice bran. Starch 70: 1800050. https://doi.org/10.1002/star.201800050

Wang, Q., Limin, L., Wang, T. and Zheng, X. 2022. A review of extrusion-modified underutilized cereal flour: chemical composition, functionality, and its modulation on starchy food quality. Food Chemistry 370: 131361. https://doi.org/10.1016/j.foodchem.2021.131361

Wang, Y., Chen, L., Yang, T., Ma, Y., McClements, D.J., Ren, F. et al. 2021. A review of structural transformations and properties changes in starch during thermal processing of foods. Food Hydrocolloids 113: 106543. https://doi.org/10.1016/j.foodhyd.2020.106543

Yagci, S., Caliskan, R., Gunes, Z.S., Capanoglu, E. and Tomas, M. 2022. Impact of tomato pomace powder added to extruded snacks on the *in vitro* gastrointestinal behaviour and stability of bioactive compounds. Food Chemistry 368: 130847. https://doi.org/10.1016/j.foodchem.2021.130847

Yang, Z., Zhou, Y., Xing, J.-J., Guo, X.-N. and Zhu, K.-X. 2021. Influence of extrusion on storage quality of dried oat noodles: Lipid degradation and off-flavours. Journal of Cereal Science 101: 103316. https://doi.org/10.1016/j.jcs.2021.103316

Zamuz, S., Sichetti Munekata, P.E., Meiselman, H.L., Zhang, W., Xing, L. and J.M Lorenzo. 2021. Chapter 2 – Consumer and Market Demand for Sustainable Food Products. pp. 23–35. *In*: Lorenzo, J.M., Sichetti Munekata, P.E. and Barba, F.J. (eds.). Sustainable Production Technology in Food. Academic Press. https://doi.org/10.1016/B978-0-12-821233-2.00008-3

Zhang, B., Dhital, S., Flanagan, B.M., Luckman, P., Halley, P.J. and Gidley, M.J. 2015. Extrusion induced low-order starch matrices: Enzymic hydrolysis and structure. Carbohydrate Polymers 134: 485–496. https://doi.org/10.1016/j.carbpol.2015.07.095

Zhang, R., Khan, S.A., Chi, J., Wei, Z., Zhang, Y., Deng, Y. et al. 2018. Different effects of extrusion on the phenolic profiles and antioxidant activity in milled fractions of brown rice. LWT – Food Science and Technology 88: 64–70. https://doi.org/10.1016/j.lwt.2017.09.042

Part III

Sustainability Emerging Technologies in Food Industries

12

Ultrasound

Wangang Zhang,*[,1] *Dacheng Kang*[2] and *Lujuan Xing*[1]

At present, the increasing demands of consumers for innovative texture-building, high-quality, or customized foods are actuating the food industry to use innovative technology to build sustainable food supply chains. One of the key solutions of the industry is to select the appropriate advanced methods and techniques for sustainable food manufacturing. Among these, the emerging technology of high-power ultrasound has been widely applied in food science and technology since the late 1990s (Knorr et al. 2011). Ultrasound is now considered a mild and green technique that can provide innovative products at moderate or low cost. It has been proved that ultrasound can modify the physical properties of some foods, offering novel ingredients for the development of these kinds of products (Kang Dacheng et al. 2021). More importantly, ultrasound technology does not only introduce preservatives during food processing but it can also remove some undesirable compounds from foods, such as allergens (Arzeni et al. 2012; Capote and de Castro 2007). Ultrasound can improve the quality of the product after the process (better texture and/or color) under the right processing conditions. In addition, due to the fact that ultrasound is a cheap, fast, and portable non-destructive technique, the high convenience for online measurements and quality control during food processing is attracting more attention (Bermúdez-Aguirre et al. 2011).

Because of these advantages, ultrasound has been introduced to some basic unit operations in food engineering, such as drying, freezing, or extraction (Ashokkumar et al. 2008; Awad et al. 2012; Sun et al. 2019). During the past five years, a lot of research has been attempted regarding the manufacture of ultrasound equipment that is compatible with the food industry. Thus ultrasound is a promising and sustainable technology that could be applied in the food industry. This chapter is to illustrate the basic principle of ultrasound during food processing and gather relevant information on current ultrasound applications in different food processes. Additionally, the advantages and drawbacks, as well as the new and potential applications, are evaluated and described from industry and sustainable point of view.

1. Fundamentals of Ultrasound

Sound can be simply ascribed to a variation in pressure or vibration that moves through the environment (Tang and Ko 2003). Mechanical action is the main reason for the variation in pressure or vibration. The parameter of the frequency of an acoustic wave is determined by the frequency of the electric field applied, which is the important index to divide the application area of the acoustic field (Fig. 1). In ultrasonics, frequencies ranging from 20 to 40 kHz (above the frequency of human hearing)

[1] Nanjing Agricultural University, Nanjing, Jiangsu, China.
[2] Linyi University, Linyi, Shandong, China.
* Corresponding author: wangang.zhang@njau.edu.cn

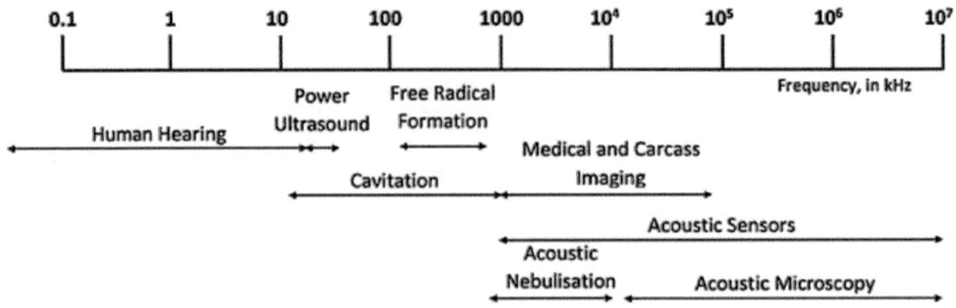

Figure 1. The Frequency Range of Application of Ultrasound Technology and Acoustic Phenomena. Cited From Dalmoro et al. (2012).

are common for food processing, such as microbial inactivation, meat tenderization, emulsification, and substance extraction. The generated high-power sound waves can produce significant shear fields and lead to extremely high temperatures and pressures within the small area of liquid, which is called "acoustic cavitation" (Ashokkumar 2011). The cavitation generated by ultrasound in liquid can lead to producing microstirring and microchannels on interfaces; meanwhile, the free radicals are appeared and produced (Kuijpers et al. 2002; Tao and Sun 2015; Turantas et al. 2015) (Fig. 2). These phenomena are responsible for the modification of molecular reactions and the characteristics and microstructure of food (Gulseren et al. 2007; Li K. et al. 2014).

Alternatively, ultrasound with low energy (low intensity) or high frequencies (> 1 MHz) is used for imaging applications, such as the detection of foreign matter in cans and the fat in animal carcasses (Alles et al. 2021; Hopkins et al. 2007) and measurements for determining food composition in acoustic velocity (Jiménez et al. 2017). During these applications, any damage to the texture, nutritional, flavor, and color value of foods are avoided.

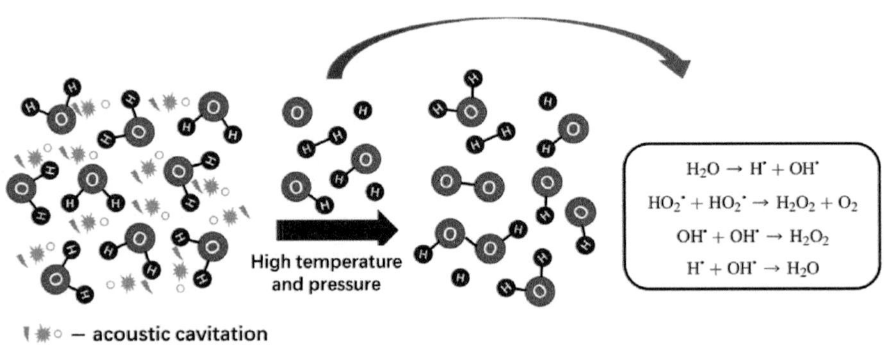

Figure 2. The Mechanism of Acoustic Cavitation in Liquid.

All in all, the approach of ultrasound provides a mechanism to modify food substances and control the progress of food processing in a manner that consumers can accept. Further, it usually operates at ambient temperature, which means that there can be less damage to native proteins and other bioactive, which will be damaged by heat. However, it is recognized as an energy-intensive technology to a certain extent and the users must be aware of the potential cost when scaling from a laboratory to a full-scale operation.

2. Principles of Ultrasound Applications

2.1 High Frequency (Low-Intensity) Ultrasound Applications

High-frequency ultrasound applications are mainly indicating the exploitation of the sensing of ultrasonic signals during their propagation. In fact, due to its short wavelengths, high-frequency ultrasound can discriminate the very small differences in the mechano-elastic constitutive characteristics of the medium (Bard 2017). Meanwhile, any changes in the structure of the medium can result in a change of the wave propagation characteristics, including reflection, attenuation, refraction, and velocity (Liao and Li 2009). Thus, the basic principle of high-frequency applications of ultrasound is to accurately measure these characteristics and their variations under various conditions. Among these applications, imaging, process control, and non-destructive detections are the main application fields in foods, including cereals, meat products, fruits and vegetables, and fats and emulsions (Ashokkumar 2015).

2.1.1 Cereal and Oil Products

Owing to the relationship between dough quality and product texture, the major object of study in cereal-based products is bread (Coupland 2004). Studies have shown that high-frequency ultrasound is a fast, most suitable, and nondestructive technique for online quality control of doughs and the textural property evaluation of bread products (Petrauskas 2007). Ross et al. (2004) exanimated the mixed extent of three varied flour doughs by using ultrasonic and conventional rheology techniques. The observed relationship between the velocity and attenuation parameters of ultrasound and the physical and rheology properties of dough provide the proof of application of ultrasound in the quality control of bread. Meanwhile, when combining the monitoring of proteins and gelatinization of starch in dough, the information of the parameters above could help to monitor changes in wheat flour dough consistency, which allows the determination of critical kneading times (Lee and Campanella 2013; Skaf et al. 2009). Another study conducted by Elmehdi et al. (2003) introduced the application of low-intensity ultrasound in the characterization of the fermentation of dough during bread preparation. In addition, to optimize the texture and rheology of the bread, air incorporation during the mixing of bread dough is necessary. Thus, the number of bubbles or the porosity is an important mechanical-physical property of a dough that directly influences the quality of bakery products (Jha et al. 2017). It has been illustrated that the air bubbles in the dough can affect sound velocity and attenuation (Elmehdi et al. 2005; Petrauskas 2007). The physical properties of cakes and batters including symmetry, density, volume, and rheology properties can be monitored and controlled by high frequency ultrasound. Therefore, for the evaluation of the textural properties of products of cereal, the low frequency, and the direct ultrasonic measurement process is suitable.

As for the crispiness of cereal products, such as biscuits and noodles, the crispiness characteristic can be assessed by the behavior of displacement/force of products and acoustic attenuation, which are measured by an acoustic envelop detector (Chen et al. 2020). Fox et al. (2004) described the design and application of an ultrasound system (2.25 MHz and 15 mm diameter transducer) at a low cost to monitor the batters as they were mixed for their specific purpose. Through monitoring the changes in compressibility of the doughs and the acoustic impedance, a substantial relationship was attained between the consistency of batters and acoustic impedance.

With respect to the oil product analysis, the method of ultrasound velocity measurement has been introduced for the evaluation of the composition, purity, and quality of the oil. After evaluating the ultrasonic velocity at a frequency of 3 MHz and the density of some refined and unrefined edible oils, including sunflower, castor, coconut, and groundnut, a series of the physical parameters of oil, such as molar compressibility, specific volume, intermolecular free length, and adiabatic compressibility could be better estimated (Sankarappa et al. 2005). Recently, through measuring ultrasound velocity and attenuation, a pulse-echo technique was successfully applied to determine the acoustical characteristics of olive oil, argan oil, and commercial cooking oil. The data on the acoustical characteristics of these oils included the density, the acoustical impedance, the celerity, the absorption losses coefficient, the dynamic viscosity, and the compressibility modulus (Awad et al. 2012).

2.1.2 Meat Products

The major application of low-intensity ultrasound to meat products is the analysis of the composition of pork, beef, poultry, and fish products by determining the behaviors of the sound wave at various frequencies (McClements and Gunasekaran 1997; Welti-Chanes et al. 2017). As same as for human pregnancy, the images of tissues such as internal organs, fat, and muscle in a carcass can be obtained according to the vibration reflection of sound waves (Awad et al. 2012). Many studies have confirmed the advantages of low-intensity ultrasound as a nondestructive and rapid method in meat analysis.

As for the studies of the composition of poultry and fish, Ghaedian et al. (1998) employed a frequency scanning pulse echo reflectometer to evaluate the relationship between temperature and sound velocity in fish analogs with various concentrations of water, oil, and solids without fat. An excellent correlation between the values predicted and the actual measured values was observed. With regard to the chicken/ analog products, the same method was successfully applied to determine the solid fat content of chicken fat and the composition of the products (Chanamai and McClements 1999). Meanwhile, by means of the relationship between temperature profiles and ultrasonic velocity, the composition of Atlantic mackerel (*Scomber scombrus*) tissues including solids, fat content, and non-fat content was also determined (Sigfusson et al. 2001). However, though no significant correlations between the fat content in mackerel tissue and the attenuation coefficient were observed, the authors indicated that this method may be useful for monitoring structural degradation of the mackerel tissue during post-mortem.

Another method that is applied to determine the composition of raw meat mixtures is the ultrasound velocity measurements (Benedito et al. 2001). The ratios of protein, fat, and moisture in lean tissue and fats can be accurately predicted based on the behavior of ultrasound velocity at different temperatures. By comparing the melting behavior of fat and the solid/liquid fat content from animals of different feeding regimes and breeds, the established temperature-ultrasound velocity plots were also used to classify and characterize back fat in pork (Niñoles et al. 2007). During the cheese processing, Telis-Romero et al. (2011) proposed an approach to assess cheese composition by ultrasound velocity measurements. It has been proved that this method was an effective quality control tool to detect process anomalies online during the cooling process after curdling processing. Contact or non-contact low-intensity ultrasound is also used to detect defects and internal objects in cheeses (Leemans and Destain 2009; Nassar et al. 2010) and skinless poultry breasts (Cho and Irudayaraj 2003).

In addition, these technologies have also been used in the following fields: (1) analyzing the fat properties of pork meat products (Niñoles et al. 2007) and estimation of the intramuscular fat percentage of beef (Ribeiro et al. 2008); (2) as a selection and replacement tool for genetic improvement of herds (Houghton and Turlington 1992; Williams 2002); (3) in research and management of the quality of animal carcasses through the online ultrasound testing system (Fiore et al. 2020; Jakaria et al. 2017).

2.1.3 Fruits and Vegetables

In contrast with meat products, fruits and vegetables are considered attenuating materials, which makes ultrasound technology unsuitable for evaluating the quality of tissues because of the scattering of sound from pores and voids (Awad et al. 2012; Povey 1998). Nevertheless, the quality control of fresh vegetables and fruits in both pre- and post-harvest can be monitored by high frequency ultrasound. It was reported that the parameters of high frequency ultrasound (velocity, amplitude, or attenuation) together with the plant physicochemical indexes (dry weight percentage, oil content, total soluble solids, and acidity) were related to the assessment of firmness, ripeness, and composition of fruits and vegetables (Mizrach 2008). These relationships can be employed to determine the proper harvesting time or shelf life of plant products. For instance, Mizrach et al. (1991) observed that when the color of fruit peels changed from green to yellow, the amplitude of the transmitted ultrasound waves would increase. They reported a good correlation between fruit ripeness and acoustic attenuation. Moreover, the firmness of tomatoes and plums could be well evaluated through the measurement of attenuation of ultrasound (Mizrach 2004; 2007). These results implied the importance of the attenuation parameter of ultrasound.

In other works, the method of ultrasound velocity measurements has been applied to determine the content of different sugar species in drinks and fruit juices (Contreras et al. 1992). With the help of a non-contact ultrasonic system, Kuo et al. (2008) measured the viscosity and sugar content of reconstituted orange juice. The pulse-echo of ultrasound was exhibited to have a good exponential correlation with viscosity, while a linear correlation with sugar contents. In addition, based on pulse transmission induced by an ultrasonic measuring system, microbial growth and contamination in fruit juices could be detected in a non-invasive method (Elvira et al. 2014). It is clear that the growth of microorganisms can affect the time of wave flight in liquid. That is, applying ultrasonic measuring to assess the microbial contamination viability in juice quality control while decreasing detection times and reducing economical waste.

2.1.4 Other Foods

Like aerated foods including ice cream, confectionery, whipped cream and desserts, the air bubbles distributed in viscoelastic liquid or solid matrix may influence the transmission and attenuation of sound waves. It is well known that the quality of aerated products is closely related to the characteristics of the dispersed gas phase, such as the mean bubble size, the uniformity of the bubble size distribution, and the bubble morphology. However, owing to the optically opaque or easily damaged delicate structures of these products, the analytical techniques of ultrasound are capable to provide information about bubble characteristics without damaging the structure of aerated foods (Labbafi et al. 2007). Dough, to a certain extent, can be recognized as aerated materials. Leroy et al. (2008) combined the measurements of ultrasound attenuation and velocity to investigate the bubble size distribution in the dough. The results showed that the time after mixing and the bubble concentration in dough could affect the position and shape of the attenuation peaks and the velocity of the sound waves. In another work, an aerated agar gel system in which bubbles and embedded polystyrene beads were employed to analyze the image characteristic by using ultrasonic spectroscopy (Strybulevych et al. 2007). The result indicated a good correlation between ultrasound parameters and image analysis. In addition, though adding the comparable size of foreign matters, both theory and experiment demonstrated that the ultrasonic signatures of solid inclusions and bubbles can be distinguished.

Due to its activities as an antimicrobial against food-borne microorganisms, honey is a popular natural sweetening agent that has been applied as a preservative and functional food additive (Bogdanov et al. 2008). Nevertheless, honey can be easily adulterated by adding amounts of starch, gelatins, water, sucrose, chalk, commercial glucose, and other substances. To detect fraud and ensure the quality of honey products, a list of analytical methods, such as HPLC (Wang Qian et al. 2020), NMR spectroscopy (Yong et al. 2022), and GC-MS systems (Alghamdi et al. 2020) have been proved authenticity for analyzing honey composition. Moreover, the high frequency ultrasound has also been applied in determining the mechanical and physical properties of honey. Singh and Dwivedi (1995) illustrated that the physical properties of adulterated honey, such as homogeneity viscosity and density, would induce changes in ultrasound velocity. Ultrasound velocity measurements can also be applied as an effective method to distinguish the place of origin of honey by measuring the high frequency dynamic shear rheology, viscosity, and moisture content (Wang Qin et al. 2005).

Freezing is one of the most widespread industrial methods for food preservation which represents a point of sharp discontinuity in the relationship between temperature and the stability and sensory properties of foods (Kim et al. 2018). The preserving effect of freezing is not only because of the lower temperature but is due to the depression of water activity as a result of the conversion of part of the water to ice. Sigfusson et al. (2004) measured the time of flight of an ultrasonic pulse moving parallel to the direction of heat flux during freezing. The percentage of food frozen can be calculated through a function of time, which provided a prediction of the complete freezing time for foods. More importantly, the temperature of food and the ice content can be determined by measuring the speed of sound waves (Aparicio et al. 2008). In another study, in frozen sucrose solutions, the velocity of ultrasonic increased linearly with the ice content approximately (Gülseren and Coupland 2008). Therefore, the high frequency ultrasound is a quick and suitable method for online monitoring of freezing and thawing systems, while the material and geometrical shape of containers and water contents of foods are less influenced by the measurement results.

2.2 Low Frequency (High Intensity) Ultrasound Applications

As mentioned above, ultrasound with high intensity can influence the physical, chemical, and biochemical properties of food products through its cavitations in a liquid medium. Therefore, the high intensity ultrasound has been used in research for more than 10 years, which mainly focused on the microorganism inactivation, food processing, and extraction of the active ingredient in foods (Chemat et al. 2011a; Haughton et al. 2012; McClements 1995).

2.2.1 Microorganism Inactivation

As far as the antimicrobial effect of ultrasound is concerned, ultrasound has been demonstrated to contribute to increased microbial safety and extended product shelf life (Elizaquível et al. 2012). Nowadays, many reviews have systematically summarized the application of ultrasonic technology in sterilization and food preservation. The main mechanism for ultrasound-induced microbial inhibition is due to the heating and free radical produced from the cavitation of ultrasound, which results in the thinning and destruction of cell membranes (M. Sango et al. 2014). Especially, hydrogen peroxide formed by hydroxyl radicals can also lead to the formation of pores, erosion, and deposition of cell components on the cell surface and DNA damage at high temperatures and pressure (Chandrapala et al. 2012) (Fig. 3). Moreover, recent studies have shown that the combination of ultrasound with physical (heat or ultra-high pressure) or chemical (lactic acid, chlorine solution, or red wine) methods are more efficient for microbial inactivation (Alarcon-Rojo et al. 2015; Cichoski et al. 2015; Evelyn and Silva 2016). However, Smith (2011) found no significant effects of ultrasound for reducing *Salmonella* and *E. coli* in broiler meat samples, suggesting that the combination of greater ultrasound power and a longer treatment period might achieve better effects for the reduction of microorganisms.

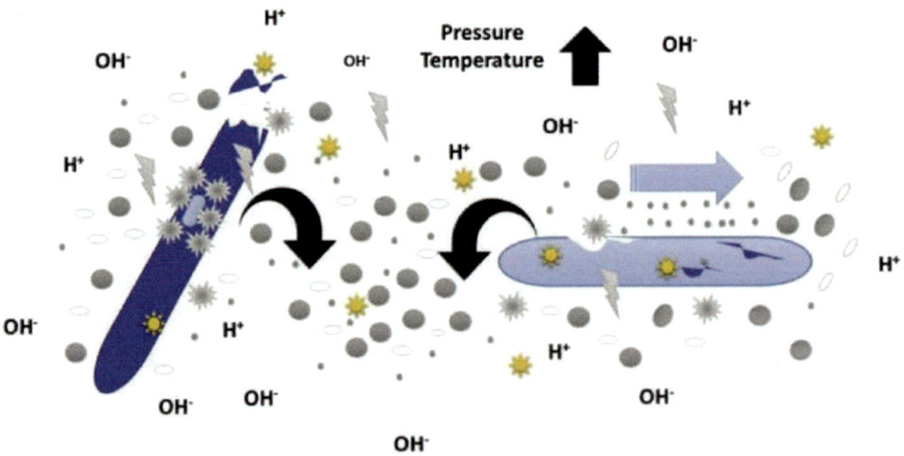

Figure 3. Mechanisms of Microbial Inactivation Under Cavitation of Ultrasound Cited by (Welti-Chanes et al. 2017).

2.2.2 Extracting

Extraction is an old technology that has been applied in many industries. However, time consuming, requiring heating for long periods, and large volumes of solvents are the disadvantage of the conventional methods for extraction processes. Recently, the ultrasound-assisted extraction method has been studied and proven as feasible, whether alone or in combination with other methods, including hydro-distillation and Clevenger (Dzah et al. 2020; Singla and Sit 2021). Owing to the cavitation which induced high shear forces, cell disruption, and microjet or microbubble formation, the extraction processing can be enhanced through fragmentation, surface erosion, and mass transfer resulting in a high yield of extracted materials. Therefore, the ultrasonic process is less time consuming, thus being an alternative for lower energy and solvent consumption. The major application of high-intensity ultrasound is for accelerating the extraction process of bioactive ingredients from plant and animal resources as well as food components, such as herbal, polysaccharides, protein, and oil (Vilkhu et al. 2008).

As reported by many previous studies, the major advantages of ultrasound are minimum effect on extractable materials and avoidance of organic solvents, which can potentially enhance the extraction of heat-sensitive bioactive and food components at lower processing temperatures (Ashokkumar 2015; Welti-Chanes et al. 2017). Like ultrasound-combined microorganism inactivation, the combination of ultrasound with supercritical fluid extraction is more efficient for material extraction. It is well known that supercritical fluid extraction based on CO_2 is environmentally friendly and has higher extraction potential along with its inexpensive nature. The mass transfer of the target components from the solid phase to the extracting media can be enhanced through the ultrasound treatment. Applying the method of ultrasound pretreatment combined with supercritical CO_2 extraction, Liu et al. (2020) reported an improved yield of *iberis amara* seed oil by more than 25% as well as improving antioxidant activity and physiochemical properties of the oil. In addition, the higher expenses of the equipment and the complexity of separating polar components without the aid of CO_2 modifiers are certain disadvantages of supercritical fluid extraction. It has been illustrated that ultrasound can allow a wide range of compounds to be extracted using polar or non-polar solvents and much simpler equipment. For instance, ethanol is less efficient than methanol in classical extraction. However, under the background of ultrasound-incorporated extraction, ethanol showed similar or higher yields compared to methanol. Moreover, ethanol is less toxic and sometimes classified as a green solvent that can be applied in food processing (Rodrigues et al. 2015).

2.2.3 Thermal Processing

Many food materials are denatured when exposed to moderate heat treatments (80–90°C; 1 hour or less). Extensive denaturation of food ingredients often results in damage to the functional properties and product qualities. Thermal processing has been used to improve the sensory property of foods, which can result in the cleavage of disulfide bonds, unfolding, aggregation, and formation of dimers and larger oligomers of proteins. In addition to protein modifications, chemical reactions such as glycation, Maillard reactions, oxidation, and deamidation (acid treatment at high temperatures or enzymatic treatment with transglutaminases) may occur during heat treatment (Aryee and Boye 2017).

In conventional cooking, the outside of food may be overcooked while the interior is insufficiently cooked when exposed to elevated temperatures. This will lead to a reduction in the quality of the product. The improved rate of heat transfer by high power ultrasound could provide a solution to avoid such problems because of the cavitation effect of ultrasound. Thus, high intensity ultrasound combined with thermal has been utilized in the thermal processing of foods (Chemat et al. 2011a). Pohlman et al. (1997) first confirmed that the high-intensity ultrasound-assisted cooking (20 kHz; 1,000 W) for *Longissimus* and *pectoralis* muscles of beef could result in a greater cooking speed, moisture retention, and efficiency of energy consumption, while during less cooking loss, the product yield increased two times than conventional cooking. Recently, Zou et al. (2018) further evaluated the effect of high-intensity ultrasound (0, 400, 600, 800, and 1,000 W; 20 kHz) assisted cooking (80, 100, 120 minutes) on quality trials of spiced beef. The authors discovered that ultrasound could significantly improve the hardness compared with the control. However, the effects on the springiness, the chewiness and the resilience of the spiced beef were not significant. In addition, Deng et al. (2018) investigated the effect of ultrasound assisted thermal processing on selected sensorial and physicochemical properties of beer. The 2.7 W/mL volumetric power of ultrasound was applied to a typical Chinese lager for 2 minutes at 40°C, 50°C, and 60°C, respectively. Results indicated that the quality attributes of the beer including bitterness, pH, original extract, and ethanol content were affected by these ultrasound treatments scarcely. As for the flavor profiles, beer treated by thermo-sonication at 50°C had a similar flavor stability to the untreated sample. Therefore, technique of thermo-sonication is a promising way for beer processing almost without affecting the original properties.

2.2.4 Freezing and Thawing

As mentioned above, low intensity ultrasound is mainly applied to monitor the process of freezing. During freezing processing, the denaturation of food ingredients especially protein is the main side effect of frozen foods, which is attributed to physical damage resulting from the formation and accretion of ice crystals and chemical processes associated with dehydration and concentration of solutes in the food matrix (Jiang Qingqing et al. 2019). It has been established that such damage is minimized in the case of quick freezing (Gambuteanu et al. 2014).

Different from low intensity ultrasound, acoustic cavitation generated by high-intensity ultrasound can result in fine ice crystals and shorten the time between the onset of crystallization and the complete formation of ice, thus reducing the damage to cellular structure. The cavitation also acts as nuclei for crystal growth or disruption of nuclei already present, which enhances both the nucleation rate and rate of crystal growth in a saturated or supercooled medium throughout the ultrasonic exposure (Chemat et al. 2011b). Sun et al. (2019) evaluated the high-intensity ultrasound-assisted immersion freezing, air freezing, and immersion freezing on the protein thermal stability and physicochemical properties of common carp (*Cyprinus carpio*) muscle. The results showed that high-intensity ultrasound-assisted immersion freezing was more efficient in keeping the muscle tissue integrity while reducing the thawing and cooking loss of samples. In addition, high-intensity ultrasound-assisted immersion freezing treatment inhibited the protein denaturation and fat oxidation of common carp muscle more than the other two treatments. Similar research conducted by Zhang et al. (2018b) revealed that porcine *longissimus* muscles treated with the HPUS condition of 180 W and 30 kHz could achieve a significant reduction in thawing loss compared with other treatments, which meant that high intensity ultrasound-assisted immersion freezing at certain powers could increase the freezing rate of muscle samples and improve meat quality. For vegetable products, Schössler et al. (2012) proposed a contact ultrasound system for the integration into freeze-drying processes for vegetables. Owing to the partial transformation of sound energy into heat, moisture removal during secondary drying was accelerated in combination with further ultrasound effects improving heat and mass transfer. The ultrasound treatment reduced the drying time required to reach a final moisture content of 10%. Therefore, the high-intensity ultrasound assisted freezing has a great application prospect in maintaining food quality and energy efficient.

Thawing is the reverse process of freezing. Major losses of sensory and nutritional qualities of frozen food are usually taking place during the faster thawing rate at room temperature as well as the damaged food matrix. Meanwhile, conventional thawing with low temperatures takes a long time and that is not suitable for the efficiency of food processing (Kalichevsky et al. 1995). Studies have shown that the thawing rate had a great influence on food quality. Quick thawing of food was desirable to assure the quality of frozen vegetables, but slow thawing was better for fish and meat which allows reabsorbing much of the water from the melting ice crystals, thus leading to less drip loss of products (Singha and Muthukumarappan 2015). Since high intensity ultrasound could accelerate the rate of heat/mass transfer, high intensity ultrasound-assisted thawing might achieve rapid thawing at low temperatures and improve the quality of food when the appropriate frequency and power of ultrasound were selected. Nevertheless, the negative aspects of localized heating and high energy requirement hindered the application of this method (Li Bing and Sun 2002). Miles et al. (1999) found that ultrasound with a frequency 500 kHz and intensity of 0.5 W/cm^2 had the minimum impact on the protein denaturation because of surface heating, and beef, pork, and cod samples were thawed to a depth of 7.6 cm within about 2.5 h. Similar research by Gambuteanu et al. (2014) discovered that ultrasound-assisted thawing (0.6 W/cm^2) of pork *Longissimus dorsi* muscle had no significant difference from the method of water immersion thawing in terms of chemical and quality characteristics. However the thawing time could be shortened, and meat thawed in ultrasound could result in a better structure by rheological determination. Moreover, the relaxation mechanism could also explain the accelerated thawing rate, which owned to the acoustic energy absorbed by frozen foods when a frequency in the relaxation frequency range of ice crystals in the food was applied (D. Kissam et al. 2006). However, there is no uniform relaxation frequency to reference when applying high-intensity ultrasound to meat thawing because of the complex composition of meat staff. Acoustic thawing still is a promising technology in the food industry if proper frequencies and high intensity ultrasound power are chosen.

2.2.5 Emulsification

Several natural and processed foods, such as milk, butter, sausage, or cakes, are emulsion-type products where protein plays an important role as an emulsifier. The proteins needed to be unfolded and dispersed spontaneously and form a protein adsorption layer at the oil-water or gas-water interface to stabilize the emulsion. Therefore, it is meaningful to the food enterprise that selects the protein with well emulsification ability or adapts the new technology to improve the quality of products.

It has been proved that high intensity ultrasound could decrease droplet size and disorder structures of proteins that results in enhanced emulsion activity. In addition, an increase in the number of hydrophobic groups exposed because of sonication can improve the emulsifying property of proteins (Yanjun et al. 2014; Zhang Longtao et al. 2018a).

Studies by de Figueiredo Furtado et al. (2017) investigated the effects of high intensity ultrasound on the emulsifying properties of sodium caseinate (1%; w/v) and lactoferrin (1%; w/v), including the protein structure, conformation, and hydrophobicity. They found that the sodium caseinate with a power of 300 W for 2–6 minutes of sonication, a significant reduction in the protein diameter, while an opposite effect was observed for lactoferrin. This effect was attributed to the unlike proteins conformation that showed opposite behavior when exposed to the ultrasound cavitation. It was suggested that after the ultrasound treatment, the lactoferrin could be dispersed into smaller droplet sizes in an oil-water solution and that it could stabilize the emulsion system. Therefore, in order to obtain a stable emulsion during high intensity ultrasound processing, the sources and molecular structure of the protein should be taken into account comprehensively; the appropriate ultrasound conditions should be selected as required for food and technology.

In other studies, combinations of high intensity ultrasound with thermal or pH-shifting treatments for the emulsifying properties of proteins have also been studied. As for the study of ultrasound/temperature for the physicochemical and emulsifying properties of thermally aggregated whey protein isolate (WPI), Shen et al. (2017) adopted an ultrasonic probe (20 kHz) for 20 minutes with pre- and post-thermal treatment (85°C for 30 minutes). They found that WPI solutions obtained the highest emulsion stability with the HPUS intensity of 31 W/cm^2 for 20 minutes. The researcher believed that HPUS treatment on post-thermal aggregation showed an improving effect on the physiochemical and emulsifying properties of whey protein with the potential for industrial applications. On the other hand, the high-intensity ultrasound was combined with pH-shifting to evaluate the effect on the functional properties and structure of pea protein isolate (PPI) (Jiang Shanshan et al. 2017). The PPI dispersion (30 mg/mL; pH 7.0) was adjusted to pH 2, 4, 10, or 12 and then sonicated using a probe (13 mm diameter) with the acoustic power density at 68 W/100 mL for 5 minutes. The treated protein solution adjusted the pH back to 7 after HPUS treatment. A better result was obtained for pH-shifting at pH 12 and HPUS (pH12-US) combined treatment with improved emulsifying properties. The alkaline condition of a solution could partially unfold and expose the hydrophobic cores to the surface of the molecule of pea protein. The cavitation effect of HPUS further broke the intermolecular interactions or disulfide bonds between proteins, thus producing the smaller molecular proteins resulting in better emulsifying capacity and stability for emulsion systems.

2.2.6 Drying

Ultrasound-assisted drying procedures can accelerate the rate of drying for fruits, vegetables, meat, and fish. The obvious advantages are the decreasing drying period, maintaining the characteristics of the product, and improving the mass transfer rate of water (Kowalski et al. 2017). Ultrasound-assisted drying is an excellent substitute for traditional drying. During product drying, ultrasound also influences the removal of moisture from the surface through air vibration (Yao 2016).

Kowalski et al. (2017) discovered that when apples were put at 0 to 200 W for 15–30 min, microchannels inside the cell were formed that give rise to impairment of the tissue. The treatments of grapes, plums, and eggplant with ultrasound prior to drying were also observed for lowering the drying time and improving the amount of dehydration (Cruz et al. 2016; Rodríguez et al. 2014). Additionally, the gain or loss of moisture of products may occur during foods are treated with ultrasound. This can be explained by the combined effect of gradient concentration between liquid media, ultrasound, and substance (Zhang Youwei et al. 2020). The rate of drying may vary across different studies by ultrasound application. Huang et al. (2020) reported that ultrasound processing could lower the absorption of water, improve the color of the substance, and reduce the loss of nutrients, such as vitamin C, flavonoid content, total phenolic content, and antioxidant activity.

2.2.7 Meat Tenderization and Curing

Tenderness is the most important quality attribute of meat in determining meat quality. The traditional way that makes the meat more palatable is usually to adapt mechanical pounding. Recently, the high intensity ultrasound has been proven to be effective to improve meat tenderness. In general, the basic principles of ultrasound for meat tenderization include two sides: enhancing enzymatic reactions and breaking the integrity of muscular cells (Chemat et al. 2011b). Ultrasonic tenderization has been successfully achieved in beef, veal, and poultry meat. Wang et al. (2018) found that ultrasound pretreatment improved the tenderness of beef by accelerating the activation and autolysis of μ-calpain during postmortem aging. The increased degradation of desmin and troponin-T was also evidence that confirmed the further tenderization of beef.

As for the investigations of meat processing, such as curing by ultrasound, the additional effect is tenderization. Therefore, the tenderization of meat is an inevitable phenomenon during meat processing. Among these, in order to enhance the shelf life, flavor, juiciness, and tenderness, curing is a widely used method in the processing and preservation of meat and meat products (Alarcon-Rojo et al. 2015). The conventional salt-brining processes mainly include three drawbacks: a high NaCl content is needed; the potential lack of control in contamination; and the marinating process with extended time may lead to textural damage. Recently, most studies have shown that high intensity ultrasound could reduce bringing time. Samples that received both sonication and salt treatment were superior in all qualities, such as water-binding capacity, tenderness, and cohesiveness (Alarcon-Rojo et al. 2015; Barretto et al. 2018; McDonnell et al. 2014a). Kang and other colleagues reported that the intensity of ultrasound, treatment time, and salt concentration were the main factors influencing the rate of mass transfer and meat quality during ultrasound-assisted curing (Kang Dacheng et al. 2016; Leal-Ramos et al. 2011). In addition, ultrasound-assisted curing could reduce the amount of NaCl supplemented in brining and shorten the processing time. Furthermore, the geometric parameters of ultrasound systems (probe size, distances between the transducer and meat, and the thickness of the sample) could also have a strong influence on the efficiency of NaCl uptake in meat (Inguglia et al. 2018).

In addition, McDonnell et al. (2014b) first explored the application of ultrasound (0, 10.7, 17.1, and 25.4 W/cm^2) in a pilot scale, which assessed the potential scale-up of this technology and the possible detrimental effects on meat quality. The study achieved up to a 50% reduction in processing times with no adverse effects on quality in the production of premium wet-cured cooked hams. It is suggested that further study is to implement ultrasound into industrial processes and perform optimization, cost analysis, and comparisons to conventional techniques.

3. Conclusions

As sustainable development is demanded, the processing, production, and consumption patterns of foods around the world may be changed because of the shortage of resources and increasing environmental awareness of the people. Ultrasound is now considered a mild and green technology that leads to changes in new regulations or even new definitions of food processes, such as ultrasound-assisted extraction. Since it is a fast, cheap, and portable technique, high frequency ultrasound is considered a non-destructive method for online measurements and food processing control. Moreover, as for low frequency or high intensity ultrasound, it has been proved that ultrasound can modify the physical properties of foods, offering novel ingredients for the development of these kinds of products. The future application of ultrasound is mainly focusing on the development of the simplicity, portability, safe and low cost of devices in laboratories, pilot plants, and food factories. The amount of research into this topic will continue to grow quickly in the coming years and encourage us to pursue this promising project.

References

Alarcon-Rojo, A.D., Janacua, H., Rodriguez, J.C., Paniwnyk, L. and Mason, T.J. 2015. Power ultrasound in meat processing. Meat. Sci. 107: 86–93.

Alghamdi, B.A., Alshumrani, E.S., Saeed, M.S.B., Rawas, G.M., Alharthi, N.T., Baeshen, M.N., Helmi, N.M., Alam, M.Z. and Suhail, M. 2020. Analysis of sugar composition and pesticides using HPLC and GC–MS techniques in honey samples collected from Saudi Arabian markets. Saudi Journal of Biological Sciences 27: 3720–3726.

Alles, E.J., Mackle, E.C., Noimark, S., Zhang, E.Z., Beard, P.C. and Desjardins, A.E. 2021. Freehand and video-rate all-optical ultrasound imaging. Ultrasonics 116: 106514.

Aparicio, C., Otero, L., Guignon, B., Molina-García, A.D. and Sanz, P.D. 2008. Ice content and temperature determination from ultrasonic measurements in partially frozen foods. Journal of Food Engineering 88: 272–279.

Aryee, A.N.A. and Boye, J.I. 2017. Comparative Study of the Effects of Processing on the Nutritional, Physicochemical and Functional Properties of Lentil. Journal of Food Processing and Preservation 41: e12824.

Arzeni, C., Martínez, K., Zema, P., Arias, A., Pérez, O.E. and Pilosof, A.M.R. 2012. Comparative study of high intensity ultrasound effects on food proteins functionality. Journal of Food Engineering 108: 463–472.

Ashokkumar, M. 2011. The characterization of acoustic cavitation bubbles—An overview. Ultrason Sonochem 18: 864–872.

---. 2015. Applications of ultrasound in food and bioprocessing. Ultrason Sonochem 25: 17–23.

Ashokkumar, M., Sunartio, D., Kentish, S., Mawson, R., Simons, L., Vilkhu, K. and Versteeg, C. 2008. Modification of food ingredients by ultrasound to improve functionality: A preliminary study on a model system. Innovative Food Science and Emerging Technologies 9: 155–160.

Awad, T.S., Moharram, H.A., Shaltout, O.E., Asker, D. and Youssef, M.M. 2012. Applications of ultrasound in analysis, processing and quality control of food: A review. Food Research International 48: 410–427.

Bard, R.L. 2017. High-Frequency Ultrasound Examination in the Diagnosis of Skin Cancer. Dermatologic Clinics 35: 505–511.

Barretto, T.L., Pollonio, M.A.R., Telis-Romero, J. and da Silva Barretto, A.C. 2018. Improving sensory acceptance and physicochemical properties by ultrasound application to restructured cooked ham with salt (NaCl) reduction. Meat. Sci. 145: 55–62.

Benedito, J., Carcel, J.A., Rossello, C. and Mulet, A. 2001. Composition assessment of raw meat mixtures using ultrasonics. Meat. Sci. 57: 365–370.

Bermúdez-Aguirre, D., Mobbs, T. and Barbosa-Cánovas, G.V. 2011. Ultrasound Applications in Food Processing 65–105.

Bogdanov, S., Jurendic, T., Sieber, R. and Gallmann, P. 2008. Honey for Nutrition and Health: A Review. Journal of the American College of Nutrition 27: 677–689.

Capote, F.P. and de Castro, M.D.L. 2007. Ultrasound in analytical chemistry. Analytical And Bioanalytical Chemistry 387: 249–257.

Chanamai, R. and McClements, D.J. 1999. Ultrasonic Determination of Chicken Composition. Journal of Agricultural and Food Chemistry 47: 4686–4692.

Chandrapala, J., Oliver, C., Kentish, S. and Ashokkumar, M. 2012. Ultrasonics in food processing - Food quality assurance and food safety. Trends In Food Science and Technology 26: 88–98.

Chemat, F., Zill-e-Huma and Khan, M.K. 2011a. Applications of ultrasound in food technology: Processing, preservation and extraction. Ultrason Sonochem 18: 813–835.

Chemat, F., Zill, e H. and Khan, M.K. 2011b. Applications of ultrasound in food technology: Processing, preservation and extraction. Ultrason Sonochem 18: 813–835.

Chen, F., Zhang, M. and Yang, C.-h. 2020. Application of ultrasound technology in processing of ready-to-eat fresh food: A review. Ultrason Sonochem 63: 104953.

Cho, B.-K. and Irudayaraj, J.M.K. 2003. Foreign Object and Internal Disorder Detection in Food Materials Using Noncontact Ultrasound Imaging. Journal of Food Science 68: 967–974.

Cichoski, A.J., Rampelotto, C., Silva, M.S., de Moura, H.C., Terra, N.N., Wagner, R., de Menezes, C.R., Flores, E.M.M. and Barin, J.S. 2015. Ultrasound-assisted post-packaging pasteurization of sausages. Innovative Food Science & Emerging Technologies 30: 132–137.

Contreras, N.I., Fairley, P., McClements, D.J. and Povey, M.J.W. 1992. Analysis of the sugar content of fruit juices and drinks using ultrasonic velocity measurements. International Journal of Food Science & Technology 27: 515–529.

Coupland, J.N. 2004. Low intensity ultrasound. Food Research International 37: 537–543.

Cruz, L., Clemente, G., Mulet, A., Ahmad-Qasem, M.H., Barrajón-Catalán, E. and García-Pérez, J.V. 2016. Air-borne ultrasonic application in the drying of grape skin: Kinetic and quality considerations. Journal of Food Engineering 168: 251–258.

D. Kissam, A., W. Nelson R, Ngao, J. and Hunter, P. 2006. Water-Thawing of Fish Using Low Frequency Acoustics 47: 71–75.

Dalmoro, A., Barba, A.A., Lamberti, G. and d'Amore, M. 2012. Intensifying the microencapsulation process: Ultrasonic atomization as an innovative approach. European Journal of Pharmaceutics and Biopharmaceutics 80: 471–477.

de Figueiredo Furtado, G., Mantovani, R.A., Consoli, L., Hubinger, M.D. and da Cunha, R.L. 2017. Structural and emulsifying properties of sodium caseinate and lactoferrin influenced by ultrasound process. Food Hydrocolloids 63: 178–188.

Deng, Y., Bi, H., Yin, H., Yu, J., Dong, J., Yang, M. and Ma, Y. 2018. Influence of ultrasound assisted thermal processing on the physicochemical and sensorial properties of beer. Ultrason Sonochem 40: 166–173.

Dzah, C.S., Duan, Y., Zhang, H., Wen, C., Zhang, J., Chen, G. and Ma, H. 2020. The effects of ultrasound assisted extraction on yield, antioxidant, anticancer and antimicrobial activity of polyphenol extracts: A review. Food Bioscience 35: 100547.

Elizaquível, P., Sánchez, G., Selma, M.V. and Aznar, R. 2012. Application of propidium monoazide-qPCR to evaluate the ultrasonic inactivation of Escherichia coli O157:H7 in fresh-cut vegetable wash water. Food Microbiology 30: 316–320.

Elmehdi, H.M., Page, J.H. and Scanlon, M.G. 2003. Using ultrasound to investigate the cellular structure of bread crumb. Journal of Cereal Science 38: 33–42.

Elmehdi, H.M., Page, J.H. and Scanlon, M.G. 2005. Investigating the porosity of dough using ultrasound. pp. 188–192. *In*: Cauvain, S.P., Salmon, S.S. and Young, L.S. (eds.). Using Cereal Science and Technology for the Benefit of Consumers, Woodhead Publishing.

Elvira, L., Durán, C.M., Urréjola, J. and Montero de Espinosa, F.R. 2014. Detection of microbial contamination in fruit juices using non-invasive ultrasound. Food Control 40: 145–150.

Evelyn, Silva F.V.M. 2016. High pressure processing pretreatment enhanced the thermosonication inactivation of *Alicyclobacillus acidoterrestris* spores in orange juice. Food Control. 62: 365–372.

Fiore, E., Fabbri, G., Gallo, L., Morgante, M., Muraro, M., Boso, M. and Gianesella, M. 2020. Application of texture analysis of b-mode ultrasound images for the quantification and prediction of intramuscular fat in living beef cattle: A methodological study. Research in Veterinary Science 131: 254–258.

Fox, P., Smith, P.P. and Sahi, S. 2004. Ultrasound measurements to monitor the specific gravity of food batters. Journal of Food Engineering 65: 317–324.

Gambuteanu, C., Patrascu, L. and Alexe, P. 2014. Effect of freezing-thawing process on some quality aspects of pork Longissimus dorsi muscle. Romanian Biotechnological Letters 19: 8916–8924.

Ghaedian, R., Coupland, J.N., Decker, E.A. and McClements, D.J. 1998. Ultrasonic determination of fish composition. Journal of Food Engineering 35: 323–337.

Gülseren, İ. and Coupland, J.N. 2008. Ultrasonic properties of partially frozen sucrose solutions. Journal of Food Engineering 89: 330–335.

Gulseren, I., Guzey, D., Bruce, B.D. and Weiss, J. 2007. Structural and functional changes in ultrasonicated bovine serum albumin solutions. Ultrason Sonochem 14: 173–183.

Haughton, P.N., Lyng, J.G., Morgan, D.J., Cronin, D.A., Noci, F., Fanning, S. and Whyte, P. 2012. An evaluation of the potential of high-intensity ultrasound for improving the microbial safety of poultry. Food and Bioprocess Technology 5: 992–998.

Hopkins, D.L., Stanley, D.F. and Ponnampalam, E.N. 2007. Relationship between real-time ultrasound and carcass measures and composition in heavy sheep. Australian Journal of Experimental Agriculture 47: 1304–1308.

Houghton, P.L. and Turlington, L.M. 1992. Application of ultrasound for feeding and finishing animals: a review. Journal of Animal Science 70: 930–941.

Huang, D., Men, K., Li, D., Wen, T., Gong, Z., Sunden, B. and Wu, Z. 2020. Application of ultrasound technology in the drying of food products. Ultrason Sonochem 63: 104950.

Inguglia, E.S., Zhang, Z., Burgess, C., Kerry, J.P. and Tiwari, B.K. 2018. Influence of extrinsic operational parameters on salt diffusion during ultrasound assisted meat curing. Ultrasonics 83: 164–170.

Jakaria, J., Khasanah, H., Priyanto, R., Baihaqi, M. and Fakhrul Ulum, M. 2017. Prediction of meat quality in Bali cattle using ultrasound imaging. J. Indonesian Trop. Anim. Agric. 42: 59–65.

Jha, P.K., Chevallier, S., Cheio, J., Rawson, A. and Le-Bail, A. 2017. Impact of resting time between mixing and shaping on the dough porosity and final cell distribution in sandwich bread. Journal of Food Engineering 194: 15–23.

Jiang, Q., Jia, R., Nakazawa, N., Hu, Y., Osako, K. and Okazaki, E. 2019. Changes in protein properties and tissue histology of tuna meat as affected by salting and subsequent freezing. Food Chemistry 271: 550–560.

Jiang, S., Ding, J., Andrade, J., Rababah, T.M., Almajwal, A., Abulmeaty, M.M. and Feng, H. 2017. Modifying the physicochemical properties of pea protein by pH-shifting and ultrasound combined treatments. Ultrason Sonochem 38: 835–842.

Jiménez, A., Rufo, M., Paniagua, J.M., Crespo, A.T., Guerrero, M.P. and Riballo, M.J. 2017. Contributions to ultrasound monitoring of the process of milk curdling. Ultrasonics 76: 192–199.

Kalichevsky, M.T., Knorr, D. and Lillford, P.J. 1995. Potential food applications of high-pressure effects on ice-water transitions. Trends In Food Science and Technology 6: 253–259.

Kang, D.-c., Wang, A.-r., Zhou, G.-h., Zhang, W.-g., Xu, S.-m. and Guo, G.-p. 2016. Power ultrasonic on mass transport of beef: Effects of ultrasound intensity and NaCl concentration. Innovative Food Science & Emerging Technologies 35: 36–44.

Kang, D., Zhang, W., Lorenzo, J.M. and Chen, X. 2021. Structural and functional modification of food proteins by high power ultrasound and its application in meat processing. Critical Reviews in Food Science and Nutrition 61: 1914–1933.

Kim, H.-W., Kim, J.-H., Seo, J.-K., Setyabrata, D. and Kim, Y.H.B. 2018. Effects of aging/freezing sequence and freezing rate on meat quality and oxidative stability of pork loins. Meat. Sci. 139: 162–170.

Knorr, D., Froehling, A., Jaeger, H., Reineke, K., Schlueter, O. and Schoessler, K. 2011. Emerging technologies in food processing. Annual Review of Food Science and Technology 2: 203–235.

Kowalski, S.J., Mierzwa, D. and Stasiak, M. 2017. Ultrasound-assisted convective drying of apples at different process conditions. Drying Technology 35: 939–947.

Kuijpers, M.W.A., Kemmere, M.F. and Keurentjes, J.T.F. 2002. Calorimetric study of the energy efficiency for ultrasound-induced radical formation. Ultrasonics 40: 675–678.

Kuo,F.-J., Sheng, C.-T. and Ting, C.-H. 2008. Evaluation of ultrasonic propagation to measure sugar content and viscosity of reconstituted orange juice. Journal of Food Engineering 86: 84–90.

Labbafi, M., Thakur, R.K., Vial, C. and Djelveh, G. 2007. Development of an on-line optical method for assessment of the bubble size and morphology in aerated food products. Food Chemistry 102: 454–465.

Leal-Ramos , M.Y., Alarcon-Rojo, A.D., Mason, T.J., Paniwnyk, L. and Alarjah, M. 2011. Ultrasound-enhanced mass transfer in Halal compared with non-Halal chicken. J. Sci. Food Agric. 91: 130–133.

Lee, S. and Campanella, O. 2013. Impulse viscoelastic characterization of wheat flour dough during fermentation. Journal of Food Engineering 118: 266–270.

Leemans, V. and Destain, M.-F. 2009. Ultrasonic internal defect detection in cheese. Journal of Food Engineering 90: 333–340.

Leroy, V., Fan, Y., Strybulevych, A.L., Bellido, G.G., Page, J.H. and Scanlon, M.G. 2008. Chapter 5 - Investigating the Bubble Size Distribution in Dough Using Ultrasound. pp. 51–60. *In*: Campbell, G.M., Scanlon, M.G., Pyle, D.L. (eds.). Bubbles in Food 2, AACC International Press.

Li, B. and Sun, D.-W. 2002. Novel methods for rapid freezing and thawing of foods—a review. Journal of Food Engineering 54: 175–182.

Li, K., Kang, Z.L., Zhao, Y.Y., Xu, X.L. and Zhou, G.H. 2014. Use of high-intensity ultrasound to improve functional properties of batter suspensions prepared from pse-like chicken breast meat. Food and Bioprocess Technology 7: 3466–3477.

Liao, A.-H. and Li, P.-C. 2009. The Role of High Frequency Ultrasound in Multimodality Small Animal Imaging for Cancer Research. Journal of Medical Ultrasound 17: 86–97.

Liu, X., Ou, H., Xiang, Z. and Gregersen, H. 2020. Ultrasound pretreatment combined with supercritical CO_2 extraction of Iberis amara seed oil. Journal of Applied Research on Medicinal and Aromatic Plants 18:100265.

M. Sang, D., Abela, D., McElhatton, A. and Valdramidis, V.P. 2014. Assisted ultrasound applications for the production of safe foods. Journal of Applied Microbiology 116: 1067–1083.

McClements, D.J. 1995. Advances in the application of ultrasound in food analysis and processing. Trends in Food Science and Technology 6: 293–299.
McClements, D.J. and Gunasekaran, S. 1997. Ultrasonic characterization of foods and drinks: Principles, methods, and applications. Critical Reviews in Food Science and Nutrition 37: 1–46.
McDonnell, C.K., Allen, P., Morin, C. and Lyng, J.G. 2014a. The effect of ultrasonic salting on protein and water-protein interactions in meat. Food Chemistry 147: 245–251.
McDonnell, C.K., Lyng, J.G., Arimi, J.M. and Allen, P. 2014b. The acceleration of pork curing by power ultrasound: A pilot-scale production. Innovative Food Science and Emerging Technologies 26: 191–198.
Miles, C.A., Morley, M.J. and Rendell, M. 1999. High power ultrasonic thawing of frozen foods. Journal of Food Engineering 39: 151–159.
Mizrach, A. 2004. Assessing plum fruit quality attributes with an ultrasonic method. Food Research International 37: 627–631.
Mizrach, A. 2007. Nondestructive ultrasonic monitoring of tomato quality during shelf-life storage. Postharvest Biology and Technology 46: 271–274.
Mizrach, A. 2008. Ultrasonic technology for quality evaluation of fresh fruit and vegetables in pre- and postharvest processes. Postharvest Biology and Technology 48: 315–330.
Mizrach, A., Galili, N., Rosenhouse, G. and Teitel, D.C. 1991. Acoustical, mechanical, and quality parameters of winter-grown melon tissue. Transactions of the ASABE 34: 2135–2138.
Nassar, G., Lefbvre, F., Skaf, A., Carlier, J., Nongaillard, B. and Noêl, Y. 2010. Ultrasonic and acoustic investigation of cheese matrix at the beginning and the end of ripening period. Journal of Food Engineering 96: 1–13.
Niñoles, L., Clemente, G., Ventanas, S. and Benedito, J. 2007. Quality assessment of Iberian pigs through backfat ultrasound characterization and fatty acid composition. Meat. Sci. 76: 102–111.
Petrauskas, A. 2007. Evaluation of porous food products by using ultrasonic methods. Ultragarsas (Ultrasound) 62: 1392–2114.
Pohlman, F.W., Dikeman, M.E., Zayas, J.F. and Unruh, J.A. 1997. Effects of ultrasound and convection cooking to different end point temperatures on cooking characteristics, shear force and sensory properties, composition, and microscopic morphology of beef longissimus and pectoralis muscles. Journal of Animal Science 75: 386–401.
Povey, M.J.W. 1998. Ultrasonics of food. Contemporary Physics 39: 467–478.
Ribeiro, F.R.B., Tedeschi, L.O., Stouffer, J.R. and Carstens, G.E. 2008. Technical note: A novel technique to assess internal body fat of cattle by using real-time ultrasound. Journal of Animal Science 86: 763–767.
Rodrigues, S., Fernandes, F.A.N., de Brito, E.S., Sousa, A.D. and Narain, N. 2015. Ultrasound extraction of phenolics and anthocyanins from jabuticaba peel. Industrial Crops and Products 69: 400–407.
Rodríguez, Ó., Santacatalina, J.V., Simal, S., Garcia-Perez, J.V., Femenia, A. and Rosselló, C. 2014. Influence of power ultrasound application on drying kinetics of apple and its antioxidant and microstructural properties. Journal of Food Engineering 129: 21–29.
Ross, K.A., Pyrak-Nolte, L.J. and Campanella, O.H. 2004. The use of ultrasound and shear oscillatory tests to characterize the effect of mixing time on the rheological properties of dough. Food Research International 37: 567–577.
Sankarappa, T., Prashant kumar, M. and Ahmad, A. 2005. Ultrasound velocity and density studies in some refined and unrefined edible oils. Physics and Chemistry of Liquids 43: 507–514.
Schössler, K., Jäger, H. and Knorr, D. 2012. Novel contact ultrasound system for the accelerated freeze-drying of vegetables. Innovative Food Science & Emerging Technologies 16: 113–120.
Shen, X., Fang, T., Gao, F. and Guo, M. 2017. Effects of ultrasound treatment on physicochemical and emulsifying properties of whey proteins pre- and post-thermal aggregation. Food Hydrocolloids 63: 668–676.
Sigfusson, H., Decker, E.A. and McClements, D.J. 2001. Ultrasonic characterization of Atlantic mackerel (Scomber scombrus). Food Research International 34: 15–23.
Sigfusson, H., Ziegler, G.R. and Coupland, J.N. 2004. Ultrasonic monitoring of food freezing. Journal of Food Engineering 62: 263–269.
Singh, V.R., Dwivedi S. 1995. Ultrasonic detection of adulteration in fluid foods. pp. 1/73–71/74. Proceedings of the First Regional Conference, IEEE Engineering in Medicine and Biology Society and 14th Conference of the Biomedical Engineering Society of India. An International Meet.
Singha, P. and Muthukumarappan, K. 2015. Quality changes and freezing time prediction during freezing and thawing of ginger. Food Science & Nutrition 4: 521–533.
Singla, M. and Sit, N. 2021. Application of ultrasound in combination with other technologies in food processing: A review. Ultrason Sonochem 73: 105506.
Skaf, A., Nassar, G., Lefebvre, F. and Nongaillard, B. 2009. A new acoustic technique to monitor bread dough during the fermentation phase. Journal of Food Engineering 93: 365–378.
Smith, D. 2011. Effect of ultrasonic marination on broiler breast meat quality and *salmonella* contamination. International Journal of Poultry Science 10.
Strybulevych, A., Leroy, V., Scanlon, M.G. and Page, J.H. 2007. Characterizing a model food gel containing bubbles and solid inclusions using ultrasound. Soft Matter 3: 1388–1394.

Sun, Q., Sun, F., Xia, X., Xu, H. and Kong, B. 2019. The comparison of ultrasound-assisted immersion freezing, air freezing and immersion freezing on the muscle quality and physicochemical properties of common carp (Cyprinus carpio) during freezing storage. Ultrason Sonochem 51: 281–291.

Tang, S.K. and Ko, N.W.M. 2003. Basic sound generation mechanisms in inviscid vortex interactions at low Mach number. Journal of Sound and Vibration 262: 87–115.

Tao, Y. and Sun, D.W. 2015. Enhancement of food processes by ultrasound: A review. Critical Reviews In Food Science And Nutrition 55: 570–594.

Telis-Romero, J., Váquiro, H.A., Bon, J. and Benedito, J. 2011. Ultrasonic assessment of fresh cheese composition. Journal of Food Engineering 103: 137–146.

Turantas, F., Kilic, G.B. and Kilic, B. 2015. Ultrasound in the meat industry: General applications and decontamination efficiency. International Journal of Food Microbiology 198: 59–69.

Vilkhu, K., Mawson, R., Simons, L. and Bates, D. 2008. Applications and opportunities for ultrasound assisted extraction in the food industry—A review. Innovative Food Science & Emerging Technologies 9: 161–169.

Wang, A., Kang, D., Zhang, W., Zhang, C., Zou, Y. and Zhou, G. 2018. Changes in calpain activity, protein degradation and microstructure of beef M. semitendinosus by the application of ultrasound. Food Chemistry 245: 724–730.

Wang, Q., Rademacher, B., Sedlmeyer, F. and Kulozik, U. 2005. Gelation behaviour of aqueous solutions of different types of carrageenan investigated by low-intensity-ultrasound measurements and comparison to rheological measurements. Innovative Food Science & Emerging Technologies 6: 465–472.

Wang, Q., Zhao, H., Zhu, M., Zhang, J., Cheng, N. and Cao, W. 2020. Method for identifying acacia honey adulterated by resin absorption: HPLC-ECD coupled with chemometrics. LWT 118: 108863.

Welti-Chanes, J., Morales-de la Peña, M., Jacobo-Velázquez, D.A. and Martín-Belloso, O. 2017. Chapter 19 - Opportunities and Challenges of Ultrasound for Food Processing: An Industry Point of View. pp. 457–497. *In*: Bermudez-Aguirre, D., ed. Ultrasound: Advances for Food Processing and Preservation, Academic Press.

Williams, A.R. 2002. Ultrasound applications in beef cattle carcass research and management. Journal of Animal Science 80: E183–E188.

Yanjun, S. et al. 2014. Effect of power ultrasound pre-treatment on the physical and functional properties of reconstituted milk protein concentrate. Journal of Food Engineering 124: 11–18.

Yao Y. 2016. Enhancement of mass transfer by ultrasound: Application to adsorbent regeneration and food drying/ dehydration. Ultrason Sonochem 31: 512–531.

Yong, C.-H., Muhammad, S.A., Aziz, F.A., Nasir, F.I., Mustafa, M.Z., Ibrahim, B., Kelly, S.D., Cannavan, A. and Seow, E.-K. 2022. Detecting adulteration of stingless bee honey using untargeted 1H NMR metabolomics with chemometrics. Food Chemistry 368: 130808.

Zhang, L., Pan, Z., Shen, K., Cai, X., Zheng, B. and Miao, S. 2018a. Influence of ultrasound-assisted alkali treatment on the structural properties and functionalities of rice protein. Journal of Cereal Science 79: 204–209.

Zhang, M., Haili, N., Chen, Q., Xia, X. and Kong, B. 2018b. Influence of ultrasound-assisted immersion freezing on the freezing rate and quality of porcine longissimus muscles. Meat Sci. 136: 1–8.

Zhang, Y., Li, T., Shen, Y., Wang, L., Zhang, H., Qian, H. and Qi, X. 2020. Extrusion followed by ultrasound as a chemical-free pretreatment method to enhance enzymatic hydrolysis of rice hull for fermentable sugars production. Industrial Crops and Products 149: 112356.

Zou, Y., Zhang, W., Kang, D. and Zhou, G. 2018. Improvement of tenderness and water holding capacity of spiced beef by the application of ultrasound during cooking. International Journal of Food Science & Technology 53: 828–836.

13

Electrolyzed Water

An Innovative Alternative in the Food Industry

Yasmim Sena Vaz Leães, Carla Cristina Bauermann Brasil, Paulo Cezar Bastianello Campagnol and *Alexandre José Cichoski**

1. Introduction

Foodborne transmissible diseases (FTDs) are caused by eating contaminated food or water. According to the World Health Organization (WHO), FTDs are responsible for the death of more than two million people worldwide with an average of 700 outbreaks recorded per year in Brazil (Pereira et al. 2021).

Methods used during food manufacturing aim to maintain the hygienic and sanitary conditions of the food and processing environment. Many of these methods use heat treatment or chemical products, generating residues in the environment, which can cause damage to the health of the food handler. Among the main sanitizers used in the industry is chlorine, which is usually in the form of sodium hypochlorite; although effective, it has disadvantages related to environmental pollution, the risk to the handler due to its toxicity, and the need to control its use and storage. As an alternative to these methods, electrolyzed water (EW) has potential characteristics for use in the food industry, such as being an industrial disinfectant, being effective in cleaning equipment, and reducing the microbial load of products of plant and animal origin (Wang et al. 2019; Athayde et al. 2018).

Recently, studies aimed at extracting bioactive compounds and protein isolates, as well as improving or modifying technological, physical-chemical, and sensory characteristics in different food matrices have been evidenced. It also explores its antimicrobial and biofilm elimination effect (Leães et al. 2021; Li et al. 2021; Leães et al. 2020; Lin et al. 2020; Watanabe et al. 2020; Soquetta et al. 2020).

Thus, the purpose of this chapter is to present information on the main characteristics of EW, microbial inactivation mechanisms, and possible applications in the food industry.

2. General Features of EW

The EW is produced by the electrolysis of a saline solution (usually NaCl) in an electrolytic cell, which has a negatively charged electrode (cathode) and a positively charged electrode (anode).

Department of Food Science and Technology (DTCA), Universidade Federal de Santa Maria, Santa Maria CEP 97105-900, Rio Grande do Sul, Brazil.

Emails: yasmimsvl@hotmail.com; carlacristinab@gmail.com; paulocampagnol@gmail.com; cijoale@gmail.com

* Corresponding author: cijoale@gmail.com

During electrolysis, the water molecule (H_2O) and the NaCl present in the solution dissociate, hence characterizing each type of EW (Yatao et al. 2012; Hricova et al 2008; Huang et al. 2008). Thus, according to the redox potential (ORP) and pH characteristics, the EW can be divided into basic electrolyzed water (BEW), acidic electrolyzed water (AEW), and slightly acidic electrolyzed water (SAEW) (Table 1).

The AEW formed at the anode contains dissolved Cl_2, hypochlorous acid (HOCl), and dilute hydrochloric acid (HCl), while the BEW (cathode) contains small amounts of sodium hydroxide (NaOH), hypochlorite (OCl^-), and H_2. SAEW has chlorine mostly in the form of HOCl (about 95%), which makes it less corrosive and more stable compared to AEW (Afari and Hung 2018; Athayde et al. 2018).

There are two types of electrolysis equipment: those with a diaphragm/separation membrane, giving rise to two distinct cells, which on one side produces AEW and on the other, BEW, which is based on the migration of Na^+ to the cathode side (+), and from Cl_2 to the anode side; the other type of equipment is the one that does not have a membrane (chambers with a single cell), which produce AEW and BEW (Afari and Hung 2018, Rahman et al. 2016) (Fig. 1).

Table 1. Characterization of EW Types in Terms of ORP and pH Values. (Adapted From Guentzel et al. 2008; Huang et al. 2008; Rahman et al. 2010; Forghani and Oh 2013).

Classification	ORP	pH
BEW	< -800 mV	> 11
AEW	$\geq 1{,}000$ mV	2.0 to 3.0

(Adapted From Guentzel et al. 2008, Huang et al. 2008, Rahman et al. 2010, Forghani and Oh 2013)

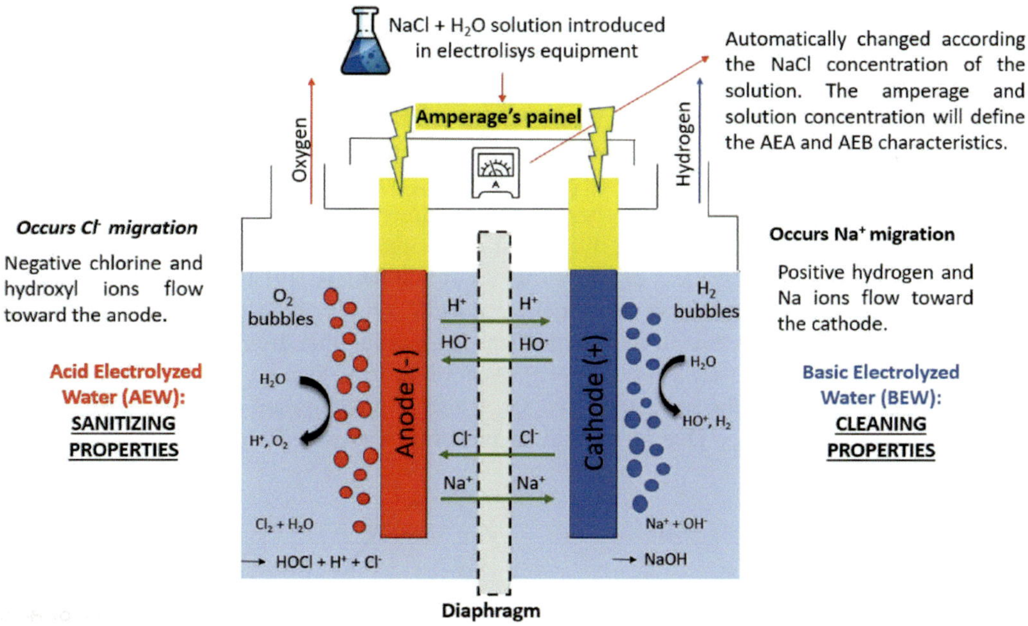

Figure 1. Water electrolysis system: on the left, AEW formation; on the right, BEW formation (Adapted from Shiroodi and Ovissipour, 2018 and Rahman, Khan and Oh 2016).

2.1 Main Advantages and Limitations of EW

EW is a non-thermal treatment and can be used in foods without causing changes in chemical and sensory properties (Zhang et al. 2012; Huang et al. 2008). It is considered a technology that provides safety for the environment and the food handler since it can be produced from a NaCl solution. Compared to other conventional disinfectants (hypochlorite), EW does not cause irritation or toxicity to the employee's skin as it is easy to use and apply. In this way, it does not present risks in transport and storage, giving lower costs to the food industry. It requires less contact time with the food or surface to reduce the number of pathogenic and spoilage microorganisms, thus preventing them from developing resistance (Huang et al. 2008; Al-Haq et al. 2005).

However, several factors limit the daily use of AEW, such as the rapid loss of the bactericidal effect because of the evaporation of chlorine (Forghani and Oh 2013; Rahman et al. 2010; Cui et al. 2009; Huang et al. 2008) and contact with organic matter (Park et al. 2009). Studies show that AEW presented problems related to the corrosion of metals and the release of chlorine and hydrogen gases into the environment, requiring adequate personal protective equipment for its use (Ayebah and Hung 2005). Given this, studies show the advantages of using SAEW because it has similar or even superior bactericidal activity when compared to AEW, despite having lower FCC. The SAEW is characterized by being more stable during storage and safer for the health of the handler (Athayde et al. 2018).

2.2 Antimicrobial Activity of EW

The mechanisms of action of EW against microorganisms are discussed by different authors. The ORP is one of the main factors involved in the bactericidal activity of EW (Park et al. 2009; Kim et al. 2000). The ORP from 200 to 800 mV favors the development of aerobic microorganisms, while anaerobes between –200 to –400 mV (Jay 2006). The ORP of AEW affects *Escherichia coli* O157:H7, damaging internal and external membranes (Liao et al. 2007). In addition, high ORP interferes with the flow of electrons in the cell of microorganisms, causing metabolic changes and the production of ATP (Huang et al. 2008).

Values of pH between 4.0 and 9.0 are ideal for bacterial multiplication. Consequently, controlling the pH of the EW becomes important to guarantee its bactericidal action since the acidic pH can sensitize the microbial cell, allowing the entry of HOCl (Mcpherson 1993) Additionally, the pH of the medium is decisive for the balance of the species of chlorine present, which is considered an important factor for the action of EW against microorganisms (Hao et al. 2012) (Fig. 2).

Evaluating ices made with AEW, starting from NaCl solutions with different concentrations (0.75 to 1.75 g/L), Li et al. (2014) suggested that the free chlorine concentration (FCC) is the main factor involved in the action of AEW against microorganisms followed by the pH and POR. However, studying the effectiveness of AEW (pH 3.74) and SAEW (pH 6.23) with the same FCC (1.4 mg/L) against *Escherichia coli,* it was noticed that AEW completely lost its disinfection capacity, while SAEW was able to destroy all microorganisms, suggesting that the free chlorine form in EW is more important than FCC (Hao et al. 2012). According to Cui et al. (2009), the bactericidal activity of SAEW is due to the presence of HOCl, which is predominant in the solution when the pH is in the range of 5.0–6.5. Due to its cellular penetration power, HOCl has greater sanitizing

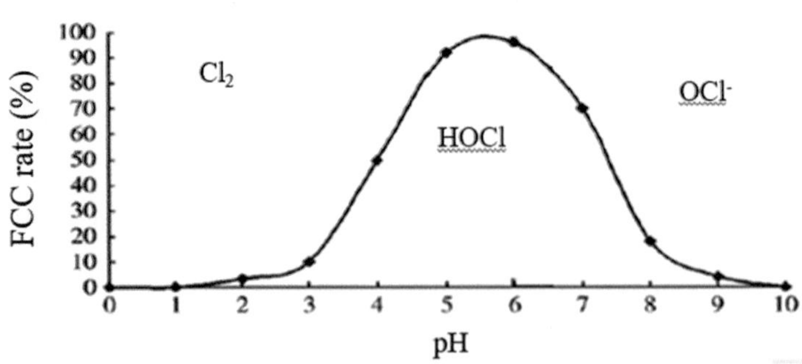

Figure 2. Free chlorine concentration (FCC) and its compounds according to EW pH (Adapted from Rahman et al. 2010).

power compared to other free chlorine fractions; it is 80 times more effective than OCl- against *E. coli,* when evaluated at the same concentration and exposure time (Anonymous 1997; White 2010). Chlorine oxidizes sulfide groups of enzymes that participate in carbohydrate metabolism, which is responsible for the oxidation of glucose in microorganisms, as well as destroying key enzymes and breaking down nucleic acids, tus promoting oxidative damage (White 2010).

Studies showed that SAEW with ORP 500 mV, pH 6.2, and 5 mg/L of FCC showed similar activity and even superior to AEW with ORP 1,100 mV, pH 2.5, and 50 mg/L of FCC against E. *coli*, *S. aureus*, *L. monocytogenes*, and *Salmonella* (Rahman et al 2010). Rahman et al. (2012a) observed similar antimicrobial activity of SAEW with 5 mg/L and 7 mg/L of FCC and AEW with 50 mg/L of FCC against *E. coli* 0157:H7 and *L. monocytogenes*. Additionally, SAEW with 10 mg/L was more effective in reducing logarithms against these microorganisms, increasing with exposure time (30, 60, and 90 seconds).

3. Application of Electrolyzed Water in Food Processing

3.1 The Application of EW on Food Processing in Surfaces

The use of EW as a sanitizer on different types of surfaces in food industries has been studied (Monnin et al. 2012; Guentzel et al. 2008) since surfaces and utensils can be sources of cross-contamination (Goh et al. 2014; Phuvasate and Su 2012) can corrode depending on the chemical used for sanitization (Waters et al. 2014).

The contact of AEW containing 50 mg/L of FCC for 5 minutes eliminated microorganisms on stainless steel surfaces used in fish handling (Phuvasate and Su 2012). Evaluating the action of AEW, SAEW, chlorinated water, and deionized water on stainless steel, aluminum, carbon, and copper for 8 days, Ayebah and Hung (2005) showed that AEW was the most corrosive with stainless steel being more resistant to oxidation. Due to pH being close to neutrality, SAEW showed less corrosion in utensils and equipment used in the industry compared to AEW (Waters et al. 2014; Huang et al. 2008).

Guentzel et al. (2008) demonstrated the effectiveness of SAEW in reducing microbial growth in household appliances present in the production area. Arevalos-Sánchez et al. (2013) verified the potential of SAEW with 70 mg/L of FCC as an alternative disinfectant on stainless steel surfaces and materials used in food handling by inhibiting the formation of *Listeria* biofilms after 3 minutes of contact. The SAEW with 120 mg/L of FCC showed a similar action to sodium hypochlorite solution with 100 mg/L of FCC in reducing the population of *Listeria* and *E. coli* in utensils used in food preparation (Monnin et al. 2012).

3.2 Electrolyzed Water in Foods of Plant Origin

In fruits and vegetables, the presence of pathogenic bacteria is relatively low, but their ingestion may represent a potential risk to human health (Artés et al. 2009; Doyle and Erickson 2008; Gómez-López et al. 2008). Outbreaks of diseases transmitted by ready-to-eat vegetables included cases of *E. coli* O157:H7 (Denmark in 2010, the Netherlands in 2007, and Sweden in 2005) and *E. coli* O104:H4 in Germany (Wu et al. 2011).

Since washing fruits and vegetables with tap water alone is not considered ideal for reducing food contamination, chemical compounds such as sodium hypochlorite (Akbas 2007), chlorine dioxide (Kim et al. 2009), organic acids (Zhang et al. 2012), calcium chloride (Izumi and Watada 1994), acidified sodium chloride (Liao 2009), and ozone (Nagashima and Kamoi 1997) are used. However, they can pose a risk to human health and the environment. Thus, the use of EW in the decontamination of fresh products has been extensively researched.

Issa-Zacharia et al. (2010) suggested that AEW could be used as an alternative to conventional hygiene methods. This is because it significantly reduced mesophilic aerobic bacteria in Chinese celery, lettuce, and turnip by 2.7, 2.5, and 2.45 logs CFU/g, respectively, compared to the conventional process that uses chlorine.

The AEW (pH 2.8, ORP 1,125 mV, and FCC 48 mg/L) reduced the incidence of spoilage in blueberries, conferred greater skin firmness, and increased the amounts of anthocyanins and total phenolics, proving to be a viable alternative for increasing the shelf life of blueberries (Chen et al 2019a). The AEW (FCC 80 mg/L; pH 2.5 for 10 minutes) in longan fruits, slowed the onset of diseases and reduced the presence of reactive oxygen species during storage (6 days) in a controlled atmosphere (25 °C and 90 % RH) (Tang et al. 2021). Apples sliced and immersed in AEW (pH 3.54) showed a reduction of enzymatic browning after 14 days of storage at 8ºC, maintenance of firmness, and total phenolic content and antioxidant activity higher than the control (Plesioanu et al. 2021).

Although AEW has an action against microorganisms, SAEW has been studied in this sense as it has greater effectiveness and less corrosiveness. The use of SAEW in chopped cabbage stored in a modified atmosphere improved the visual characteristics and increased the shelf life to 5 days (Gómez-López et al. 2007). Compared to conventional sanitizers, the use of SAEW (pH 6.3 ± 2, ORP 520 ± 20, and 5 ± 0.1mg/L of FCC) in spinach leaves inoculated with *E. coli* O157:H7 and *L. monocytogenes* was promising, in addition to not polluting the environment (Rahman et al. 2010).

The application of SAEW (pH 6.5, ORP > 450, and 50 mg/L of FCC) reduced the development of *E. coli* O157:H7 in lettuces that were packaged in a modified atmosphere and stored at 4ºC (Posada-Izquierdo et al. 2014). However, the use of SAEW had negative effects on the browning of minimally processed lettuce, loss of turgidity, and mineral content concerning conventional sanitizers (Rico et al. 2008).

The SAEW with 25 mg/L FCC extended the shelf life without compromising the physicochemical quality and freshness of 20-day cherries (Hayta and Aday 2015). The concentration of 60 mg/L for 3 minutes reduced by up to 50% the incidence of rot in tomatoes (*Solanum Lycopersicum* L.) inoculated with *Fusarium oxysporum*, *Galactomyces geotrichum,* and *Alternaria sp* when compared to the control, thus prolonging the shelf life of tomatoes (Vázques-López et al. 2016). The SAEW in spray form (pH 7.1–9.4 and ORP 733.7 to 906.3) reduced infections in avocados in the field and during storage (Hassan and Dann 2019).

The SAEW (pH 5.42, ORP 818–854 mV, and FCC 30 mg/L) combined with fumaric acid and calcium oxide applied to apples, tangerines, and tomatoes for 14 days of storage reduced the colonies of *E. coli* O157:H7 and L. monocytogenes and improved sensory characteristics (Chen et al. 2019b).

Graça et al. (2020) evaluated the effect of AEW and SAEW in a mixture of yeasts (*Candida sake, Hanseniaspora uvarum, Pichia fermentans,* and *Metschnikowia pulcherrima*) applied to fresh-cut apples stored at 4ºC control with NaOCl, thus being a promising alternative to be used. The combination of SAEW with fumaric acid (FA) and antioxidant solution, totally inhibited *Salmonella* sp., in yellow and red peppers, increasing in 9 days of shelf life at 4ºC (Saravanakumar et al. 2021). The SAEW with 4 mg/L at 80°C was shown to be effective in disinfecting organic carrots, inactivating *E. coli* and *S. typhimurium* (Liu et al. 2019). The SAEW with FCC of 50 mg/L for 45 seconds was sufficient to reduce 4 logs CFU/g of *Salmonella* spp. without affecting the texture and flavor of carrots (Cap et al. 2021).

Other EW applications aimed at cleaning or conservation but combined with other technologies, showed encouraging results. Persimmon paste intended for wine production and soaked in AEW (FCC 531.2 mg/L) showed a significant improvement in their sensory characteristics when compared to the control. There was also an increase in the total amino acid content, which improved the flavor of the wine produced (Zhu et al. 2016). The SAEW (pH 6.2 and ORP 297 mV) combined with the ultrasonic probe provided higher values of phenolic compounds, total flavonoids, and antioxidant power with a reduction in extraction time of 87.5% compared to the traditional method (Soquetta et al. 2019).

The combination of AEW and isoelectronic precipitation (IP) increased the protein content (65.1%) and amino acids (76.6%) in rice bran protein isolate and increased the recovery rate of the concentrate by 50% after processing. extraction when compared to the standard method (Watanabe et al. 2019). The SAEW (60°C, FCC 80 mg/L, and 15 minutes) combined with the ultrasound (US) proved to be an effective alternative for removing biofilms from two strains of *Bacillus cereus* in spinach, beetroot, and lettuce leaves, reducing by 3.0–3.4 logs CFU/cm^2 and 4 logs CFU/cm^2 (Hussain et al. 2019). The combination of SAEW (FCC 60 mg/L) and ultraviolet light-emitting diode (UV-LED) for 5 minutes considerably reduced *Salmonella* and *E. coli* counts (2.27 and 2.43 log CFU/g, respectively) in coriander (Jiang et al. 2020).

3.3 Electrolyzed Water in Foods of Animal Origin

The EW can also be an alternative method for the control of pathogens during the slaughter of cattle, swine, and chickens, that is being able to act against microorganisms, such as *Salmonella* (Yang et al. 2013; Bunic and Sofos 2012; Rahman et al. 2012a; Fabrizio et al. 2002), *Campylobacter jejuni* (Fabrizio and Cutter 2003; Park et al. 2002), *Listeria monocytogenes* (Arevalos-Sánchez et al. 2013; Rahman et al. 2012a), and *E. coli.* (Panglori and Hung 2013; Yang et al. 2013; Hao et al. 2011).

Studies have shown that washing bovine leather with AEW in the form of a spray was effective in reducing *E. coli* and *Salmonella* and minimized cross-contamination of carcasses in the slaughter line (Jadeja and Hung 2014; Bosilevac et al. 2005). Fabrizio and Cutter (2003) demonstrated the effectiveness of AEW in reducing mesophilic microorganisms in fresh pork meat after 15 seconds of washing. Fabrizio and Cutter (2004) suggested that the combined use of EW with organic acids, such as calcium lactate, improves the sensory and microbiological quality of pork meat. Kim et al (2005) suggested that washing chicken carcasses with EW in the form of a spray can reduce cross-contamination during the slaughter line by removing fecal material and other pathogens, showing positive effects against *Campylobacter jejuni*, *Salmonella*, and *E. coli* (Northcutt et al. 2007). Rahman et al. (2012b) demonstrated that immersion of chicken breasts in SAEW with 10 mg/L of FCC was effective in reducing *Listeria monocytogenes* and *Salmonella.*

The application of AEW ice in shrimp was satisfactory in microbiological preservation and the inhibition of the polyphenol oxidase enzyme (Lin et al. 2013; Wang et al. 2014). Also in the form of ice, SAEW reduced the microbial load in fish and the water from the melting of storage ice, as well as cross-contamination (Feliciano et al. 2010). Furthermore, the application of AEW decreased the number of histamine-producing bacteria in fish (Park et al. 2002).

The combination of AEW, SAEW, and BEW (0.05%) applied in spray form reduced the count of mesophilic and psychrotrophic bacteria, especially BEW reduced lipid and protein oxidation in pork loin and increased its shelf life (Athayde et al. 2017). Similar results were obtained in chicken carcasses after the evisceration step from the combination of AEW and SAEW (Wang et al. 2018), as well as an increase of 2 days of shelf life when applying SAEW together with peracetic acid (Hamidi et al. 2020).

In beef steak, SAEW (pH 6.29 ± 1.33, ORP 870–900 mV, FCC 40 mg/L) promoted a reduction in microbial counts and improved sensory attributes of color, flavor, texture, and odor, extending shelf life for another 8 days at 4ºC (Sheng et al. 2018). The SAEW applied (FCC 30 mg/L for 2.5 minutes) in combination with black teapolyphenols in beef, increased by 9 days the shelf life at 4ºC because of reduced microbial development and lipid oxidation (Bing et al. 2022). The antimicrobial action of SAEW is due to the presence of HOCl, and the antioxidant action is due to the presence of H_2 in BEW, which reacts with free radicals (Misir and Koral 2019).

The EW has also been studied for cleaning surfaces and utensils in the meat industry. The joint action of US and SAEW (pH 6.50; ORP 853.53), and BEW (pH 11.04; ORP 371.30 mV) for 20 minutes applied at 25ºC, reduced ($p < 0.05$) the counts of mesophilic bacteria, enterobactéria, and *S. aureus* when compared to the values recommended by Brazilian legislation (Brasil et al. 2020). Furthermore, the combination of SAEW (FCC 30 mg/L) with ultraviolet light in 15 minutes times reduced to 4.5 log CFU/cm$_2$ the *Vibrio parahaemolyticus* count on contact surfaces in the processing of shrimp and crabs and eliminating biofilms (Roy et al. 2021).

The EW has been incorporated as a processing step and also in the formulation for the reduction of additives and/or in the process of extracting protein isolates, usually combined with the US. The combination of US and SAEW (pH 6.0, ORP 800–850 mV, and FCC 5 mg/L) reduced ($p < 0.05$) the counts of mesophilic, psychrotrophic, and enterobacterial bacteria in chicken breasts during the chilling stage (Cichoski et al. 2019).

Regarding the reduction of additives in meat products, BEW (0.1%, pH 11.6, and ORP –301 mV) enabled a 50% reduction in polyphosphate without compromising the technological quality of catfish fillets. Improved water holding capacity (WRC), texture, and color (Lin et al. 2020). The combination of US and BEW (0.01%, pH 10.91, and ORP −330 mV) allowed a 30% reduction in NaCl content in meat emulsions without affecting yield, stability, color, and texture when compared to the control. The possible effect of BEW, in this case, was attributed to the presence of NaOH, acting to reduce the fat globule and increase the pH, which distanced itself from the isoelectric point (IP) of the protein (Leães et al. 2020). Made with this emulsion, the BEW reduced ($p < 0.001$) the count of lactic acid bacteria about the control during the 90 days of storage (Leães et al. 2021). The BEW (pH 12.3 and ORP –850 mV) in combination with the US increased the extraction of Antarctic krill crustacean proteins by 9.4% when compared to the standard method, improved solubility, reduced particle size, and did not affect oxidation protein (Li et al. 2021).

In dairy products, EW has been studied aiming at the disinfection of contact surfaces and equipment. The combination of BEW (100 and 300 mg NaOH/L at 30°C for 10 and 30 minutes) and SAEW (FCC 40 mg/L for 3 minutes) reduced the total bacterial count by up to 3.90 logs CFU/cm^2 in steel with no modifications (Jiménez-Pichardo et al. 2016). While SAEW was applied to stainless steel, sealing rubber and PVC hose in a clean-in-place milking system was effective in disinfection (Liu et al. 2020). Dev et al. (2014) in a pilot-scale CIP milking system used BEW (pH 11.5, ORP –850 mV, and 58.8°C) for cleaning and AEW (pH 2.6, ORP 1.150 mV, FCC 80 mg/L and 39.3°C) for disinfection and obtained 100% microbial destruction. Application of BEW (50 mg NaOH/L at 30°C for 10 minutes) in the cleaning step and SAEW (50 mg/L FCC at 20°C for 5 minutes) in the disinfection step on stainless steel plates in the dairy industry was effective in preventing the formation of biofilms of lactic acid bacteria, yeasts, and *L. monocytogenes* (Jiménez-Pichardo et al. 2021). Moradi and Tajik (2017) applied SAEW with FCC of 75 $\mu g\ ml^{-1}$ for 5 minutes, which was effective in destroying *E. coli*, *L. monocytogenes*, and *P. aeruginosa* biofilms in the UHT pasteurization system.

In dairy products (kashk that consists of a powdered cereal of cracked wheat fermented with milk), AEFA (pH 5.3–5.5, Eh 545–600 mV, and CCL 20–22 mg/L) combined with the US provided a reduction of 1.87, 1.67, 1.71, and 1.91 log CFU/mL in the counts of *S. aureus*, *B. cereus*, *E. coli*, and *A. fumigatus*, respectively, proving to be effective in decontamination of this product (Forghani et al. 2015).

In eggs, SAEW stood out as a promising alternative to increase microbial safety and extend shelf life. The SAEW (pH 6.0–6.5 and ORP 238.4–265.2) with FCC of 4 mg/L for 2 minutes inactivated *Salmonella enteritidis* in eggs with contaminated shells (Cao et al. 2009). The same was observed in eggshells contaminated with *L. monocytgenes* in a study carried out by Rivera-Garcia et al (2019), using SAEW with pH 6.86, ORP 872 mV, and FCC of 46 mg/L, obtaining a reduction of 2.18 log^{10} CFU/mL without compromising the mineral concentration of the bark. Zang et al. (2019) obtained satisfactory results regarding the inactivation of *S. enteritidis* and *E. coli* using SAEW under conditions of pH 6.37–6.53, ORP 645.5–675.9 mV, FCC of 26 mg/ L at times of 3 and 4 minutes, respectively, and with different FCC (10, 18, and 26 mg/L). In addition, during storage at 25°C, weight loss minimization (5.52%), and albumen and yolk preservation were observed. Sheng et al. (2021) combined SAEW (pH 6.37, ORP 980, and FCC 30 mg/L) and chitosan as egg coating and observed an increase in egg shelf life.

3.5 Other Applications of Electrolyzed Water

New fields of application of EW are related to the removal of pesticides, such as acephate, omethoate, dimethyl dichlorovinyl phosphate from the surface of vegetables when immersed in AEW with FCC of 70 mg/L, and BEW when compared to immersion in water or detergent solution (Hao et al. 2012). In one study, BEW showed efficiency in removing pesticides in pepper, and AEW in cabbage and broccoli (Liu et al. 2021).

In addition, the application of AEW against fungi such as *Candida albicans* (Zeng et al., 2011) and *Aspergillus flavus* were positive (Xiong et al. 2010). Aflatoxin B1 was 85% eliminated from contaminated peanuts after 15 minutes of contact with AEW containing FCC of 70 mg/L at 20°C without causing damage to food (Zhang et al. 2012).

The addition of SAEW (pH 7.64 and 1.28 mg/L of FCC) in gutters used for the hydration of dairy sheep reduced the growth of mesophilic bacteria, total coliforms, and streptococci without affecting the health or performance of the animals (Bodas et al. 2013). The SAEW (pH 5.0–6.5; FCC of 12 mg/L) significantly reduced the incidence of clinical and subclinical cases of *Pseudomonas mastitis*, making it viable to control the disease (Kawai et al. 2017).

4. Conclusion

The EW has shown to be a promising alternative in several processes that go beyond disinfection, improving technological characteristics in meat products, extraction of bioactive compounds and protein isolates with differentiated functional characteristics.

Despite the existence of several reports in the literature about its use in foods, studies still need to be conducted to evaluate the real effects of the treatment with EW on the biochemical and sensorial properties of foods, as well as the improvement of the critical factors of the process to reach desirable levels. of microbial inactivation without affecting the quality of the product.

The optimization in the use of EW, to require shorter treatment times, still needs to be designed at an industrial level. It is suggested that the application of EW in the food industry should be compared to other non-thermal treatments, to increase the effectiveness of its use.

References

Afari, G.K. and Hung, Y.C. 2018. A meta-analysis on the effectiveness of electrolyzed water treatments in reducing foodborne pathogens on different foods. Food Control. 93: 150–164.

Al-Haq, M.I., Sugiyama, J. and Isobe, S. 2005. Applications of electrolyzed water in agriculture and food industries. Food Sci Technol. 11: 135–150.

Anonymous. 1997. Principle os formation of electrolytic water.Hoshizac Eletric Co Ltd. Sakae, Toyoake. Aichi. Japan.

Arevalos-Sánchez, M., Regalato, C., Martin, S.E., Meas-Vong. Y., Cadena-Moreno, E. and García-Almendárez, B.E. 2013. Effect of neutral electrolyzed water on lux-tagged Listeria monocytogenes EGDe biofilms adhered to stainless steel and visualization with destructive and non-destructive microscopy techniques. Food Control. 34: 472–477.

Artés, F., Gómez, P., Aguayo, E., Escalona, V. and Artés-Hernández, F. 2009. Sustainable sanitation techniques for keeping quality and safety of fresh-cut plant commodities. Postharvest Biol. Technol. 51: 287–96.

Athayde, D.R., Flores, D.R.M.; Silva, J.S., Silva, M.S., Genro, A.L.G., Wagner, R. et al. 2018. Characteristics and use of electrolyzed water in food industries. Int. Food Res. Journal. 25: 11–16.

Athayde, D.R., Flores, D.R.M., Silva, J.S., Genro, A.L.G., Silva, M.S. and Klein, B. et al. 2017. Application of electrolyzed water for improving pork meat quality. Food Res. Int. 100: 757–763.

Ayebah, B. and Hung, Y.E. 2005. Electrolyzed water and its corrosiveness on various surface materials commonly found in food processing facilities. J. Food Process Eng. 28: 247–264.

Bing, S., Zang, Y., Li, Y., Zhang, B., Mo, Q. and Zhao, X. et al. 2022. A combined approach using slightly acidic electrolyzed water and tEWpolyphenols to inhibit lipid oxidation and ensure microbiological safety during beef preservation. Meat Sci. 183: 108643.

Bodas, R., Bartolomé, D.J., Tabernero-De-Paz, M.J., Posado, R., García, J.J. and Rodríguez, L. et al. 2013. Electrolyzed water as novel technology to improve hygiene of drinking water for dairy ewes. Res. Vet. Sci. 95(3): 1169–70.

Bosilevac, J.M., Shackelford, S.D., Brichta, D.M. and Koohmaraie, M. 2005. Efficacy of Ozonated and Electrolyzed Oxidative Waters To Decontaminate Hides of Cattle before Slaughter. J. Food Prot. 68: 1393–1398.

Brasil, C.C.B., Menezes, C.R., Jacob-Lopes, E., Barin, J.S., Zepka, L.Q. and Campagnol, P.C.B. et al. 2020. Combined application of electrolysed water and ultrasound to improve the sanitation of knives in the meat industry. Int. J. Food Science and Technology 55: 1136–1144.

Bunic, S. and Sofos, J. 2012. Interven tions to control Salmonella contamination during poultry, cattle and pig slaughter. Food Res. Int. 45: 641–655.

Cap, M., Rojas, D., Fernandez, M., Fulco, M., Rodriguez, A., Soteras, T. et al. 2020. Effectiveness of short exposure times to electrolyzed water in reducing *Salmonella* spp and Imidacloprid in lettuce. LWT. 28:109496.

Cao, W., Zhu, Z.W., Shi, Z.X., Wang, C.Y. and Li, B.M. 2009. Efficiency of slightly acidic electrolyzed water for inactivation of *Salmonella enteritidis* and its contaminated shell eggs. Int. J. Food Microbiol. 130: 88–93.

Cichoski, A.J., Flores, D.R.M., Menezes, C.R., Jacob-Lopes, E., Zepla, L.Q. and Wagner, R. et al. 2019. Ultrasound and slightly acid electrolyzed water application: An eficiente combination to reduce the bacterial counts of chicken breast during pre-chilling. Int. J. Food Microbiol. 301: 27–33.

Chen, X., Tango, C.H., Daliri, E.B.-M., Oh, S.-Y. and Oh, D.-H. 2019a. Disinfection efficacy of slightly acidic electrolyzed water combined with chemical treatments on fresh fruits at the industrial scale. Foods 8: 1–19.

Chen, Y., Hung, Y.-C., Chen, M., Lin, M. and Lin, H. 2019b. Enhanced storability of blueberries by acidic electrolyzed oxidizing water application may be mediated by regulating ROS metabolism. Food Chem. 270: 229–235.

Cui, X., Shang, Y., Shi, Z., Xin, H. and Cao, W. 2009. Physicochemical properties and bactericidal efficiency of neutral and acidic electrolyzed water under different storage conditions. J. Food Eng. 91: 582–586.

Dev, S.R.S., Demirci, A., Graves, R.E. and Puri, V.M. 2014. Optimization and modeling of an electrolyzed oxidizing water based Clean-In-Place technique for farm milking systems using a pilot-scale milking system. J. Food Eng. 135: 1–10.

Doyle, M.P. and Erickson, M.C. 2008. Summer meeting 2007-the problems with fresh produce: an overview. J. App. Microbiol. 105: 317–30.

Fabrizio, K.A. and Cutter, C.N. 2004. Comparison of electrolyzed oxidizing water with other antimicrobial interventions to reduce pathogens on fresh pork. Meat. Sci. 68: 463–468.

Fabrizio, K.A. and Cutter, C.N. 2003. Stability of Electrolyzed Oxidizing Water and Its Efficacy against Cell Suspensions of *Salmonella Typhimurium* and *Listeria monocytogenes*. J. Food Prot. 66: 1379–1384.

Fabrizio, K.A., Sharma, R.R., Demirci, A. and Cutter, C.N. 2002. Comparison of Electrolyzed Oxidizing Water with Various Antimicrobial Interventions to Reduce Salmonella Species on Poultry. Poult. Sci. 81: 1598–1605.

Forghani, F., Eskandardi, M. and Oh, D.-H. 2015. Application of slightly acidic electrolyzed water and ultrasound for microbial decontamination of kashk. Food Sci. Biotechnol. 24: 1011–1016.

Forghani, F. and Oh, D. 2013. Hurdle enhancement of slightly acidic electrolyzed water antimicrobial efficacy on Chinese cabbage, lettuce, sesame leaf and spinach using ultrasonication and water wash. Food Microbiol. 36: 40–45.

Goh, S.G., Leili, A-H., Kuan, C.H., Loo, Y.Y., Lye, Y.L. and Chang, W.S. et al. 2014 Transmission of *Listeria monocytogenes* from raw chicken meat to cooked chicken meat through cutting boards. Food Control. 37: 51–55.

Gómez-López, V.M., Ragaert, P., Debevere, J. and Devlieghere, F. 2008. Decontamination methods to prolong the shelf life of minimally processed vegetables, state-of-theart. Crit. Rev. Food Sci. Nutr. 48: 487–495.

Gómez-López, V.M., Ragaert, P., Ryckeboer, J., Jeyachchandran, V., Debevere, J. and Devlieghere, F. 2007. Shelf-life of minimally processed cabbage treated with neutral electrolyzed oxidizing water and stored under equilibrium modified atmosphere. Int. J. Food Microbiol. 117: 91–98.

Graça, A., Santos, D., Pires-Cabral, P. and Quintas, C. 2020. The effect of UV-C and electrolyzed water on yeasts on fresh-cut apple at 4°C. J Food Eng. 282:110034.

Guentzel, J.L., Lam, K.L., Callan, M.A, Emmons, S.A. and Dunham, V.L. 2008. Reduction of bacteria on spinach, lettuce, and surfaces in food service areas using neutral electrolyzed water. Food Microbiol. 25: 36–41

Hamidi, R.M., Shekarforoush, S.S., Hosseinzadeh, S. and Basiri, S. 2020. Evaluation of the effect of neutral electrolyzed water and peroxyacetic acid alone and in combination on microbiological, chemical, and sensory characteristics of poultry meat during refrigeration storage. Food Sci. Technol. 27: 499–507.

Hao, J., Qiu, S., Li, H., Chen, T., Liu, H. and Li, L. 2012. Roles of hydroxyl radicals in electrolyzed oxidizing water (EOW) for the inactivation of Escherichia coli. Int. J. Food Microbiol. 155: 99–104.

Hao, J., Wuyundalai, L.H., Chen, T., Zhou, Y. and Su, Y.C. et al. 2011. Reduction of Pesticide Residues on Fresh Vegetables With Electrolyzed Water Treatment. J. Food Sci. 76: C520–C524.

Hassan, M.K. and Dann, E. 2019. Effects of treatment with electrolyzed oxidizing water on postharvest diseases of avocado. Agriculture 9:1-15.

Hayta, E. and Aday, M.S. 2015. The effect of different electrolyzed water treatments on the quality and sensory attributes of sweet cherry during passive atmosphere packaging storage. Postharvest. Biol. Technol. 102: 32–41.

Hricova, D., Stephan, E. and Zweifel, C. 2008. Electrolyzed Water and Its Application in the Food Industry. J. Food Protect. 71: 1934–1947.

Hussain, M.S., Kwon, M., Park, E-Ji, Seheli, K., Huque, R. and Oh, D. 2019. Disinfection of *Bacillus cereus* biofilms on leafy green vegetables with slightly acidic electrolyzed water, ultrasound and mild heat. LWT. 116: 108582.

Huang, Y.R., Hung, Y.C., Hsu, S.Y. and Hwang, D.F. 2008. Application of electrolyzed water in the food industry. Food Control. 19: 329–45.

Issa-Zacharia A., Kamitano, Y., Muhimbula, H.S. and Ndabikunze, B.K. 2010. A review of microbiological safety of fruits and vegetables and the introduction of electrolysed water as an alternative to sodium hypochlorite solution. African J. Food Sci. 4: 778–789.

Izumi, H. and Watada, A.E. 1994. Calcium treatments affect storage quality of shredded carrots. J Food Sci. 59:106–109.

Jadeja, R. and Hung, Y.C. 2014. Efficacy of near neutral and alkaline pH electrolyzed oxidizing waters to control Escherichia coli O157:H7 and Salmonella Typhimurium DT 104 from beef hides. Food Control. 41: 17–20.

Jay, J.M. 2006. Microbiologia de alimentos. 6th ed. Artmed, Porto Alegre, Brazil, 711p.

Jiang, Y., Ai, C., Liao, X., Donghong, L. and Tiong, D. 2020. Effect of slightly acidic electrolyzed water (SAEW) and ultraviolet light illumination pretreatment on microflora inactivation of coriander. LWT. 132: 109898.

Jiménez-Pichardo, R., Hernández-Martínez, I., Regalado-González, C., Santos-Cruz, J., Meas-Vong, Y. and Wacher-Rodarte, M.C. et al. 2021. Innovative Control of Biofilms on Stainless Steel Surfaces Using Electrolyzed Water in the Dairy Industry. Foods 10: 103.

Jiménez-Pichardo, R., Regalado, C., Castaño-Tostado, E., Meas-Vong, Y., Santos-Cruz, J. and García-Alméndarez, B.E. 2016. Evaluation of electrolyzed water as cleaning and disinfection agent on stainless steel as a model surface in the dairy industry. Food Control. 60: 320–328.

Kim, Y.J., Kim, M.H. and Song, K.B. Efficacy of aqueous chlorine dioxide and fumaric acid for inactivating pre-existing microorganisms and Escherichia coli O157:H7, Salmonella typhimurium, and Listeria monocytogenes on broccoli sprouts. Food Control. 20: 1002–1005.

Kim, C., Hung, Y.C., Russellt, S.M. 2005. Efficacy of Electrolyzed Water in the Prevention and Removal of Fecal Material Attachment and Its Microbicidal Effectiveness During Simulated Industrial Poultry Processing. Poult. Sci. 84: 1778–1784.

Kim, C., Hung, Y.C. and Brackett, R.E. 2000. Roles of oxidation-reduction potential in electrolyzed oxidizing and chemically modified water for the inactivation of food-related pathogens. J. Food Prot. 63: 19–24.

Leães, Y.S.V., Silva, J.S., Robalo, S.S., Pinton, M.B., Santos, S.P. and Wagner, R. et al. 2021. Combined effect of ultrasound and basic electrolyzed water on the microbiological and oxidative profile of low-sodium mortadellas. Int. J. Food Microbiol. 353: 109310.

Leães, Y.S.V., Pinton, M.B., Rosa, C.T.A., Robalo S.S., Wagner, R. and Menezes, C.R. et al. 2020. Ultrasound and basic electrolyzed water: A green approach to reduce the technological defects caused by NaCl reduction in meat emulsions. Ultrasonics Sonochem. 61: 104830.

Li, Y., Zeng, Q., Liu, G., Peng, Z., Wang, X., Zhu, Y. et al. 2021. Effects of ultrasound-assisted basic electrolyzed water (BEW) extraction on structural and functional properties of Antarctic krill (*Euphausia superba*) proteins. Ultrasonics Sonochem. 71: 105364.

Liao, C.H. 2009. Acidified sodium chlorite as an alternative to chlorine for elimination of Salmonella on Alfalfa Seeds. J. Food Sci. 74: 159–164.

Liao, L.B., Chen, W.M. and Xiao, X.M. 2007. The generation and inactivation mechanism of oxidation–reduction potential of electrolyzed oxidizing water. J. Food Eng. 78: 1326–1332.

Lin, T., Wang, J.J., Li, J.B., Liao, C., Pan, Y.J. and Zhao, Y. et al. 2013. Use of acidic electrolyzed water ice for preserving the quality of shrimp. J. Agric. Food Chemistry 61: 8695–8702

Lin, H., Hung, Y. and Deng, S. 2020. Effect of partial replacement of polyphosphate with alkaline electrolyzed water (AEW) on the quality of catfish fillets. Food Control. 112: 107117.

Liu, Y., Wang, J., Zhu, X., Liu, Y., Cheng, M. and Xing, W. et al. 2021. Effects of electrolyzed water treatment on pesticide removal and texture quality in fresh-cut cabbage, broccoli, and color pepper. Food Chem. 353: 129408.

Liu, Y., Wang, C., Shi, Z. and Li, B. 2020. Optimization and Modeling of Slightly Acidic Electrolyzed Water for the Clean-in-Place Process in Milking Systems. Foods 9: 1685.

Liu, Q., Jin, X., Feng, X., Yang, H. and Fu, C. 2019. Inactivation kinetics of *Escherichia coli* O157:H7 and *Salmonella Typhimurium* on organic carrot (*Daucus carota L.*) treated with low concentration electrolyzed water combined with short-time heat treatment. Food Control. 106: 106702.

Mcpherson, L.L. 1993. Understanding ORP's role in the disinfection process. Water Eng. Manag. 140: 29–31.

Misir, G.B. and Koral, S. Effects of ultrasound treatment on structural, chemical and functional properties of protein hydrolysate of rainbow trout (Oncorhynchus mykiss) by-products. Ital. J. Food Sci. 31: 205–223.

Monnin, A., Lee, J. and Pascall, M.A. 2012. Efficacy of neutral electrolyzed water for sanitization of cutting boards used in the preparation of foods. J. Food Eng. 110: 541–546.

Moradi, M. and Tajik, H. 2017. Biofilm removal potential of neutral electrolysed water on pathogen and spoilage bacteria in dairy model systems. J App Microbiol. 123: 1429–1437.

Nagashima, T. and Kamoi, I. 1997. Sterilization and preservation of vegetables by ozonated water treatment. Food Preserv Sci. 23: 127–131.

Northcutt, J., Smith, D., Ingram, K.D., Hinton, A.J. and Musgrove, M. 2007. Recovery of bacteria from broiler carcasses after spray washing with acidified electrolyzed water or sodium hypochlorite solutions. Poult. Sci. 86: 2239–44.

Park, E.J., Alexander, E., Taylor, G.A., Costa, R. and Kang, D.H. 2009. The decontaminative effects of acidic electrolyzed water for *Escherichia coli* O157:H7, *Salmonella typhimurium*, and *Listeria monocytogenes* on green onions and tomatoes with differing organic demands. Food Microbiol. 26: 386–390.

Park, H., Hung, Y.C and Brackett, R.E. 2002. Antimicrobial effect of electrolyzed water for inactivating Campylobacter jejuni during poultry washing. International J. Food Microb. 72: 77–83.

Pereira, M.B., Freitas, R.S.G., Stedefeldt, E. and Cunha. D.T. 2021. Conflicts, informality and risk perception about foodborne diseases in small family and non-family run restaurants. J. Foodserv. Bus. Res. 24: 683–708.

Plesioanu, A.M., Nour, V., Tutulescu, F. and Ionica, M.E. 2021. Quality of fresh-cut apples as affected by dip wash treatments with organic acids and acidic electrolyzed water. Food Sci. Technolg. 1–8.

Phuvasate, S. and Su, Y.C. 2010. Effects of electrolyzed oxidizing water and ice treatments on reducing histamine-producing bacteria on fish skin and food contact surface. Food Control. 21: 286–291.

Rahman, S.M.E., Khan, I. and Oh, D. 2016. Electrolyzed Water as a Novel Sanitizer in the Food Industry: Current Trends and Future Perspectives. Comp. Rev. Food Sci. Food Saf. 15: 471–490.

Rahman, S.M.E., Park J.H., Wang, J. and Oh, D.H. 2012a. Stability of low concentration electrolyzed water and its sanitization potential against food borne pathogens. J. Food Eng. 113: 548–553.

Rahman, S.M.E., Park, J., Song, K.B., Naif, A.A.H. and Oh, D.H. 2012b. Effects of slightly acidic low concentration electrolyzed water on microbiological, physicochemical, and sensory quality of fresh chicken breast meat. J. Food Sci. 71: 35–41.

Rahman, S.M.E., Ding, T. and Oh, D. 2010. Inactivation effect of newly developed low concentration electrolyzed water and other sanitizers against microorganisms on spinach. Food Control. 21: 1383–1387.

Rico, D., Martín-Diana, A., Barry-Ryan, C., Frías, J., Henehan, G. and Barat, J.M. 2008. Use of neutral electrolyzed water (EW) for quality maintenance and shelf-life extension of minimally processed lettuce. Innovat. Food Sci. Emerg. Technol. 9: 37–48.

Rivera-Garcia, A., Santos-Ferro, L., Ramirez-Orejel, J., Agredano-Moreno, L.T., Jimenez-Garcia, L.F. and Paez-Esquiliano, D. et al. 2019. The effect of neutral electrolyzed water as a disinfectant of eggshells artificially contaminated with Listeria monocytogenes. Food Sci. Nutr. 7: 2252–2260.

Roy, P.K., Mizan, M.F.R., Hossain, M.I., Han, N., Nahar, S. and Ashrafudoulla, M. 2021. Elimination of Vibrio parahaemolyticus biofilms on crab and shrimp surfaces using ultraviolet C irradiation coupled with sodium hypochlorite and slightly acidic electrolyzed water. Food Control. 128: 108179.

Saravanakumar, K., Sathiyaseelan, A., Mariadoss, A.V.A., Chelliah, R., Shin, S. and Park et al. 2021. Slightly acidic electrolyzed water combination with antioxidants and fumaric acid treatment to maintain the quality of fresh-cut bell peppers. LWT 147: 111565.

Sheng, X., Shu, D., Li, Y., Zhan, Z., Yuan, X. and Sanfeng L. et al. 2021. Combined approach consisting of slightly acidic electrolyzed water and chitosan coating to improve the internal quality of eggs during storage. J. Sci. Food Agric. 101: 2355–2361.

Sheng, X., Shu, D., Tang, X. and Zang, Y. 2018. Effects of slightly acidic electrolyzed water on the microbial quality and shelf life extension of beef during refrigeration. Food Sci. Nut. 6: 1975–1981.

Shiroodi, S.G. and Ovissipour, M. 2018. Electrolyzed Water Application in Fresh Produce Sanitation. pp. 67–89. *In:* Siddiqui, M.W. (ed.). Postharvest Disinfection of Fruits and Vegetables, Academic Press, London, UK.

Soquetta, M.B., Tonato, D., Quadros, M.M., Boeira, C.P.; Cichoski, A.J. and Terra, L.M. et al. 2019. Ultrasound extraction of bioactive compounds from *Citrus reticulata* peel using electrolyzed water. J. Food Process Preser. 43: 14236.

Tang, J., Chen, H., Lin, H., Hung, Y-C., Xie, H. and Chen, Y. 2021. Acidic electrolyzed water treatment delayed fruit disease development of harvested longans through inducing the disease resistance and maintaining the ROS metabolism systems. Postharvest. Biol. Technol. 171: 111349.

Vásquez-López, A., Villarreal-Barajas, T. and Rodríguez-Ortis, G. 2016. Effectiveness of Neutral Electrolyzed Water on Incidence of Fungal Rot on Tomato Fruits (*Solanum lycopersicum* L.). J. Food Prot. 79: 1802–1806.

Xiong, K., Liu, H., Liu, R. and Li, L. 2010. Differences in fungicidal efficiency against Aspergillus flavus for neutralized and acidic electrolyzed oxidizing waters. Int J Food Microbiol. 137(1): 67–75.

Wang, H., Duan, D., Wu, Z., Xue, S., Xu, X. and Zhou, G. 2019. Primary concerns regarding the application of electrolyzed water in the meat industry. Food Control. 95: 50–56.

Wang, H., Qi, J., Duan, D., Dong, Y., Xu, X. and Zhou, G. 2018. Combination of a novel designed spray cabinet and electrolyzed water to reduce microorganisms on chicken carcasses. Food Control. 86: 200–206.

Wang, J.J., Lin, T., Li, J.B., Liao, C., Pan, Y.J. and Zhao, Y. 2014. Effect of acidic electrolyzed water ice on quality of shrimp in dark condition. Food Control. 35: 207–212.

Watanabe, M., Yamada, C., Maeda, I., Techapun, C., Kuntyia, A. and Leksawasdi, N. et al. 2019. Evaluating of quality of rice bran protein concentrate prepared by a combination of isoelectronic precipitation and electrolyzed water treatment. LWT. 99: 262–267.

Waters, B.W., Tatum, J.M. and Hung, Y.C. 2014. Effect of chlorine-based sanitizers properties on corrosion of metals commonly found in food processing environment. J. Food Eng. 121: 159–165.

White, G.C. 2010. Chemistry of aqueous chlorine. pp. 68–172. *In*: White, G.C. (ed.). White's handbook of chlorination and alternative disinfectants. 5th ed. Black & Veatch Corporation. John Wiley & Sons, Inc. New Jersey, USA.

Wu, C.J., Hsueh, P. and Ko, W. 2011. A new health threat in Europe: Shiga toxin producing *Escherichia coli* O104:H4 infections review article. J. Microbiol. Immunol. Infect. 44: 390–393.

Yang, H., Feirtag, J., Diez-Gonzalez, F. 2013. Sanitizing effectiveness of commercial "active water" Technologies on Escherichia coli O157:H7, Salmonella enterica and Listeria monocytogenes. Food Control. 33: 32–38.

Yatao, H., Fade, L., Shifu, Z., Lijun, W., Haijie, L. and Lite, L. et al. 2012. Effect of preparing condition on physicochemical property of slightly acidic electrolyzed functional water. Transac Chinese Society Agricul Eng. 28: 232–40.

Zang, Y.T., Bing, S., Li, Y.J., Shu, D.Q., Huang, A.M. and Wu, L.T. et al. 2019. Efficacy of slightly acidic electrolyzed water on the microbial safety and shelf life of shelled eggs. Poultry Sci. 98: 5932–5939.

Zeng, X., Ye, G., Tang, W., Ouyang, T., Tian, L., and Ni, Y. et al. 2011. Fungicidal efficiency of electrolyzed oxidizing water on Candida albicans and its biochemical mechanism. J. Biosci. and Bioengin. 112: 86–91.

Zhang, Q., Xiong, K., Tatsumi, E., Li, L. and Liiu, H. 2012. Elimination of aflatoxin B1 in peanuts by acidic electrolyzed oxidizing water. Food Control. 27: 16–20.

Zhu, W., Zhu, B., Li, Y., Zhang, Y., Zhang, B. and Fan, J. 2016. Acidic electrolyzed water efficiently improves the flavour of persimmon *(Diospyros kaki L. cv. Mopan)* wine. Food Chem. 197: 141, 149.

14

High Hydrostatic Pressure Processing

Zamantha Escobedo-Avellaneda, Génesis Vidal Buitimea-Cantúa, Magdalena de Jesús Rostro-Alanis, Amado Gutierrez-Sánchez, Jorge Navarro-Baez* and *Jorge Welti-Chanes*

1. Introduction

High hydrostatic pressure (HHP), as a method to inactivate microorganisms, was first proposed in the late 1980s, but it was not until the end of the 20th century that HHP gained attention as a preservation method in the food industry (Navarro-Baez et al. 2022). The interest in this technology has been motivated by consumer demand for more nutritious and fresh food. Although thermal treatments are highly used in food preservation to inactivate pathogenic and spoilage microorganisms as well as enzymes to ensure food safety and quality, the negative impact of heat on the nutritional, sensory, and physicochemical characteristics of foods has motivated the development of nonthermal technologies. HHP is an innovative technology highly accepted by consumers (Cardello 2003; Cardello et al. 2007; Evans and Cox 2006). It has been demonstrated that this non-thermal technology is adequate in extending the shelf life of foods while minimizing the loss of general quality (Bermúdez-Aguirre and Barbosa-Cánovas 2011; Soto-Caballero et al. 2022). HHP consists of applying pressures in a range of 100–1,000 MPa, more frequently from 100–600 MPa into a chamber at refrigeration or room temperature, which makes it a non-thermal technology. In HHP, the sample, which can be either solid or liquid, is placed in a chamber filled generally with water that acts as the pressure-transmitting medium (Escobedo-Avellaneda et al. 2011; Navarro-Baez et al. 2022). Indirect pressurization obtained by pumping water in the chamber is the most common method to reach high pressures, but it can also be obtained with direct pressurization by reducing the chamber volume using a piston (Ghafoor et al. 2020).

The almost instantaneous and uniform effect of the application of HHP facilitates the scaling of processes from the laboratory level to the industrial level, thus representing an important commercial advantage of this technology. Other advantages are the inactivation of microorganisms and enzymes, the higher retention of nutrients and functional compounds, and the better sensory quality when compared with thermal treatments. Moreover, HHP reduces or eliminates the use of additives and does not produce residues because only water is involved in the entire process (Navarro-Baez et al. 2022). Nevertheless, some enzymes and bacterial spores are baroresistant, requiring the combination of pressure and temperature to be inactivated. This last method known as pressure-assisted thermal processing (PATP) has the approval of the US Food and Drug Administration (Escobedo-Avellaneda

Escuela de Ingeniería y Ciencias, Tecnologico de Monterrey, Eugenio Garza Sada 2501, Monterrey 64700, Mexico.
* Corresponding author: zamantha.avellaneda@itesm.mx

et al. 2011). In addition, residual enzyme activity and dissolved oxygen could degrade some food compounds, and most high pressurized products require refrigeration (Srinivas et al. 2018).

There are in international markets, a variety of foods treated by HHP such as meat (sliced ham, turkey or chicken cuts, and whole pieces of cured ham), fruit and vegetable products (juices, purees, smoothies, beverages, sliced products, guacamole, and avocado), ready-to-eat products, and seafood. In addition to food preservation, HPP applications include modification of enzymatic activity, starch gelatinization, shucking of shellfish, enhancement of mass transfer phenomena to increase extraction yields of high-added value compounds, and microbial stimulation to reduce fermentation time among others (Ghafoor et al. 2020). The specific effects of HHP will depend on pressure, temperature, and holding time. Parameters such as the come-up time (CUT) could also have a significant effect (Escobedo-Avellaneda et al. 2015; Soto-Caballero et al. 2022).

2. High Hydrostatic Pressure (HHP) Fundamentals

The effect of HHP in microorganisms, chemical reactions, and phase changes is governed mainly by three principles: the isostatic or Pascal's law, the adiabatic heat of compression, and the Le Chatelier-Braun. The isostatic principle states that pressure is uniformly transmitted regardless of the composition, size, and shape of the food as well as equipment (Escobedo-Avellaneda et al. 2011; Ghafoor et al. 2020). The second principle indicates that the increment in pressure will be accompanied by an increment in temperature because of the compression heating that will depend on the pressure level, as well as on the type, and initial temperature of the pressure-transmitting fluid. For water, it has reported increments from 3–5°C; for silicone oil, the increase is above 20°C, while for oils it ranges from 3–9°C per each 100 MPa. Once the pressure is released the temperature decreases. During HHP, it is important to consider the initial temperature of the pressurization fluid and its increment because it could influence microbial and enzyme inactivation as well as the final product quality. The heat transfer from the pressurization fluid to the food to be pressurized will depend on its composition. Products with high fat content will have higher temperature increment than products with a high water content (Ghafoor et al. 2020).

The Le Chatelier-Braun Principe states that under equilibrium conditions, any chemical reaction, phase transformation, or molecular transformation that implies a decrease in activation volume (Va) will be favored by pressure increments, while those accompanied by an increase will be inhibited (Escobedo-Avellaneda et al. 2011). The activation volume defined as the difference between the partial molar volume of the active state and that of the reactants is calculated by evaluating the effect of pressure (P) at constant temperature (T) on the chemical reaction rate constant (k), and it is obtained by linear regression of ln k versus pressure P, according to the following equation:

$$\ln k = \ln Ko - (Va/RT)\,P$$

Where, R is the universal gas constant (8.314 cm^3.MPa/K.mol) and *ko* is a constant of the model. The greater the magnitude of Va (positive or negative) the higher the sensitivity to pressure while reactions with V = 0 are independent of the pressure level.

The lack of effect of HHP on covalent bonds at pressure from 1,000–2,000 MPa reduces the impact of nutritious compounds of foods; therefore, the structure of aromatic compounds, primary protein structures, lipids, and vitamins will not be affected. However, the effects of pressure on weak bonds, such as hydrogen bonds, hydrophobic bonds, and electrostatic interactions result in changes in the secondary, tertiary, and quaternary structure of proteins, explaining the inactivation of microorganisms and enzymes and the denaturation of proteins. In addition, the effect of noncovalent bonds will influence other large molecules, such as polysaccharides (Navarro-Baez et al. 2022; Baranda and Montes 2020). All these changes could be characterized by a specific Va value where for example in the case of microorganims, negative values mean that pressure increment favor inactivaction, positive values that pressure increment will not favor inactivation, and values of zero indicate that the inactivation if not affected by presure changes (Serment-Moreno et al. 2014).

The effect of holding time at constant pressure and temperature on k can be described with the first-order kinetic model:

$$ln \frac{C}{C_0} - kt$$

Here, C is the change in concentration with respect to time (t), and Co is the initial concentration. C could also refer to another parameter to be evaluated with respect to time, such as a change in enzymatic activity or microbial load (Escobedo-Avellaneda et al. 2011).

3. Effects of HHP on Microorganisms

Microbial inactivation is one of the main applications of HPP. HHP causes a series of changes on microorganisms, mainly affecting the cell morphology, the biological functions essential for cell maintenance, and the genetic material (Abe 2007; Huang et al. 2014). The progressive damage to the microorganism caused by different pressure levels is represented in Fig. 1.

Diffrent morphological changes in microorganims caused by HHP have been reported. Among these are included change in the permeability of the membrane because of increased pressure, which causes loss of integrity, followed by swelling and separation of the cell wall that ultimately leads to cell death (Huang et al. 2014). Another reported change is the collapse of the vacuoles because of gas compression in this organelle (Osumi et al. 1992). Cell elongation and filament formation (Zobell and Cobet 1962), as well as cell wall contraction with pore formation, have also been reported (Téllez-Luis et al. 2001). Table 1 describes some morphological changes for specific microorganisms according to the treatment conditions used.

Once the cell membrane is damaged, the biological functions of the microorganism are compromised. The normal metabolic pathway of the microorganism is affected and the absorption of nutrients and the elimination of accumulated waste within the cell are hindered (Torres and Velazquez 2005; Huang et al. 2014). Affectation of key functions has also been reported, such as the inhibition of ATP synthesis, as well as the denaturation of proteins responsible for replication processes, regulation of metabolic activities, and maintenance of cell integrity (Tholozan et al. 2000; Simpson and Gilmour 1997). HHP also causes reversible or irreversible structural modifications of proteins, mainly damaging tertiary and quaternary structures. It has been reported that when the applied pressure is less than 300 MPa, the reaction is reversible; while at pressures above 300 MPa the damage is irreversible (Rastogi et al. 2007; Yaldagard et al. 2008). Finally, HPP can affect processes, such as the replication of genetic material and gene transcription, as well as degrade chromosomal DNA (Alpas et al. 2003; Dubins et al. 2001; Kaletun et al. 2004).

These changes can cause cell damage or death of the microorganism. However, it has been reported that sublethal stress induces the expression of repair systems; this can generate the adaptation of microorganisms to stress during food processing, which constitutes a potential risk that is important to consider (Lado and Yousef 2002).

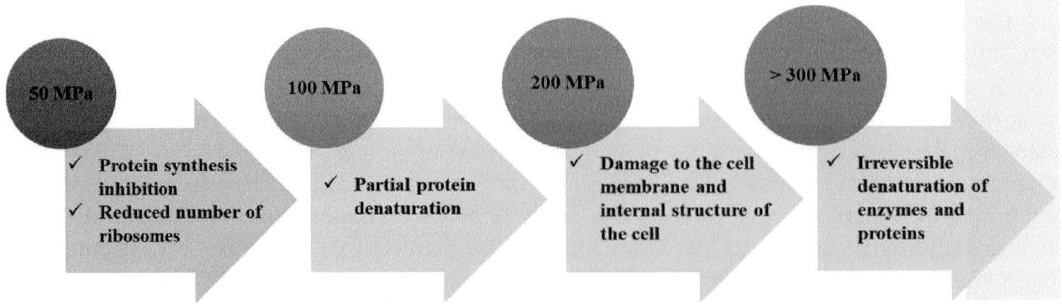

Figure 1. Progressive Effect of Damage to Microorganisms Caused by High Hydrostatic Pressure.

Table 1. Morphological and Biochemical Changes of Microorganisms After Treatment With High Hydrostatic Pressure.

Microorganism	Treatment (conditions)	Effects	Reference
Bacillus subtilis	400 MPa at 55°C	Generates depressions and ruptures of the surface of the spore, which causes a partial leak of intracellular substances.	Furukawa et al. 2003
Leuconostoc mesenteroides	500 MPa	Internal and external changes in the cell structure. Loss of chain morphology and blistering on the surface of cells. Compaction of the cytoplasm. Denaturation of ribosomes.	Kaletun et al. 2004
Escherichia coli O157:H7	200 MPa	Modifications in the properties of the membrane such as denaturation of transmembrane proteins, permeability alterations and phase transition of the lipid bilayer.	Pilavtepe-Çelik et al. 2008
Listeria monocytogenes	400 MPa	Aggregation of the cytoplasm by general denaturation of proteins.	Mohamed et al. 2012
Escherichia coli *Staphylococcus aureus*	500 MPa	Wrinkling and distortion of the membrane and cell wall. Generation of partial voids in the cytoplasm and leakage of cellular fluids due to cell ruptures.	Yang et al. 2012
Vibrio parahaemolyticus	300 MPa	The number of pimples and swelling causing some cells to be crushed and others to shattered increase progressively.	Wang et al. 2013

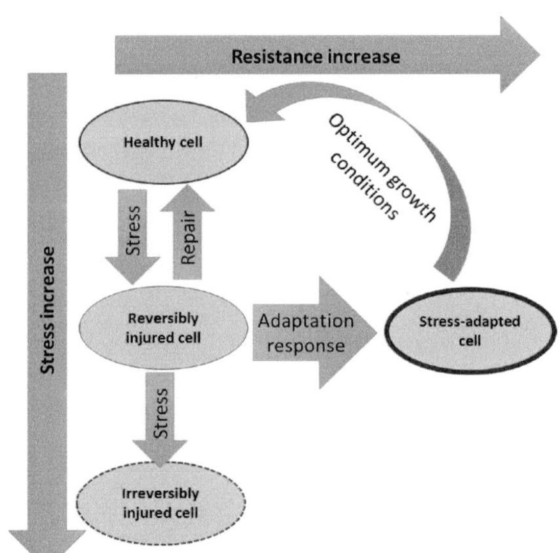

Figure 2. Microbial Stress, Cell Damage, Stress Adaptation, and Resistance to Processing; Adapted From Lado and Yousef 2002.

The stress tolerance and the survival of microorganisms depend on several factors that can be categorized in the physiology of the microorganism, the environmental or process conditions, and the properties or characteristics of the food (Lado and Yousef 2002; Huang et al. 2014). Figure 2 shows a scheme of sublethal stress and the induction of repair systems that cause the adaptation of microorganisms to the stress generated during processing.

Regarding microorganism, the factors include the type of microorganism, the growth phase, and their concentration in the food (Evrendilek 2012). The wide variety of microorganisms with different physiological characteristics may exhibit different characteristics of resistance to pressure. Gram-positive bacteria have greater resistance to HHP compared to Gram-negative bacteria; this is because the former have teichoic acid in their cell wall, which is characterized as a polysaccharide that provides rigidity to the cell wall (Toepfl et al. 2006; Quiroz-Gonzalez et al. 2018). Small coccoid-shaped bacteria are generally more resistant to HPP than large rod-shaped ones. Stationary phase cells are less susceptible to HPP because they have a stronger cytoplasmic membrane that gives them greater tolerance to treatment (Bartlett 2002). While cells in the exponential phase are less resistant to pressure because they suffer irreversible damage to the structure of the membrane and proteins, as well as loss of osmotic response and release of RNA into the medium (Garcia-Graells et al. 1999). It has been reported that pressures around 300 and 600 MPa achieve the inactivation of yeasts, molds, and most vegetative bacteria, while the inactivation of bacterial spores requires pressures greater than 1,000 MPa (Toepfl et al. 2006; Quiroz-Gonzalez et al. 2018).

Regarding environmental or process conditions, the pressure level, the temperature, and the treatment time, are inlcuded. While the factors related to the food include its composition such as fat and protein content, pH, and aw. This latter has proven to be a challenge for microbiological inactivation by HPP (Patterson 1995; 2005).

It has been reported that HPP treatment can successfully inactivate different types of organisms in different types of food (Simpson and Gilmour 1997). Currently, this technology is used to preserve products, such as ready-to-eat dishes, sausages, juices, seafood, and guacamole among others, thus eliminating or reducing the need for preservatives (Mújica-Paz et al. 2011). In the meat sector, processing with HPA has shown its potential to inactivate microorganisms and extend shelf life, improving the safety of cooked or cured processed meat products. In the case of milk and dairy products, it has been shown that the application of pressures between 300 and 600 MPa is effective for the inactivation of microorganisms, such as *Escherichia coli, Listeria monocytogenes, Salmonella typhimurium*, and *Staphylococcus aureus* (Dogan and Erkmen 2004). In the shellfish sector, APH technology has been successful in eliminating bacteria, such as *Vibrio parahaemolyticus* and *Vibrio vulnificus* in oysters without affecting their sensory characteristics (Cook 2003; Deng et al. 2015). Also, in the beverage sector, HPP has stopped bacterial and yeast spoilage in fruit juices (Jordan et al. 2001; Basak et al. 2002; Hiremath et al. 2012).

Foods treated with HPP increase every year, today the production value exceeds 10 billion dollars. However, food regulations are beginning to require certain levels of confidence that the risk of pathogens has been reduced to an acceptable level of probability (Alpas et al. 1999; Huang et al. 2014). For this, indicator strains have been selected based on the type of food, for example, *Escherichia coli* O157:H7 and *Listeria monocytogenes* require 5 log reduction to guarantee consumer health and safety (USDA 2012). The safety of emerging food preservation technologies depends on a reliable estimate of its efficacy against pathogenic and spoilage microorganisms (Huang et al. 2014). For this reason it is critical to assess the impact of variability in microbial inactivation data to make safe processing decisions.

4. Effects of HHP on Proteins and Polysaccharides

The HHP also affects macromolecules such as proteins and polysaccharides. According to the treatment conditions, HHP can change the physicochemical and structural characteristics of these biopolymers, and these changes can be either positive or negative depending on the nature of the macromolecule. For example, it has been reported that HHP produces protein denaturation, inactivation/activation of enzymes, gelatinization, and modification of the structure of polysaccharides (Michel and Autio 2001). These changes improve the biological or technological properties of proteins and polysaccharides (Amsasekar et al. 2022). The effect of HHP on these macromolecules is relevant because they are the major nutritional components in food, and they can influence texture and functional properties of the food system (Guo et al. 2020).

Proteins are complex molecules composed of 20 different amino acids linked between them substituted amide bonds (Tahergorabi and Hosseini 2017). Protein structure is divided into primary, secondary, tertiary, and quaternary structures. The primary structure refers to the amino acid sequence in which aminoamides are linked by covalent bonds. The secondary structure is stabilized by intra- or intermolecular hydrogen bonds; the tertiary structure is related to folding into a three-dimensional configuration by non-covalent interactions between amino acids; and finally, the quaternary structure refers to the spatial arrangement of subunits by non-covalent bonds (Hendrickx et al.1998). The effects of HHP of proteins structures depends on treatment conditions (Farjami et al. 2021) (Fig. 3). At pressure lower to 300 MPa a reversible denaturation can occur because of the effects on quaternary and tertiary structure, while at pressure higher than 300 MPa an irreversible denaturation is presented because of the effects on secondary structure. Denaturation produces an effect on protein functionality, it has been demonstrated that HHP improved protein digestibility and bioactive peptide yield and profile, enhancing the generation of antioxidant peptides (Perreault et al. 2017).

Some proteins have catalytic properties, thus, the effect on enzymes is also presented in this section. The effect of the HHP on enzymes depends on the enzyme type, as well as on the pressure level applied. Enzymes such as polyphenol oxidase (PPO), peroxidase (PO), pectin methylesterase (PME), and lipoxygenase (LOX) may cause undesirable changes on colour, texture, and flavour of foods (Hu et al. 2013), thus, they inactivation enhance food shelf-life (Jadhav et al. 2021). These enzymes can be inactivated at pressures from 400-800 MPa, however, pressure resistant enzymes like PPO and PO in some food matrices could require higher pressures of 600-900 MPa or even the combination with temperature (Weemaes et al. 1998; Hu et al. 2013). On the other side, some enzyme activities are desired, i.e., recently it was reported that the enzymatic activity of beta-D-glucosidase, the enzyme responsible for glucovanillin hydrolysis to form vanillin in Vanilla planifolia beans during the curing process, was incremented during the HHP-assisted curing (Buitimea-Cantúa et al. 2022). It has been reported that the activation of enzymes occurs at pressures lower to 200 MPa attributed to reversible protein unfolding. The mechanism for enzyme activation is still unclear, however, it has been speculated that HHP, in addition, to exposing the catalytic sites, also produces tissue decompartmentalization and induces the interactions between substrates and enzymes accelerating the enzymatic reactions (Morild 1981; Escobedo-Avellaneda et al. 2011; Escobedo-Avellaneda et al. 2017; Suwal et al. 2019; Buitimea-Cantúa et al. 2022). Enzyme inactivation occurs at pressures higher than 200 MPa, attributed to the irreversible unfolding, which produces a loss of the native three-dimensional configuration (Chakraborty et al., 2014; Augusto et al. 201). It is important to note, that enzymes in foods matrix present great variations either in the same food

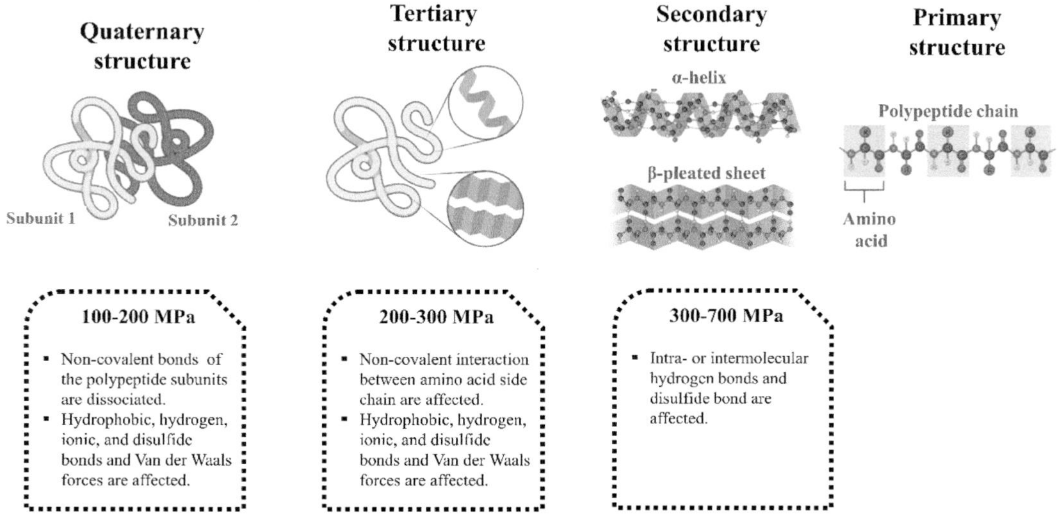

Figure 3. Effect of the High Hydrostatic Pressure on Protein Structure. Figure Made in BioRender (https://app.biorender.com).

matrix, as well as different resistance to the HHP, making it impossible to generalize and predict their behaviours after HHP processing (Buckow et al. 2009), requiring studies for each type of matrix and processing conditions. However, HHP can modify the functional properties of proteins increasing their applications in the food industry (Cao et al. 2017).

Other important food biopolymers are polysaccharides, which interact with protein and directly or indirectly influence the characteristics of food (Parmar et al. 2021). The effects of HHP on starch have been highly reported. HHP induces starch gelatinization. Gelatinization of starch has relevant applications in the food industry. It has been reported that gelatinization increases with increasing pressure (Yang et al. 2016). This effect has been attributed to the modifications of non-covalent intermolecular interactions which produces morphological and structural changes in starch (Balny 2001; Zang et al. 2022). Although the exact mechanism is still unclear, it has been attributed to the formation of intra- and inter-molecular associations, where hemiacetal oxygen, hydroxyl, or methyl groups of the sugar residues of the polysaccharides contribute to hydrogen bonding or van der Waals forces of attraction (Tako et al. 2014). Thus, it has been suggested that the transition of starch crystalline structures could occur during HHP (Katopo et al. 2002). HHP does not accelerate chemical reactions in polysaccharides, but rather causes physical changes in molecules. Among other things, molecules' movements are limited while hydrogen bonds are broken and the molecules are packed together toward filling the gaps between them, resulting in the production of hydrogels with different rheological and mechanical properties compared to the ones obtained by the conventional method (Yamamoto 2017).

5. Effects of HHP on Sensory Attributes

The sensory quality of foods is crucial for consumer acceptance (van Eck and Stieger 2020). Buyers are looking for increasing "naturalness" in the food they consume, such a trait is derived from farming practices all the way to the final product processing (Battacchi et al. 2020). Organoleptic properties, such as taste, smell, and appearance, must be preserved during food processing. Unlike pasteurization and other more heat-intensive processing techniques, HPP does not affect covalent bonds, rather HHP affects non-covalent bonds, which are associated with negative Va values. This allows a milder alteration of the food structure and hence its sensorial properties since small covalent molecules associated with a color and flavor are not directly affected (Oey et al. 2008). However organoleptic properties are interrelated with enzymes, pH, and different chemical reactions that may occur in specific cell types (Oey et al. 2008). Because of the effect that HPP has on the noncovalent bonds, enzyme activity, lipid stability, and even cell membrane integrity may be affected depending on the processing conditions (Gómez-Maqueo et al. 2020; Tribst et al. 2016; Medina-Meza et al. 2014).

Volatile compounds are responsible for food flavor (Zang et al. 2013; Garicano et al. 2020). Along with visual appearance, the aroma of a food product is the first quality evaluation parameter that consumers have. Because of this, special care is procured to preserve the aromatic qualities of food. Volatile organic compounds (VOCs) present in food, such as alcohols, aldehydes, ketones, esters, lactones, and terpenes, are part of the aroma profile of foods (Lomelí-Martín et al. 2021). As VOCs are in the norm small covalently bound molecules, it is well accepted that HPP up to 1,500 MPa at ambient temperature does not affect the molecular structure (Martinez-Monteagudo et al. 2014). Nevertheless, there is cumulative evidence that even non-thermal processing techniques like HHP could produce a variation of the volatile profile detected through headspace GC in different food samples (Xia et al. 2020). Analysis of VOCs in mango juice showed reductions of around 30% of major mango volatile compounds (β-myrcene, butanoic acid ethyl ester, (E, Z)-2,6-nonadienal, and o-cymene) after HPP treatment, but the overall aroma profile was not significantly altered (Pan et al. 2021). Red plum purée samples treated at 400 and 600 MPa did not exhibit significant variation of VOCs with exception of some ester compounds (González-Cebrino et al. 2016). A comparative quality study between raw, thermally, and HPP-treated milk did not find significant differences in volatile species' concentrations between samples (Liu et al. 2020). In general terms, a specific mechanism and relation of the effect that HPP has over the VOCs concentration profile is yet to be determined. Although VOCs variation has been detected for HPP-treated samples, it usually does not represent an impact on sensorial properties, which implies that the technique is appropriate to preserve the food qualities desired in a commercial application (Liu et al. 2020; Trejo Araya et al. 2009; Zhang et al. 2021).

Another important factor regarding the sensorial desirability of food is its texture, which is intrinsically related to the overall structural composition of foods. Food texture analysis can be classified into three main groups of properties: mechanical (reaction to stress, e.g., chewing), geometrical (tactile perception of particles), and moisture (perception of water, oil, and fat through tactile means) (Meilgaard et al. 2016). Crystal and gel structures that constitute foods are usually networks of carbohydrates, proteins, or a mixture of both (Foegeding et al. 2015). For instance, as stated earlier, HPP can exert changes to protein structures because of the disruption of hydrophobic, ionic, and hydrogen bonds which in consequence will alter the secondary, tertiary, and quaternary structure of proteins. Depending on cell type these conformational changes could lead to alterations in enzyme activities, gelation properties of the protein, and cell membrane permeability. High protein and water-content foods are arguably the most susceptible because of the structural changes of proteins and the water-holding capacity of cells (Foegeding et al. 2015; Grossi et al. 2016; Teixeira et al. 2013; Zhang 2017). This is especially important for meat products that contain a high degree of myofibrillar protein. Studies have also been made on the effect that HPP has on dairy products, such as milk and cheese. Milk-produced organoleptic properties are affected by the presence of protein and lipid compounds, such as casein micelles and milk fat globules; special attention must be given to HPP parameters to prevent the denaturation of these structures. A study performed on Camembert cheese quality and safety revealed that may or may not affect the texture and general appearance of the product depending on the conditions used (Batty et al. 2019).

On the other hand, fruits and vegetables are less susceptible to changes through the mechanisms stated above. However, enzymatic, and non-enzymatic biochemical reactions are facilitated via the decompartmentalization of substrates and enzymes. Pectin Methylesterase (PME) is an enzyme associated with several vegetable cell-wall structural functions (Pelloux et al. 2007). HPP treatments within a 200–600 MPa range have exhibited PME inactivation in relation to different experiment temperatures (Torres et al. 2016; Boulekou et al. 2010).

Food's visual appearance is a determinant factor for consumer acceptance since it is the first impression of buyers towards quality, ripening, and other desirable traits. Plant-based products owe their color to different biomolecules of which chlorophylls, carotenoids, and anthocyanins are the main ones (Clydesdale et al. 2009). Although these molecules are not sensible to HPP by themselves, special attention must be given to parameters above 200 MPa that could induce membrane damage and facilitate enzyme-catalyzed reactions that alter color, such as browning by the baroresistant PPO (Queiroz et al. 2008; Keenan et al. 2010). The effect on color variation between thermal and HPP was evaluated in blueberry purée, L*, a*, and b* (CIE color space) values did not differ from control values at 200 and 400 MPa, which was not the case for heat-treated samples. At higher pressures of 600 and 800 MPa lower L* values were obtained for blueberry purée and asparagus juice, mainly attributed to the liberation of chlorophyll compounds. HPP in spinach purée retained green color for longer because of reduced chlorophyllase activity (Zhang et al. 2021; Chen et al. 2015).

Meat product color is mainly attributed to the myoglobin (MG) in, and hemoglobin content and the Fe oxidation state found in the heme group (Liu et al. 2021; Liu et al. 2021). Due to the protein nature of MG, it is susceptible to conformational changes because of HPP (Gupta et al. 2018). The exact denaturation mechanism of HPP on MG is yet to be determined, but there are several studies on its chromatic effects in meat. A comprehensive review on the topic has been made in which the loss of a* (redness) in different meat products becomes an issue above 200 MPa (Bak et al. 2017). Similarly, the denaturation of casein micelles and fat globules can alter the way light is diffracted by dairy products, thus altering its color.

6. Effects of HHP on Bioactive Compounds

Bioactive compound, such as vitamins, flavonoids, tannins, carotenoids, tannins, terpenoids, anthocyanins, phenolics, mono and poly unsaturated fatty acids (Biesalski et al. 2009; Mahadevan and Karwe 2016); exhibits a beneficial effect on human health, and they could be present in fruits, vegetables, grains, herbal, and meat products. HHP have shown higher stability in bioactive compounds than thermal treatments, this could be attributed to HPP low effect in covalent bonds (Rastogi et al. 2007). In some cases HHP technology have shown a higher extraction of bioactive compounds, compared to non-treated samples, along with the beneficial effects of preservation (Navarro-Baez et al. 2022; Gómez-Maqueo et al. 2020). Table 2 summarizes some reports on the effect of high hydrostatic pressure on bioactive compounds.

The different phenomena that can occur depend on the level of pressure. When the pressure is less than 100 MPa, stress reactions can be induced leading to changes in the concentration of secondary metabolites. While when pressure is higher than 100 MPa usually the mechanism for biosynthesis is not activated, due to the high stress the cells are inactivated, above this pressure, the release of bioactive compounds is promoted extraction from cellular compartments. High pressure damage the cell membrane, which increases the mass transfer, improving permeability and facilitating the

Table 2. Effect of High Hydrostatic Pressure on Some Bioactive Compounds.

Bioactive compound	Treatment condition	Food	Results	Reference
Vitamins	500 MPa 5 min 35°C	Orange juice	Higher retention of Vitamin C compared to pasteurization	Polydera et al. 2003
	100–600 MPa 1–20 min 30 °C	Mango pulp	The HHP treatments retained a maximum of 90% of its original ascorbic acid at 600 MPa	Kaushik et al. 2014
	400–600 MPa 1.5 and 3 min 20°C	Strawberry puree and juice	Vitamin C content is not affected.	Leyva-Daniel et al. 2017
	600 MPa 5 min	Egg yolk	Folate content was unaffected after HHP	Naderi et al. 2017
	600 MPa 2–3 min < 37°C	Honey	HHP treatments did not induce significant changes on vitamin C content	Aaby et al. 2018
Anthocyanins	400, 500, 600 MPa 15 min 10–30°C	Blackberry and strawberry purées	Preservation of the color, anthocyanins content and antioxidant activity of purées	Patras et al. 2009
	400–500 MPa 5–10 min 10–25°C	Red fruit derivatives	Enhanced extraction of anthocyanins	Ferrari et al. 2011
Carotenoids	100, 300, and 500 MPa 20 min 20°C	Carrot and spinach	The stability of carotenoids in HHP-treated products was greater than that in thermally treated samples	Lae-Seunget et al. 2013
	450-650 MPa 3 min 20°C	Orange, papaya, melon and milk smoothies	Increase of carotenoids' (lycopene, β- and α-carotene) content	Andrés et al. 2016
	550 MPa 6 min	Carrot juice	Increased isomerization and oxidation of carotenoids	Zhang et al. 2016
	200–400 MPa 5–15 min	Bee-pollen paste Bee-pollen-based beverage	Pressure and time had a positive effect by increasing the extraction of total carotenoids.	Zuluaga et al. 2016

release of secondary metabolites (Gómez-Maqueo et al. 2020; Prasad et al. 2009; Huang et al. 2013; Aganovic et al. 2021). The increments of bioactive compounds such as phenolics and vitamins are associated with two mechanisms: 1) the release from extracellular compartments because of stress generated from high pressure treatments and 2) enhance of biosynthesis of bioactive compounds related to immediate stress response of plants by generating signaling molecules to regulate gene expression, which is associated to higher enzymatic activity (Navarro-Baez et al. 2022; Gómez-Maqueo et al. 2019; Jacobo-Velázquez et al. 2017).

7. HHP Equipment and Packaging

The first high pressure food processing equipment to treat foods was used to pressurize milk (Hite 1899). It was a vertical small HHP machine and a product container made of a collapsible tin tube where the milk was packed (Elamim et al. 2015). It was after almost 100 years when the firts commercial high-pressure equipment came out in Japan in 1990 and this technology started to grow (Yaldagard et al. 2008). Since then, HHP technology has been established gradually in other countries. Now, the industrial application of HHP has increased and in 2019 there were about 500 industrial units installed worldwide representing a volume for production of about 125000 L (Hiperbaric). Although, many HHP units have been installed in America and Mexico (67%), Europe (18%), Asia (8%), Latin America (3%), Oceania (3%), and Africa (1%), the high cost of the equipments is the reason for the late commercial spread of the technology (Ghafoor et al. 2020). The design of HHP machines has progressed reaching higher pressures and having larger vessel volumes. In addition, they include automated baskets for product loading, unloading in a horizontal configuration. This horizontal configuration increases productivity because of the faster loading and unloading, reduces the risk of confusion between processed and non-processed products, and decreases equipment height facilitating the installation process and reducing costs. Until 2019, HHP processing required that the product was processed on their final package (in-pack), but now the new in-bulk equipment does not require that the product be treated in their final package which has increased production capacity with a consequent improvement in industry acceptance (Tonello-Samson et al. 2020). The maximum pressure, pressurization rate, temperature, chamber volume, and temperature control depends on equipment design. Regarding packaging materials, they should be water-resistant and flexible to support about 15% reduction in the product volume, withstand this compression, and retain their original geometry upon decompression. LDPE and HDPE pouches; PET and HDPE bottles among other options can be used for HHP processing (Hiperbaric).

References

Abe, F. 2007. Exploration of the effects of high hydrostatic pressure on microbial growth, physiology, and survival: perspectives from piezophysiology. Biosci. Biotechnol. Biochem. 71: 2347–2357

Alpas, H., Kalchayanand, N., Bozoglu, F., Sikes, A., Dunne, C.P. and Ray, B. 1999. Variation in resistance to hydrostatic pressure among strains of food-borne pathogens. Appl. Environ. Microbiol. 65: 4248–4251.

Alpas, H., Lee, J., Bozoglu, F. and Kaletunc, G. 2003. Evaluation of high hydrostatic pressure sensitivity of Staphylococcus aureus and Escherichia coli O157: H7 by differential scanning calorimetry. Int. J. Food Microbiol. 87: 229–237.

Amsasekar, A., Mor, R.S., Kishore, A., Singh, A. and Sid, S. 2022. Impact of high pressure processing on microbiological, nutritional and sensory properties of food: a review. Nutr Food Sci.

Augusto, P.E., Tribst, A.A. and Cristianini, M. 2018. High hydrostatic pressure and high-pressure homogenization processing of fruit juices. pp. 393–421. *In*: Fruit juices. Academic Press.

Bak, K.H., Bolumar, T., Karlsson, A.H., Lindahl, G. and Orlien, V. 2019. Effect of high pressure treatment on the color of fresh and processed meats: A review. Crit. Rev. Food Sci. Nutr. 59: 228–252.

Balci, A.T. and Wilbey, R.A. 1999. High pressure processing of milk-the first 100 years in the development of a new technology. Int. J. Dairy Technol. 52: 149–155.

Balny, C. 2002. High pressure and protein oligomeric dissociation. Intl. J. High Press Res. 2: 737–741.

Baranda, A.B. and Montes, P. 2020. HPP for improving preservation of vitamin and antioxidant contents in vegetable matrices. pp. 15–70. *In*: Present and Future of High Pressure Processing. Elsevier.

Bartlett, D.H. 2002. Pressure effects on *in vivo* microbial processes. Biochim Biophys Acta (BBA)-Protein Structure and Molecular Enzymology 1595: 367–381.

Battacchi, D., Verkerk, R., Pellegrini, N., Fogliano, V. and Steenbekkers, B. 2020. The state of the art of food ingredients' naturalness evaluation: A review of proposed approaches and their relation with consumer trends. Trends Food Sci. Technol. 106: 434–444.

Batty, D., Meunier-Goddik, L. and Waite-Cusic, J.G. 2019. Camembert-type cheese quality and safety implications in relation to the timing of high-pressure processing during aging. Int. J. Dairy Sci. 102: 8721–8733.

Biesalski, H.K., Dragsted, L.O., Elmadfa, I., Grossklaus, R., Müller, M., Schrenk, D. et al. 2009. Bioactive compounds: Definition and assessment of activity. Nutr. 25: 1202–1205.

Boulekou, S.S., Katsaros, G.J. and Taoukis, P.S. 2010. Inactivation kinetics of peach pulp pectin methylesterase as a function of high hydrostatic pressure and temperature process conditions. Food Bioproc. Tech 3: 699–706.

Buckow, R., Weiss, U. and Knorr, D. 2009. Inactivation kinetics of apple polyphenol oxidase in different pressure–temperature domains. Innov. Food Sci. Emerg. Technol. 10: 441–448.

Buitimea-Cantúa, G.V., Welti-Chanes, J. and Escobedo-Avellaneda, Z. 2022. Metabolite transformation and β-D-glucosidase activity during the high hydrostatic pressure assisted curing process of vanilla beans (Vanilla planifolia) to improve phenolic compounds formation. Food Chem. 132497.

Cao, B., Fang, L., Liu, C., Min, W. and Liu, J. 2018. Effects of high hydrostatic pressure on the functional and rheological properties of the protein fraction extracted from pine nuts. Food Sci. Technol. Int. 24: 53–66.

Cardello, A.V. 2003. Consumer concerns and expectations about novel food processing technologies: Effects on product liking. Appetite 40: 217–233.

Cardello, A.V., Schutz, H.G. and Lesher, L.L. 2007. Consumer perceptions of foods processed by innovative and emerging technologies: A conjoint analytic study. Innovative Food and Emerging Technologies 8: 73–83.

Chakraborty, S., Kaushik, N., Rao, P.S. and Mishra, H.N. 2014. High-pressure inactivation of enzymes: a review on its recent applications on fruit purees and juices. Compr. Rev. Food Sci. Food Saf. 13: 578–596.

Chen, X., Qin, W., Ma, L., Xu, F., Jin, P. and Zheng, Y. 2015. Effect of high pressure processing and thermal treatment on physicochemical parameters, antioxidant activity and volatile compounds of green asparagus juice. LWT-Food Sci. Technol. 62: 927–933.

Clydesdale, F.M. 1993. Color as a factor in food choice. Crit. Rev. Food Sci. Nutr. 33: 83–101

Cook, D.W. 2003. Sensitivity of Vibrio species in phosphate-buffered saline and in oysters to high hydrostatic pressure processing. J. Food Prot. 66: 2277–2282.

Deng, K., Wu, X., Fuentes, C. Su, Y., Welti-Chanes, J. Paredes-Sabja, D. and Torres, J.A. 2015. Analysis of Vibrio vulnificus Infection Risk When Consuming Depurated Raw Oysters. J. Food Protec. 8: 1113–1118

Dogan, C. and Erkmen, O. 2004. High pressure inactivation kinetics of Listeria monocytogenes inactivation in broth, milk, and peach and orange juices. J. Food Eng. 62: 47–52.

Dubins, D.N., Lee, A., Macgregor, R.B. and Chalikian, T.V. 2001. On the stability of double stranded nucleic acids. J. Am. Chem. Soc. 123: 9254–9259.

Elamin, W.M., Endan, J.B., Yosuf, Y.A., Shamsudin, R. and Ahmedov, A. 2015. High Pressure Processing Technology and Equipment Evolution: A Review. J. Eng. Sci. Technol. 8: 75–83.

Escobedo-Avellaneda, Z., Gutiérrez-Uribe, J., Valdez-Fragoso, A., Torres, J.A. and Welti-Chanes, J. 2015. High hydrostatic pressure combined with mild temperature for the preservation of comminuted orange: effects on functional compounds and antioxidant activity. Food and Bioprocess Technology 8(5): 1032–1044.

Escobedo-Avellaneda, Z., Moure, M.P., Chotyakul, N., Torres, J.A., Welti-Chanes, J. and Lamela, C.P. 2011. Benefits and limitations of food processing by high-pressure technologies: effects on functional compounds and abiotic contaminants Beneficios y limitaciones del procesamiento de alimentos por tecnologías de alta presión: efectos en componentes funcionales y contaminantes abióticos. CyTa- J. Food 9: 351–364.

Escobedo-Avellaneda, Z., Pérez-Simón, I., Lavilla-Martín, M., Baranda-González, A. and Welti-Chanes, J. 2017. Enzymatic and phytochemical stabilization of orange–strawberry–banana beverages by high hydrostatic pressure and mild heat. Food Sci. Technol. Int. 23: 185–193.

Evans, G. and Cox, D.N. 2006. Australian consumers' antecedents of attitudes towards foods produced by novel technologies. British Food Journal 108: 916–930.

Evrendilek, G.A. 2012. Microbial stress response to high pressure processing. pp. 95e108. *In*: Stress response of foodborne microorganisms. Nova Science Publishers, Inc.

Farjami, T., Babaei, J., Nau, F., Dupont, D. and Madadlou, A. 2021. Effects of thermal, non-thermal and emulsification processes on the gastrointestinal digestibility of egg white proteins. Trends Food Sci. Technol. 107: 45–56.

Foegeding, E.A., Vinyard, C.J., Essick, G., Guest, S. and Campbell, C. 2015. Transforming structural breakdown into sensory perception of texture. J. Texture. Stud. 46: 152–170.

Furukawa, S., Shimoda, M. and Hayakawa, I. 2003. Mechanism of the inactivation of bacterial spores by reciprocal pressurization treatment. J. Appl. Microbiol. 94: 836e841.

Galanakis, C.M. 2021. Functionality of food components and emerging technologies. Foods 10: 102–128.

Garcia-Graells, C., Masschalck, B. and Michiels, C.W. 1999. Inactivation of Escherichia coli in milk by high hydrostatic pressure treatment in combination with antimicrobial peptides. J. Food Prot. 62: 1248–1254.

Ghafoor, K., Gavahian, M., Marszalekc, K., Barba, F.J., Xia, Q. and Denoya, G.I. 2020. An overview of the potential applications based on HPP mechanism. pp. 3–14. *In*: Present and Future of High Pressure Processing: A Tool for

Developing Innovative, Sustainable, Safe and Healthy Foods. Francisco J. Barba, Carile Tonello-Samson, Eduardo Puértolas and María Lavilla (eds.). Elsevier.

Gómez-Maqueo, A., Welti-Chanes, J. and Cano, M.P. 2020. Release mechanisms of bioactive compounds in fruits submitted to high hydrostatic pressure: A dynamic microstructural analysis based on prickly pear cells. Food Res. Int. 130: 108909.

González-Cebrino, F., García-Parra, J. and Ramírez, R. 2016. Aroma profile of a red plum purée processed by high hydrostatic pressure and analysed by SPME-GC/MS. Innov. Food. Sci. Emerg. Technol. 33: 108–114.

Grossi, A., Olsen, K., Bolumar, T., Rinnan, Å., Øgendal, L.H. and Orlien, V. 2016. The effect of high pressure on the functional properties of pork myofibrillar proteins. Food Chem. 196: 1005–1015.

Guo, X., Chen, M., Li, Y., Dai, T., Shuai, X., Chen, J. and Liu, C. 2020. Modification of food macromolecules using dynamic high pressure microfluidization: A review. Trends. Food Sci. Technol. 100: 223–234.

Gupta, J., Bower, C.G., Cavender, G.A. and Sullivan, G.A. 2018. Effectiveness of different myoglobin states to minimize high pressure induced discoloration in raw ground beef. LWT-Food Sci. Tech. 93: 32–35.

Hendrickx, M., Ludikhuyze, L., Van den Broeck, I. and Weemaes, C. 1998. Effects of high pressure on enzymes related to food quality. Trends Food Sci. Technol. 9: 197–203.

Hiremath, N. and Ramaswamy, H.S. 2012. High-pressure destruction kinetics of spoilage and pathogenic microorganisms in mango juice. J Food Process Preserv 36: 113–125.

Hite, B.H. 1899. The effect of pressure in the preservation of milk, Bull. West Virginia Univ. Agric. Exper. Stn., 58: 15–35.

Hu, W., Zhou, L., Xu, Z., Zhang, Y. and Liao, X. 2013. Enzyme inactivation in food processing using high pressure carbon dioxide technology. Crit Rev Food Sci Nutr. 53: 145–161.

Huang, H. Hsiao-Wen Huang, Hsiang-Mei Lung, Binghuei Barry Yang and Chung-Yi Wang. 2014. Responses of microorganisms to high hydrostatic pressure processing. Food Cont. 40: 250–259.

Huang, H.W., Hsu, C.P., Yang, B.B. and Wang, C.Y. 2013. Advances in the extraction of natural ingredients by high pressure extraction technology. Trends. Food Sci. Technol. 33: 54–62.

Jacobo-Velázquez, D.A., del Rosario Cuéllar-Villarreal, M., Welti-Chanes, J., Cisneros-Zevallos, L., Ramos-Parra, P.A. and Hernández-Brenes, C. 2017. Nonthermal processing technologies as elicitors to induce the biosynthesis and accumulation of nutraceuticals in plant foods. Trends. Food Sci. Technol. 60: 80–87.

Jadhav, H.B., Annapure, U.S. and Deshmukh, R.R. 2021. Non-thermal technologies for food processing. Front. Nutr. 8.

Jaenicke, R. 1991. Protein stability and molecular adaptation to extreme conditions. EJB Reviews 1991: 291–304.

Jordan, S.L., Pascual, C., Bracey, E. and Mackey, B.M. 2001. Inactivation and injury of pressure-resistant strains of E. coli 0157 and Listeria monocytogens in fruit juices. J Appl Microbiol 91: 1–14.

Kaletun, G., Lee, J., Alpas, H. and Bozoglu, F. 2004. Evaluation of structural changes induced by high hydrostatic pressure in Leuconostoc mesenteroides. Appl. Environ. Microbiol. 70: 1116–1122.

Katopo, H., Song, Y. and Jane, J.L. 2002. Effect and mechanism of ultrahigh hydrostatic pressure on the structure and properties of starches. Carbohydr. Polym. 47: 233–244.

Keenan, D.F., Brunton, N.P., Gormley, T.R., Butler, F., Tiwari, B.K. and Patras, A. 2010. Effect of thermal and high hydrostatic pressure processing on antioxidant activity and colour of fruit smoothies. Innov. Food Sci. Emerg. Technol. 11: 551–556.

Lado, B.H. and Yousef, A.E. 2002. Alternative food-preservation technologies: efficacy and mechanisms. Microbes and Infect 4: 433–440.

Li, X., Zhang, Z.H., Qi, X., Li, L., Zhu, J., et al. 2021. Application of nonthermal processing technologies in extracting and modifying polysaccharides: A critical review. Compr. Rev. Food Sci. Food Saf. 20: 4367–4389.

Liu, G., Carøe, C., Qin, Z., Munk, D. M., Crafack, M., Petersen, M.A. et al. 2020. Comparative study on quality of whole milk processed by high hydrostatic pressure or thermal pasteurization treatment. LWT-Food Sci. Tech. 127: 109370.

Lomelí-Martín, A., Martínez, L.M., Welti-Chanes, J. and Escobedo-Avellaneda, Z. 2021. Induced changes in aroma compounds of foods treated with high hydrostatic pressure: A review. Foods 10(4): 878.

Lovegrove, A., Edwards, C.H., De Noni, I., Patel, H., El, S.N., Grassby, T. and Shewry, P.R. 2017. Role of polysaccharides in food, digestion and health. Crit. Rev. Food Sci. Nutr. 57: 237–253.

Mahadevan, S. and Karwe, M.V. 2016. Effect of high-pressure processing on bioactive compounds. pp. 479–507. *In*: High Pressure Processing of Food. Springer, New York, NY.

Martinez-Monteagudo, S.I. and Saldana, M.D. 2014. Chemical reactions in food systems at high hydrostatic pressure. Food Eng. Rev. 6: 105–127.

Medina-Meza, I.G., Barnaba, C. and Barbosa-Cánovas, G.V. 2014. Effects of high pressure processing on lipid oxidation: A review. Innov Food Sci Emerg Technol 22: 1–10.

Meilgaard, M. C., Carr, B. T. and Civille, G. V. 1999. Sensory evaluation techniques. CRC press.

Michel, M. and Autio, K. 2001. Effects of high pressure on protein-and polysaccharide-based structures. pp. 189–214. *In*: Ultra high pressure treatments of foods. Springer, Boston, MA.

Mohamed, H.M.H., Diono, B.H.S. and Yousef, E.Y. 2012. Structural changes in Listeria monocytogenes treated with gamma radiation, pulsed electric field and ultra-high pressure. J. Food Saf. 32: 66–73.

Morild, E. 1981. The theory of pressure effects on enzymes. Adv. Protein Chem. 34: 93–166.

Mújica-Paz, H., Valdez-Fragoso, A., Samson, C.T., Welti-Chanes, J. and Torres, J.A. 2011. Highpressure processing technologies for the pasteurization and sterilization of foods. Food and Bioproc. Tech. 4: 969.

Navarro-Baez, J.E., Martínez, L.M., Welti-Chanes, J., Buitimea-Cantúa, G.V. and Escobedo-Avellaneda, Z. 2022. High Hydrostatic Pressure to Increase the Biosynthesis and Extraction of Phenolic Compounds in Food: A Review. Molecules 27: 1502.

Oey, I., Lille, M., Van Loey, A. and Hendrickx, M. 2008a. Effect of high-pressure processing on colour, texture and flavour of fruit-and vegetable-based food products: a review. Trends Food Sci. Technol. 19: 320–328.

Oey, I., Van der Plancken, I., Van Loey, A. and Hendrickx, M. 2008b. Does high pressure processing influence nutritional aspects of plant based food systems?. Trends. Food Sci. Technol. 19: 300–308.

Osumi, M., Yamada, N., Sato, M., Kobori, H., Shimada, S. and Hayashi, R. 1992. Pressure effects on yeast cell ultrastructure: changes in the ultrastructure and cytoskeleton of the dimorphic yeast, Candida tropicalis. p. 9. *In*: Balny et al. (eds.). High Pressure and Biotechnology. John Libbey Eurotext, Montrouge.

Pan, X., Wu, J., Zhang, W., Liu, J., Yang, X., Liao, X. et al. 2021. Effects of sugar matrices on the release of key aroma compounds in fresh and high hydrostatic pressure processed Tainong mango juices. Food Chem. 338: 128117.

Parmar, P., Goraya, R.K. and Pathania, S. 2021. Macromolecules Interactions in Food Formulations. Food Formulation: Novel Ingredients and Processing Techniques 49–70.

Patterson, M.F. 2005. Microbiology of pressure-treated foods. J. Appl. Microbiol. 98: 1400–1409.

Patterson, M.F., Quinn, M., Simpson, R. et al. 1995. Sensitivity of Vegetative Pathogens to High Hydrostatic Pressure Treatment in Phosphate-Buffered Saline and Foods. J. Food Prot. 61: 524–529.

Pelloux, J., Rusterucci, C. and Mellerowicz, E.J. 2007. New insights into pectin methylesterase structure and function. Trends. Plant Sci. 12: 267–277.

Pilavtepe-Çelik, M., Balaban, M.O., Alpas, H. and Yousef, A.E. 2008. Image analysis based quantification of bacterial volume change with high hydrostatic pressure. J. Food Sci. 73: 423–429.

Prasad, K. N., Yang, B., Zhao, M., Wei, X., Jiang, Y. and Chen, F. 2009. High pressure extraction of corilagin from longan (Dimocarpus longan Lour.) fruit pericarp. Sep. Purif. Technol. 70: 41–45.

Queiroz, C., Mendes Lopes, M.L., Fialho, E. and Valente-Mesquita, V.L. 2008. Polyphenol oxidase: characteristics and mechanisms of browning control. Food Rev. Int. 24: 361–375.

Quiroz-González, B., Rodríguez-Martínez, V., García-Mateos M., de R., Torres, J.A. and Welti-Chanes, J. 2018. High hydrostatic pressure inactivation and recovery study of Listeria innocua and Saccharomyces cerevisiae in pitaya (Stenocereus pruinosus) juice. Innov. Food Sci. Emerg. Technol. 50: 169–173.

Rastogi, N.K., Raghavarao, K.S.M.S., Balasubramaniam, V.M., Niranjan, K. and Knorr, D. 2007. Opportunities and challenges in high pressure processing of foods. Crit. Rev. Food Sci. Nutr. 47: 69–112.

Silva, J.L. and Weber, G. 1993. Pressure stability of proteins. Annu. Rev. Phys. Chem. 44: 89–113.

Simpson, R.K. and Gilmour, A. 1997. The effect of high hydrostatic pressure on the activity of intracellular enzymes of Listeria monocytogenes. Lett. Appl. Microbiol. 25: 48–53.

Soto-Caballero, M.C., Cano-Monge, E.E., Cano-Monge, S.M., Welti-Chanes, J. and Escobedo-Avellaneda, Z. 2021. Effect of high hydrostatic pressures on microorganisms, total phenolic content and enzyme activity of mamey (Pouteria sapota) nectar. Journal of Food Science and Technology 1–6.

Srinivas, M.S., Madhu, B., Srinivas, G. and Jain, S.K. 2018. High pressure processing of foods: A Review. The Andhra Agriculture Journal 65: 467–476.

Suwal, S., Perreault, V., Marciniak, A., Tamigneaux, É., Deslandes, É., Bazinet, L. et al. 2019. Effects of high hydrostatic pressure and polysaccharidases on the extraction of antioxidant compounds from red macroalgae, Palmaria palmata and Solieria chordalis. J. Food Eng. 252: 53–59.

Tahergorabi, R. and Hosseini, S.V. 2017. Tahergorabi, R. and Hosseini, S.V. 2017. Proteins, peptides and amino acids. pp. 15–38. *In*: Nutraceutical and functional food components.

Tako, M., Tamaki, Y., Teruya, T. and Takeda, Y. 2014. The principles of starch gelatinization and retrogradation. Food Nutr. Sci. 5: 280–291

Tao, Y., Sun, D. W., Hogan, E. and Kelly, A. L. 2014. High-pressure processing of foods: An overview. Emerging Technologies for Food Processing, 3–24.

Teixeira, B., Fidalgo, L., Mendes, R., Costa, G., Cordeiro, C., Marques, A. et al. 2013. Changes of enzymes activity and protein profiles caused by high-pressure processing in sea bass (Dicentrarchus labrax) fillets. J. Agric. Food Chem. 61: 2851–2860.

Tholozan, J. L., Ritz, M., Jugiau, F., Federighi, M. and Tissier, J.P. 2000. Physiological effects of high hydrostatic pressure treatments on Listeria monocytogenes and Salmonella typhimurium. J. Appl. Microbiol. 88: 202–212.

Tonello-Samson, C., Queirós, R.P. and González-Angulo, M. 2020. Advances in high-pressure processing in-pack and in-bulk commercial equipment. pp. 297–316. *In*: Present and Future of High Pressure Processing.

Torres, E.F., González-M, G., Klotz, B. and Rodrigo, D. 2016. Effects of high hydrostatic pressure and temperature increase on Escherichia coli spp. and pectin methyl esterase inactivation in orange juice. Food Sci. Tech. Int. 22: 173–180.

Torres, J.A. and Velazquez, G. 2005. Commercial opportunities and research challenges in the high pressure processing of foods. J. Food Eng. 67: 95–112.

Tribst, A.A.L., Júnior, B.R.D.C.L., de Oliveira, M.M. and Cristianini, M. 2016. High pressure processing of cocoyam, Peruvian carrot and sweet potato: Effect on oxidative enzymes and impact in the tuber color. Innov. Food Sci. Emerg. Technol. 34: 302–309.

United States Department of Agriculture. 2012. High pressure processing (HPP) and inspection program personnel (IPP) verification responsibilities. Food Safety and Inspection Service, 6120.2.

van Eck, A. and Stieger, M. 2020. Oral processing behavior, sensory perception and intake of composite foods. Trends Food Sci. Technol. 106: 219–231.

Wang, C.Y., Huang, H.W., Hsu, C.P., Shyu, Y.T. and Yang, B.B. 2013. Inactivation and morphological damage of Vibrio parahaemolyticus treated with high hydrostatic pressure. Food Control. 32: 348–353.

Weemaes, C., Ludikhuyze, L., Van den Broeck, I. and Hendrickx, M. 1998. High pressure inactivation of polyphenoloxidases. J. Food Sci. 63: 873–877.

Xia, Q., Zheng, Y., Liu, Z., Cao, J., Chen, X., Liu, L. et al. 2020. Nonthermally driven volatilome evolution of food matrices: The case of high pressure processing. Trends. Food Sci. Technol. 106: 365–381.

Yaldagard, M., Mortazavi, S.A. and Tabatabaie, F. 2008. The principles of ultra high pressure technology and its application in food processing/preservation: A review of microbiological and quality aspects. African Journal of Biotechnology 7: 2739–2767.

Yamamoto, K. 2017. Food processing by high hydrostatic pressure.

Yang, B., Shi, Y., Xia, X., Xi, M., Wang, X., Ji, B. et al. 2012. Inactivation of foodborne pathogens in raw milk using high hydrostatic pressure. Food Control 28: 273–278.

Yang, Z., Gu, Q., Lam, E., Tian, F., Chaieb, S. and Hemar, Y. 2016. *In situ* study starch gelatinization under ultra-high hydrostatic pressure using synchrotron SAXS. Food Hydrocoll 56: 58–61.

Zang, M., Wang, L., Zhang, Z., Zhang, K., Li, D., Li, X. et al. 2020. Changes in flavour compound profiles of precooked pork after reheating (warmed-over flavour) using gas chromatography–olfactometry–mass spectrometry with chromatographic feature extraction. Int. J. Food Sci. Tech. 55: 978–987.

Zhang, S., Zheng, C., Zeng, Y., Zheng, Z., Yao, X., Zhao, Y. et al. 2021a. Mechanism of colour change of carambola puree by high pressure processing and its effect on flavour and physicochemical properties. Int. J. Food Sci. Tech. 56: 5853–5860.

Zhang, W., Shen, Y., Li, Z., Xie, X., Gong, E.S., Tian, J. et al. 2021b. Effects of high hydrostatic pressure and thermal processing on anthocyanin content, polyphenol oxidase and β-glucosidase activities, color, and antioxidant activities of blueberry (Vaccinium Spp.) puree. Food Chem. 342: 128564.

Zhang, Z., Yang, Y., Tang, X., Chen, Y. and You, Y. 2015. Chemical forces and water holding capacity study of heat-induced myofibrillar protein gel as affected by high pressure. Food Chem. 188: 111–118.

Zhang, Z., Yang, Y., Zhou, P., Zhang, X. and Wang, J. 2017. Effects of high pressure modification on conformation and gelation properties of myofibrillar protein. Food Chem. 217: 678–686.

15

Application of Microwave Heating in Food Processing

Current Trends, Challenges and Prospects

Mohammad Uzzal Hossain Joardder,[1,3,*] *Abdul Mojid Parvej,*[1]
Md Bakhtier Khalzi,[1] *Golam Kibria M. Hasanuzzaman*[2] and *Azharul Karim*[3]

1. Introduction

Heating is the basic treatment in food processing, which is used to increase shelf life. There are various types of heating options to process foods. The common disadvantages of conventional heating methods are higher energy and time consumption. These limitations of conventional heating can be overcome using advanced technologies, including microwave heating and infrared heating.

Microwave heating offers a range of benefits over traditional heating methods, particularly in terms of the energy economy. In a conventional thermal process, heat is transmitted slowly from the outer surface of the food to the inside through convection and conduction modes of heat transfer. Eventually, it also requires higher energy usage and longer processing time, whereas microwave heating generates heat inside the food within a short period by transmitting a microwave into it (Pozar 2009). Due to this internal volumetric heating nature, a relatively lower temperature difference between the inside and surface of food materials prevails (Decareau 1985; Witkiewicz and Nastaj 2010).

Although there are numerous articles available regarding MW heating in food processing, this chapter aims to capture the most important aspects of MW heating in a brief and simplified approach. Firstly, the underlying physics during microwave heating along with the dielectric properties of different food materials have been presented. Then, the application of MW heating in different types of food processing along with the quality aspects of processed foods have been discussed briefly. Following this, the challenges associated with MW heating have been pointed out. Finally, the challenges of MW heating and approaches to achieving uniform heating have been discussed.

[1] Department of Mechanical Engineering, Rajshahi University of Engineering & Technology, Rajshahi, Bangladesh.
[2] Department of Electrical & Electronic Engineering, Rajshahi University of Engineering & Technology, Rajshahi, Bangladesh
[3] School of Mech., Medical & Process Engineering, Queensland University of Technology, QLD, Australia.
* Corresponding author: muhjoardder@gmail.com

2. Underlying Physics of MW Heating

Microwave does not fall into the category of thermal radiation in the electromagnetic spectrum. It means that the MW does not carry heat and irradiate on the outer surface. Rather it generates heat while interacting with the materials having conducible dielectric properties. Microwaves are electromagnetic waves, having a frequency ranging from 300 MHz to 300 GHz; whereas, 2,450 MHz (having a wavelength of about 12 cm) is the most frequently used frequency in a typical MW oven (Metaxas and Meredith 1983).

As the term 'electromagnetic' implies, both electrical and magnetic field propagates simultaneously. In microwaves, electromagnetic waves are directed toward a dielectric medium to generate heat. The permanent dipoles in the dielectric materials including food reorient by the application of the electrical field are due to orientation polarization (Curtis 1999). For microwave applications, this reorientation allows dielectric materials to store electric energy causing volumetric heat generation (Balanis 1999). This volumetric nature of heating offers faster heating of food while taking lower energy.

2.1 MW Energy Distribution

The overall energy distribution within food during MW heating is affected by the heat generation because of microwave interaction, heat and mass transfer rate within food materials, as well as dielectric properties of food materials. MW generates heat in the food sample in the presence of water molecules. Maxwell's or Lambert's equations can be used to calculate generated heat during MW propagation (Kumar et al. 2018). In the case of modeling microwave distribution and heat generation, Maxwell's equation is more appropriate while taking higher computational time than Lambert's equation.

2.1.1 Maxwell's Equation

Electromagnetic field distribution in the computational domain can be visualized by applying Maxwell's equations which can be expressed in the frequency domain as (Chen et al. 2014):

$$\nabla \times \left(\frac{1}{\mu} \nabla \times \vec{E} \right) - \frac{(2\pi f)^2}{c} (\varepsilon' - i\varepsilon'')\vec{E} = 0 \tag{1}$$

Here, $\vec{E}$ is the strength of the electric field (V/m); μ is the electromagnetic permeability; *f* is the frequency of MW (Hz); *c* is the speed of light (m/s); ε' and ε'' are is the dielectric constant and dielectric loss factor, respectively.

Heat generation Q_m (W/m^3) because of MW can be calculated using the distribution of $\vec{E}$, which is written as follows (Kumar et al. 2018; Metaxas 1996; Wentworth 2006):

$$Q_m = \pi f \varepsilon_0 \varepsilon'' |E|^2 \tag{2}$$

2.1.2 Lambert's Law

Many research reported using Lambert's law for predicting MW energy distribution for food processing including drying (Kumar et al. 2016a). The energy absorbed by the food materials from the microwave can also be calculated using Lambert's law as expressed in Equation 3.

$$P(x) = P_0\, exp^{(-2ax)} \tag{3}$$

Here, P_o and α are the incident power at the surface and attenuation constant, respectively. The attenuation constant is expressed by the following equation:

$$\alpha = \frac{2\pi}{\lambda}\sqrt{\varepsilon'\left[\frac{\left(1+\left(\frac{\varepsilon''}{\varepsilon'}\right)^2\right)-1}{2}\right]} \quad (4)$$

Furthermore, the sample size being processed is dictated by the microwave penetration depth, which may be calculated using the formula below.

$$d_p = \frac{1}{2\alpha} \quad (5)$$

Penetration depth is inversely related to the frequency of MW. The penetration depth is defined as the distance from the product's surface to the point within the food at which the power density is dropped by 37% (Feynman et al. 2011).

Knowledge about penetration depth is important to make sure the complete heating of the food. Thinner food is preferable to heat thoroughly under the MW. The heating of a dish having different ingredients is sometimes problematic as the variation of their dielectric properties and dimensions results in more non-uniform heating.

2.2 Dielectric Properties of Different Foods

Prediction of MW energy distribution requires through modelling the dielectric property of food materials. The degree of interaction between the applied microwave field and the sample is determined by the material's dielectric properties. In other words, the amount of heat energy that can be attained from an MW depends on the dielectric properties (Chandrasekaran et al. 2013).

Dielectric permittivity (ε) is a vital property of food materials that expresses the susceptibility of food materials in absorption, reflection, and dissipation of MW. The dielectric nature of solid foods is dominated by their physical structure and water contents, whereas liquid foods are determined mainly by the concentration and mobility of solutes in water. Energy conversion during microwave heating is governed by the following equation (Chandrasekaran et al. 2013).

$$\varepsilon^* = \varepsilon' - j\varepsilon''$$

The dielectric constant (ε') and the dielectric loss factor (ε'') for a specific material determine the transformation of electromagnetic energy into heat when absorption takes place.

The dielectric constant (ε') is a measurement of a material's capability to hold energy after being polarized by an external field (such as microwaves) (Kingston and Jassie, 1986). The dielectric loss factor (ε''), on the other side, indicates the material's capability to disperse absorbed electromagnetic energy by transforming it to heat. The dissipation factor () can give the idea of the potential of transformation of electromagnetic energy into heat energy. The dissipation factor can be determined from the ratio of the dielectric loss factor (ε'') and the dielectric constant (ε') as expressed in Equation 6.

$$\tan\delta = \varepsilon'/\varepsilon' \quad (6)$$

The food composition, moisture content, and structure of foodstuffs all have an impact on dielectric properties.

2.2.1 Composition of Food

Food composition plays a critical role in determining dialectic properties. Among the compositions, water, salt, sugar, and oil are more conducive to MW heating.

The dielectric constant, dielectric loss factor, and dissipation factor of different materials are shown in Table 1.

Table 1. Dielectric Properties of Different Materials.

	Material	ε′	ε″	tan δ	T(°C)	f (MHz)	Reference
Meats, Fishes, and Seafood	Fish	2.1	0.5	0.24	90	10,000	Kent (1987)
	Chicken Breast	59	18.3	0.31	20	915	Basaram et al. (2010)
	Turkey Breast	56.3	18.0	0.32	–	2,450	Lyng et al. (2005)
	Lamb (leg)	49.4	15	0.30	72.6	2,450	
Eggs and egg products	Precooked egg white	53.0	34.6	0.65	80	915	Wang et al. (2009)
	Liquid egg white	50.5	33.3	0.66	80	915	
Dairy products	Butter (unsalted)	22.5	4.6	0.20	50	915	Guo et al. (2010)
	Milk	68.5	12.6	0.18	22	2,450	
Fresh fruits and vegetables	Bean (12% MC)	2.8	0.8	0.29	19	9,000	Torrealba-Meléndez et al. (2014)
	Broccoli	15.7	8.3	0.53	21.5	915	Kristiawan et al. (2011)
	Apple	57.0	8.0	0.14	23	915	Venkatesh and Raghavan (2004)
	Carrot	59.0	18.0	0.31	23	915	
	Cucumber	71.0	11.0	0.15	23	915	
	Grape	69.0	15.0	0.22	23	915	
	Lemon	73.0	15.0	0.21	23	915	
	Mango	64.0	13.0	0.20	23	915	
	Orange	73.0	14.0	0.19	23	915	
	Potato	62.0	22.0	0.35	23	915	
	Radish	68.0	20.0	0.29	23	915	
Liquid foods	Water	79.2	7.9	0.10	20	1,700	Nelson and Trabelsi (2012)
	Tomato Juice	81.3	33.6	0.41	35	2,450	Ghanem (2010)
	Strawberry Juice	92.6	23.6	0.25	35	2,450	
	Apple Juice	72.7	10.9	0.15	35	2,450	Zhu et al. (2012)
	Orange Juice	72.3	13.0	0.18	35	2,450	
	Grape Juice	70.4	13.5	0.19	35	2,450	

From the table, it is depicted that food materials of different origins posses a wide range of dielectric properties. For liquid food, sugar content is crucial in dielectric properties.

2.2.2 Moisture Content and Temperature

Moisture content is the most crucial factor affecting the dielectric properties of food materials. Water presented in either free or bound form can influence the dielectric properties of food materials (Tiras et al. 2019). Therefore, during the mass transfer process, the dielectric properties of food material significantly change.

Dielectric properties of food materials can be correlated with their moisture contents and processing temperature. Containing higher water makes the food more conducive to microwave heating. Therefore, fresh fruits and vegetables are heated up quickly, whereas dried food takes a longer time to get heated. In the literature, there are numerous empirical relations of dielectric characteristics of various materials. For example, moisture temperature dependant dielectric constant and dielectric loss components for potatoes are as follows (Kumar et al. 2016ab):

$$\varepsilon' = 64.5876 + 0.0056M - 0.2223T - 0.0046MT \tag{8}$$

$$\varepsilon'' = -8.2227 + 0.2360M + 0.2041T \tag{9}$$

Here, M is the moisture content and T is the temperature.

From the equation, it is clear that dielectric constant decreases with the increase in temperature, whereas the dielectric loss factor increases with the increase in temperature.

2.2.3 Porosity

The porosity of food materials reflects the presence of air in food materials (Joardder er al. 2015). The heating rate under MW power can be affected by the macro- and microstructure of food materials (Martín-Esparza et al. 2006). Porous materials show lower dielectric strength as it has a higher amount of trapped air. Air has the lower dielectric constant of 1 and the lowest dielectric loss factor of 0. As a result, the more porous the material gets during heating or drying, the more trapped air it contains, resulting in poor dielectric characteristics. In addition to this, the specific surface area of the food sample affects the dielectric properties (Jones and Or 2002).

Apart from the mentioned factors, there are other factors such as MW frequency, humidity, sample shape, and size that can affect the dielectric nature of food materials. In order to attain maximum benefit from MW heating, the proper knowledge of the dielectric properties of food material is vital.

3. MW Heating in Food Processing

The application of MW is dominant in domestic and commercial food processing. Microwave is commonly used in a wide range of heat-associated food processing application, such as thawing, blanching, cooking, tempering, drying, tempering, pasteurization, sterilization, and baking. From many perspectives, MW heating is more advantageous than conventional heating. Some applications of microwave heating in food processing are discussed briefly in the following sections.

3.1 MW Blanching

Blanching is a pretreatment of food materials that are used to enhance food quality by inactivating enzymes (Binsi et al. 2014). Steam, hot water, or hot solutions are commonly used in traditional blanching techniques to deactivate enzymes through convective and conductive heat transfer, which may lead to vitamin and another soluble component leaching into the water or solutions (Chávez-Reyes et al. 2013). The shortcomings associated with the traditional technique can be solved by using microwave-assisted blanching (MAB) (Ramesh et al. 2002; Giami 1991). Microwave blanching does not need a large amount of water to transmit heat efficiently into food, which minimizes the loss of nutrients by leaching (Lin and Brewer 2005). MAB allows volumetric heating yielding high temperatures at a short time, which helps to inactivate the enzyme at a short time and lowest energy consumption. In microwave blanching, Patricia et al. (2011) compared the broccoli blanched in the microwave and hot water with fresh broccoli in terms of vitamin C, protein, iron, phosphorus and ashes where they found that the properties of microwave blanched broccoli was very close to the fresh broccoli.

Furthermore, it was reported that when green peas (150 g) with water (100 ml) blanched in a microwave retained more carotene than those blanched in hot water (Chung et al. 1981).

3.2 MW Cooking

Microwave cooking has recently become the most versatile way of cooking all around the world. Microwave ovens are extremely common kitchen tools appliance for food preparation. Microwave-cooked food retains more nutrition than conventional oven-cooked foods (Zia-ur-Rehman et al. 2003).Daomukda et al. (2011) compared the brown rice cooked in a microwave oven with the brown rice cooked by boiling or steaming, and they found that when rice cooked in MW the protein, fat, and ash content was kept at higher levels. Moreover, MW cooking takes less energy (12–24%) in comparison to regular cooking (Lakshmi et al. 2007).

3.3 MW Baking

Baking is a thermal technique that alters the physicochemical properties of dough substantially. The baking process is divided into three phases: the first phase begins with dough expansion and moisture loss; the second phase is when dough expansion and moisture loss rates are at their highest; the third phase is when dough expansion and moisture loss rates are at their lowest. The reason behind the minimum moisture loss rate in the third phase is due to the breakdown of the air structure in the dough, which rises the vapor pressure (Mondal and Datta 2008). Therefore, the application of MW at the final stage of baking is preferable.

In microwave baking, high temperature is attainable within a very short time, which results in lesser loss of nutrients. Baking methods have a substantial impact on the nutritional content of protein in foods. Megahey et al. (2005) used a microwave oven of power 250 W and a convection oven of temperature 200°C to investigate the effects of baking conditions on the quality of cake in terms of texture, and they found that microwave-baked cake had higher springiness along with low firmness.

3.4 MW Roasting

Roasting is a cooking method where heat is employed on every side of food uniformly at a minimum temperature of 150°C, which may come from an oven, open flame, or other heat sources. Roasting is accomplished at a high temperature within a very short time, where various chemical reactions take place. There are several roasting methods like sand roasting, microwave roasting, and oven roasting, which have substantially influenced the roasted food's quality. However, microwave roasting has expanded the range of availability of roasted products globally because of its advantages such as being cost-effective, quick, and simple (Joshi et al. 2014).

Although hot air roasting is simpler and more economic, it necessitates a lengthy roasting time, whereas MW reduces the roasting time. In general, utilizing a better roasting procedure involving microwaves and high pressure may have a positive effect on the good impact on the product's physicochemical qualities of of the final product (Yoshida et al. 2001).

3.5 MW Drying

Drying is simultaneous heat and mass transfer process, which drives moisture out from food materials. The water molecules of food products absorb microwave radiation, which results in the heating and evaporation of water. The dielectric property of food product polar ion possesses microwave irradiation, which absorbed energy and rise the temperature of food products. This is susceptible of accelerates the evaporation of rate of intracellular liquid water. Non-uniform heating during MW drying causes hot and cold spot problems in food materials. It is basically a combination of underheating and overheating takes place in the food simultaneously. To accelerate the drying process and mitigation of the formation of crest hardening, minimize uneven heating problems and incorporate MW in with a conventional drying system. A combination of other drying approaches, including hot air, freeze, vacuum, and infrared drying with MW drying, provides better quality dried foods.

3.5.1 MW Hot-Air Drying

The surface temperature during convective drying requires high evaporation of moisture in the core. Therefore, internal heat generation during MW propagation better leads evaporation of water present at the core of food materials. The vapor intensity in the material is due to pressure difference, which causes higher moisture exchange.

The limitation of convective drying can be eliminated by combined drying of the convective-microwave method. The combined drying raises the drying kinetics and quality of the dried product. The microwave power is absorbed by moisture, which causes quick evaporation from the internal part of the material; and the same time convective air removes surface moisture. The microwave in the drying process reduces drying time and provides volumetric heat transfer (Joardder and Karim 2022).

The microwave convective drying process reduces the drying time by about 80–90% in comparison to convective drying and produces better-dried product quality. The drying cost of the microwave was reduced by combined microwave convective drying (MCD) because of the lower cost of convective drying (Masud et al. 2022; Andrés and Bilbao, 2004) .

3.5.2 MW Freeze-Drying

Freeze-drying (FD) is one of the supreme drying processes for heat-sensitive products (Hammami and René 1997). However, the FD is a lengthy and energy-consuming process as the drying takes place at low pressure. The microwave freeze drying (MFD) process is the combination of microwaves in the FD system. In the MFD process, the matrix materials drying happened layer by layer. The microwave energy enhances the performance of the MFD system by engrossing the moisture of the product directly (Fan et al. 2019).

The drying rate of raspberry puree is considerably raised by the modification of FD by adding microwave energy (Ozcelik et al. 2019). MFD drying techniques provide the same or better quality than conventional FD (Ambros et al. 2018). The problem associated with the non-uniform spread of heating is diminished by introducing microwave energy in the FD process and reducing energy requirements (Vadivambal and Jayas, 2010). Moreover, MFD takes less drying time reducing about 50–75% in comparison with FD (Torringa et al. 2001).

3.5.3 MW Vacuum Drying

Microwave vacuum drying (MVD) system combines the advantages of both microwave and vacuum effects, which maintain food quality with less power consumption (Wang et al. 2013). The microwave helps to accumulate water vapor on the surface and in that time, vacuum condition decreases the boiling point of water that allow drying at lower temperature by reducing pressure, which maintains the quality, color, and texture of food materials (Zhao et al. 2018).

The MVD drying process removes the moisture from food materials and maintains the chemical properties of food by performing lower-temperature drying (Viji et al. 2019). Eventfully, the vacuum condition creates boiling point depression causing water to rapidly vaporize. Low thermal energy is required to evaporate the moisture of food samples. The drying cost and time required for crispy materials are decreased in MVD technique (Dhital et al. 2018). It was observed that the rehydration time, color, and texture of mushrooms were found during the MVD technique in comparison to hot air drying (Giri and Prasad, 2007). VMD techniques reduce drying time by about 70–90% with respect to HAD and FD (Lin et al. 1998).

3.5.4 MW Infrared Drying

The water molecules allow near-infrared (0.78 μm to 1.4 μm) absorbance, whereas the longer wavelength is absorbed by the surface (Nuthong et al. 2011). IR enters in quite a depth of moisture of porous matrix, which permits moisture removal from the product. The MW heating accumulate moisture at the surface of the material, which made the surface moist without making it crispy. The problem is overcome by introducing IR techniques associated with MW, which reduce the drying time and maintain quality. The microwave IR drying techniques remove the surface water of food materials, which relieves the wetness of dried food.

Agricultural product longevity, reduction of dying time, and food quality can be achieved through microwave IR drying. The MW heating accumulates moisture from the core to the surface, whereas IR vaporizes it rapidly from surface to atmosphere resulting in faster drying (Heydari et al. 2020). The combined effect of halogen lamp-microwave on carrots reduced the drying time remarkably (Tireki et al. 2006).

Although MW-associated drying system has the potential to reduce drying time along with maintaining high product quality compared to conventional drying, the non-uniform nature of temperature distribution during MW heating results in hot and cold spots. Intermittent application of MW can eliminate the non-uniform heating problem as discussed in Section 4.2.1 (Kumar et al. 2018).

3.6 MW Heating for Liquid Food Processing

Microwave radiation is effective for pasteurization, sterilization, tempering, evaporation, and leaching of liquid foods. There is much research work on microwave heating for liquid food processing and preservation such as fruit (Franco et al. 2017) juices (Mendes-Oliveira et al. 2020), milk (Zhu et al. 2014), and sugarcane juice (Marszalek et al. 2015). It is reported that MW heating technology shows higher inactivation efficiencies of pathogens and ensures higher nutritional quality than those of conventional processing. Moreover, MW heating results in higher preservation of thermo-sensitive nutritional compounds, such as vitamin C and flavonoids (Martins et al. 2021).

The inactivation of enzymes and microbial during MW heating can be described as a set of selective heating, magnetic field coupling, and cell membrane fission mechanisms (Kozempel et al. 1998). According to selective heating theory, MW heating destroys the microorganisms through a high heating effect in the microorganisms' area. The conventional heating of liquid food is processed through a heat exchanger, which causes excess heating to contact the surface and consumes a long time to process food, which degrades nutritional composition, texture, and color (Marszalek et al. 2015). The presence of excess ion mobility and viscous effect rises the factor of loss of liquid food with respect to solid matrices (Coronel et al. 2008). MW heating destroys microorganisms more quickly than the traditional thermal system (Salazar-González et al. 2012).

4. Challenges and Recommendations

The innovation of MW heating changed the ways of food processing to a great extent. Despite posing several advantages compared to conventional heat-related processing of foods, the industrial use of MW heating is not still widespread. The main reason of limited application of MW are listed below:

- As discussed in the previous section, uneven heating is one of the major problems in applying Mw heating in food processing. A set of options regarding achieving uniform heating during MW application has been discussed in the following section.
- The initial investment in MW heating is still high in comparison with the conventional heating options.
- Monitoring and manipulation of food processing, such as cooking, is not feasible under MW heating conditions.
- In the case of cooking or heating, utensils of selective materials are suitable. For example, reflective materials and materials with a low dissipation factor are not appropriate for microwave heating. Similarly, plastic materials are not preferable for MW heating because of the chance of contamination of food with plastic particles. Using acceptable utensils during MW heating is a crucial issue in order to avoid any unexpected hazardous incident.
- A microwave oven is preferable for cooking or processing small quantities of food, whereas scaling up is an expensive option. The proper size of the food sample is vital in order to maintain a more uniform heating.
- Microwave heating necessitates the use of electrical energy, which is the most expensive type of energy. However, it is far more energy efficient than conventional heating systems. The selection of appropriate MW power and frequency can save extra energy consumption.
- Direct contact with MW rays in the human body may lead to heat body tissue in the same manner that it can heat food. The special arrangement must be ensured to hinder any linkage of the MW heating system. Especially, MW leakage is a great concern when any modification is required in an existing MW oven. Special observation and test of Mw leakage must be ensured prior to using the modified MW.

- Moreover, the inadequacy of the data on the dielectric properties of different foods is one of the main hindrances in applying MW heating at the industrial level. Open-source data bank generation can assist in the wider application of MW in food processing.

4.1 Methods of Achieving Uniform Heating

Many different solutions can be adapted to mitigate uneven heating problems during the application of MW. MW oven turntable and the recent approach of the double magnetron in domestic level MW ovens are the measures of making MW heating more uniform. However, industry-level application microwave in food processing needs to take further research and developments for mitigating the uneven heating problem. Following are the promising approaches in order to mitigate the uneven temperature distribution under MW heating conditions.

4.1.1 Intermittent Application of MW

The uneven heating problem of microwave heating can be resolved by intermitting the microwave, which allows the materials to distribute the heat and avoid hot and cold spots (Kumar et al. 2014ab). Intermittent application of MW involves heat generation during ON time, whereas heat redistribution prevails during OFF time. Eventually, the non-uniform heat energy in hot and cold can easily achieve thermal equilibrium. One of the major issues in this approach is the selection of intermittency or MW pulse ratio. Several experimental observations and mathematical modeling can ensure the optimum ON/OFF condition of the MW application in a food processing system.

4.1.2 Proper Power Selection

Some microwave ovens include power settings, which aid in more even heating of food. Selection of the right MW power is vital to minimize overheating or even underheating of food. Selection of MW power depends on several factors, including moisture content, size and shape of food samples as well as shape and geometry of container, and dielectric properties of container and food among others. For many instances, selective MW power would be suitable. In other words, different MW power at different stages of processing can ensure a better quality of food. For example, high MW power is appropriate at the early stage of processing, whereas lower MW power is suitable for the sample at the final stage of the processing. The power level and application time need to be selected properly after conducting comprehensive research for a particular food processing.

4.1.3 Multi-Magnetron

The uniform temperature distribution can be achieved by using multiple magnetrons in the MW-heating system. In many households, MW oven is now equipped with dual magnetrons to mitigate non-uniform problems. The placement of multiple magnetrons is a crucial issue in this case. The number of magnetrons and their placement must be designed using advanced-numerical modeling of temperature distribution during MW-heating

One of the major issues in this approach is the selection of the right condition including optimum MW power, proper intermittency of MW power, and exact location of the magnetron in case of use of multiple magnetron incorporation. Several experimental observations and mathematical modeling can ensure the optimum condition of the MW application in a food processing system.

5. Conclusion

Microwaves are electromagnetic radiation, which propagates through space using time-varying electric and magnetic fields. Microwaves penetrate the material and interacted with water molecules resulting in volumetric heating, eventually facilitating higher evaporation of moisture from inside of the material. Food materials with higher moisture content are suitable for MW heating as water molecules are bipolar in nature. The application of MW heating become popular in food processing at both domestic and industrial levels because of its volumetric heating capability. The application of MW can significantly reduce food processing time along with maintaining a high quality of processed food. Despite the existing advantages of MW, there are some challenges including uneven heating associated with MW application in food processing. Although microwave-assisted food processing seems promising, should be seen as preliminary, as there are still some limitations that need to be overcome for food industry application. Control of thermal non-uniformity, enriching dielectric database, and inventing easy safety measures should be studied extensively to make MW heating more accessible and reliable in food processing.

References

[Coronel, P., Simunovic, J., Sandeep, K.P. and Kumar, P. 2008. Dielectric Properties of Pumpable Food Materials at 915 MHz. Int. J. Food Prop. 11(3): 508–518.

Ambros, S., Mayer, R., Schumann, B. and Kulozik, U. 2018. Microwave-Freeze Drying of Lactic Acid Bacteria: Influence of Process Parameters on Drying Behavior and Viability. Innov. Food Sci. Emerg. Technol. 48: 90–98. https://doi.org/https://doi.org/10.1016/j.ifset.2018.05.020.

Andrés, A. and Bilbao, P.F.C. 2004. Drying Kinetics of Apple Cylinders under Combined Hot Air–Microwave Dehydration. J. Food Eng. 63(1): 71–78.

Balanis, C.A. 1999. Advanced Engineering Electromagnetics; John Wiley & Sons.

Basaran, P., Basaran-Akgul, N. and Rasco, B.A. 2010. Dielectric Properties of Chicken and Fish Muscle Treated with Microbial Transglutaminase. Food Chem. 120(2): 361–370.

Binsi, P.K., Ninan, G., Zynudheen, A.A., Neethu, R., Ronda, V. and Ravishankar, C.N. 2014. Compositional and Chill Storage Characteristics of Microwave-blanched Sutchi Catfish (Pangasianodon Hypophthalmus) Fillets. Int. J. Food Sci. Technol. 49(2): 364–372.

Chandrasekaran, S., Ramanathan, S. and Basak, T. 2013. Microwave Food Processing—A Review. Food Res. Int. 52(1): 243–261.

Chávez-Reyes, Y., Dorantes-Alvarez, L., Arrieta-Baez, D., Osorio-Esquivel, O., Ortiz-Moreno, A. Polyphenol Oxidase Inactivation by Microwave Oven and Its Effect on Phenolic Profile of Loquat (Eriobotrya Japonica) Fruit. Food Nutr. Sci., 2013, 4 (09), 87.

Chen, J., Pitchai, K., Birla, S., Negahban, M., Jones, D. and Subbiah, J. 2014. Heat and Mass Transport during Microwave Heating of Mashed Potato in Domestic Oven—Model Development, Validation, and Sensitivity Analysis. J. Food Sci. 79(10): E1991–E2004.

Chung, S.Y., Morr, C.V. and Jen, J.J. 1981. Effect of Microwave and Conventional Cooking on the Nutritive Value of Colossus Peas (Vigna Uniguiculata). J. Food Sci. 46(1): 272–273.

Curtis, J. Experimental Verification for Microwave Processing of Materials in a Single Mode Rectangular Resonant Cavity. Virginia Tech 1999.

Daomukda, N.; Moongngarm, A.; Payakapol, L.; Noisuwan, A. Effect of Cooking Methods on Physicochemical Properties of Brown Rice. In 2nd International Conference on Environmental Science and Technology IPCBEE; IACSIT Press Singapore, 2011; Vol. 6.

Decareau, R.V. Microwaves in the Food Processing Industry; Academic Press, 1985.

Dhital, S., Baier, S.K., Gidley, M.J., Stokes, J.R. 2018. Microstructural Properties of Potato Chips. Food Struct. 16: 17–26. https://doi.org/https://doi.org/10.1016/j.foostr.2018.03.001.

Fan, K., Zhang, M. and Mujumdar, A.S. 2019. Recent Developments in High Efficient Freeze-Drying of Fruits and Vegetables Assisted by Microwave: A Review. Crit. Rev. Food Sci. Nutr. 59(8): 1357–1366. https://doi.org/10.1080/10408398.2017.1420624.

Feynman, R.P., Leighton, R.B. and Sands, M. 2011. The Feynman Lectures on Physics, Vol. I: The New Millennium Edition: Mainly Mechanics, Radiation, and Heat; Basic books Vol. 1.

Franco, A.P., Tadini, C.C. and Wilhelms Gut, J.A. 2017. Predicting the Dielectric Behavior of Orange and Other Citrus Fruit Juices at 915 and 2450 MHz. Int. J. Food Prop. 20(sup2): 1468–1488.

Ghanem, T.H. 2010. Dielectric Properties of Liquid Foods Affected by Moisture Contents and Temperatures. Misr J. Agric. Eng. 27(2): 688–698.

Giami, S.Y. 1991. Effects of Pretreatments on the Texture and Ascorbic Acid Content of Frozen Plantain Pulp (Musa Paradisiaca). J. Sci. Food Agric. 55(4): 661–666.

Giri, S.K. and Prasad, S. 2007. Drying Kinetics and Rehydration Characteristics of Microwave-Vacuum and Convective Hot-Air Dried Mushrooms. J. Food Eng. 78(2): 512–521. https://doi.org/https://doi.org/10.1016/j.jfoodeng.2005.10.021.

Guo, W., Zhu, X., Liu, H., Yue, R. and Wang, S. 2010. Effects of Milk Concentration and Freshness on Microwave Dielectric Properties. J. Food Eng. 99(3): 344–350.

Hammami, C.; René, F. Determination of Freeze-Drying Process Variables for Strawberries. J. Food Eng., 1997, 32 (2), 133–154. https://doi.org/https://doi.org/10.1016/S0260-8774(97)00023-X.

Heydari, M.M., Kauldhar, B.S. and Meda, V. 2020. Kinetics of a Thin-layer Microwave-assisted Infrared Drying of Lentil Seeds. Legum. Sci. 2(2): e31.

Joardder, M.U.H., Karim, A., Kumar, C. and Brown, R.J. 2015. Porosity: Establishing the Relationship between Drying Parameters and Dried Food Quality; Springer, 2015.

Joardder, M.U.H. and Karim, M.A. 2022. Drying Kinetics and Properties Evolution of Apple Slices under Convective and Intermittent-MW Drying. Therm. Sci. Eng. Prog. 101279.

Jones, S.B. and Or, D. 2002.Surface Area, Geometrical and Configurational Effects on Permittivity of Porous Media. J. Non. Cryst. Solids 305(1–3): 247–254.

Joshi, N.D., Mohapatra, D., Joshi, D.C., Sutar, R.F. 2014.Puffing Characteristics of Parboiled Milled Rice in a Domestic Convective–Microwave Oven and Process Optimization. Food Bioprocess Technol. 7(6): 1678–1688.

Kent, M. 1987. Electrical and Dielectric Properties of Food Materials. A Bibliography and Tabulated Data. A COST 90bis Production. Sci. Technol. Publ. Hornchurch.

Kingston, H.M. and Jassie, L.B. 1986. Microwave Energy for Acid Decomposition at Elevated Temperatures and Pressures Using Biological and Botanical Samples. Anal. Chem. 58(12): 2534–2541.

Kozempel, M.F., Annous, B.A., Cook, R.D., Scullen, O.J. and Whiting, R.C. 1998. Inactivation of Microorganisms with Microwaves at Reduced Temperaturas. J. Food Prot. 61(5): 582–585.

Kristiawan, M., Sobolik, V., Klíma, L. and Allaf, K. 2011. Effect of Expansion by Instantaneous Controlled Pressure Drop on Dielectric Properties of Fruits and Vegetables. J. Food Eng. 102(4): 361–368.

Kumar, C., Joardder, M.U.H., Farrell, T.W. and Karim, M.A. 2018. Investigation of Intermittent Microwave Convective Drying (IMCD) of Food Materials by a Coupled 3D Electromagnetics and Multiphase Model. Dry. Technol. 36(6): 736–750. https://doi.org/10.1080/07373937.2017.1354874.

Kumar, C., Joardder, M.U.H., Farrell, T.W., Millar, G.J. and Karim, M.A. 2016a. Mathematical Model for Intermittent Microwave Convective Drying of Food Materials. Dry. Technol. 34(8): 962–973.

Kumar, C., Joardder, M.U.H., Farrell, T.W. and Karim, M.A. 2016b. Multiphase Porous Media Model for Intermittent Microwave Convective Drying (IMCD) of Food. Int. J. Therm. Sci. C(104): 304–314. https://doi.org/10.1016/J.IJTHERMALSCI.2016.01.018.

Kumar, C., Joardder, M.U.H., Karim, A., Millar, G.J. and Amin, Z. 2014a. Temperature Redistribution Modelling during Intermittent Microwave Convective Heating. Procedia Eng. 90: 544–549.

Kumar, C., Karim, M.A. and Joardder, M.U.H. 2014b. Intermittent Drying of Food Products: A Critical Review. J. Food Eng. 121: 48–57.

Lakshmi, S., Chakkaravarthi, A., Subramanian, R. and Singh, V. 2007. Energy Consumption in Microwave Cooking of Rice and its Comparison with Other Domestic Appliances. J. Food Eng. 78(2): 715–722.

Lin, S. and Brewer, M.S. 2005. Effects of Blanching Method on the Quality Characteristics of Frozen Peas. J. Food Qual. 28(4): 350–360.

Lin, T.M., Durance, D. and Scaman, T. 1998. C. H. Characterization of Vacuum Microwave, Air and Freeze Dried Carrot Slices. Food Res. Int. 31(2): 111–117. https://doi.org/https://doi.org/10.1016/S0963-9969(98)00070-2.

Lyng, J.G., Zhang, L. and Brunton, N.P.A. 2005. Survey of the Dielectric Properties of Meats and Ingredients Used in Meat Product Manufacture. Meat Sci. 69(4): 589–602.

Marszałek, K., Mitek, M. and Skąpska, S. 2015. Effect of Continuous Flow Microwave and Conventional Heating on the Bioactive Compounds, Colour, Enzymes Activity, Microbial and Sensory Quality of Strawberry Purée. Food Bioprocess Technol. 8(9): 1864–1876.

Martín-Esparza, M.E., Martínez-Navarrete, N., Chiralt, A. and Fito, P. 2006. Dielectric Behavior of Apple (Var. Granny Smith) at Different Moisture Contents: Effect of Vacuum Impregnation. J. Food Eng. 77(1): 51–56.

Martins, C.P.C., Cavalcanti, R.N., Cardozo, T.S.F., Couto, S.M., Guimarães, J.T., Balthazar, C.F., Rocha, R.S., Pimentel, T.C., Freitas, M.Q. and Raices, R.S.L. 2021. Effects of Microwave Heating on the Chemical Composition and Bioactivity of Orange Juice-Milk Beverages. Food Chem. 345: 128746.

Masud, M.H., Joardder, M.U.H., Ananno, A.A. and Nasif, S. 2022. Feasibility Study and Optimization of Solar-Assisted Intermittent Microwave–Convective Drying Condition for Potato. Eur. Food Res. Technol. 1–15.

Megahey, E.K., McMinn, W.A.M. and Magee, T.R.A. 2005. Experimental Study of Microwave Baking of Madeira Cake Batter. Food Bioprod. Process. 83(4): 277–287.

Mendes-Oliveira, G., Deering, A.J., San Martin-Gonzalez, M.F. and Campanella, O.H. 2020. Microwave Pasteurization of Apple Juice: Modeling the Inactivation of Escherichia Coli O157: H7 and Salmonella Typhimurium at 80–90 C. Food Microbiol. 87: 103382.

Metaxas, A.C. and Meredith, R.J. 1983. Industrial Microwave Heating; IET.

Metaxas, A.C. Fundamentals of Electroheat: A Unified Approach. John Wiley & Sons Ltd.: Chichester, England 1996.

Mondal, A. and Datta, A.K. 2008. Bread Baking–A Review. J. Food Eng. 86(4): 465–474.

Nelson, S.O. and Trabelsi, S. 2012. Factors Influencing the Dielectric Properties of Agricultural and Food Products. J. Microw. Power Electromagn. Energy 46(2): 93–107.

Nuthong, P., Achariyaviriya, A., Namsanguan, K. and Achariyaviriya, S. 2011. Kinetics and Modeling of Whole Longan with Combined Infrared and Hot Air. J. Food Eng. 102(3): 233–239.

Ozcelik, M., Ambros, S., Heigl, A., Dachmann, E. and Kulozik, U. 2019. Impact of Hydrocolloid Addition and Microwave Processing Condition on Drying Behavior of Foamed Raspberry Puree. J. Food Eng. 240: 83–91. https://doi.org/https://doi.org/10.1016/j.jfoodeng.2018.07.001.

Patricia, C.M., Bibiana, D.Y. and José, P.M. 2011. Evaluation of Microwave Technology in Blanching of Broccoli (Brassica Oleracea L. Var Botrytis) as a Substitute for Conventional Blanching. Procedia Food Sci. 1: 426–432.

Pozar, D.M. 2009 Microwave Engineering USA: John Wiley & Sons.

Ramesh, M.N., Wolf, W., Tevini, D., Bognar, A. 2002. Microwave Blanching of Vegetables. J. Food Sci. 67(1): 390–398.

Salazar-González, C., Martín-González, M.F.S., López-Malo, A. and Sosa-Morales, M.E. 2012. Recent Studies Related to Microwave Processing of Fluid Foods. Food Bioprocess Technol. 5(1): 31–46. https://doi.org/10.1007/s11947-011-0639-y.

Tıraş, B., Dede, S. and Altay, F. 2019. Dielectric Properties of Foods. 2019.

Tireki, S., Şumnu, G. and Esin, A. 2006. Production of Bread Crumbs by Infrared-Assisted Microwave Drying. Eur. Food Res. Technol. 222(1): 8–14.

Torrealba-Meléndez, R., Sosa-Morales, M.E., Olvera-Cervantes, J.L. and Corona-Chávez, A. 2014. Dielectric Properties of Beans at Ultra-Wide Band Frequencies. J. Microw. Power Electromagn. Energy 48(2): 104–112.

Torringa, E., Esveld, E., Scheewe, I., van den Berg, R. and Bartels, P. 2001. Osmotic Dehydration as a Pre-Treatment before Combined Microwave-Hot-Air Drying of Mushrooms. J. Food Eng. 49(2–3): 185–191.

Vadivambal, R. and Jayas, D.S. 2010. Non-Uniform Temperature Distribution during Microwave Heating of Food Materials—A Review. Food bioprocess Technol. 3(2): 161–171.

Venkatesh, M.S., Raghavan, G.S.V. 2004. An Overview of Microwave Processing and Dielectric Properties of Agri-Food Materials. Biosyst. Eng. 88(1): 1–18.

Viji, P., Shanmuka Sai, K.S., Debbarma, J., Dhiju Das, P.H., Madhusudana Rao, B. and Ravishankar, C.N. 2019. Evaluation of Physicochemical Characteristics of Microwave Vacuum Dried Mackerel and Inhibition of Oxidation by Essential Oils. J. Food Sci. Technol. 56(4): 1890–1898. https://doi.org/10.1007/s13197-019-03651-7.

Wang, J., Tang, J., Wang, Y. and Swanson, B. 2009. Dielectric Properties of Egg Whites and Whole Eggs as Influenced by Thermal Treatments. LWT-Food Sci. Technol. 42(7): 1204–1212.

Wang, Y., Zhang, M., Mujumdar, A.S., Mothibe, K.J. and Roknul Azam, S.M. 2013. Study of Drying Uniformity in Pulsed Spouted Microwave–Vacuum Drying of Stem Lettuce Slices with Regard to Product Quality. Dry. Technol. 31(1): 91–101.

Wentworth, S.M. Fundamentals of Electromagnetics with Engineering Applications; John Wiley & Sons, 2006.

Witkiewicz, K. and Nastaj, J.F. 2010. Simulation Strategies in Mathematical Modeling of Microwave Heating in Freeze-Drying Process. Dry. Technol. 28(8): 1001–1012.

Yoshida, H., Hirakawa, Y. and Abe, S. 2001. Roasting Influences on Molecular Species of Triacylglycerols in Sunflower Seeds (Helianthus Annuus L.). Food Res. Int. 34(7): 613–619.

Zhao, Y., Jiang, Y., Zheng, Y., Li, Z., Zhang, Y., Zheng, B., Lo, Y.M., Miao, S., Tian, Y. 2018. Effects of Microwave Vacuum Drying on Macroscopic Properties and Microstructure of Lotus (Nelumbo Nucifera Gaertn.) Seeds. Int. J. Food Eng. 14(3). https://doi.org/doi:10.1515/ijfe-2017-0313.

Zhu, X., Guo, W. and Jia, Y. 2014. Temperature-Dependent Dielectric Properties of Raw Cow's and Goat's Milk from 10 to 4,500 MHz Relevant to Radio-Frequency and Microwave Pasteurization Process. Food Bioprocess Technol. 7(6): 1830–1839.

Zhu, X., Guo, W. and Wu, X. 2012. Frequency-and Temperature-Dependent Dielectric Properties of Fruit Juices Associated with Pasteurization by Dielectric Heating. J. Food Eng. 109(2): 258–266.

Zia-ur-Rehman, Z., Islam, M. and Shah, W.H. 2003. Effect of Microwave and Conventional Cooking on Insoluble Dietary Fibre Components of Vegetables. Food Chem. 80(2): 237–240.

16

Ohmic Heating

Design, Thermal Performance, and Applications in Food Processing

Asaad Rehman Al-Hilphy[1,*] and *Amin Mousavi Khaneghah*[2]

1. Introduction

In Ohmic (OH), the electric current passes within the food, is converted into electrical resistance, and then is heated to transform the electrical energy into heating energy. OH has several names, such as heat by direct electrical resistance, Joule heating, heat by electrical resistance, electrical heating, and heating by electroconductive. The heat inside the food distributes evenly and quickly. The direction of the heat is from the inside to out, in contrast to the use of conventional heated surfaces in which the direction of the heating is from the outside to the inside slowly (Shirsat et al. 2004; Icier and Ilicali 2005; Leizerson and Shimoni 2005). Therefore, food contains ions, such as acids and salts. The electrical current passes within the food, generating heat in it (FDA 2000). OH's efficiency depends on the food's electrical conductivity (Zoltai and Swearingen 1996). The OH technology is also environmentally friendly (Sakr and Liu 2014). It applies widely in the bleaching, fermentation, evaporation, peeling, cooking, sterilization, extraction, water distillation, dehydration, and pasteurization (Icier 2012; Bozkurt and Icier 2010; Al-Hilphy et al. 2015; Makroo 2017; Makroo et al. 2020; Abdulstar et al. 2020; Al-Hilphy et al. 2020). The OH technique is especially suitable for liquid and viscous foods with appropriate electrical conductivity (Sakr and Liu 2014). The present chapter aims to investigate applying OH in the food industry as an alternative heating technology.

2. OH Fundamentals

The electrical energy flows to the food (Fig. 1) because the food has high moisture content and dissolved salts, which conduct the electricity through electrolytic conduction. The electricity converts into heat energy dispersed by an electrical conductor (food). The food is in the form of electrical resistance through passing the alternating electric current in the food, known as the Joule effect, which causes volumetric heating. It raises food temperature with an efficiency of more

[1] Department of Food Science, College of Agriculture, University of Basrah, Iraq.
[2] Department of Fruit and Vegetable Product Technology, Prof. Wacław Dąbrowski Institute of Agricultural and Food Biotechnology – State Research.
* Corresponding author: aalhilphy@yahoo.co.uk

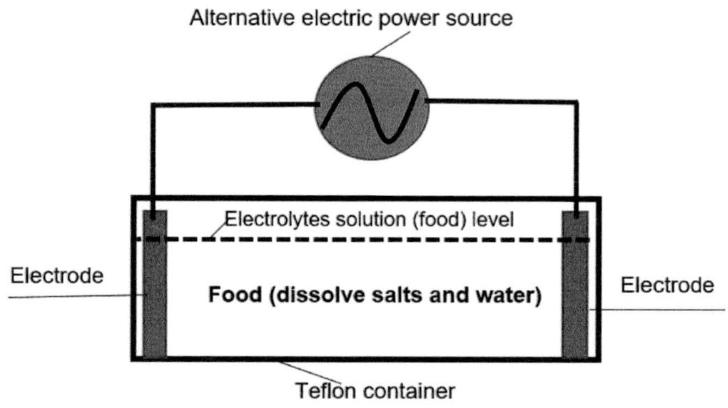

Figure 1. Principle of Ohmic Heating.

than 90% because the heat is transmitted from inside the food to the outside, in contrast to the traditional heating method (in traditional heating, the heat transfer is through thermal gradients or hot surfaces to the food. Ohmic heating is fast and homogeneous heating because of the oscillation of the motion of ions when exposed to electric fields, and ions collide with each other. Thus, the dynamic motion of ions and dipole moments results in heating because of the increased kinetic energy. The heat generated is immediately related to the squared electrical field intensity and the electrical conductivity and induces electrical current (Cappato et al. 2017; Goullieux and Pain 2014; Singh and Heldman 2014; Silva et al. 2017; Indiarto and Rezaharsamto 2020).

Volumetric heat generation is given by Equation 1 as follows (Samprovalaki et al. 2007):

$$Q = \sigma E^2 = I^2 \ R \qquad (1)$$

Here, is the volumetric heat generation (W/m^3), is the electrical conductivity (S/m), is electrical field intensity (V/m), I is the current (A), and R is the resistance (ohm).

3. Advantages and Disadvantages of OH

3.1 Advantages

In OH, heat energy is transported from the inside of the treated food material to the outside, unlike traditional heating in which heat energy transmits from the treated food surface to the inside. OH is fast and homogeneous, does not require hot surfaces, and heat damage and loss of nutrients in food are low (high quality). Also, the formation of fouling is low. In addition, the maintenance costs are low. OH is environmentally friendly and safe technology. The process of heating food is done by the generation of heat energy inside the food. Moreover, the temperature of the treated particles is higher than that of the liquid, and this condition is not present in conventional heating (Kim et al. 1996; Al-Hilphy 2018 a,b).

The energy efficiency and volumetric heating of OH are high. Alternating voltages are applied to the electrodes that surround the food. Furthermore, the heating rate relies on the electric field intensity and the square of the electric conductivity. The electric field intensity varies with the gap between the electrodes and the voltage used. Some foods have high temperatures and others low because of the complex composition of the food. The electrodes get corroded, and they need to be cleaned constantly. Some processes require pretreatment, such as blanching (Zareifard et al. 2003; Goullieaux and Pain 2014a). Fouling formation on the electrodes when using high voltages to heat milk is due to the denaturation of whey proteins during OH at high voltages (Al-Hilphy et al. 2018).

3.2 Disadvantages

The initial costs of setting up an OH system are high. Information about OH is insufficient. The volumetric heat generation rate in heating heterogeneous foods and the complex correlation between the distribution of the electric field and temperature, heat channeling, the shape of the particles, and the orientation. The application of OH differs from one product to another, which can increase development costs (De Halleux et al. 2005). OH works in a narrow range of frequencies (Muhammad et al. 2019).

4. Design of OH System

Designs of OH apparatuses are different according to process type, such as pasteurization, thawing, sterilization, hamburger treatment, drying food, and evaporation). When designing an OH device, the type of product to be pasteurized must be considered, along with the qualities and characteristics of that product, the electrical conductivity of that material, and the heating coefficient (Lima 2007). Kong et al. (2008) have designed an Ohmic heater that comprises two main parts: the OH unit, which includes a power transducer, stainless steel electrodes, and a Teflon cylinder, and the other part is the data collection system, which consists of a digital conduction meter and thermocouple temperature sensors. The main components of the OH apparatus are an alternative electric generator and heating chamber as follows:

4.1 Electrodes

Pair of electrodes must be presented in the heating chamber. Electrodes are made of stainless steel, such as 304 and 316 types of platinum-coated titanium electrodes. Samaranayake and Sastry (2005) showed that the use of electrodes made of stainless steel are electrochemically reactive materials during OH in resisting all pH values; Berthou et al. (2001) and Stirling (1987) used platinum-coated titanium electrodes or lithium to prevent electrolysis that occurs during the low frequency alternative current of 50 Hz or 60 Hz. Reznik (1996) used electrodes made of stainless steel for a frequency higher than 100 kHz or graphite electrodes, and Reznik (1996) added electrolytes between the electrodes and the food to block food pollution. The characteristics of electrodes are high conductivity and made of non-corrosive materials to prevent migration to food, like Mo, Ni, Mn, Fe, and Cr. The electrodes must directly contact the food so as to pass alternative electrical to the food. The gap between the electrodes is changed depending on the size of the OH system. The distance between electrodes affects the electrical field intensity value.

4.2 Heating Chamber

The heating chamber must be made of insulator materials, such as Teflon. Al-Hilphy et al. (2020) used Teflon material to manufacture the heating chamber of the OH system to extract bioactive compounds from the wheat barn. In addition, Abdulstar et al. (2020) used a heat-resistant plastic cylinder as a heating chamber for evaporating water using OH.

5. Types of OH Systems

There are three types of OH systems (Goullieaux and Pain 2014b; Indiarto and Rezaharsamto 2020):

5.1 Batch Configuration

The food is static in the cylinder heating. This system is ideal for laboratory use and testing development of the new product. This system is easy to calculate the electrical conductivity, time of heating, and process uniformity. Also, it can be used to monitor the influence of processing on the final food quality and identify the composition of the optimum initial product. This system is a simple design.

5.2 Transvers Configuration

This configuration is classified as a continuous system of OH. The product flow is parallel with electrodes and perpendicular to the electrical field intensity. The two types of electrodes used in this system are coaxial and flat. Straightforward type is used in this configuration. The disadvantage of this configuration is that a part of the current leaves the system with material (leaking current) because the electrodes are close to inlet and outlet pipes, and the electrical current density leads to flow product unequal around the electrode edge.

5.3 Collinear Configuration

In this configuration, the food flows from one electrode to another, and the electrical field is parallel with the food flow.

6. OH Process Parameters

6.1 Electrical Conductivity

Electrical conductivity is considered the main factor in designing an efficient OH system (Kaur and Singh 2015). Electrical conduction is the movement of charges through a medium (Al-Hilphy et al. 2018b), and it is a critical property influencing the OH process (Palaniappan and Sastry 1991). Electrical conduction is considered one of the crucial matters for the success of OH technology, as it depends on the basic components of food so that the electrical conductivity rises in the case of the presence of mineral salts, acids, and moisture. The electrical conductivity decreases when fat and alcohol are present in the food (Henningsson et al. 2005). The electrical conductivity is given by Equation 2 (Wang and Sastry 1997; Icier et al. 2008):

$$\sigma\left(\frac{s}{m}\right)=\frac{IL}{VA} \tag{2}$$

Here, is the current (A), is the gap between electrodes (m), is the voltage (V), and is the section area (m^2).

Also, the electrical conductivity of the material is calculated from cell constant, current, and voltage, which is calculated from Equation 3:

$$\sigma = k\frac{I}{V} \tag{3}$$

Here, refers to the cell constant (1/m) that is calculated from the length (the gap between electrodes) (m) of the sample divided by cross-section area (m^2) (electrodes area).

The area can be calculated from Equation 4 (Darvishi et al. 2015) during evaporation by OH because of the change of the sample volume and the contact surface area between electrodes and sample during the evaporation process:

$$A_i = \frac{m_i}{\rho_i L} \tag{4}$$

Here, is the change in the touch area between food and the electrodes (m^2), is the food mass at any time (kg), and is the food density at any time (kg/m^3).

Electrical conductivity relies on the food microstructure materials, ionic dissociation, temperature, and electrical field intensity (Lima and Sastry 1999). Electrical conductivity linearly increases as temperature increases at constant electrical field intensity (Castro et al. 2004). Preheating leads to an increase in the electrical conductivity of materials because of increasing temperature. OH is ineffective when electrical conductivity is less than 0.01 and above 10 S/m because of the need for huge currents or voltages (Knirsch et al. 2010). Al-Hilphy et al. (2020) depicted that the electrical conductivity decreased as the electrical field excessed—i.e., when electrical field intensity increased from 4.28–15.70, the electrical conductivity of the mixture (water+ wheat bran) reduced from 2.02–1.02 S/m, respectively. This change is due to the polluting of electrodes and the reactions of electrolysis. Also, Abdulstar et al. (2020) found that the electrical conductivity of water treated by OH decreased as the electrical field intensity increased. Some published papers indicated that the electrical conductivity excessed with rising electrical field intensity (Al-Hilphy et al. 2015; Al-Hilphy et al. 2014; Altemimi et al. 2018; Darvishi et al. 2011; Castro et al. 2004).

The temperature has a significant effect on the electrical conductivity, as presented in Equation 5, which presents the relevance between electrical conductivity and temperature as follows (Palaniappan and Sastry 1991):

$$\sigma_t = \sigma_{ref.}\ 1 + m(t-t_{ref.} \qquad (5)$$

Here, is the electrical conductivity at temperature T (S/m), $\sigma_{ref.}$, is the electrical conductivity at reference temperature (S/m), and *m* is the temperature coefficient.

The content of fat excesses electrical conductivity in food, such as meat, is explained in Equation 6:

$$\sigma = \sigma_s + \sigma_o T + c(f\%)^n \qquad (6)$$

Here, σ_s is the electrical conductivity of the sample at a reference temperature of zero°C (S/m), σ_o is the electrical conductivity (S/m °C), *T* is the temperature dependency constant for the electric conductivity (S/m per fat %), and *n* is dimensionless.

Al-Hilphy et al. (2018) cited an empirical equation for the electrical conductivity of milk treated by OH at 80 volts as follows:

$$\sigma = 0.002T + 0.778 \qquad (7)$$

Here, T is the temperature (°C).

Electric conductivity changes with the change of TSS% as shown in Equation 5 (Sabanci and Icier 2017):

$$\sigma = d + TSS\% + e \qquad (8)$$

Here, and are constants, and is the total soluble solid (%).

6.2 Frequency and Waveform

The typical frequency in ohmic heating ranges from 50–60Hz (Knirsch et al. 2010). The heating rate is affected by changing the frequency and waveform of the alternative current (Lima et al.1999). Reducing the frequency of alternative current leads to increased oil extraction (Lakkakula et al. 2004). Enzyme stabilization is enhanced using low frequency (Lima and Sastry 1999). The influence of electroporation at low frequency during OH is more significant than the higher frequency because the electrical conductivity increases suddenly, and the product texture is affected by that—however, the required time to reach the specified temperature increases (Shynkaryk et al. 2010). As for the waveforms, the sinusoidal and triangular increase the electricity higher than 4 Hz quadratic waveforms (Gavahian et al. 2019).

6.3 Electrical Field Intensity

Increasing electrical field intensity (voltage gradient) increases the heating rate (Casaburi et al. 2021) because the electrical field intensity increases fluid motion through capillaries, which have a direct effect on the electrical conductivity (Halden et al. 1990). Electrical conductivity increases as the electrical field intensity increases, and hence heating rate rises (Icier et al. 2008). Electrical field intensity (*E*) is given in Equation 9 (Floury et al. 2006):

Here, is the applied voltage (V), and is the gap between electrodes (cm).

$$E = \frac{V}{L} \tag{9}$$

6.4 Particle Size and Location

Particle size directly affects electrical conductivity, where increasing particle size leads to a decline in electrical conductivity (Palaniappan and Sastry 1991). The particle size of material lower than 5 mm impacts the electrical conductivity. Large particle has a significant influence on the electrical conductivity and heating rate (McKenna et al. 2006). The magnitude of particle geometry is demonstrated when the solid particle width ratio to height is away from unity (Laekin and Spinkin 1996). De Alwis et al. (1989) reported that when the particle positions with its axis parallel to the electric field, it heats up more slowly than the liquid.

Nevertheless, when the particle is placed with its longer axis vertical to the electrical field, the particle will heat up much quicker than the liquid. Benabderrahmane and Pain (2000) depicted that the particle diameter increases, which results in decreasing efficiency of liquid heating. As for the location of a particle, particle location and orientation play a significant role in OH. Davis et al. (1999) showed the impact of the particle's direction on each other. Zareifard et al. (2003) investigated the OH conduct of particles' mass according to their electrode location. The surface of the electrode is wholly or partially connected to the Ohmically heated food. They found that in the parallel state, the heating of the liquid phase is faster than that of the solid state; while in the series case, the opposite is noticed.

6.5 Heat Capacity

Food with lower heat capacity heats faster (Varghese et al. 2014). Materials have high specific heat, and gravity requires longer to arrive at the specific temperature (Silva et al. 2017).

6.6 Viscosity

Materials have higher viscosity inclined to heat rapidly compared to lower viscosity (Marcotte et al. 2000). Higher viscous liquids increase the heating rate compared to lower viscous liquids (Ghnimi et al. 2008).

6.7 Concentration

The heating rate of materials with higher concentrations is faster than the low concentration because the electrical current pushes through more particles. Thus the heat generation in the particles is faster (Varghese et al. 2014). Sastry (1991) stated that the concentration of particles is a critical parameter in determining the rate of heating in the two stages. Castro et al. (2003) Noted that strawberry pulp electrical conductivity declined as strawberry pulp concentration was excessed. The pulp of initial pH value of 4.0 and a Brix value of 14.5° deg. Moreover, 2.5% of starch content has a higher electrical conductivity than pulp with pH 4.0, 26.5° Brix value, and without starch. Zareifard et al. (2003) concluded that the required heating time to raise the temperature from 20°C to 80°C was twice as considerable compared to the lower solid's concentration, where the bulk density declined as the particle size increased.

6.8 Concentration of Ionic

The increment in the electrical conductivity occurs during heating of biological tissues because of the rise of the ionic mobility, which is due to changes in the structure of tissues like breaking down of cell wall protopectin, nonconductive gas bubbles expulsion, and softening and reducing the hydrous stage viscosity (Bean et al. 1960; Sasseon and Monselis 1977). In general, the heating rate increases with the increase of ionic constriction.

7. Heat Performance of OH Systems

7.1 Temperature and Heating Rate

The temperature rate of wheat bran treated by OH significantly ($p < 0.05$) increased with the excess of electrical field intensity because passing higher electrical current in the material led to increasing consumed electrical energy and then converted it to thermal energy, which raised temperature (Al-Hilphy et al. 2020). Increasing electrical field intensity led to a rising temperature of treated material by OH (Darvishi et al. 2020).

The following equations (10–12) can be used for determining the temperature of liquid leaving the OH unit (Berk 2009):

$$\frac{aT+b}{aT_o+b} = e^{\left|\frac{a\pi d_c L}{mc_p}\right|} \tag{10}$$

$$a = \frac{|\Delta V|^2 d_c \sigma_{om^n}}{4} - U\backslash \tag{11}$$

$$b = \frac{dc\,|\Delta V|^2 \sigma_o}{4} + U\backslash \tag{12}$$

ΔV is the voltage gradient along the heating tube (V/m), is the electrical conductivity of liquid at 0°C (S/m), dc is the heating tube diameter (m), L is the heating tube length (m), and is the constants related to electrical conductivity, according to the following equation:

$$\sigma_L = \sigma_0(1 + m^n T) \tag{13}$$

σ_L is the electrical conductivity of milk (S/m) at temperature T, and is the total heat transfer coefficient (W/m^2.°C).

As for the heating rate, Al-Hilphy (2014) found that the heating rate during the extraction of essential oil using OH was excessed with the rise of the applied voltage. Al-Hilphy et al. (2012) depicted that when the applied voltage increased from 80–220 V, the heating rate increased from 10.0–13.84 °C/min, respectively. Al-Hilphy et al. (2020) disclosed that the heating rate significantly rises with the increment of electrical field intensity. For instance, when the electrical field intensity was 4.28, 7.85, and 15.71 V/cm, the heating rate reached 5.83°C/min, 11.66°C/min, and 17.50°C/min, respectively. The authors observed that the heating rate was faster than conventional heating because of the volumetric heating in OH, which quickly increased temperature. Cho et al. (2017) disclosed that the heating rate is faster in OH than in traditional heating. The heating rate is estimated according to Equation 14 (De Halleux et al. 2005):

$$H_r = \frac{T}{t} \tag{14}$$

Here, T is the temperature (°C), and t is the time (min).

7.2 Type and Shape of Electrodes

Increasing thick electrodes leads to reduce temperature increase rate—i.e., when the thickness of electrodes was 20 mm and 2 mm, the temperatures at the surface electrodes were 44.3°C and 61.8°C, respectively, after the 90 seconds of OH. In addition, the surface temperature of the titanium electrode is slightly higher than the stainless steel electrode by 3°C after the 90s of OH because the density of titanium is lower than stainless steel. Using an aluminum foil electrode gave a higher interface temperature, which reached 83.3°C. The 1 mm thick platinized titanium electrode gave an interface temperature of 68.8°C. Stainless steel inclines to electrolysis (Zell et al., 2011). Platinized titanium electrodes do not use electrolysis at 50 Hz frequency during heating food by OH (Tzedakis et al. 1999). To reduce product contamination by reaction between the product and platinized titanium electrodes, bipolar pulses at a frequency of up to 10 kHz are used (Samaranayake et al. 2005). Abdulstar et al. (2019) demonstrated that the shape of the electrode had a significant influence on decreasing the needed time to arrive at the boiling point using OH. The researchers illustrated that the OH of water using vertical electrodes is higher than the horizontal.

7.3 Specific Ohmic Energy (SOE)

Specific ohmic energy is calculated from Equation 15 as follows (Raso et al. 2016):

$$SOE = \frac{1}{m}\int_0^{\infty} V(t)\,.\,I(t)dt \tag{15}$$

Here, is the specific energy consumption (kJ/kg), m is the mass of evaporated water in the case of evaporation and drying and the mass of heated product during OH, V is the voltage (V), I is the current (A), and t is the time (s). Equation 5 is divided into 1,000 to convert J to kJ.

Al-Hilphy et al. (2020) declared that the SOE increased by 113.88% as the electrical field intensity increased from 4.28 to 15.71. Energy consumption of OH is lower than conventional heating by 4.6–5.3 times (Alkanan et al. 2021).

7.4 Energy Efficiency and System Performance Coefficient

Energy efficiency for heating food is calculated according to Equation 16 (Kasturi and Ayalur Kannappan 2020):

$$E_{eff} = \frac{mc_p(T_f - T_i)}{\Sigma\ VI\Delta t} \tag{16}$$

Also, energy efficiency can be calculated for Ohmic evaporation according to Equation 17 (Hosainpour et al. 2014):

$$E_{eff} = \frac{mc_p(T_f - T_i) + m_w\lambda_w}{\Sigma\ VI\Delta t} \times 100 \tag{17}$$

Here, is the energy efficiency (%), is the mass of the sample (kg), is the specific heat capacity (kJ/kg. K), IS the final temperature (°C), is the initial temperature (°C), is the mass of evaporated water (kg), is the latent heat of water evaporation (kJ/kg).

Al-Hilphy et al. (2020) stated that the energy efficiency of the Ohmic bioactive compound extractor was decreased by 53.11% as the electrical field intensity increased from 4.28 to 15.71). This is due to an increase the heat loss. As for system performance coefficient (SPC), it is generally utilized to assess the OH energy effectiveness (Darvishi et al. 2012, 2013; Icier and Ilicali 2005; Park et al. 20017). SPC is given by Equation 5 (Icier and Ilicali 2005):

$$SPC = \frac{Q_t}{E_g} \tag{18}$$

$$E_g = Q_t + E_{loss} = \sum\ \Delta VIt \tag{19}$$

$$Q_t = mC_p(T_f - T_i) \tag{20}$$

Here, is the system performance coefficient, is the amount taken heat energy (J), is the given energy (J), is the final temperature (°C), and is the initial temperature (°C). Energy efficiency can be calculated from Equation 21 depending on the SPC as follows:

$$E_{eff.} = SPC \times 100 \tag{21}$$

Al-Hilphy (2014) reported that SPC decreased as the voltage gradient increased. It was found that when voltage decreased from 80–60 V, the SPC decreased from 0.977–0.990, respectively. This means that 2.29–3.69% of electrical energy was not converted into heat energy during essential oil extraction from leaves of eucalyptus using OH. Al-Hilphy et al. (2012) demonstrated that the SPC of the ohmic milk pasteurizer was significantly decreased with the increase of applied voltage—i.e., SPC was reduced from 0.82 to 0.59 when applied voltage increased from 80 V to 220 V, respectively.

7.5 Heat Generation

Volumetric heat generation during OH is determined by Equation 22 (Icier 2003):

$$H_g = \sigma|\Delta V|^2 \tag{22}$$

Volumetric heat generation (J) also can be calculated according to Equation 23

$$H_g = kI/V|V|^2 \tag{23}$$

Compensation of Equation 2 in Equation 6 to determine total volumetric heat generation dose (J) as given by Equation 24 (Jo and Park 2019):

is the total volumetric heat generation dose (J), and v is the sample volume (m³).

The heat generation rate is calculated by Equation 27

$$E_{vd} = \int_{ti}^{tf} H_g dt \tag{24}$$

$$E_{vd} = \int_{ti}^{tf} k\frac{I}{v}|\nabla V|^2 dt \tag{25}$$

$$\begin{aligned} E_{vd} = \frac{k}{2}\Bigg[&\left(\frac{I_0}{v_0}|\nabla V_0|^2 + \frac{I_1}{V_1}|\nabla V_1|^2\right)\Delta t_{0\sim1} + \left(\frac{I_1}{V_1}|\nabla V_1|^2 + \frac{I_2}{v_2}|\nabla V_2|^2\right)\Delta t_{1\sim2} \\ &+ \left(\frac{I_2}{v_2}|\nabla V_2|^2 + \frac{I_3}{v_3}|\nabla V_3|^2\right)\Delta t_{2\sim3} \ldots\ldots + \left(\frac{I_{n-1}}{v_{n-1}}|\nabla V_{n-1}|^2 \right. \\ &\left. + \frac{I_n}{v_n}|\nabla V_n|^2\right)\Delta t_{n-1\sim n}\Bigg] \end{aligned} \tag{26}$$

Here, is the volumetric heat generation (W/m³), is the electrical field intensity (electrical field intensity) (V/cm), is the heat generation rate (W), and is the sample volume (m³).

Sabanci and Icier (2017) demonstrated that the increase in heat generation rate during OH of cheery sour juice depends on applied electrical field intensity (voltage gradient). It was also found that the heat generation decreased with TSS content because electrical conductivity linearly decreased during the Ohmic vacuum evaporation process. At the beginning of evaporation, the heat generation rate increased but fell at the end of the process. Jo and Park (2019) found that total volumetric heat generation increases with OH's temperature comes up. The researchers presented no significant differences among electrical field intensities in the volumetric heat generation—i.e., when electrical field intensity of 10 V/cm, 12.5 V/cm, 15 V/cm, and 17.5 V/cm, total volumetric heat generation reached 49,793 J, 48,944 J, 50,859 J, and 51,460 J, respectively.

8. Applications of OH in Food Processing

OH had many applications in food processing, like extraction, pasteurization, sterilization, dehydration, water distillation, enzyme stabilization, starch gelatinization, fermentation, blanching, thawing, and microbial inactivation.

8.1 Extraction

OH is used to extract many bioactive compounds, dye, and oil, such as sucrose extraction from sugar beets (Katrokha et al. 1984), soymilk from soybean (Kim and Pyun 1995), OH-assisted mechanical juice extraction to increase the yield of juice apple (Lima and Sastry 1999), an increase of diffusion of dye in beet (Halden et al. 1990), extraction of lipids from the bran of rice (Lakkakula et al. 2004), and bioactive extraction compound from wheat bran (Al-Hilphy et al. 2014). Al-Hilphy (2014) designed an essential oil ohmic extractor from eucalyptus leaves and found that the yield of essential oil using OH was higher than the conventional extraction because of the increasing permeability of oil cells presented in eucalyptus leaves by the electrical field because of pass electrical current. Al-Hilphy et al. (2020) designed a pilot-scale ohmic extractor of wheat bran bioactive compound. The researchers declared that OH gave a higher yield quantity of bioactive compounds (total phenolic and antioxidant effectiveness) than conventional extraction by 110 to 460 ppm and 65–84%, respectively.

8.2 Pasteurization and Sterilization

Al-Hilphy et al. (2012) found that the test of alkaline phosphatase enzyme gave positive results in fresh milk and a negative ones in pasteurized milk, whether by OH at all voltages or conventional pasteurization (HTST). Also, the pasteurization efficiency using OH is higher than traditional pasteurization. Moreover, Ali et al. (2012) illustrated that higher voltage causes a large denaturation of whey proteins. As for heating at a voltage of 80 V, the denaturation of whey proteins is almost nonexistent. Pereira et al. (2010) showed that temperature affects whey proteins when using OH, as the rise in temperature with an exceeds in the heating time means more significant denaturation of whey proteins. Leizerson and Shimoni (2005) reported that the concentration of flavor compounds was higher in Ohmic-heated orange juice and had a shelf life twice as long as conventional pasteurized juice. Elzubier et al. (2009) sterilized guava juice by OH. Castro et al. (2004) mentioned that the ascorbic acid decomposition did not affect an electric field during their studies on the decomposition of Vitamin C in pasteurized products of strawberries using OH and conventional heating.

8.3 Drying

There are many studies about using OH technology in drying some food products where Kemp and Fryer (2007) have investigated the influence of OH on the drying rate of red grapes; the results showed that the drying rate of grapes treated by OH increased significantly, especially when using low electrical frequencies. Zhong and Lima (2003) also studied the impact of OH on the rate of vacuum dehydration of sweet potato tissue, comparing the treated potato slices by Ohmic heating with untreated potato slices. The results illustrated that the dehydration rate of Ohmic-treated samples was quicker than the untreated samples. Ali et al. (2010) investigated the influence of OH on the osmosis dehydration kinetics of strawberries. The researchers found that the transfer of mass and efficient diffusion rate significantly increased and improved sugar and water transfer during osmosis dehydration.

8.4 Water Distillation

Abdulstar et al. (2020) designed OH apparatus to distillate water. It was found that OH can be used to eliminate Ca^{+2} and other metals from brine. Moreover, the researchers reported that the maximum productivity was 500 mL/h at 11 V/cm, and electrical conductivity and TDS were significantly ($p < 0.05$) lower than in untreated water. Assiry et al. (2010) studied water desalination using OH and investigated wastewater electrical conductivity after evaporation. Also, the researchers showed that the OH method creates heat in seawater in trying to use it in the distillation process as an alternate treatment method instead of utilizing a steam boiler. Distillate water using OH is faster than conventional water distillation (Seidi Damyeh and Niakousari 2017). The application of OH in the process of desalination has expectant interests like decreasing the repair and chemical efficacies and enhancing plant reliability and period (Assiry et al. 2011; Seidi Damyeh and Niakousari 2017). The application of OH in the process of desalination has a unique feature at the rising rate of heating because of increasing scaling outside the boiler tube in conventional seawater heating in MSF, which leads to a reduction in the coefficient of heat transfer as the wear and corrosion produced in the boiler tube. However, there are no restrictions on heat transfer (Malik et al. 2001).

8.5 Thawing

Şeyhun et al. (2014) compared thawing a piece of beef using Ohmic thawing and traditional thawing using a water bath where the meat was frozen at a temperature of -18°C. The results illustrated that the Ohmic thawing took 160 minutes while conventional thawing, it took 450 minutes. Ohmic thawing relies on the food components, the type of material manufactured for the electrodes, and the gap between the electrodes, where the food is put between two electrodes, and the food acts as an electrical conductor. Kim et al. (2006) studied the impact of Ohmic thawing on the physical and chemical characteristics of frozen beef burgers at -18°C at different electric field strength values (10 V/cm, 20 V/cm, 30 V/cm, 40 V/cm, 50 V/cm) were used at the frequency of 50 Hz. The results depicted that the higher the electric field intensity values, the faster thawing, and that the thawing rate of ground beef at 50 V/cm was faster as it affected the physical and chemical properties of minced meat; it caused the oxidation of burger fat.

8.6 Blanching

OH is utilized as a substitutional method for blanching vegetables (Mezrahi 1996). Ohmic Blanching decreased the solid leaching extent and blanching time compared to traditional blanching (Mizrah 1996). Iier et al. (2006) blanched pea puree ohmically using electrical field intensity range from 20 to 50 V/cm, and conventionally at 100°C. The results depicted that the required time to inactivate peroxide enzyme at 30V/cm electrical field intensity and above was lower than the conventional. Blanching

Moreover, blanching using OH at 50 V/cm presented the best quality of color. Sensoy and Sastry (2004) depicted that ohmic blanching offered a higher solid content than the conventional. Moreover, weight loss was not affected by the waveform and frequency.

8.7 Enzyme and Microbial Inactivation

OH significantly reduced polyphenol activity (PPO) (Delfiya and Thangavel 2016). Abedelmaksoud (2018) revealed the minimum PPO activity in coconut ohmically heated at 80°C for 3 minutes at an electrical field intensity of 20V/cm. Treatment with OH leads to inactivation of the enzyme mainly as a result of thermal effects. Also, enzymes like PPO, phosphatase, and lipoxygenase, containing a metallic prosthetic group, are nonthermally inhibited by an electric field (Makroo et al. 2020). Urease activity was highly decreased by using OH at a frequency between 50 kHz to 10 kHz and the voltage between 160 V to 220 V compared to conventional (Li et al. 2015). As for microbial inactivation, electrical field intensity significantly influences the decreasing survival number in juice oranges (Lee et al., 2012).

8.8 Starch Gelatinization

Fa-De Li et al. (2004) reported that the electrical conductivity during starch gelatinization suspension treated by OH (90°C, 100 V AC, and 50Hz) was reduced because of the migration of charged particles. Commercial starch heated by OH gave the highest decrease in enthalpy, reducing energy requirements (An and King, 2007).

8.9 Fermentation

OH decreases the fermentation time of cheese, yogurt, and wine due to a decrease in the lag of fermentation Bactria (Cho et al. 1996). OH, enhanced the bioactive compound generation and sensory attributes. OH is an interesting technology to use in milk to manufacture probiotic fermented milk (Silva et al. 2021). OH is more appropriate for the bread fermentation process. Significantly, OH and an electric field of appropriate strength can help fermentation through effects of nonthermal and thermal. The electrolyte can decrease the fermentation lag stage of S. *thermophiles*, L. acidophilus, and *S. cerevisiae*. The electricity can alter the concentration and type of fermentation products to speedily set the temperature, releasing micronutrients from the substrate and enhancing cell permeability (Gavahian and Tiwari 2020). Besides, the gas cells are produced from the fermentation process, the dough of bread conductivity declines linearly as the porosity increases, and the electrode is at most accountable for the product temperature gradient, which supplies perfect results for the plain numeral pattern of the OH of the dough of bread (Gally et al. 2016). OH, has supremacy in precise control of temperature, enzymes, and microorganisms in the fermentation process, which is helpful in dereliction the time of fermentation (Wang et al. 2021).

9. Pulsed OH

In OH, the electrochemical reactions, such as chemicals at the interfaces between the electrode and solution caused by the electrical current, are likely unfavorable (Samaranayake et al. 2005). In OH, electrode corrosion and partial electrolysis of the heating medium occurs in low frequency (50 and 60 Hz) alternative current (Amatore et al. 1998; Tzedakis et al. 1999). Therefore, high frequency pules alternative current in OH will reduce the corrosion of electrodes and the medium electrolysis. Samaranayake et al. (2005) declared that utilizing stainless steel, titanium, and platinized titanium electrodes with a pulse electrical field, short pulse width, and higher frequency (10 kHz) can reduce the electrochemical reactions compared to conventional heating. Kim (2018) used pulsed OH with a high frequency (0.06–1kHz) to treat tomato juice. The results showed that the foodborne pathogen was inactivated effectively, and electrode corrosion was not observed.

Acknowledgments

The authors are thankful to the Department of Food Sciences, College of Agriculture, University of Basrah for available electronic references.

References

Abdulstar, A.R., Altemimi, A., Al-Hilphy, A.R., Watson, D.G. and Lakhssassi, N. 2020. Water distillation using an ohmic heating apparatus. Int. J. Ambient Ener. 1: 11.

Abedelmaksoud, T.G., Mohsen, S.M., Duedahl-Olesen, L., Elnikeety, M.M. and Feyissa, A.H. 2018. Optimization of ohmic heating parameters for polyphenoloxidase inactivation in not-from-concentrate elstar apple juice using RSM. Journal of Food Science & Technology 55(7): 2420–2428.

Al-Hilphy, A.R.S. 2018a. Engineering intervention for extraction of essential oils from plants. pp. 51–85. *In*: Verma, D.K. and Goyal, M.R. (eds.). Advances, engineering intervention in foods and plants. Apple Academic Press, Inc., AAP. pp.325. ISBN: 978-1-77188-596-6.pp314.

Al-Hilphy, A.R.S. 2018b. Technology of Ohmic heating in Meat processing. pp. 57–70. *In*: Goyal, M.R. and Verma, D.K. (eds.). Advances, Innovations and Emerging Technologies for Agricultural and Biological Science. Apple Academic Press, Inc., AAP. pp. 325. ISBN: 978-1-77188-556-0.

Al-Hilphy A.R.S., Ali, H.I. and Mohsin, G.F. 2018. Technology of Ohmic heating for the pasteurization of milk. pp. 3–46. *In*: Goyal, M.R., Kumar, A. and Gupta, A.K. (eds.). Novel dairy processing Technologies. Apple Academic Press. pp. 298. USA. ISBN: 978-1-77188-612-3.

Al-Hilphy, A.R.S. 2014. A Practical Study for new Design of Essential Oils Extraction Apparatus Using Ohmic Heating. Int. J. l Agric. Sci. 4(12): 351–366.

Al-Hilphy, A.R.S., AlRikabi, A.K.J. and Al-Salim, A.M. 2015. Extraction of phenolic compounds from wheat bran using ohmic heating. Food Sci. and Quality Manag. 43.

Al-Hilphy, A.R., Al-Musafer, A.M. and Gavahian, M. 2020. Pilot-scale ohmic heating-assisted extraction of wheat bran bioactive compounds: Effects of the extract on corn oil stability. Food Res. Int. 137: 109649.

Al-Hilphy, A.R., Ali, H. and Mohsin, G.F. 2012. Designing and Manufacturing Milk Pasteurization Apparatus by Ohmic Heating and studying Its Efficiency. J. Basrah Res. (Sciences), 38(4B).

Ali, H.I., Al-Hilphy, A.R.S. and Mohsin, G.F. 2012. Effect of Ohmic Heating in the Elimination of Microorganisms. J. Kerbala Univ. The second conference of Agriculture College, pp. 1128–1137.

Alkanan, Z.T., Altemimi, A.B., Al-Hilphy, A.R., Watson, D.G. and Pratap-Singh, A. 2021. Ohmic Heating in the Food Industry: Developments in Concepts and Applications during 2013–2020. Applied Sciences 11(6): 2507.

All, H., Marchal, L. and Vorobiev, E. 2010. Blanching of strawberries by ohmic heating: effects on the kinetics of mass transfer during osmotic dehydration. Food and Bioprocess Technology 3(3): 406–414.

Altemimi, A., Ali, H.I., Al-Hilphy, A.R.S., Lightfoot, D.A. and Watson, D.G. 2018. Electric Field Applications on Dried key Lime Juice Quality with Regression Modeling. J. Food Process Preserv 42(6): e13637. doi:10.1111/jfpp.13637.

An, H.J. and King, J.M. 2007. Thermal characteristics of ohmically heated rice starch and rice flours. Journal of Food Science 72(1): C084–C088.

Assiry, A.M. 2011. Application of Ohmic heating technique to approach near-ZLD during the evaporation process of seawater. Desalination. 280: 217–23.

Assiry, A.M., Gaily, M.H., Alsamee, M. and Saifuddin, A. 2010. Electrical conductivity of seawater during Ohmic heating. Desalination. 260: 9–17.

Amatore, C., Berthou, M. and Hebert, S. 1998. Fundamental principles of electrochemical ohmic heating of solutions. J. Electroanal Chem. 457: 191–203.

Bean, E.C., J.P. Rasor and G.C. Porter. 1960. Changes in electrical conductivities of avocados during ripening. Year Book of Californian Avocado Society 44: 75 –78.

Benabderrahmane, Y. and J.P. Pain. 2000. Thermal behaviour of solid-liquid mixture in an ohmic heating sterilise-slip phase model. Chem. Eng. Sci. 55: 1371 –1384.

Berk, Z. 2009. Food Engineering and technology. Food Science and Technology international series, AP.

Berthou, M., Chopard, F. and Aussudre, C. 2001. Heater for resistive heating of a fluid, fluid-treatment apparatus incorporating such a heater, and a method of treating a fluid by resistive heating. Word Patent No 0141509.

Bozkurt, H. and Icier, F. 2010. Ohmic cooking of ground beef: Effects on quality. Journal of Food Engineering 96: 481–490.

Cappato, L. P. et al. 2017. Ohmic heating in dairy processing: Relevant aspects for safety and quality,‖ Trends Food Sci. Technol. 62: 104–112.

Casaburi, O., Brondi, C., Romano, A. and Marra, F. 2021. Ohmic Heating of Basil-based Sauces: Influence of the Electric Field Strength on the Electrical Conductivity. Chemical Engineering Transactions 87: 343–348.

Castro, A., Teixeira, J.A., Salengke, S., Sastry, S.K. and Vicente, A.A. 2004. Ohmic Heating of Strawberry Products: Electrical Conductivity Measurements and Ascorbic Acid Degradation Kinetics. Innov. Food Scie. and Emerg. Technol. 5: 27–36.

Castro, I., Teixeira, J.A., Salengke, S., Sastry, S.K. and Vicente, A.A. 2003. The influence of field strength, sugar and solid content on electrical conductivity of strawberry products. Journal of Food Process Engineering 26(1): 17–29.

Cho, W.I.l., Kim, E.J., Hwang, H.J., Cha, Y.H., Cheon, H.S., Choi, J.B. and Chung, M.S. 2017. Continuous ohmic heating system for the pasteurization of fermented red pepper paste. Innovative Food Sci. Emerg. Technol. 42: 190–196. https://doi.org/10.1016/j.ifset.2017.07.020

Darvishi, H., Hosainpour, A. and Nargesi, F. 2012. Ohmic heating behaviour and electrical conductivity of tomato paste. J. Nutr. Food Sci. 2(167): 1–5.

Darvishi, H., Hosainpour, A., Nargesi, F. and Fadavi, A. 2015. Exergy and energy analyses of liquid food in an Ohmic heating process: A case study of tomato production. Innovative Food Sci and Emerg. Technol. 31: 73–82 (Retrieved from) HTTP:// www.sciencedirect.com/science/article/pii/S1466856415001216

Darvishi, H., Hosainpour, A., Nargesi, F., Khoshtaghaza, M.H. and Torang, H. 2011. Ohmic Processing: Temperature-Dependent Electrical Conductivities of Lemon Juice. Modern Appl. Sci. 5(1): 209.

Darvishi, H., Khostaghaza, M.H. and Najafi, G. 2013. Ohmic heating of pomegranate juice: electrical conductivity and pH change. J. Saudi Soc. Agri. Sci. 12(2): 101–108.

Darvishi, H., Salami, P., Fadavi, A. and Saba, M.K. 2020. Processing kinetics, quality and thermodynamic evaluation of mulberry juice concentration process using Ohmic heating. Food and Bioproducts Processing 123: 102–10.

Davies, L.J., Kemp, M.R. and Fryer, P.J. 1999. The geometry of shadows: effects of inhomogeneities in electrical field processing. J. Food Eng. 40(4): 245–258.

De Alwis, A.A.P., Halden, K. and Fryer, P.J. 1989, Shape and conductivity effects in the ohmic heating of foods. Chem. Eng. Res. & Design 67(2): 159–168.

De Alwis, A.A.P., Halden, K. and Fryer, P.J. 1989. Shape and conductivity effects in the ohmic heating of foods. Chemical Engineering Research Design. 67: 159–168.

De Halleux, D., Piette, G., Buteau, M.L. and Dostie, M. 2005. Ohmic cooking of processed meats: energy evaluation and food safety considerations. Canadian Biosys. Eng. 47(3): 341–347.

Delfiya, D.S.A. and Thangavel, K. 2016. Effect of ohmic heating on polyphenol oxidase activity, electrical and physicochemical properties of fresh tender coconut water. International J. Food Eng. 12(7): 691–700.

Elzubier, A.S., Thomas, C.S.Y., Sergie, S.Y., Chin, N.L. and Ibrahim, O.M. 2009 The effect of buoyancy force in computational fluid dynamics simulation of a two dimensional continuous ohmic heating process. Am. J. Appl. Sci. 6(11): 1902–1908.

Fa-De-L., Te-Li, L., Zaigui, L. and Eizo, T. 2004. Determination of starch gelatinization temperature by ohmic heating. J. Food Eng. 62: 113–120.

Floury J., Grosset, N., Leconte, N., Pasco, M. and Madec, M. 2006. Continuous raw skim milk processing by pulsed electric field at non-lethal temperature: Effect on microbial inactivation and functional properties. Lait. 86: 43–57.

Food and Drug Administration-Center for Food Safety and Applied Nutrition (FDACFSAN) 2000. Kinetics of Microbial Inactivation for Alternative Food Processing Technologies-Ohmic and Inductive heating. www.cfsan.fda.gov/~comm/ift-ohm.html

Gally, T., Rouaud, O., Jury, V. et al. 2016. Bread baking using ohmic heating technology; a comprehensive study based on experiments and modelling. J. Food Eng. 190: 176–184.

Gavahian, M., B.K. Tiwari. 2020. Moderate electric fields and ohmic heating as promising fermentation tools, Innov. Food Sci. Emerg. Technol. 64102422:

Gavahian, M., Chu, Y.H. and Farahnaky, A. 2019. Effects of ohmic and microwave cooking on textural softening and physical properties of rice. J. Food Eng. 243: 114–124.

Ghnimi, S., Flach-Malaspina, N., Dresch, M., Delaplace, G. and Maingonnat, J.F. 2008. Design and performance evaluation of an Ohmic heating unit for thermal processing of highly viscous liquids. Chem. Eng. Des. 86: 626–32.

Goullieaux, A. and Pain, J.P. 2014a. Recent Development in Microwave Heating. Sun, D. Ed., Emerging Technologies for Food Processing, pp. 361–377.

Goullieux, A. and Pain, J.P. 2014b. Chapter 22-Ohmic Heating,‖ D.-W. B. T.-E. T. for F. P. (Second ed. Sun, Ed. San Diego: Academic Press, pp. 399–426.

Halden K., De Alwis, A.A.P. and Fryer, P.J. 1990. Changes in electrical conductivity of foods during ohmic heating. Int. J. Food Sci. Technol. 25: 9–25.

Henningsson, M., Östergren, K. and Dejmek, P. 2005. The electrical conductivity of milk—the effect of dilution and temperature. International J. Food Property 8(1): 15–22.

Hosainpour, A., Nargesi, F., Darvishi, H. and Fadavi, A. 2014. Ohmic pre-drying of tomato paste. Food Sci. Technol. Int. 20: 193–204.

Icier, F. 2012. Novel thermal and non-thermal technologies for fluid foods: Chapter 11: Ohmic heating of fluid foods. Elsevier Inc. doi: 10.1016/B978-0-12-381470-8.00011-6.

Icier, F. and Ilicali, C. 2005. The effects of concentration on electrical conductivity of orange juice concentrates during ohmic heating. Eur. Food Res. Technol. 220(3): 406–14.

Icier, F., Yildiz, H. and Baysal, T. 2008. Polyphenol oxidase deactivation kinetics during ohmic heating of grape juice. J. Food Eng. 85(3): 410–417.

Indiarto, R. and Rezaharsamto, B. 2020. A review on ohmic heating and its use in food. Int. J. Sci. Technol. Res. 9(2): 485–490.

Jaeger, H., Roth, A., Toepfl, S., Holzhauser, T., Engel, K.H., Knorr, D. and Steinberg, P. 2016. Opinion on the use of ohmic heating for the treatment of foods. Trends in Food Sci. Technol. 55: 84–97.

Jo, Y.J. and Park, S.H. 2019. Evaluation of energy efficacy and texture of ohmically cooked noodles. Journal of Food Engineering 248: 71–79.

Kasturi, S. and Ayalur, K.B. 2020. Energy diagnosis and savings potential in knitted fabric based wet processing units. Energy Sust. Develop. 54: 85–100. https://doi.org/10.1016/j.esd.2019.10.003.

Katrokha, I., Matvienko, A., Vorona, L., Kupchik, M.. and Zaets, V. 1984. Intensification of sugar extraction from sweet sugar beetcossettes in an electric field. Sakharnaya Promyshlennost 7: 28–31.

Kaur, N. and Singh, A.K. 2016. Ohmic heating: concept and applications—a review. Critical Reviews in Food Sci. Nutr. 56(14): 2338–2351.

Kemp, M.R and Fryer, P.J. 2007. Enhancement of diffusion through foods using alternating electric fields. Innovative Food Science and Emerging Technologies 8: 143–153.

Kim, H.J., Choi, Y.M., Yang, T.S., Taub, I.A. and Tempest, P. 1996. Validation of ohmic heating quality enhancement of food products: ohmic heating for thermal processing of foods: government, industry, and academic perspectives. Food Technol. (Chicago) 50(5): 253–261.

Kim, J. and Pyun, Y. 1995. Extraction of Soy Milk Using Ohmic Heating. Abstract, 9th Congress of Food Sci. and Tech., Budapest, Hungary.

Kim, J.Y., Hong, G.P., Park, S.H., Spiess, W.E. and S.G. Min. 2006. Effect of ohmic thawing on Physico-chemical properties of frozen hamburger patties. Food Science of Animal Resources 26(2): 223–228.

Kim, S. 2018. Application of Pulsed Ohmic Heating for Inactivation of Foodborne Pathogens (Doctoral dissertation, Department of Agricultural Biotechnology College of Agriculture and Life Sciences Seoul National University.

Knirsch, M.C., Alves dos Santos, C., Martins de Oliveira Soares Vicent, A.A. and Vessoni Penna, T.C. 2010. Ohmic heating —a review, ‖ Trends Food Sci. Technol. 21(9): 436–441.

Kong,Y.Q., Dong, L.i., Wang, L.J., Bhandari, B., Chen, D.X. and Mao, Z.H. 2008. Ohmic Heating Behavior of Certain Selected Liquid Food Materials. Intern. J.l Food Eng. 4(3).

Lakkakula R.N., Lima, M. and Walker, T. 2004. Rice bran stabilization and rice bran oil extraction using ohmic heating. Bioresource Technol. 92: 157–161.

Larkin, J.W. and Spinak, S.H. 1996. Safety considerations for ohmically heated, aseptically processed, multiphase low-acid food products. Food Technol. 50(5): 242–245.

Lee, S., Sagong, H., Ryu, S. and Kang, D. 2012. Effect of continuous ohmic heating to inactivate Escherichia coli O157:H7, Salmonella Typhimurium and Listeria monocytogenes in orange juice and tomato juice. J. Appl. Microbiol. 112(4): 723–731.

Leizerson, S. and Shimoni, E. 2005. Stability and sensory shelf life of orange juice pasteurized by continuous ohmic heating. J. Agr. Food Chem. 53(10): 4012–4018.

Lima M. and Sastry, S.K. 1999. The effects of ohmic heating frequency on hot-air drying rate and juice yield. J. Food Eng., 41: 115–119.

Lima, M. 2007. Ohmic Heating: Quality Improvements, Encyclop. Agric., Food and Biolog. Eng. 1: 1–3.

Makroo, H.A., Rastogi, N.K. and Srivastava, B. 2017. Enzyme inactivation of tomato juice by ohmic heating and its effects on Physico-chemical characteristics of concentrated tomato paste. Journal of Food Process Engineering 40(3): e12464.

Makroo, H.A., Rastogi, N.K. and Srivastava, B. 2020. Ohmic heating assisted inactivation of enzymes and microorganisms in foods: A review. Trends Food Sci. Technol. 97: 451–465. doi:10.1016/j.tifs.2020.01.015

Malik, A.U., Al-Ghamdi, M.A., N.K. and Hodhan, A.H. 2001. Investigation on the boiler tubes of Boiler #4, Jeddah-testing for microstructure and life prediction. In: Ninth middle east corrosion conference and exhibition in Bahrain, Feb.12–14.

Marcotte, M., Trigui, M. and Ramaswamy, H.S. 2000. Effect of salt and citric acid on electrical conductivities and ohmic heating of viscous liquids,‖ J. Food Process. Preserv. 24: 389–406.

McKenna, B.M., Lyng, J., Brunton, N. and Shirsat, N. 2006. Advances in radio frequency and ohmic heating of meats, ‖ J. Food Eng. 772: 215–229.

Mizrahi, S. 1996. Leaching of soluble solids during blanching of vegetables by ohmic heating. Journal of Food Engineering, 29: 153–166. doi:10.1016/0260-8774(95)00074-7.

Muhammad, A.I., Shitu, A. and Tadda, M.A. 2019. Ohmic Heating as Alternative Preservation Technique—A Review. Arid Zone J. Eng. Technol. Environ. 15: 268–277.

Nimratbir K. and Singh, A.K. 2015. Ohmic Heating: Concept and Applications—A Review, Critical Revi. Food Sci. Nutr. 56(14): 2338–2351 DOI: 10.1080/10408398.2013.835303

Palaniappan, S. and Sastry, S.K. 1991. Electrical conductivity of selected juices: Influences of temperature, solids content, applied voltage, and particle size. J. Food Process Eng. 14: 247–260.

Palaniappan, S. and Sastry, S.K. 1991. Electrical conductivity of selected juices: influences of temperature, solids content, applied voltage, and particle size 1. J. Food Process Eng. 14(4): 247–260.

Park, I.K., Ha, J.W. and Kang, D.H. 2017. Investigation of optimum ohmic heating conditions for inactivation of Escherichia coli O157:H7, Salmonella enterica serovar Typhimurium, and Listeria monocytogenes in apple juice. BMC Microbiol. 17(1): 117.

Pereira, R.N., Teixeira, J.A. and Vicente, A.A. 2010. Denaturation of Whey Proteins of milk during Ohmic heating, IBB-Institute for Biotechnology and Bioengineering, Centre for Biological Engineering, University of Minho Campus de Gualtar, 4700-035 Braga, Portugal.

Raso, J., Frey, W., Ferrari, G., Pataro, G., Knorr, D., Teissie, J. and Miklavčič, D. 2016. Recommendations guidelines on the key information to be reported in studies of application of PEF technology in food and biotechnological processes. Innovative Food Science & Emerging Technologies 37: 312–321.

Reznik, D. 1996. ohmic heating of fluid foods: various parameters affect the performance of ohmic heating devices used to heat fluid food products. Food Technol. 5: 251–260.

Sabanci, S. and Icier, F. 2017. Applicability of ohmic heating assisted vacuum evaporation for concentration of sour cherry juice. Journal of Food Engineering 212: 262–270.

Sakr, M. and Liu, S. 2014. A comprehensive review on applications of ohmic heating (OH). Renewable and Sustainable Energy Revi. 39: 262–269.

Samaranayake, C.P. and Sastry, S.K. 2005. Electrode and pH effects on electrochemical reactions during Ohmic heating. J. Electroanal Chem. 577: 125–35.

Samprovalaki, K., Bakalis, S. and Fryer, P.J. 2007. Ohmic heating: models and measurements. pp. 159–186. *In*: Yanniotis, S. and Sunden, B. (eds.). Heat transfer in food processing. WIT Press, Southampton.

Sasson, A. and Monselise, S.P. 1977. Electrical conductivity of Shamouti orange peel during fruit growth and postharvest senescence. J. American Soci. Horticultural Sci. 102: 142 –144.

Sastry, S.K. 1991. Ohmic sterilisation and related safety issues. International winter meeting of ASAE. Paper No. 916616.

Seidi Damyeh, M. and Niakousari, M. 2017. Ohmic Hydrodistillation, an Accelerated Energy-Saver Green Process in the Extraction of Pulicaria Undulata Essential oil. Industrial Crops and Products 98: 100–107. doi:10.1016/j.indcrop.2017.01.029.

Sensoy, İ.L.K.A.Y. and Sastry, S.K. 2004. Ohmic blanching of mushrooms. Journal of Food Process Engineering 27(1): 1–15.

Şeyhun, A. 2014. Islamist thinkers in the late Ottoman Empire and early Turkish Republic. Brill.

Shirsat, N., Lyng, J.G., Brunton, N.P. and McKenna, B. 2004. Ohmic processing: Electrical conductivities of pork cuts. Meat Sci. 67: 507– 514.

Shynkaryk, M.V., Ji, T., Alvarez, V.B. and Sastry, S.K. 2010. Ohmic Heating of Peaches in the Wide Range of Frequencies (50 Hz to 1 MHz),‖ J. Food Sci. 75(7): E493–E500.

Silva, A.B., Scudini, H., Ramos, G.L.P., Pires, R.P., Guimarães, J.T., Balthazar, C.F. and Cruz, A.G. 2021. Ohmic heating processing of milk for probiotic fermented milk production: Survival kinetics of Listeria monocytogenes as contaminant post-fermentation, bioactive compounds retention and sensory acceptance. International Journal of Food Microbiology 348: 109204.

Silva, V.L., Santos, L.M. and Silva, A.M. 2017. Ohmic heating: An emerging concept in organic synthesis. Chem.–A Eur. J. 23(33): 7853–7865.

Singh, R.P. and Heldman, D.R. 2014. Introduction to food engineering, 5th ed. Academic Press Incorporation, Orlando, Florida.

Stirling, R. 1987. Ohmic heating—a new process for the food industry. Power Eng. J. 1: 365–371.

Tzedakis, T., Basseguy, R. and Comtat, M. 1999. Voltammetric and coulometric techniques to estimate the electrochemical reaction rate during ohmic sterilization. J. Appl. Electrochem. 29(7): 819–826.

Varghese, K.S., Pandey, M.C., Radhakrishna, K. and Bawa, A.S. 2014. Technology, applications and modelling of ohmic heating: a review,‖ J. Food Sci. Technol. 51(10): 2304–2317.

Wang, W. and Sastry, S.K. 1997. Changes in electrical conductivity of selected vegetables during multiple thermal treatments. J Food Process Eng. 20: 449–516.

Wang, Z., Ma, S.L., Li, Sun, B., Wang, F., Huang, J. and Wang, X. 2021. Novel heating technologies and potential methods to improve fermentation efficiency and quality in wheat products: A short review. Grain & Oil Sci. Technol. 4(2): 81–87.

Zareifard, M.R., Ramaswamy, H.S., Trigui, M. and Marcotte, M. 2003. Ohmic heating behavior and electrical conductivity of two-phase food systems. Innov. Food Sci. Emerg. Technol. 4: 45–55.

Zell, M., Lyng, J.G., Morgan, D.J. and Cronin, D.A. 2011. Minimizing heat losses during batch ohmic heating of solid food. Food Bioprod Process 89: 128–134.

Zoltai, P. and Swearingen, P. 1996. Product development consideration for ohmic processing. Food Technol. 50: 263–266.

17

Microencapsulation of Functional Foods

Simara Somacal and *Cristiano Ragagnin de Menezes**

1. Introduction

Microencapsulation technology plays a promising role in the food industry, as the consumer's concern to acquire products that provide healthiness is increasing (Martins et al. 2019). This technique is a tool that allows the fine coating of solid particles, liquid drops, and dispersions. This technique can solve limitations in the use of food ingredients, as it can suppress or attenuate undesirable flavors, reduce volatility and reactivity to external factors, and increase their stability under adverse environmental conditions, such as the changes in temperature, presence of light, oxygen, and extreme pH (Đorđević et al. 2014).

Encapsulation protects a bioactive material, which can release its contents at controlled rates under specific conditions, preventing inadequate exposure. The material used to protect the composite is called a shell, support, or wall material (Budinčić et al. 2021).

The microencapsulation process can produce microspheres or microcapsules. Microspheres can be defined as a matrix in which the encapsulated substance and polymer are a uniform mixture. In microcapsules, the encapsulated substance is dispersed in the polymeric cavities by agglomerates, forming a nucleus (Wang et al. 2013). The morphology of the microcapsules will depend on the core material, the wall material, and the technique used. As can be seen in Fig. 1, microcapsules can be of the mononuclear type (microcapsules), where the wall material covers a single core and can also be called a reservoir or polynuclear, which is defined as particles that have many coated cores as a wall material (microparticles); or also of the matrix type, which is polynucleated and has the active agent dispersed in the wall material nucleus (Wang et al. 2013). Capsules can be classified by size into three categories: macro- (> 5,000 μm), micro- (0.2–5,000 μm), and nanocapsules (< 0.2 μm).

Microencapsulation has numerous applications where we can highlight the pharmaceutical, food, and cosmetic areas, such as the encapsulation of essential oils, dyes, flavorings, vitamins, amino acids, polyunsaturated fatty acids, probiotics, and prebiotics. The main objective is to protect the encapsulated material from adverse environmental conditions and release it at its site of action in quantity and at the appropriate time to perform the desired function (Poletto et al. 2019; FJ et al. 2020; Castejón et al. 2021a).

Department of Food Science and Technology, Federal University of Santa Maria, 97105-9003, Santa Maria, RS, Brazil.
* Corresponding author: cristiano.ufsm@gmail.com

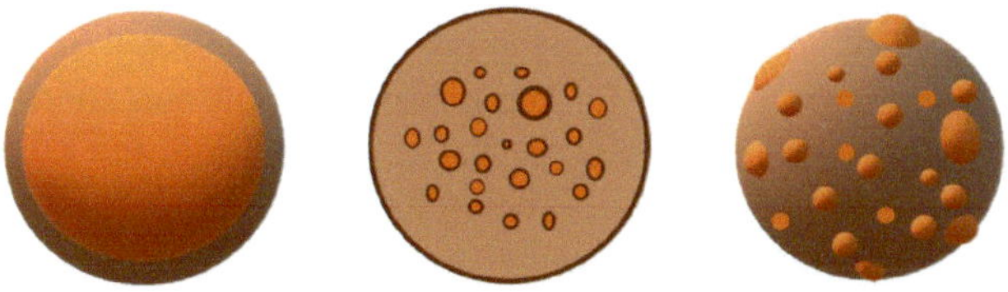

Figure 1. Morphology of Different Particles Obtained by the Microencapsulation Process. Adapted From Singh et al. (2010).

2. Functional Foods

In recent decades, consumers are increasingly convinced that certain foods can make a major contribution to their overall health status. This is related to advances in research that have increased understanding of the impact that certain foods or food compounds have on nutrition and health and resulted in the advancement of the concept of functional foods (Nystrand and Olsen 2020; Fu et al. 2022).

Functional foods include natural and processed foods that have one or more bioactive substances in their composition, which can generate beneficial physiological effects and even improve health and well-being or reduce the risk of developing diseases (Yeung et al. 2018; Granato et al. 2020). Bioactive substances found in these foods can be grouped into probiotics and prebiotics, polyunsaturated fatty acids, sulfur and nitrogen foods, pigments and vitamins, phenolic compounds, fiber, and others. Functional foods can be characterized into three different groups as shown in Table 1 (Granato et al. 2020):

It is estimated that the category of foods with functional properties will have an expansion rate of 6.7% in the next five years (Foods Market Functional 2021). The SARS-CoV-2 pandemic notably impacted the demand for functional foods, with a portion of consumers adopting changes in their eating habits, prioritizing homemade preparations of natural foods as well as consuming foods that bring health benefits. On the other hand, another portion of consumers increased unhealthy eating patterns, reporting greater consumption of ultra-processed foods and reduced consumption of vegetables (Rodrigues et al. 2021; Janssen et al. 2021; Lima et al. 2021).

Despite the estimated growth rate, the development of foods and functional ingredients is challenging for industries because of the difficulty in ensuring that these products arrive biologically active and meet consumer expectations (Granato et al. 2020). Foods and functional ingredients can undergo degradation during the processing, distribution, and storage caused by moisture, light, temperature, oxygen, pH, microorganisms, and contaminants among other factors (Marrapu et al. 2020). In this context, the use of technology to protect against adverse conditions, such as microencapsulation, can be used as a strategy to prolong the stability of these compounds. Microencapsulation technology is based on encapsulating labile substances that need protection or that are intended to modify their release in a specific environment, such as the gastrointestinal tract (Suave et al. 2006). The characteristics of size reduction involve better sensory qualities of the food matrix and increased viability of functional compounds over time (Lu et al. 2021).

Table 1. Characterization of Functional Foods.

Classification		
I	II	III
Conventional foods	Foods that have been modified	Synthesized or derived food ingredients
Contains naturally occurring bioactive compounds	Obtained through enrichment with bioactive substances	Added to conventional foods
Examples: Carrots (containing the carotenoid β-carotene); dietary fiber	Examples: Margarine containing added phytosterols; probiotics	Examples: Prebiotics;

2.1 Probiotics, Prebiotics, and Symbiotics

Currently, the food supply is challenging for the pharmaceutical and food industries because of the appeal of health promotion (Canizales et al. 2018). In addition, the pointed relevance of the intestinal microbiota in human health and well-being has led to an increase in the development of new products, such as probiotics, prebiotics, and symbiotics.

Probiotics are live microorganisms that when administered in adequate amounts, induce health benefits to the host (Hill et al. 2014). To be considered probiotics, such as those belonging to the *Bifidobacterium* and *Lactobacillus* genera, microorganisms must survive an acidic environment and exposure to bile salts found in the digestion stage of the human body, have a good absorption capacity in the intestine, and maintain a clear link with health markers (Papizadeh et al. 2017; Champagne et al. 2018).

The health benefits of probiotics are acquired through the daily ingestion of a probiotic product that contains 10^8–10^9 CFU g^{-1}, as per the legislation, or the food must contain at least 106–10^7 CFU g^{-1} per unit dose (WGO 2017). A series of health benefits are attributed to probiotics, related to the improved immune response to pathogens and inflammatory diseases, such as neurodegenerative diseases, anti-infective properties against *Salmonella* enterica, prevention of inflammatory bowel disease, modulation of dysbiosis of the intestinal microbiota, therapeutic effect against depression, or anxiety and anti-diabetes therapeutic effect with reduced inflammation among others (Roobab et al. 2020).

Prebiotics are non-digestible food ingredients that have the ability to improve survival, growth, metabolism, and the health-promoting activities of probiotics in the digestive system (Ashaolu 2020). Prebiotics are natural food constituents and the best known are fructooligosaccharides (FOS), galactooligosaccharides (GOS), inulin, and xylooligosaccharides (XOS) with FOS and GOS mainly stimulating the proliferation of *Lactobacillus* and *Bifidobacterium* (Ashaolu 2020).

Symbiotics, on the other hand, are a combination of probiotics and prebiotics in which the intention is to increase the benefits of probiotics, efficiently improving the survival of beneficial microorganisms present in the intestinal microbiota (Mohanty et al. 2018; Quigley 2019).

The commercial offer of probiotics, prebiotics, and symbiotics is currently as supplements or are incorporated as ingredients in dairy products, such as yogurts, fermented milk, dairy drinks, dairy desserts, and cheeses (Krumbeck et al. 2018; Champagne et al. 2018).

2.1.1 Polyunsaturated Fatty Acids (PUFAs)

Lipids play an important role in the quality of food products, particularly concerning the organoleptic properties that make them desirable. On the other hand, they add nutritional value to foods, constituting an important source of metabolic energy, essential fatty acids (linoleic, linolenic, and arachidonic acids), and fat-soluble vitamins (A, D, E, and K) (Damodaran et al. 2010).

The major constituents of lipids are triacylglycerols, that is, fatty acids esterified with glycerol. Fatty acids are classified according to the type of carbon chain where they can be saturated (SFA) or unsaturated (UFA), which in turn can be monounsaturated (MUFAs) or PUFAs (Shahidi and Ambigaipalan 2019).

PUFAs are called essential fatty acids and must be supplied by the diet. These fatty acids are mainly composed of the omega 3 (n-3) and omega 6 (n-6) families. PUFAs are associated with many health benefits; there is a well-known association between a high proportion of PUFAs (particularly long-chain n-3 PUFAs) and reduced risk of cardiovascular diseases (Shahidi and Ambigaipalan 2018) and depression (Liao et al. 2019). In addition, n-3 fatty acids bring benefits to human health, such as lowering blood triglyceride levels and anti-inflammatory effects (Kaushik et al. 2015).

The current strategy to develop functional foods containing PUFAs focuses primarily on adding one or a combination of PUFAs from natural sources to foods to increase their individual and total content (Heck et al. 2017). However, the incorporation of PUFAs into foods is a challenge, as this group of fatty acids has a high number of unsaturations in its carbon chain, which makes these fatty acids more susceptible to the lipid oxidation process and can negatively affect the sensory properties of foods where PUFAs were added (Jiménez-Martín et al. 2016). Thus, the feasibility of incorporating these compounds as functional ingredients has to overcome challenges related to their stability during food processing, storage, and passage through the gastrointestinal tract to ensure that they exert the desired health benefits even after ingestion.

Microencapsulation technology is an attractive strategy to increase the stability of functional foods. Proper selection of encapsulation methods or core materials and carriers can influence most of the desired properties of the final food product. (Canizales et al. 2018).

3. Techniques for Microencapsulation of Functional Compounds

Different microencapsulation techniques can be used to help the incorporation of bioactive compounds into foods or beverages efficiently and thus collaborate in the development of new functional foods. Microcapsules containing bioactive compounds can be produced by methods, such as spray drying, internal ionic emulsification/gelation, external ionic extrusion/gelation, and complex coacervation (Holkem et al. 2017; Timilsena et al. 2019; Solomando et al. 2020; Somacal et al. 2020). The success of a microencapsulation system is related to the microencapsulation method, the type of active material, the encapsulating agent, the desired release mechanism for its action, and the characteristics of the food or other product where the microcapsules will be incorporated. The encapsulation efficiency and particle size of microcapsules also depend on the encapsulating agent and the encapsulation method used, as well as on the physicochemical characteristics of the functional compounds to be microencapsulated (Table 2). Therefore, it is essential to know the different techniques that can be used in the production of microcapsules containing bioactive agents and all other variables that interfere in the process to produce quality microcapsules with good yield and efficiently protect the compounds of interest.

Table 2. Different Methods Used to Microencapsulate Functional Compounds.

Encapsulation Method	Functional Compounds	Encapsulating Agent	Encapsulation Efficiency (%)	Size (μm)	References
Spray drying	PUFAs	Sodium caseinate + lactose	58.8–76.9	1.5–30	(Castejón et al. 2021a)
Spray drying	Probiotic	Gum arabic + maltodextrin + glycerol	79.73–84.6	4.85–8.75	(Nunes et al. 2018)
Complex coacervation	Probiotic	Gelatin + arabic gum	77.6–87.53	127–227	(Silva et al. 2019)
Extrusion/external ionic gelation	PUFAs	Pectin	95.27–97.58	862–1,165	(Menin et al. 2018)
Extrusion/external ionic gelation	Probiotic + prebiotic	Sodium alginate	92	149–167	(Raddatz et al. 2022)
Extrusion/external ionic gelation	Probiotic	Sodium alginate	96.75	127–234	(Poletto et al. 2019)
Extrusion/external ionic gelation	PUFAs	Sodium alginate	80	105	(Somacal et al. 2020)
Emulsification/ internal ionic gelation	Probiotic	Pectin	91.24	24.4	(Poletto et al. 2019)
	Probiotic + prebiotic	Pectin + inulin	90.59	462	

3.1 Spray Drying

The spray drying microencapsulation method (Fig. 2) is performed in two steps. In the first step, the substance to be encapsulated is dispersed or dissolved in an aqueous solution containing the encapsulating agent after which it is atomized into fine droplets using a stream of hot air. In this step,

Figure 2. Laboratory Scale Spray Dryer.

the solvent present in the droplets is immediately dried, giving rise to a fine, dry powder, which is the microcapsules containing the substance that was encapsulated (Nunes et al. 2018).

Microencapsulation by spray drying is an efficient method because it produces dry microparticles ready to be applied to the final product, which reduces the total processing time. This technique is also energy efficient and has low operating costs compared to other advanced particle engineering techniques, such as FD (Sosnik and Seremeta 2015).

Formulation parameters, droplet size, and thermal conditions inside the dryer are the main factors that affect the size and morphology of particles produced by spray drying. The concentration and properties of the solvent in the formulation influence the surface tension, viscosity, and density of the solution, which impact the particle size (Us-Medina et al. 2018; Solomando et al. 2020). Castejón et al. (2021b) studied the effects of the spray drying microencapsulation process on the omega-3 fatty acid profile. The microencapsulation of PUFAs by this method showed that it did not affect the omega-3 fatty acid profile, but the encapsulation efficiency varied for the different PUFAs sources; the best microencapsulation efficiency was achieved for *Salvia hispanica* L. (76.9%), while the lowest was obtained for *Camelina sativa* L oil.

3.2 Extrusion/External Ionic Gelation

The external ion extrusion/gelation technique is a simple, fast, and low-cost encapsulation method based on the ability of polyelectrolytes to cross-link in the presence of counterions, forming a three-dimensional mesh structure (Menin et al. 2018). This method is based on incorporating the material to be encapsulated in a sodium alginate solution, and then the mixture undergoes drop-by-drop extrusion—through a small-bore pipette, a syringe, or an extruder nozzle—into a solution of calcium chloride ($CaCl_2$), where it must remain for about 20 minutes in contact with this ionic solution to achieve stability and mechanical strength (Etchepare et al. 2015; Raddatz et al. 2022).

The external ion extrusion/gelation technique is promising for the encapsulation of bioactive compounds because the production of microparticles is done at room temperature and does not use organic solvents, minimizing environmental impacts without compromising the efficiency of microencapsulation (Menin et al. 2018). This technique has been successfully applied for the encapsulation of probiotics, as demonstrated by Raddatz et al. (2022), who obtained an encapsulation efficiency of < 92% and observed that the microcapsules containing the probiotic *Lactobacillus casei* remained stable during 90 days of storage. Furthermore, they found that microencapsulation increased the survival of the probiotic during simulated gastrointestinal conditions and provided a controlled release of the probiotics, signaling that microparticles produced by this technique can promote the extended viability and stability of the probiotic *L. casei* under different conditions. An excellent encapsulation efficiency of probiotic cells using the external ionic extrusion/gelation technique was also observed by Poletto et al. (2019), which obtained 96.75%, and Vaziri et al. (2018) reached 99.9% encapsulation efficiency.

Promising effects of microencapsulation by extrusion/external ionic gelation were also reported by Menin et al. (2018) that observed a high encapsulation efficiency (up to 98%) of linseed oil by this technique. In addition, the authors evaluated the oxidative stability of free and microencapsulated linseed oil since this oil had a content of almost 50% of alpha-linolenic acid (n-3) and observed that the microencapsulated oil was 13 times more stable than free oil, confirming that this technique is a viable approach to preserving functional compounds, such as PUFAs, in vegetable oils.

3.3 Emulsification/Internal Ionic Gelation

Microencapsulation by emulsification/internal ionic gelation produces particles formed from the water-in-oil emulsion. It is an advantageous technique as it does not use heat, retains a high amount of the ingredient to be encapsulated, has a low cost as it does not require sophisticated equipment and can be performed on a laboratory bench, and does not use expensive reagents. In this technique, an aqueous polysaccharide solution is dispersed in the oil phase to form a water-in-oil emulsion, followed by the addition of a $CaCl_2$ solution and an organic acid which reacts to form microparticles (Etchepare et al. 2015).

Studies show that the technique has the potential to protect functional ingredients, as reported by Holkem et al. (2017), that analyzed the stability of microencapsulated or not microencapsulated probiotics under gastrointestinal conditions and during storage and observed that microencapsulation protected *Bifidobacterium*. BB-12 cells when compared to free cells. The authors also observed that microcapsules containing probiotics and stored at freezing temperature (–18°C) had counts of 7.31 log CFU g^{-1}, even after 120 days. Raddatz et al. (2020) also evaluated different storage temperatures of microcapsules produced by emulsification/internal ionic gelation and containing probiotics for 120 days. The authors also added a prebiotic (inulin) to the probiotic, which further improved the viability of the microencapsulated microorganism at freezing temperatures (–18°C) and refrigeration (7°C).

3.4 Complex Coacervation

The complex coacervation microencapsulation method is a highly promising technique in the pharmaceutical, food, and textile industries (Timilsena et al. 2019). The structure formed in this microencapsulation technique is called core-shell and is formed through the use of oppositely charged hydrocolloid solutions, such as proteins and polysaccharides, which through interactions and precipitations form a complex layer that will envelop the material to be encapsulated (core). These electrostatic interactions are caused by changing the pH of the aqueous phase (Timilsena et al. 2019; Silva et al. 2019). The steps of the microencapsulation process by coacervation are represented in Fig. 3.

Due to the high efficiency of encapsulation and processing at room temperature, complex coacervation is advantageous for microencapsulating a wide variety of ingredients as the core is protected by the encapsulating agent and thus external damage, such as moisture, light, and oxidation is avoided. However, a limiting factor for the commercial application of the technology is its pH sensitivity and ionic strength, as complex coacervation occurs within a very narrow pH range.

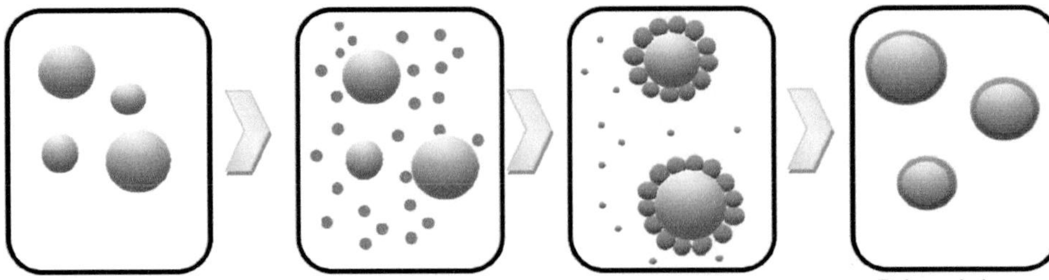

Figure 3. Schematic Representation of the Steps of the Microencapsulation Process by Coacervation.

(Timilsena et al. 2017; 2019). The high encapsulation efficiency of this technique was recently reported by Silva et al. (2019), which achieved an encapsulation efficiency for *Lactobacillus acidophilus* La-5 microcapsules of 87.53%. The authors also reported that when the microcapsules encountered adverse conditions in the gastrointestinal tract, the microorganisms were protected by the wall material, but the viable cell count was below the recommended daily intake, exposing that the success of microencapsulation is dependent on the ingredient a be encapsulated, the microencapsulation technique used, as well as the processing conditions.

4. Final Considerations

Microencapsulation is an emerging alternative for packaging functional compounds in a matrix that can provide the desired release characteristic and a physical barrier against adverse environmental conditions. The characteristics of microencapsulation methods and encapsulating agents are extremely important in the preservation of various nutritional, sensory components, microorganisms, enzymes, dyes, etc., protecting the compound of interest against adverse conditions, such as those that occur during processing, storage, or digestion.

Spray drying is an efficient and fast method, having the advantage of producing dry microparticles. However, the use of high temperatures requires the use of thermo-protectors during processing.

Microencapsulation by extrusion/external ionic gelation is a simple method that, even being carried out at room temperature and without the use of organic solvents, has a good encapsulation efficiency and is a promising technique from an environmental point of view.

The advantage of using the microencapsulation technique by emulsification/internal ionic gelation is its low cost, the absence of heat, and sophisticated equipment during processing, but this technique uses organic solvents.

Microencapsulation by complex coacervation has the advantage of using room temperature during processing and with high encapsulation efficiency; however, this technology is sensitive to pH and ionic strength.

References

Al-Sheraji, S.H., Ismail, A., Manap, M.Y. et al. 2013. Prebiotics as functional foods: A review. J Funct Foods 5:1542–1553. https://doi.org/10.1016/J.JFF.2013.08.009

Ashaolu, T.J. 2020. Immune boosting functional foods and their mechanisms: A critical evaluation of probiotics and prebiotics. Biomed Pharmacother 130: 110625. https://doi.org/10.1016/J.BIOPHA.2020.110625

Budinčić, J.M., Petrović, L., Đekić, L. et al. 2021. Study of vitamin E microencapsulation and controlled release from chitosan/sodium lauryl ether sulfate microcapsules. Carbohydr. Polym. 251: 116988. https://doi.org/10.1016/J.CARBPOL.2020.116988

Canizales, J.R., Rodríguez, G.R.V., Avila, J.A.D. et al. 2018. Encapsulation to Protect Different Bioactives to Be Used as Nutraceuticals and Food Ingredients. 1–20. https://doi.org/10.1007/978-3-319-54528-8_84-1

Castejón, N., Luna, P. and Señoráns, F.J. 2021a. Microencapsulation by spray drying of omega-3 lipids extracted from oilseeds and microalgae: Effect on polyunsaturated fatty acid composition. LWT 148:. https://doi.org/10.1016/J.LWT.2021.111789

Castejón, N., Luna, P. and Señoráns, F.J. 2021b. Microencapsulation by spray drying of omega-3 lipids extracted from oilseeds and microalgae: Effect on polyunsaturated fatty acid composition. LWT 148: 111789. https://doi.org/10.1016/J.LWT.2021.111789

Champagne, C.P., Gomes da Cruz, A. and Daga, M. 2018. Strategies to improve the functionality of probiotics in supplements and foods. Curr. Opin. Food Sci. 22: 160–166. https://doi.org/10.1016/J.COFS.2018.04.008

Damodaran, S., Parkin, K.L. and Fennema, O.R. 2010. Química de alimentos de Fennema, 4th edn. Porto Alegre

Đorđević, V., Balanč, B., Belščak-Cvitanović, A. et al. 2014. Trends in Encapsulation Technologies for Delivery of Food Bioactive Compounds. Food Eng. Rev. 2015 74 7: 452–490. https://doi.org/10.1007/S12393-014-9106-7

Etchepare, M., de, A., Barin, J.S., Cichoski, A.J. et al. 2015. Microencapsulation of probiotics using sodium alginate. Ciência Rural 45:1319–1326. https://doi.org/10.1590/0103-8478cr20140938

FJ L-J, J L-S, ML C-G et al. 2020. Spray-Drying Microencapsulation of Bioactive Compounds from Lemon Verbena Green Extract. Foods (Basel, Switzerland) 9:. https://doi.org/10.3390/FOODS9111547

Foods Market Functional. 2021. Functional Foods Market Growth, Size, Share and Forecast to 2026. https://www.marketdataforecast.com/market-reports/functional-food-market. Accessed 15 Jan 2022

Fu, J., Zhang, L. Le, Li, W. et al. 2022. Application of metabolomics for revealing the interventional effects of functional foods on metabolic diseases. Food Chem. 367: 130697. https://doi.org/10.1016/J.FOODCHEM.2021.130697

Granato, D., Barba, F.J., Kovačević, D.B. et al. 2020. Functional Foods: Product Development, Technological Trends, Efficacy Testing, and Safety. https://doi.org/101146/annurev-food-032519-051708 11: 93–118. https://doi.org/10.1146/ANNUREV-FOOD-032519-051708

Heck, R.T., Vendruscolo, R.G., de Araújo Etchepare, M. et al. 2017. Is it possible to produce a low-fat burger with a healthy n − 6/n − 3 PUFA ratio without affecting the technological and sensory properties? Meat. Sci. 130: 16–25. https://doi.org/10.1016/j.meatsci.2017.03.010

Hill, C., Guarner, F., Reid, G. et al. 2014. The International Scientific Association for Probiotics and Prebiotics consensus statement on the scope and appropriate use of the term probiotic. Nat. Rev. Gastroenterol. Hepatol. 118 11: 506–514. https://doi.org/10.1038/nrgastro.2014.66

Holkem, A.T., Raddatz, G.C., Barin, J.S. et al. 2017. Production of microcapsules containing Bifidobacterium BB-12 by emulsification/internal gelation. LWT - Food Sci. Technol. 76: 216–221. https://doi.org/10.1016/J.LWT.2016.07.013

Janssen, M., Chang, B.P.I., Hristov, H. et al. 2021. Changes in Food Consumption During the COVID-19 Pandemic: Analysis of Consumer Survey Data From the First Lockdown Period in Denmark, Germany, and Slovenia. Front. Nutr. 8: 60. https://doi.org/10.3389/FNUT.2021.635859/BIBTEX

Jiménez-Martín, E., Pérez-Palacios, T., Carrascal, J.R. and Rojas, T.A. 2016. Enrichment of Chicken Nuggets with Microencapsulated Omega-3 Fish Oil: Effect of Frozen Storage Time on Oxidative Stability and Sensory Quality. Food Bioprocess Technol. 9: 285–297. https://doi.org/10.1007/S11947-015-1621-X

Kaushik, P., Dowling, K., Barrow, C.J. and Adhikari, B. 2015. Microencapsulation of omega-3 fatty acids: A review of microencapsulation and characterization methods. J. Funct Foods 19: 868–881. https://doi.org/10.1016/j.jff.2014.06.029

Krumbeck, J.A., Walter, J. and Hutkins, R.W. 2018. Synbiotics for Improved Human Health: Recent Developments, Challenges, and Opportunities. https://doi.org/101146/annurev-food-030117-012757 9:451–479. https://doi.org/10.1146/ANNUREV-FOOD-030117-012757

Liao, Y., Xie, B., Zhang, H. et al. 2019. Efficacy of omega-3 PUFAs in depression: A meta-analysis. Transl Psychiatry 2019 91 9: 1–9. https://doi.org/10.1038/s41398-019-0515-5

Lima, E.R., Silva, T.S.S., Vilela, A.B.A. et al. 2021. Implicações da pandemia de COVID-19 nos hábitos alimentares de brasileiros: revisão integrativa. Res. Soc. Dev. 10: e29810414125. https://doi.org/10.33448/rsd-v10i4.14125

Lu, W., Yang, X., Shen, J. et al. 2021. Choosing the appropriate wall materials for spray-drying microencapsulation of natural bioactive ingredients: Taking phenolic compounds as examples. Powder. Technol. 394: 562–574. https://doi.org/10.1016/J.POWTEC.2021.08.082

Marrapu, B., Ma, J., Geng, Z. et al. 2020. Chemo-enzymatic synthesis, characterization, in vitro antioxidant capacity and oxidative stability studies of novel phosphatidylcholines with ω-3/ω-6 PUFAs and phenolic acids. Food Res. Int. 131:109010. https://doi.org/10.1016/J.FOODRES.2020.109010

Martins, A.P. de, O., Bezerra, M. de, F., Marques Júnior, S. et al. 2019. Consumer behavior of organic and functional foods in Brazil. Food Sci. Technol. 40: 469–475. https://doi.org/10.1590/FST.03519

Menin, A., Zanoni, F., Vakarelova, M. et al. 2018. Effects of microencapsulation by ionic gelation on the oxidative stability of flaxseed oil. Food Chem. 269: 293–299. https://doi.org/10.1016/j.foodchem.2018.06.144

Mohanty, D., Misra, S., Mohapatra, S. and Sahu, P.S. 2018. Prebiotics and synbiotics: Recent concepts in nutrition. Food Biosci 26: 152–160. https://doi.org/10.1016/J.FBIO.2018.10.008

Nunes, G.L., Motta, M.H., Cichoski, A.J. et al. 2018. Encapsulation of Lactobacillus acidophilus La-5 and Bifidobacterium Bb-12 by spray drying and evaluation of its resistance in simulated gastrointestinal conditions, thermal treatments and storage conditions. Ciência Rural. 48:. https://doi.org/10.1590/0103-8478CR20180035

Nystrand, B.T. and Olsen, S.O. 2020. Consumers' attitudes and intentions toward consuming functional foods in Norway. Food Qual Prefer 80: 103827. https://doi.org/10.1016/J.FOODQUAL.2019.103827

Papizadeh, M., Rohani, M., Nahrevanian, H. et al. 2017. Probiotic characters of Bifidobacterium and Lactobacillus are a result of the ongoing gene acquisition and genome minimization evolutionary trends. Microb Pathog 111:118–131. https://doi.org/10.1016/J.MICPATH.2017.08.021

Poletto, G., Raddatz, G.C., Cichoski, A.J. et al. 2019. Study of viability and storage stability of Lactobacillus acidophillus when encapsulated with the prebiotics rice bran, inulin and Hi-maize. Food Hydrocoll. 95: 238–244. https://doi.org/10.1016/J.FOODHYD.2019.04.049

Quigley, A.M.M. 2019. Prebiotics and Probiotics in Digestive Health. Clin Gastroenterol Hepatol 17:333–344. https://doi.org/10.1016/J.CGH.2018.09.028

Raddatz, G.C., de Souza da Fonseca, B., Poletto, G. et al. 2020. Influence of the prebiotics hi-maize, inulin and rice bran on the viability of pectin microparticles containing Lactobacillus acidophilus LA-5 obtained by internal gelation/emulsification. Powder Technol. 362: 409–415. https://doi.org/10.1016/J.POWTEC.2019.11.114

Raddatz, G.C., Fonseca, V.R., Cichoski, A.J. et al. 2022. Viability and stability evaluation of Lactobacillus casei LC03 co-encapsulated with red onion (Allium cepa L.) peel extract. LWT 153: 112434. https://doi.org/10.1016/J.LWT.2021.112434

Rodrigues, M.B., Matos, J.D.P. and Horta, P.M. 2021. The COVID-19 pandemic and its implications for the food information environment in Brazil. Public Health Nutr. 24: 1. https://doi.org/10.1017/S1368980020004747

Roobab, U., Batool, Z., Manzoor, M.F. et al. 2020. Sources, formulations, advanced delivery and health benefits of probiotics. Curr. Opin. Food Sci. 32: 17–28. https://doi.org/10.1016/J.COFS.2020.01.003

Shahidi, F. and Ambigaipalan, P. 2019. Omega-3 Fatty Acids. In: Encyclopedia of Food Chemistry. Elsevier, pp. 465–471.

Shahidi, F. and Ambigaipalan, P. 2018. Omega-3 Polyunsaturated Fatty Acids and Their Health Benefits. Annu. Rev. Food Sci. Technol. 9: 345–381. https://doi.org/10.1146/annurev-food-111317-095850

Silva, T.M. da, Barin, J.S., Lopes, E.J. et al. 2019. Development, characterization and viability study of probiotic microcapsules produced by complex coacervation followed by freeze-drying. Ciência Rural 49:. https://doi.org/10.1590/0103-8478CR20180775

Solomando, J.C., Antequera, T. and Pérez-Palacios, T. 2020. Study on fish oil microcapsules as neat and added to meat model systems: Enrichment and bioaccesibility of EPA and DHA. LWT 120: 108946. https://doi.org/10.1016/J.LWT.2019.108946

Somacal S, Pinto VS, Vendruscolo RG, et al (2020) Maximization of microbial oil containing polyunsaturated fatty acid production by Umbelopsis (Mortierella) isabellina. Biocatal Agric Biotechnol 30:101831. https://doi.org/10.1016/j.bcab.2020.101831

Sosnik, A. and Seremeta, K.P. 2015. Advantages and challenges of the spray-drying technology for the production of pure drug particles and drug-loaded polymeric carriers. Adv. Colloid Interface Sci. 223: 40–54. https://doi.org/10.1016/J.CIS.2015.05.003

Suave, J., DALL'AGNOL EC, PEZZIN APT et al. 2006. Microencapsulação: Inovação em diferentes áreas

Timilsena, Y.P., Akanbi, T.O., Khalid, N. et al. 2019. Complex coacervation: Principles, mechanisms and applications in microencapsulation. Int. J. Biol. Macromol. 121: 1276–1286. https://doi.org/10.1016/J.IJBIOMAC.2018.10.144

Timilsena, Y.P., Wang, B., Adhikari, R. and Adhikari, B. 2017. Advances in microencapsulation of polyunsaturated fatty acids (PUFAs)-rich plant oils using complex coacervation: A review. Food Hydrocoll. 69: 369–381. https://doi.org/10.1016/j.foodhyd.2017.03.007

Us-Medina, U., Julio, L.M., Segura-Campos, M.R. et al. 2018. Development and characterization of spray-dried chia oil microcapsules using by-products from chia as wall material. Powder Technol. 334: 1–8. https://doi.org/10.1016/j.powtec.2018.04.060

Vaziri, A.S., Alemzadeh, I., Vossoughi, M. and Khorasani, A.C. 2018. Co-microencapsulation of Lactobacillus plantarum and DHA fatty acid in alginate-pectin-gelatin biocomposites. Carbohydr. Polym. 199: 266–275. https://doi.org/10.1016/J.CARBPOL.2018.07.002

Wang, L., Liu, Y., Zhang, W. et al. 2013. Microspheres and microcapsules for protein delivery: strategies of drug activity retention. Curr. Pharm. Des. 19: 6340–52

Wgo. 2017. World Gastroenterology Organisation Probióticos E Prebióticos. In: World Gastroenterol. Organ. https://www.worldgastroenterology.org/guidelines/global-guidelines/probiotics-and-prebiotics?source=post_page--------------------------. Accessed 26 Sep 2021

Yeung, A.W.K., Mocan, A. and Atanasov, A.G. 2018. Let food be thy medicine and medicine be thy food: A bibliometric analysis of the most cited papers focusing on nutraceuticals and functional foods. Food Chem. 269: 455–465. https://doi.org/10.1016/J.FOODCHEM.2018.06.139

Part IV

Sustainable Management in Food Industries

18

Wastewater Treatment in Sustainable Food Industries

*William Michelon** and *Aline Viancelli*

1. Introduction

Water contamination has become a growing problem for public and social authorities since the beginning of industrialization. Demand for freshwater is increasing dramatically as the industrial world develops and the population increases. By 2050, worldwide water demand for agriculture, industry, and cities is estimated to increase by 20–30% (Boretti and Rosa 2019). Thus, as the world's population grows, so does the demand for freshwater, food, energy, and technology, resulting in a rise in wastewater production from both domestic and industrial sources (Dinar et al. 2019). One of the implications of this development is the production of greater volumes and types of wastewater, containing a diverse range of chemical compounds and microorganisms with pathogenic potential (Rogowska et al. 2020).

Conventional wastewater treatment facilities work (physically or mechanically) mainly on the removal of suspended materials and the decrease in biological oxygen demand via activated sludge processes (Wang et al. 2017). This biodegradation involves the organic molecules' breakdown as well as inorganic compounds (mainly nitrogen and phosphorus), which are key points to preventing the eutrophication of water in rivers and lakes (Bhateria and Jain 2016). The capacity of these traditional technologies to degrade heavy metals, high nutrient loads, and xenobiotics is still limited, which can result in an increase in the concentration of these compounds in surface and groundwater (Chowdhury et al. 2016; Menció et al. 2016; Sousa et al. 2018).

To reduce the environmental impacts of these contaminations, a tertiary treatment process known as phycoremediation has attracted attention (Koul et al. 2022). Wastewater treatment using microalgal systems is a technique that has been around for more than 60 years. Oswald's pioneering work in California established the principles of wastewater treatment in that so high-rate algal ponds, that were initially designed to remove organic material and nutrients (Oswald 1973; 1962). The advantages of using microalgae-based treatment include biomass with added value, which may be recovered simultaneously with wastewater treatment (Letry et al. 2019). Based on the type of wastewater treated and the environmental conditions, pigments (Maroneze et al. 2020), nutraceuticals (Dhandayuthapani et al. 2021), fertilizers (Castro et al. 2020), fatty acids (Goswami et al. 2021), protein (amino acids and peptides) (Michelon et al. 2022), animal and fish feeds (Apandi et al. 2019;

Universidade do Contestado, Concórdia, SC, Brazil, 89711-330.
* Corresponding author: william@unc.br

Lum et al. 2013) can be obtained from collected microalgae biomass (Fig. 1). Thus, microalgae-based systems are currently recognized as sustainable wastewater treatment technologies because they have a lower energy requirement for oxygen, which is due to the photosynthetic oxygen generation when compared to conventional activated sludge processes (Posadas et al. 2017). In addition, microalgae-based processes require fewer consumable resources (such as water and soil), do not compete with food production, and also have a high potential for CO_2 mitigation (Michelon et al. 2019).

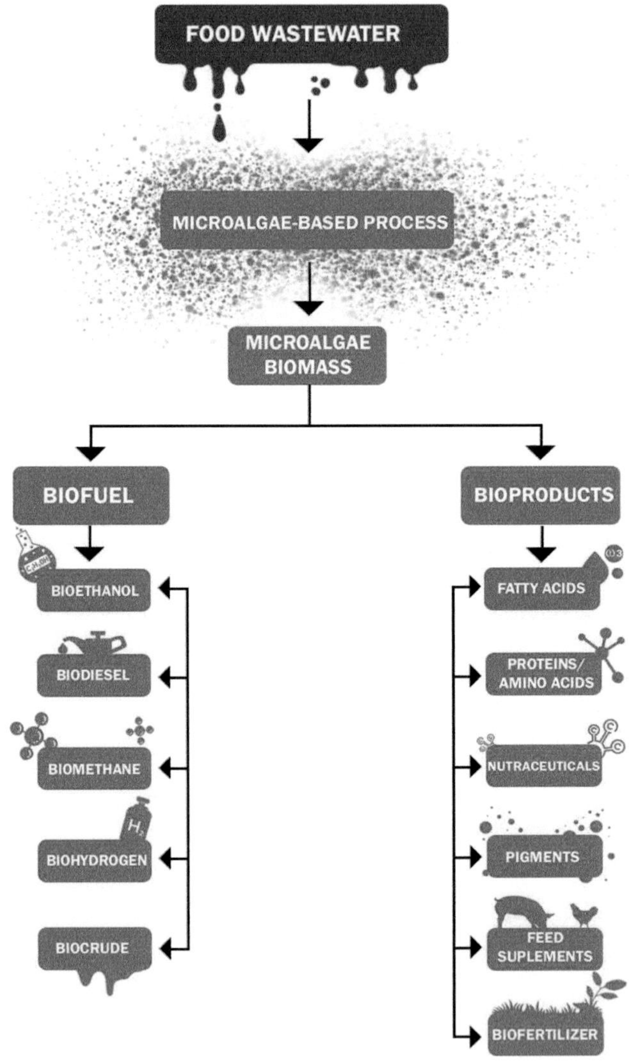

Figure 1. Potential Application of Phycoremediation on Food Industry Wastes for Value-Added Products Obtaintion.

2. Microalgae's Use in Food Wastewater Treatment

Microalgae are photosynthetic microorganisms that are both eukaryotic and prokaryotic and despite their differences, both organisms are produced and conduct photosynthesis in the same way (Samiee et al. 2019). Microalgae are fast-growing microorganisms (with a doubling rate of one day) that are cultivated in water and may be grown in a variety of settings, where the microalgae can present autotrophic and heterotrophic behavior (Mutanda et al. 2020). Thus, microalgae assimilate nutrients from food wastewater and produce new biomass, and several microalgae species have been studied for their ability to remove contaminants from different wastewater (Maryjoseph and Ketheesan 2020). As it is a green technology and is based on the circular bioeconomy, microalgae-based treatment has become increasingly appealing (Ahmad et al. 2022).

Microalgae-based treatment could be applied in many effluents generated during the food production chain, such as brewery, dairy, vinasse, aquaculture, swine, and cattle wastewater. The successful application of different microalgae species will be presented in this chapter.

2.1 Brewery Wastewater

Beer production is an important part of the economy of the countries where it is produced; however, the brewery industries consume large volumes of water and consequently, there is a significant volume of wastewater (Amenorfenyo et al. 2019). Sugars, protein (amino acids), alcohol, cellulose, nitrogen, phosphate, calcium, potassium, BOD, and COD are the main components of brewery effluent (Song et al. 2020). Song et al. (2020) studied simultaneous brewery wastewater purification and CO_2 fixation using four different microalgae strains (*Chlorella* sp. L166, *Chlorella* sp. UTEX1602, *Scenedesmus* sp. 336, and *Spirulina* sp. FACHB-439). The hybrid system's target species were *Chlorella* sp. UTEX1602 and *Scenedesmus* sp. 336, achieving a purification integrated with 15% CO_2. After 10 days of culture, the dry weight of *Scenedesmus* sp. 336 reached 1.0 g L^{-1}, and the removal rates of NH_3-N, TN, TP, and COD from brewery wastewater were 89.9%, 75.9%, 95.7%, and 73.6%, respectively. Chlorophyll, carotenoids, carbohydrates were at 20.4 mg L^{-1}, 7.5 mg L^{-1}, and 63.6 mg L^{-1}, respectively, and lipid productivity was of 38 mg $L^{-1}d^{-1}$ (Song et al. 2020). *Scenedesmus obliquus* has been widely used in the treatment of brewery effluent. After 9 days, the highest biomass growth was 0.9 g of dry biomass L^{-1}, with a relatively high decrease of wastewater pollutants (57.5% COD and 20.8% of total N after 14 days; 56.9% of total C after 13 days) (Mata et al. 2012). Pollutant removal efficiency in photobioreactors using *Scenedesmus obliquus* 10% (v v^{-1}), brewery CO_2 supplement, and brewery wastewater were 92.9%, 88.5%, 40.8%, and 61.9% for ammonia, total nitrogen, phosphorus, and COD, respectively (Ferreira et al. 2017). After treating brewery wastewater, the biomass of *Scenedesmus obliquus* was used to generate biohydrogen, bio-oil, biochar, and biogas. The yields for biohydrogen were 67.1 mL H_2 g^{-1} and 64%, 30%, and 6% for bio-oil, biochar, and biogas, respectively (Ferreira et al. 2017).

2.2 Dairy Wastewater

Milk consumption and production of milk-derived products (cheese, curd, butter, yogurt, and dry milk powder) put the dairy industry at the top of the food industry. However, during the production, cleaning, and washing of the processing equipment, large amounts of water are transformed into wastewater (Kaur 2021). Dairy effluents other than cheese effluents have an organic matter content of less than 15 g L^{-1} of COD and 5.0 g L^{-1} of BOD. Furthermore, it includes fat, nitrogen, phosphorus, potassium, chloride, and solid materials. The pH values range from 4 to 11 (Carvalho et al. 2013). When the microalgae *Scenedesmus* sp. ASK22 was grown in dairy wastewater, its lipid production was 31.16 mg L^{-1} d^{-1}, and its removal efficiency was 100% for nitrate, 98.63% for phosphate, and > 99 % for COD (Pandey et al. 2019). Kumar et al. (2019) used the microalgae *Ascochloris* sp. ADW007 to treat water from raw dairy effluents. Biomass productivity increased significantly in both the column (0.284 g L^{-1} d^{-1}) and flat plate (0.292 g L^{-1} d^{-1}) photobioreactor cultures. Furthermore, after 7 days of growth, lipid production was 33.4%. This strain produced more than 80% clean and odorless water, with reductions up to 96%, 80% and 97%, for COD and nitrate total phosphate, respectively (Kumar et al. 2019). On the 10th day of its cultivation, *Chlamydomonas polypyrenoideum* reduced the pollution load of nitrate, nitrite, phosphate, chloride, fluoride, and ammonia by 90%, 74%, 70%, 61%, 58%, and 90%, respectively. The lipid content of algal biomass produced on dairy effluent on the 10th and 15th days was 1.6 g and 1.2 g, respectively (Kothari et al. 2013). After 6 days of cultivation, the outdoor cultivation of *Chlorella zofingiensis* in a system supplemented with 6% CO_2, revealed removal rates of 51.7% for total nitrogen and 97.5% for orthophosphate (Huo et al. 2012).

Lu et al. (2015) investigated the potential of *Chlorella* sp. to remove nutrients from dairy effluent in indoor and outdoor cultures. Under indoor conditions, the COD, total nitrogen, and phosphorus removal rates obtained were 88.4 mg L^{-1} d^{-1}, 38.3 mg L^{-1} d^{-1}, and 2.0 mg L^{-1} d^{-1}, respectively, contrasting to the 41.3 mg L^{-1} d^{-1}, 6.6 mg L^{-1} d^{-1}, and 2.7 mg L^{-1} d^{-1}, respectively, obtained under outdoor conditions. In addition, the maximum biomass productivity achieved was 260 mg L^{-1} d^{-1} and 110 mg L^{-1} d^{-1} in indoor and outdoor pilot-scale cultures, respectively (Lu et al. 2015). *Chlorococcum* sp. RAP-13 produced more saturated fatty acids when grown in dairy wastewater, and the predominant fatty acid components of the microalgal oil were palmitic (16:0), oleic (18:1), stearic (18:0), linoleic (18:2), and linolenic (18:3) acids. Dairy wastewater quality increased following microalgae culture because of lower COD and BOD levels (Ummalyma and Sukumaran 2014).

2.3 Vinasse

Vinasse is the main liquid residue from distillation processes, thus it is generated during the production of sugarcane ethanol, tequila, mezcal, and others (Robles-González et al 2012). For every liter of ethanol produced, 10–20 liters of vinasse are generated (Christofoletti et al. 2013). Vinasse has different physical and chemical properties depending on the feedstock, fermentation and distillation processes, juice composition, and quality (de Godoi et al. 2019). Vinasse has significant polluting potential because of its physical and chemical characteristics, such as COD, BOD, phenols, sulfates, phosphorus, ammonia nitrogen, nitrate, calcium, iron, magnesium, potassium, and organic compounds (carbohydrates, proteins, and lipids) (Parsaee et al. 2019). COD and BOD reductions of up to 49% and 70 %, respectively, were observed, along with a maximum algal biomass yield of 10.5±0.9 g L^{-1} (Soto et al. 2021). Calixto et al. (2016) investigated the biomass chemical composition and productivity of *Chlorella* sp., *Chlamydomonas* sp., *Lagerheimia longiseta*, and *Pediastrum tetras* in sugarcane vinasse, sewage, and chicken manure. *Chlorella* sp. generated more protein, whereas *Chlamydomonas* sp. produced biomass with a higher carbohydrate content in the same culture medium. Vinasse was used as a long-term production medium of *Chlorella vulgaris* UTEX 1803; the results indicated that microalgae grew rapidly in the first 10 days, with an average biomass production of 0.87 g L^{-1}. The concentration of metabolites in the biomass began to grow on the 10th day, reaching an essentially constant amount on the 16th, presenting 4.8% of lipids, 48.9% of proteins, 2.9% of xylose, 7.8% of glucose, 4.5% of arabinose and 8.3% of fructose (Quintero-Dallos et al. 2019).

Anaerobic digestion is indeed a suitable alternative for vinasse pretreatment with the goal of using it for cultivating microalgae. The specific growth rate of biomass obtained with anaerobically treated vinasse was 0.76 d^{-1} (Marques et al. 2013). The maximum production of *Chlorella vulgaris* biomass (70 mg L^{-1} d^{-1}) was obtained when a medium prepared with anaerobic digester effluent was used and lipid production ranged from 0.5 to 17 mg L^{-1} d^{-1} (Marques et al. 2013). Silva et al. (2017) studied the heterotrophic growth of *Desmodesmus subspicatus* in vinasse and lipid extraction with supercritical CO_2. As a result, *Desmodesmus subspicatus* produced lipids at temperatures ranging from 15°C to 40°C, with the maximum lipid output (1,100 mg L^{-1} d^{-1}) obtained at 20°C. By increasing pressure from 20 MPa to 30 MPa at 60°C, lipid extraction yield improved from 23% to 45%. The isolated lipids' fatty acid profiles revealed large proportions of palmitic acid (16:0), stearic acid (18:0), oleic acid (18:1), linoleic acid (18:2), arachidic acid (20:0), and arachidonic acid (20:4) (Silva et al. 2017).

2.4 Aquaculture Wastewater

The global consumption of fish, shellfish, prawns, and algae has doubled in the past decades, making aquaculture the fastest-growing economy on the food chain. However, aquaculture presents a concern related to water pollution with wastewater (Mohd Nizam et al. 2020). The aquaculture wastewater contains a high concentration of COD and a lower concentration of nitrogen (ammonium and nitrate), phosphorus and suspension solids (Kurniawan et al. 2021). The utilization of microalgae to remediate raw effluent from brown crab aquaculture with *Chlorella vulgaris* and *Scenedesmus obliquus* in semi-continuous growth yielded biomass productivities of 879.8 mg L^{-1} d^{-1} and 811.7 mg L^{-1} d^{-1}, respectively, with efficacy removal of 100% for total nitrogen and phosphorus and more than 96% for COD. The biomass from *Chlorella vulgaris* and *Scenedesmus obliquus,* harvested after aquaculture wastewater treatment, contained 31–35% protein, 6–8% lipids, and 30–39% carbohydrates. In addition, microalgal biomass was applied as a biostimulant of seeds, with germination indexes of 175% and 48% in watercress and 84% and 98% in wheat for *Chlorella vulgaris* and *Scenedesmus obliquus*, respectively (Viegas et al. 2021a).

In another study, five microalgal species were grown in batch mode: *Chlorella vulgaris*, *Chlorococcum* sp. GD, *Parachlorella kessleri* TY, *Scenedesmus quadricauda*, and *Scenedesmus obliquus* with the potential for treating real aquaculture effluent from a fishery industry. In comparison to the other four microalgae, *Parachlorella kessleri* TY showed better growth and pollutant removal performance in aquaculture effluent, achieving removal of up to 94.4%, 96.2%, 99%, 94.3% and 95.6% of COD, ammonium, nitrite, nitrate, and phosphorus, respectively, after 3 days of cultivation (Liu et al. 2019). Tossavainen et al. (2019) reported on microalgal (*Euglena gracilis* CCAP 1224/5Z and *Selenastrum* SCCAP K-1877) growing in aquaculture wastewater from recirculating aquaculture systems in association with the synthesis of important fatty acids and tocopherols. The maximum yields were 1.5 g L^{-1} of algal biomass, 84.9 mg L^{-1} of lipid, and 877.2 g L^{-1} of tocopherol. Additionally, the optimum nutrient removal efficiency for ammonium-N and phosphate-P was 99.5% and 99.8%, respectively, while COD was decreased by 67.6% (Tossavainen et al. 2019). Thus, the proposed innovative microalgae-based aquaculture system incorporates wastewater treatment, a biomass harvesting process, and a platform for producing aquatic animals (Li et al. 2021).

2.5 Swine Wastewater

The world consumption of animal protein increased significantly in the last few years, especially demanding for pork. As a consequence, swine farms generate significant amounts of wastewater (Nagarajan et al. 2019). Swine wastewater typically comprises more than 1% total solids, the majority of which are organic, with COD concentrations ranging from 10 g L^{-1} to 100 g L^{-1}. Additionally, it contains a high concentration of nutrients (nitrogen and phosphorus) as well as other micronutrients, such as calcium, chlorine, chromium, cobalt, copper, iron, magnesium, manganese, potassium, sodium, sulfur, and zinc (Cheng et al. 2020). In the culture of *Chlorella pyrenoidosa*, diluted swine wastewater with COD concentrations < 1 g L^{-1} exhibited ammonia removal of over 90 % and the greatest phosphorus removal of 78%, which was associated with biomass increment (Wang et al. 2012). The culture of *Coelastrella* sp. QY01 in anaerobically treated swine wastewater resulted in the removal of 100% of ammonium and phosphorus, where the maximum biomass production was 57.5 mg L^{-1} d^{-1} and the highest lipid productivity was 13.4 mg L^{-1} d^{-1} after 10 days of cultivation (Luo et al. 2016).

Different microalgae species cultured in swine wastewater revealed lipid content and lipid productivities ranging from 21% to 46% and 130 mg L^{-1} d^{-1} to 1,100 mg L^{-1} d^{-1}, respectively (Cheng et al. 2019). *Chlorella spp.* microalgae harvested after treatment of nutrient-rich swine wastewater digestate exhibited a high protein and carbohydrate content but low lipids content. However, when nitrogen and phosphorus were limited, the lipid content increased from 3% to 16%. This improvement in lipid content is especially important for biodiesel industries as well as a variety of other businesses interested in the commercial potential of algae-derived oil byproducts (Michelon et al. 2015). The fatty acid profiles of microalgae *Coelastrella* sp. grown in anaerobically and aerobically treated swine wastewater was found to be mostly palmitic acid (16:0) (23.8–30.8%), linoleic acids (18:2) (11.9–20.8 %) and linoleic acids (18:3) (40.5–53.8 %) of total fatty acid (Luo et al. 2016). Furthermore, swine wastewater microalgae biomass can be converted into bioenergy, such as biomethane and biohydrogen. For instance, Ferreira et al. (2018) investigated the growth and biohydrogen production capability of *Scenedesmus obliquus* cultivated in several wastewaters and indicated that biomass collected from swine wastewater had the greatest biohydrogen productivity (390 mL H_2 gVS^{-1}). Microalgae biomass obtained following swine tertiary treatment and nitrogen and phosphorus limited generated significantly more biomethane (approximately 57%) than microalgae biomass collected after 8 days of treatment (i.e., 103.5 ± 1.7 vs 44 ± 2.5 L-CH_4 (kg biomass)$^{-1}$) (Perazzoli et al. 2016).

2.6 Cattle Manure Wastewater

Beef is a protein source of high-quality, whose demand is increasing globally. On beet cattle production farms, the manure produced by one animal every day corresponds to 5–6% of its body weight (about 5.5 kg per animal) (Font-Palma 2019). Manure from cattle farms' composition varies according to climate, feedlot surface type, and management strategies. Cattle manure includes nutrients derived from animal diets, such as COD, suspension solids, nitrogen, phosphorus, potassium, sulfur, and magnesium, as well as trace minerals (Font-Palma 2019). Microalgae have been used to remediate the liquid fraction of cattle wastewater, for example, the application of microalgae cultivation *Scenedesmus* sp. resulted in a 92.5% decrease in total nitrogen and a 51.9% decrease in phosphorus present in cattle manure (Scherer et al. 2017).

Coelastrum sp. was isolated from cattle manure leachate and applied to treat cattle wastewater, as a result, the maximal cell growth was 2.71 g L^{-1}. The maximum levels of COD, total nitrogen, nitrate, and phosphorus removal were 53.4%, 91.2%, 87.5%, and 100%, respectively; under 2,300 lux light, the maximum lipid content of the biomass was 50.8% (Mousavi et al., 2018). Three *Chlorella sorokiniana* strains were cultivated in cattle anaerobic digestate. The biomass production was approximately 280 mg L^{-1}, and nitrogen and phosphorus removal was up to 88.6% and 64.9%, respectively. The overall lipid, protein, and starch content of the three *Chlorella sorokiniana* strains were around 25–35%, 30–35%, and 20–25 %, respectively; the mostly synthesized fatty acids were palmitic acid (16:0), oleic (18:1), and linoleic acids (18:3) (Kobayashi et al. 2013).

Biomass productivities for bioremediation of cattle manure using *Chlorella vulgaris, Chlorella protothecoides*, and *Scenedesmus obliquus* were 64.4 mg L^{-1} d^{-1}, 67.7 mg L^{-1} d^{-1}, and 67 mg L^{-1} d^{-1}, respectively. The fatty acid profile of the lipid fraction indicated a dominancy of oleic and linoleic acids (18:1 and 18:2), as well as palmitic acid (16:0), which could be because of a combination of lipid molecules generated during microalgae biotransformation and adsorption of lipid molecules present in cattle wastewater. All microalgae cultivated from cattle wastewater had a high amount of oleic acid (18:1), reaching 63.5% for *Chlorella vulgaris*, 62.7% for *Chlorella protothecoides*, and 58.4% for *Scenedesmus obliquus*. Following biomass separation, the liquid wastewater may be recirculated into it to reduce water demand, and the precipitate can be utilized as crop fertilizer or soil pH corrector (Viegas et al. 2021b).

3. Challenges and Future Perspectives

Although food wastewater is regarded as a valuable source for microalgae-based biomass production, excessive quantities of nutrients and hazardous contaminants may inhibit microalgae growth; considering its feasible and cost-effective application in large-scale wastewater treatment and bioenergy generation, there is still a need to produce an efficient process that allows microalgae to grow effectively in undiluted wastewaters (Cheng et al. 2019). On the other hand, excessive growth can restrict light from penetrating the water and inhibit microalgae growth; it reduces the nutrient removal capability (Parr et al. 2002). Typically, wastewaters are sterilized using autoclaving, which enhances operational costs and limits the large-scale use of microalgae-based processes. In this respect, studies might be conducted on non-axenic and polyculture microalgae derived from food industry effluent that can generate greater biomass with stable growth, attempting to avoid the possibility of culture crash and environmental contamination, lower production costs, and improve wastewater treatment efficiency (Ummalyma et al. 2022).

4. Conclusion

Recovering resources from food industry wastes is a growing technique since it supports environmental sustainability while economically benefiting companies by commercializing value-added recovered bioproducts. In this sense, the microalgae-based process is a simple and low-cost technology for industrial facilities for treating food industry wastewater that can have polluting potential, once the use of microalgae processes in industrial wastewater treatment demands less energy and results in cheap operating and maintenance expenses.

References

Ahmad, A., W. Hassan, S. and Banat, F. 2022. An overview of microalgae biomass as a sustainable aquaculture feed ingredient: food security and circular economy. Bioengineered 13: 9521–9547. https://doi.org/10.1080/21655979.2022.2061148

Amenorfenyo, D.K., Huang, X., Zhang, Y., Zeng, Q., Zhang, N., Ren, J. and Huang, Q. 2019. Microalgae Brewery Wastewater Treatment: Potentials, Benefits and the Challenges. International Journal of Environmental Research and Public Health 16, 1910. https://doi.org/10.3390/ijerph16111910

Apandi, N.M., Mohamed, R.M.S.R., Al-Gheethi, A. and Kassim, A.H.M. 2019. Microalgal biomass production through phycoremediation of fresh market wastewater and potential applications as aquaculture feeds. Environmental Science and Pollution Research 26: 3226–3242. https://doi.org/10.1007/s11356-018-3937-3

Bhateria, R. and Jain, D. 2016. Water quality assessment of lake water: a review. Sustainable Water Resources Management 2: 161–173. https://doi.org/10.1007/s40899-015-0014-7

Boretti, A. and Rosa, L. 2019. Reassessing the projections of the World Water Development Report. npj Clean Water 2: 15. https://doi.org/10.1038/s41545-019-0039-9

Calixto, C.D., da Silva Santana, J.K., de Lira, E.B., Sassi, P.G.P., Rosenhaim, R., da Costa Sassi, C.F., da Conceição, M.M. and Sassi, R. 2016. Biochemical compositions and fatty acid profiles in four species of microalgae cultivated on household sewage and agro-industrial residues. Bioresource Technology 221: 438–446. https://doi.org/10.1016/j.biortech.2016.09.066

Carvalho, F., Prazeres, A.R. and Rivas, J. 2013. Cheese whey wastewater: Characterization and treatment. Science of The Total Environment 445–446: 385–396. https://doi.org/10.1016/j.scitotenv.2012.12.038

Castro, J. de S., Calijuri, M.L., Ferreira, J., Assemany, P.P. and Ribeiro, V.J. 2020. Microalgae based biofertilizer: A life cycle approach. Science of The Total Environment 724: 138138. https://doi.org/10.1016/j.scitotenv.2020.138138

Cheng, D.L., Ngo, H.H., Guo, W.S., Chang, S.W., Nguyen, D.D. and Kumar, S.M. 2019. Microalgae biomass from swine wastewater and its conversion to bioenergy. Bioresource Technology 275: 109–122. https://doi.org/10.1016/j.biortech.2018.12.019

Cheng, H.-H., Narindri, B., Chu, H. and Whang, L.-M. 2020. Recent advancement on biological technologies and strategies for resource recovery from swine wastewater. Bioresource Technology 303: 122861. https://doi.org/10.1016/j.biortech.2020.122861

Chowdhury, S., Mazumder, M.A.J., Al-Attas, O. and Husain, T. 2016. Heavy metals in drinking water: Occurrences, implications, and future needs in developing countries. Science of The Total Environment 569–570: 476–488. https://doi.org/10.1016/j.scitotenv.2016.06.166

Christofoletti, C.A., Escher, J.P., Correia, J.E., Marinho, J.F.U. and Fontanetti, C.S. 2013. Sugarcane vinasse: Environmental implications of its use. Waste Management 33: 2752–2761. https://doi.org/10.1016/j.wasman.2013.09.005

de Godoi, L.A.G., Camiloti, P.R., Bernardes, A.N., Sanchez, B.L.S., Torres, A.P.R., da Conceição Gomes, A. and Botta, L.S. 2019. Seasonal variation of the organic and inorganic composition of sugarcane vinasse: main implications for its environmental uses. Environmental Science and Pollution Research 26: 29267–29282. https://doi.org/10.1007/s11356-019-06019-8

Dhandayuthapani, K., Malathy, S., Mulla, S.I. and Gupta, S.K. 2021. An Insight into the Potential Application of Microalgae in Pharmaceutical and Nutraceutical Production, in: Algae. Springer Singapore, Singapore, pp. 135–179. https://doi.org/10.1007/978-981-15-7518-1_7

Dinar, A., Tieu, A. and Huynh, H. 2019. Water scarcity impacts on global food production. Global Food Security 23: 212–226. https://doi.org/10.1016/j.gfs.2019.07.007

Ferreira, A., Marques, P., Ribeiro, B., Assemany, P., de Mendonça, H.V., Barata, A., Oliveira, A.C., Reis, A., Pinheiro, H.M. and Gouveia, L. 2018. Combining biotechnology with circular bioeconomy: From poultry, swine, cattle, brewery, dairy and urban wastewaters to biohydrogen. Environmental Research 164, 32–38. https://doi.org/10.1016/j.envres.2018.02.007

Ferreira, A., Ribeiro, B., Marques, P.A.S.S., Ferreira, A.F., Dias, A.P., Pinheiro, H.M., Reis, A. and Gouveia, L. 2017. Scenedesmus obliquus mediated brewery wastewater remediation and CO 2 biofixation for green energy purposes. Journal of Cleaner Production 165: 1316–1327. https://doi.org/10.1016/j.jclepro.2017.07.232

Font-Palma, C. 2019. Methods for the Treatment of Cattle Manure—A Review. C 5, 27. https://doi.org/10.3390/c5020027

Goswami, R.K., Mehariya, S., Verma, P., Lavecchia, R. and Zuorro, A. 2021. Microalgae-based biorefineries for sustainable resource recovery from wastewater. Journal of Water Process Engineering 40: 101747. https://doi.org/10.1016/j.jwpe.2020.101747

Huo, S., Wang, Z., Zhu, S., Zhou, W., Dong, R. and Yuan, Z. 2012. Cultivation of Chlorella zofingiensis in bench-scale outdoor ponds by regulation of pH using dairy wastewater in winter, South China. Bioresource Technology 121: 76–82. https://doi.org/10.1016/j.biortech.2012.07.012

Kaur, N. 2021. Different treatment techniques of dairy wastewater. Groundwater for Sustainable Development 14: 100640. https://doi.org/10.1016/j.gsd.2021.100640

Kobayashi, N., Noel, E.A., Barnes, A., Watson, A., Rosenberg, J.N., Erickson, G. and Oyler, G.A. 2013. Characterization of three Chlorella sorokiniana strains in anaerobic digested effluent from cattle manure. Bioresource Technology 150: 377–386. https://doi.org/10.1016/j.biortech.2013.10.032

Kothari, R., Prasad, R., Kumar, V. and Singh, D.P. 2013. Production of biodiesel from microalgae Chlamydomonas polypyrenoideum grown on dairy industry wastewater. Bioresource Technology 144: 499–503. https://doi.org/10.1016/j.biortech.2013.06.116

Koul, B., Sharma, K. and Shah, M.P. 2022. Phycoremediation: A sustainable alternative in wastewater treatment (WWT) regime. Environmental Technology & Innovation 25: 102040. https://doi.org/10.1016/j.eti.2021.102040

Kumar, A.K., Sharma, S., Patel, A., Dixit, G. and Shah, E. 2019. Comprehensive evaluation of microalgal-based dairy effluent treatment process for clean water generation and other value-added products. International Journal of Phytoremediation 21: 519–530. https://doi.org/10.1080/15226514.2018.1537248

Kurniawan, S.B., Abdullah, S.R.S., Imron, M.F., Ahmad, A., Mohd Said, N.S., Mohd Rahim, N.F., Mohammad Alnawajha, M., Abu Hasan, H., Othman, A.R. and Purwanti, I.F. 2021. Potential of valuable materials recovery from aquaculture wastewater: An introduction to resource reclamation. Aquaculture Research 52: 2954–2962. https://doi.org/10.1111/are.15180

Letry, K.A., Castro, E.D., Gupta, S.K. and Kumar, M. 2019. Industrial Wastewater-Based Algal Biorefineries: Application Constraints and Future Prospects. pp. 371–392. *In*: Application of Microalgae in Wastewater Treatment. Springer International Publishing, Cham. https://doi.org/10.1007/978-3-030-13909-4_16

Li, H., Chen, S., Liao, K., Lu, Q. and Zhou, W. 2021. Microalgae biotechnology as a promising pathway to ecofriendly aquaculture: a state-of-the-art review. Journal of Chemical Technology & Biotechnology 96: 837–852. https://doi.org/10.1002/jctb.6624

Liu, Y., Lv, J., Feng, J., Liu, Q., Nan, F. and Xie, S. 2019. Treatment of real aquaculture wastewater from a fishery utilizing phytoremediation with microalgae. Journal of Chemical Technology & Biotechnology 94: 900–910. https://doi.org/10.1002/jctb.5837

Lu, W., Wang, Z., Wang, X. and Yuan, Z. 2015. Cultivation of Chlorella sp. using raw dairy wastewater for nutrient removal and biodiesel production: Characteristics comparison of indoor bench-scale and outdoor pilot-scale cultures. Bioresource Technology 192: 382–388. https://doi.org/10.1016/j.biortech.2015.05.094

Lum, K.K., Kim, J. and Lei, X.G. 2013. Dual potential of microalgae as a sustainable biofuel feedstock and animal feed. Journal of Animal Science and Biotechnology 4: 53. https://doi.org/10.1186/2049-1891-4-53

Luo, L., He, H., Yang, C., Wen, S., Zeng, G., Wu, M., Zhou, Z. and Lou, W. 2016. Nutrient removal and lipid production by Coelastrella sp. in anaerobically and aerobically treated swine wastewater. Bioresource Technology 216, 135–141. https://doi.org/10.1016/j.biortech.2016.05.059

Maroneze, M.M., Dias, R.R., Severo, I.A. and Queiroz, M.I. 2020. Microalgae-Based Processes for Pigments Production, in: Pigments from Microalgae Handbook. Springer International Publishing, Cham, pp. 241–264. https://doi.org/10.1007/978-3-030-50971-2_11

Marques, S.S.I., Nascimento, I.A., de Almeida, P.F. and Chinalia, F.A. 2013. Growth of Chlorella vulgaris on Sugarcane Vinasse: The Effect of Anaerobic Digestion Pretreatment. Applied Biochemistry and Biotechnology 171: 1933–1943. https://doi.org/10.1007/s12010-013-0481-y

Maryjoseph, S. and Ketheesan, B. 2020. Microalgae based wastewater treatment for the removal of emerging contaminants: A review of challenges and opportunities. Case Studies in Chemical and Environmental Engineering 2: 100046. https://doi.org/10.1016/j.cscee.2020.100046

Mata, T.M., Melo, A.C., Simões, M. and Caetano, N.S. 2012. Parametric study of a brewery effluent treatment by microalgae Scenedesmus obliquus. Bioresource Technology 107, 151–158. https://doi.org/10.1016/j.biortech.2011.12.109

Menció, A., Mas-Pla, J., Otero, N., Regàs, O., Boy-Roura, M., Puig, R., Bach, J., Domènech, C., Zamorano, M., Brusi, D. and Folch, A. 2016. Nitrate pollution of groundwater; all right…, but nothing else? Science of The Total Environment 539: 241–251. https://doi.org/10.1016/j.scitotenv.2015.08.151

Michelon, W., da Silva, M.L.B., Matthiensen, A., de Andrade, C.J., de Andrade, L.M. and Soares, H.M. 2022. Amino acids, fatty acids, and peptides in microalgae biomass harvested from phycoremediation of swine wastewater. Biomass Conversion and Biorefinery 12, 869–880. https://doi.org/10.1007/s13399-020-01263-2

Michelon, W., Da Silva, M.L.B., Mezzari, M.P., Pirolli, M., Prandini, J.M. and Soares, H.M. 2015. Effects of Nitrogen and Phosphorus on Biochemical Composition of Microalgae Polyculture Harvested from Phycoremediation of PiggeryWastewater Digestate. Applied Biochemistry and Biotechnology 178: 1407–1419. https://doi.org/10.1007/s12010-015-1955-x

Michelon, W., Pirolli, M., Mezzari, M.P., Soares, H.M. and da Silva, M.L.B. 2019. Residual sugar from microalgae biomass harvested from phycoremediation of swine wastewater digestate. Water Science and Technology 79, 2203–2210. https://doi.org/10.2166/wst.2019.226

Mohd Nizam, N.U., Mohd Hanafiah, M., Mohd Noor, I. and Abd Karim, H.I. 2020. Efficiency of Five Selected Aquatic Plants in Phytoremediation of Aquaculture Wastewater. Applied Sciences 10: 2712. https://doi.org/10.3390/app10082712

Mousavi, S., Najafpour, G.D., Mohammadi, M. and Seifi, M.H. 2018. Cultivation of newly isolated microalgae Coelastrum sp. in wastewater for simultaneous CO2 fixation, lipid production and wastewater treatment. Bioprocess and Biosystems Engineering 41: 519–530. https://doi.org/10.1007/s00449-017-1887-7

Mutanda, T., Naidoo, D., Bwapwa, J.K. and Anandraj, A. 2020. Biotechnological Applications of Microalgal Oleaginous Compounds: Current Trends on Microalgal Bioprocessing of Products. Frontiers in Energy Research 8. https://doi.org/10.3389/fenrg.2020.598803

Nagarajan, D., Kusmayadi, A., Yen, H.-W., Dong, C.-D., Lee, D.-J. and Chang, J.-S. 2019. Current advances in biological swine wastewater treatment using microalgae-based processes. Bioresource Technology 289: 121718. https://doi.org/10.1016/j.biortech.2019.121718

Oswald, W.J. 1973. Productivity of algae in sewage disposal. Solar Energy 15: 107–117. https://doi.org/10.1016/0038-092X(73)90013-3

Oswald, W.J. 1962. The Coming Industry of Controlled Photosynthesis. American Journal of Public Health and the Nations Health 52: 235–242. https://doi.org/10.2105/AJPH.52.2.235

Pandey, A., Srivastava, S. and Kumar, S. 2019. Sequential optimization of essential nutrients addition in simulated dairy effluent for improved Scenedesmus sp ASK22 growth, lipid production and nutrients removal. Biomass and Bioenergy 128: 105319. https://doi.org/10.1016/j.biombioe.2019.105319

Parr, L.., Perkins, R. and Mason, C. 2002. Reduction in photosynthetic efficiency of Cladophora glomerata, induced by overlying canopies of Lemna spp. Water Research 36, 1735–1742. https://doi.org/10.1016/S0043-1354(01)00395-5

Parsaee, M., Kiani Deh Kiani, M. and Karimi, K. 2019. A review of biogas production from sugarcane vinasse. Biomass and Bioenergy 122, 117–125. https://doi.org/10.1016/j.biombioe.2019.01.034

Perazzoli, S., Bruchez, B.M., Michelon, W., Steinmetz, R.L.R., Mezzari, M.P., Nunes, E.O. and da Silva, M.L.B. 2016. Optimizing biomethane production from anaerobic degradation of Scenedesmus spp. biomass harvested from algae-based swine digestate treatment. International Biodeterioration & Biodegradation 109: 23–28. https://doi.org/10.1016/j.ibiod.2015.12.027

Posadas, E., Muñoz, R. and Guieysse, B. 2017. Integrating nutrient removal and solid management restricts the feasibility of algal biofuel generation via wastewater treatment. Algal Research 22: 39–46. https://doi.org/10.1016/j.algal.2016.11.019

Quintero-Dallos, V., García-Martínez, J.B., Contreras-Ropero, J.E., Barajas-Solano, A.F., Barajas-Ferrerira, C., Lavecchia, R. and Zuorro, A. 2019. Vinasse as a Sustainable Medium for the Production of Chlorella vulgaris UTEX 1803. Water 11: 1526. https://doi.org/10.3390/w11081526

Robles-González, V., Galíndez-Mayer, J., Rinderknecht-Seijas, N. and Poggi-Varaldo, H.M. 2012. Treatment of mezcal vinasses: A review. Journal of Biotechnology 157, 524–546. https://doi.org/10.1016/j.jbiotec.2011.09.006

Rogowska, J., Cieszynska-Semenowicz, M., Ratajczyk, W. and Wolska, L. 2020. Micropollutants in treated wastewater. Ambio 49: 487–503. https://doi.org/10.1007/s13280-019-01219-5

Samiee, S., Ahmadzadeh, H., Hosseini, M. and Lyon, S. 2019. Algae as a Source of Microcrystalline Cellulose, in: Advanced Bioprocessing for Alternative Fuels, Biobased Chemicals, and Bioproducts. Elsevier, pp. 331–350. https://doi.org/10.1016/B978-0-12-817941-3.00017-6

Scherer, M.D., de Oliveira, A.C., Filho, F.J.C.M., Ugaya, C.M.L., Mariano, A.B., Vargas, J.V.C., 2017. Environmental study of producing microalgal biomass and bioremediation of cattle manure effluents by microalgae cultivation. Clean Technologies and Environmental Policy 19, 1745–1759. https://doi.org/10.1007/s10098-017-1361-x

Silva, M.A., Barbosa, G.H., Brito Codato, C., Arjonilla de Mattos, L.F., Gaspar Bastos, R. and Kieckbusch, T.G. 2017. Heterotrophic growth of green microalgae Desmodesmus subspicatus in ethanol distillation wastewater (vinasse) and lipid extraction with supercritical CO 2. Journal of Chemical Technology & Biotechnology 92: 573–579. https://doi.org/10.1002/jctb.5035

Song, C., Hu, X., Liu, Z., Li, S. and Kitamura, Y. 2020. Combination of brewery wastewater purification and CO2 fixation with potential value-added ingredients production via different microalgae strains cultivation. Journal of Cleaner Production 268, 122332. https://doi.org/10.1016/j.jclepro.2020.122332

Soto, M.F., Diaz, C.A., Zapata, A.M. and Higuita, J.C. 2021. BOD and COD removal in vinasses from sugarcane alcoholic distillation by Chlorella vulgaris: Environmental evaluation. Biochemical Engineering Journal 176, 108191. https://doi.org/10.1016/j.bej.2021.108191

Sousa, J.C.G., Ribeiro, A.R., Barbosa, M.O., Pereira, M.F.R. and Silva, A.M.T. 2018. A review on environmental monitoring of water organic pollutants identified by EU guidelines. Journal of Hazardous Materials 344: 146–162. https://doi.org/10.1016/j.jhazmat.2017.09.058

Tossavainen, M., Lahti, K., Edelmann, M., Eskola, R., Lampi, A.-M., Piironen, V., Korvonen, P., Ojala, A. and Romantschuk, M. 2019. Integrated utilization of microalgae cultured in aquaculture wastewater: wastewater treatment and production of valuable fatty acids and tocopherols. Journal of Applied Phycology 31: 1753–1763. https://doi.org/10.1007/s10811-018-1689-6

Ummalyma, S.B., Sirohi, R., Udayan, A., Yadav, P., Raj, A., Sim, S.J. and Pandey, A. 2022. Sustainable microalgal biomass production in food industry wastewater for low-cost biorefinery products: a review. Phytochemistry Reviews. https://doi.org/10.1007/s11101-022-09814-3

Ummalyma, S.B. and Sukumaran, R.K. 2014. Cultivation of microalgae in dairy effluent for oil production and removal of organic pollution load. Bioresource Technology 165: 295–301. https://doi.org/10.1016/j.biortech.2014.03.028

Viegas, C., Gouveia, L. and Gonçalves, M. 2021a. Aquaculture wastewater treatment through microalgal. Biomass potential applications on animal feed, agriculture, and energy. Journal of Environmental Management 286: 112187. https://doi.org/10.1016/j.jenvman.2021.112187

Viegas, C., Gouveia, L. and Gonçalves, M. 2021b. Bioremediation of cattle manure using microalgae after pre-treatment with biomass ash. Bioresource Technology Reports 14: 100681. https://doi.org/10.1016/j.biteb.2021.100681

Wang, H., Xiong, H., Hui, Z. and Zeng, X. 2012. Mixotrophic cultivation of Chlorella pyrenoidosa with diluted primary piggery wastewater to produce lipids. Bioresource Technology 104: 215–220. https://doi.org/10.1016/j.biortech.2011.11.020

Wang, Q., Wei, W., Gong, Y., Yu, Q., Li, Q., Sun, J. and Yuan, Z. 2017. Technologies for reducing sludge production in wastewater treatment plants: State of the art. Science of The Total Environment 587–588: 510–521. https://doi.org/10.1016/j.scitotenv.2017.02.203

19

Strategies for Food Waste Valorizations and Products

José Enrique Botello-Álvarez,[1,*] *Pasiano Rivas-García*,[2,*] *Alejandro Padilla-Rivera*,[3] *Brenda Ríos-Fuentes*,[4] *Juan Felipe Rueda-Avellaneda*[5] and *Uriel Galvan-Arzola*[5]

1. Introduction

Approximately one-third of the world's food is produced and not consumed (FAO 2018), becoming losses and waste that are often improperly managed, generating environmental, social and economic impacts. To move forward sustainability in the fields of production, transformation, and consumption of food, circular economic systems have been proposed where waste is sought to be transformed into new raw materials and products that are recovered, reused, or recycled (CEC 2017).

Nzihou (2010) defined recovery as converting waste and biomass into energy, fuel, and other useful materials through processes that contribute to sustainability objectives. The generation of value in a waste requires recognizing its potential, so that through its processing it is appreciated as a satisfier of an existing need. The recovery potential of waste depends on its composition, the amount, state, and source of its generation, the existence of the infrastructure and technologies for its processing and transformation, and the existence of a marketing channel. These factors must be considered in the development of a recovery strategy.

Papargyropoulou et al. (2014), considering the sustainability criteria, proposed a hierarchical order of food waste (FW) management alternatives: prevention, reuse, recycling, recovery, and final disposal. However, the prevailing socio-economic conditions and regulations in the different regions of the world determine the management method and its coverage.

[1] Doctorado en Ciencias de la Ingeniería, Departamento de Ingeniería Bioquímica, Tecnológico Nacional de México en Celaya, Av. Tecnológico y A. García Cubas, Celaya, Gto., 38010, México.

[2] Facultad de Ciencias Químicas, Universidad Autónoma de Nuevo León, San Nicolás de los Garza, Mexico. Centro de Investigacion en Biotecnologia y Nanotecnologia, Facultad de Ciencias Químicas, Universidad Autonoma de Nuevo León, Apodaca, Mexico.

[3] School of Architecture, Planning and Landscape, University of Calgary, 2500 University Drive NW, Calgary, AB T2N 1N4, Canada.
Email: Alejandro.padillariv@ucalgary.ca

[4] Doctorado en Ciencias de la Ingeniería, Departamento de Ingeniería Bioquímica, Tecnológico Nacional de México en Celaya, Av. Tecnológico y A. García Cubas, Celaya, Gto., 38010, México.
Email: brenda.riosf@gmail.com

[5] Facultad de Ciencias Químicas, Universidad Autónoma de Nuevo León, San Nicolás de los Garza, Mexico
Email: juan.ruedand@uanl.edu.mx; ugalvan@uanl.edu.mx

* Corresponding authors: enrique.botello@itcelaya.edu.mx; pasiano.rivasgr@uanl.edu.mx

In the ReFED report (2016), solutions are proposed and evaluated to reduce FW in the United States by at least 20%. The solutions were classified into the categories of prevention, recovery, and recycling. In prevention solutions, the following stand out—i.e., consumer education campaigns, waste tracking and analysis (waste tracking and analytics), and standardization of expiration dates (standardized date labelling). The recovery measures are donation tax incentives, standardized donation regulation, and donation matching software. Finally, the recycling solutions are composting, anaerobic digestion, and anaerobic digestion with water recovery.

In the evaluation of a solution to reduce the generation of FW, it is necessary to consider representative indicators of the benefits of each proposed solution: economic value, annual reduction of waste, potential for commercial profits, jobs generated, reduction of greenhouse gases, consumption of water, land use, and resources (ReFED 2016). Recycling and recovery are the management alternatives that are linked to recovery processes. Food recycling refers to transforming food waste into new products or intermediate raw materials. In contrast, the recovery of food waste refers to the extraction of its valuable components or its transformation into fuels or energy. The recovery alternatives are manufacturing food or food ingredients for humans, manufacturing food for livestock, composting, biofertilizers, biorefinery products, biofuel production, and obtaining energy by incineration.

In this chapter, a general strategy is proposed for the selection of an alternative for the recovery of food losses and waste. The basic stages that must be considered in these strategies are presented and discussed—i.e., characterization and possible valorization alternatives, location and generation sources, and finally the integral evaluation and selection of the valorization process.

2. Categories and Characterization of Food Waste

Food waste is classified into two main categories: food losses (FL) and food waste (FW). In the following lines, both definitions and differentiations are presented. FL refers to the decrease in edible food mass throughout the part of the supply chain that specifically leads to edible food for human consumption. FL takes place at the production, postharvest, and processing stages in the food supply chain. FL that is occurring at the end of the food chain (retail and final consumption) is rather called 'food waste', which relates to retailers' and consumers' behaviour (FAO 2011).

Foods are grouped into cereals, roots and tubers, oilseeds and pulses, fruit and vegetables, meat, fish and seafood, and dairy products (FAO 2011). Different studies about FL and FW have been identified by food source groups and sources. García et al. (2005) classified FW as meat waste, fish waste, fruit and vegetable waste, restaurant waste, and household waste. Fausto-Castro et al. (2020) classified restaurant residues as cereal waste, vegetable waste (VeW), citric waste and processed food waste. The classification of waste and its selection facilitate its recovery processes.

For certain recovery processes, it is also important to distinguish the terms FW and FL. FW are foods or their parts, considered edible, that according to their origin and handling, are in good condition, clean, and contain nutritional value so they can be processed to obtain safe products with added value. FL is food waste discarded in good condition for final disposal. In some European countries, it is prohibited to feed cattle with FW when this term is identified with FL (Salemdeeb et al. 2017).

Household waste in developing countries constitutes 38% of urban waste (MSW) and is mostly consigned into landfills, confined sites, or open dumps (Rueda Avellaneda et al. 2021; Aldana-Espitia et al. 2017). The household waste within the MSW matrix is FW that can no longer be valued even as feed for livestock, but it can be recycled in the production of biogas, compost, or in the recovery of energy through incineration (Margallo et al. 2019). In the rest of the document, we will generally refer to food losses and waste with the abbreviation FLW.

Table 1 presents the proximal analysis of some FLW. Some residues have a high content of nutritional groups, such as meat waste, which on a dry basis contains a high-fat content and can be recycled for the production of biodiesel (Toldrá-Reig et al. 2020). VeW are high in moisture

Table 1. Proximal Composition of Waste From Different Sources (% Dry Basis).

Source	Moisture	Crude Protein	Crude Fibre	Fat Gross	Carbohydrates	Ash	Reference
Vegetal waste	66.06	9.72	1.12	1.93	60.69	3.52	San Martín et al. (2016)
	88.10	11.60	13.40	1.50	65.50	7.90	García et al. (2005)
Meat waste	61.00	24.60	0.60	69.9	0.00	4.90	García et al. (2005)
Fish waste	73.90	57.00	1.20	19.10	0.00	21.80	García et al. (2005)
Restaurant waste	60.40	27.50	2.30	28.80	26.90	14.40	García et al. (2005)
	65.52	19.65	6.14	24.66	42.13	7.42	Fausto-Castro et al. (2020)
	58.20	26.0	----	16.20	52.20	5.51	Ho and Chu (2019)
Household waste	66.90	16.30	12.40	11.30	41.80	18.20	García et al. (2005)
	68.20	25.70		15.40	50.60	8.28	Ho and Chu (2019)

and simple or complex carbohydrates; these residues have been used as fodder for ruminant feeding (Quintero-Herrera et al. 2021). In addition, they contain a large quantity and diversity of nutraceuticals and can be recycled for ingredients and food supplements with high added value (Rudra et al. 2015).

Table 2 presents a summary of recovery alternatives for FLW from different types of food. In the columns, the recovery alternatives were classified by the type of food from which the FLWs come; in the rows, the valuation alternatives are classified according to the valuation product in a particular way; in each cell the FLW and the valuation product obtained are specified. In the recycling of FLW to produce food, ingredients, or food supplements for human consumption, the use of FLW from vegetables stands out because of its high content of nutrients and nutraceutical compounds.

The fruit and vegetable processing industry discards significant amounts of vegetables for aesthetic reasons that can be used to produce creams, pastes, dehydrated products, and flours that are appreciated in the production of vegan or naturopathic products since they can replace synthetic colours or flavours. Another relevant vegetable waste is the non-consumable parts such as peels, skins, and seeds that contain a high content of phytochemicals considered antioxidants.

Cereal residues are mainly made up of starch that can easily be hydrolyzed to give rise to simple sugars, which are used as substrates for the biotechnological production of different chemical products in the so-called biorefineries. These processes are limited by high production costs, but they are expected to be the basis of the green industry in the future. The production of biogas through anaerobic digestion, composting, and energy by incineration are technically viable alternatives for any of the waste or combination of these, however, they are processed with low added value that require large infrastructures with high investment costs.

3. The Amount and Source of Wastes

The FAO, through the food balances database, offers an excellent tool to learn about the food systems of different countries through a balance of basic foods (FAO 2021). Table 3 presents the different parameters considered in a balance of a food product at the national level. With this information, it is possible to have a first approximation of the generation of FLW in a country. As an example, the balances for some staple foods in Mexico are presented. White corn is the most important basic grain for the country, but yellow corn is also imported, which is the main raw material for feeding national livestock; however, about 10% of the country's total corn supply is wasted. Also, surprising is the low production and consumption of beans, a legume that was considered a basic ingredient in traditional Mexican cuisine for a long time. In recent years, Mexico has become an important exporter of vegetables, including broccoli, cauliflower, and asparagus, among others; however, in this sector there are also 10.9% losses of national availability.

On the other hand, it is essential to recognize where FLW generation occurs along the food supply chain. Figure 1 represents a representative scheme of the supply chain from agricultural production to domestic consumption (adapted from CEC 2017).

Table 2. Valorization Alternatives for FLW From Different Foods.

Valorization Alternative	Cereals	Roots and Tubers	Oilseed and Pulses	Fruit and Vegetables	Meat	Fish and Seafood	Dairy products	Food Waste
Food	Bakery waste/Beer (1)	Carrot peel waste/ phytonutrients (14) Waste in onion production / Bioactive compounds (15)	Peanut skins and peanut waste/ Potential functional food ingredient (19)	Peels, seeds, and pomances/ nanoemulsions of phytochemicals (2) Grape seed / Bioactive compound in the food industry and health food (10)	Beef tongue/ food ingredient (7)	Tilapia, tuna, salmon, and sardine waste/ fish meal (27)	Products of sheep milk waste/fermented dairy products (23)	
Feed	Cereals and corn starch/fish feed (16) Bakery waste/feed lambs (30) Bakery waste/broiler chicken feed (33)	Carrot waste/feeds in animal diets (35)	Peanut skins and peanut waste/animal feed (19)	Fruit and vegetable waste/fish feed (16)	Packaged Meat waste/pet food (5) Meat products and bone meal/ Fish feed (16)	Fish/pig feed (6) Fishmeal/fish feed (16) Fish waste/Fish feed; fish meal (17)		Food waste/chicken feed (12) Food waste/pig feed (13)
Composta	Bakery waste/ compost (31)		Soy waste / Composting (29)	Viticulture waste/ Biofertilizer (28) Mushroom wastes/ Broiler chickens (34)		Marine waste/ composting (18)		Food waste/líquid fertilizer (12) Food waste/ compost, mineral fertilizer (13)
Product of biorefineries	Bakery waste/ bioethanol (1) Maize, wheat, rice waste/ bioethanol (11) Bakery waste/ succinic acid (32) Wheat, barley, oats, waste/biopolymers (26)	Sweet potato / Bioethanol (4) Potato waste/ bioethanol (11) Sugar beet, Carrots, and onions waste/biopolymers (26)	Peanut waste / Bioethanol (21) Chia seed flour/ wood plastic composite (20) Rapeseed waste/ biopolymers (26)	Waste pomegranate fruits / Bioethanol (36)			Milk and soft goat cheese / Propionic acid (25)	Kitchen waste/ poly–3–hydroxybutyrate, bioethanol, pectinase, and 2, 3-butanediol (37)

Biogas	Bakery waste/ Biogas (1) Maize husk/biogas (22)	Carrots waste/ biogas (3)		Lettuce, cabbage, apple, banana, pear, watermelon/Biogas (3) Vegetal waste/biogas (24)	Meat and bone meal waste/ biogas (9)		Whey cheese/ biogas (24)	Food waste from restaurants/biogas (22)
Incineration	Bakery waste/ energy (1)				Manure, fat, blood, sludge/ energy (8)			Food waste/energy (13)

(1) Brancoli et al. 2020; (2) Saini et al. 2019; (3) Lin et al. 2011; (4) Weber et al. 2020; (5) Mosna et al. 2021; (6) Márquez et al. 2007; (7) Long and Mohan (2021); (8) Mofijur et al. 2021; (9) Bedoić et al. 2020; (10) Chen et al. 2020; (11) Melikoglu and Turkmen 2019; (12) Siddiqui et al. 2021; (13) Salemdeeb et al. 2016; (14) Jayesree et al. 2021; (15) Osojnik et al. 2021; (16) Mo et al. 2015; (17) Mo et al. 2018; (18) Etxabide et al. 2016; (19) Toomer 2018; (20) Dominguez et al. 2021; (21) Polachini et al. 2016; (22) Owamah et al. 2021; (23) Tribst et al. 2020; (24) Ivanchenko et al. 2021; (25) Ali et al. 2021; (26) Bolaji et al. 2021; (27) Souza et al. 2017.

Table 3. Food Balance of Mexico in 2018, Data is Presented by 10^3 t (FAO 2020; Food Balances).

Product	Corn and Products	Beans	Vegetables
Production	27,170	1,196	12,133
Import Quantity	17,279	166	244
Stock Variation	2,712	0	85
Export Quantity	1,223	51	5,363
Domestic Supply Quantity	40,514	1,311	6,909
Food	14,813	1,174	6,159
Feed	18,622	0	0
Seed	181	85	0
Losses	4,571	52	750
Residuals	26	0	0
Processing	2,028		
Other Uses (Non-food)	273		

To estimate the amount of food that is lost in each of the stages of the supply chain, the use of factors or percentages of losses is proposed (FAO 2011). These factors are estimated based on databases from local agencies, such as the USDA Economic Research Service and ERS in the United States, which collect information through specialized studies, censuses, and surveys with agricultural producers, companies, and establishments that are integrated into the stages supply chain, as well as measuring consumption trends (Buzby et al. 2009). Loss factors are estimated at the level of food groups and each stage of the supply chain of a country or region. Buzby et al. (2009) found that the loss factors for fruits and vegetables in supermarkets in the United States show large variations with respect to the average assigned by the FAO (CEC 2017). For example, for fruits and vegetables, there is a fraction of 0.12; however, for papaya, a fraction of 0.55 was estimated.

Table 4 presents a comparison of the loss factors for Canada, the United States, and Mexico; there are important differences both in the types of food and in the stages where losses occur. In Mexico, for fruits and vegetables, large loss factors were estimated at all stages of the supply chain. In Canada and the United States, the greatest loss factors are generated in the final stage of consumption.

The stage of the supply chain where FLW is generated is decisive in choosing the most convenient recovery alternative. The FW that is generated in the domestic consumption stage is mainly disposed of in sanitary landfills or open-air dumps in developing countries or is valued as compost, biogas, or to obtain energy by incineration. Other alternatives are difficult to implement

Table 4. Loos Factors From the Food Supply Chain in Canada, USA, and Mexico (CEC 2017).

Country	Food Production Preharvest	Food Production Postharvest	Processing and Packaging	Distribution	Consumption
Cereal					
Canada - USA	0.02	0.02	0.005	0.02	0.27
Mexico	0.06	0.04	0.02	0.04	0.10
Fruits and Vegetables					
Canada - USA	0.20	0.04	0.02	0.12	0.28
Mexico	0.20	0.10	0.20	0.12	0.10
Meat					
Canada - USA	0.037	0.01	0.05	0.04	0.11
Mexico	0.056	0.011	0.05	0.05	0.06

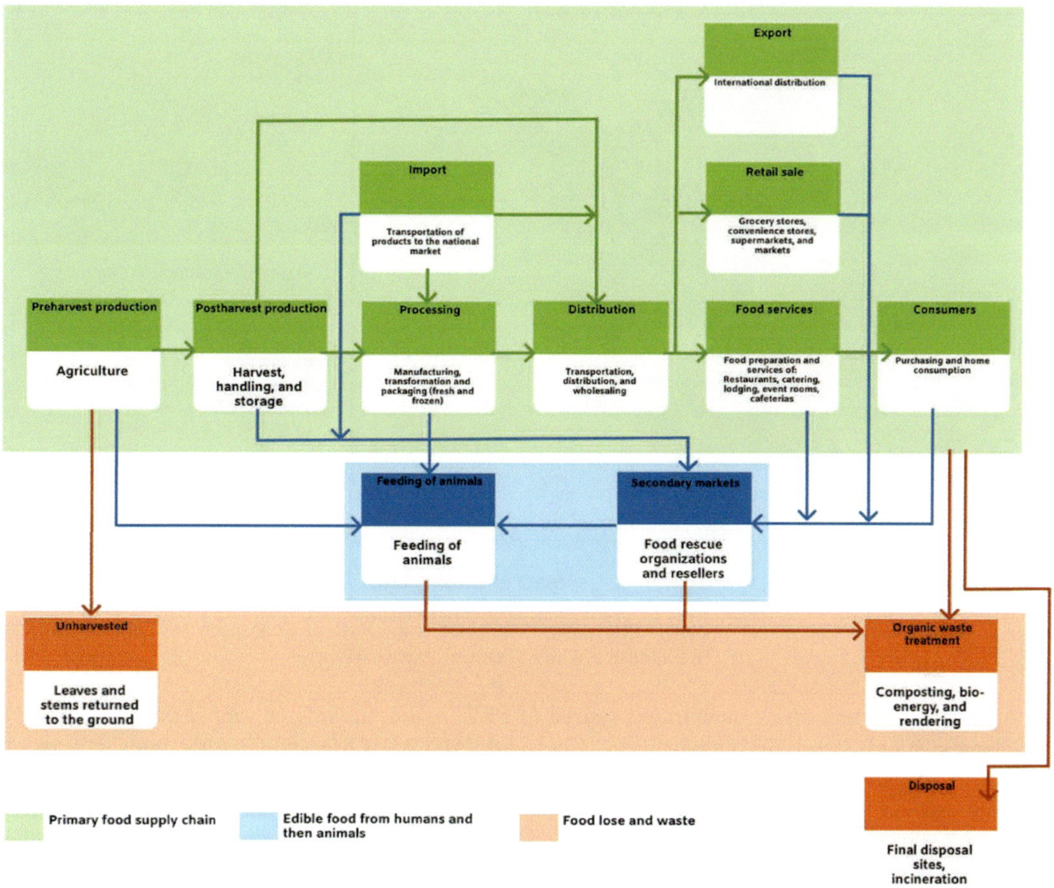

Figure 1. Food Supply Chain With FL and FW Recovery and Recycling Options (Adapted from CEC 2017).

because these residues are generated in a dispersed manner, present variations in composition, and are often in poor condition or contaminated. In food processing plants, there are both internal and government regulations, as well as certification processes, which influence the management and generation of byproducts or waste in an ideal state to be used in a recovery process. For example, fruit skins or peels are obtained fresh and clean and in large quantities to be used in the production of food products and ingredients appreciated for their high content of phytochemicals (Esparza et al. 2020).

The establishment of a processing plant for the valorization of FLW considers its abundance and availability in a delimited geographic region. Considering the information in Tables 2 and 3, in Mexico, 750,000 t of (VeW) is generated annually; in the vegetable processing stage, a loss factor of 0.20 is estimated, so this sector generates approximately 150,000 t of VeW. The state of Guanajuato is located in the centre of Mexico, it stands out for the agricultural production and processing of vegetables, and it ranks first nationally in the production of broccoli with 23,994 ha and a production of 413,084 t per year in 2019 (SIAP 2019). In Mexico, there are at least 154 companies (with 100 or more workers) dedicated to the processing of vegetables, mainly for export. Figure 2 shows the geographical contour of the state of Guanajuato in which several plants of this type can be seen in an agro-industrial corridor of around 100 km, which is closely related to the high agricultural production of vegetables (DENUE 2021). Aviles-Rios et al. (2009) estimated the annual generation of 15,200 t of broccoli and cauliflower residues in this region with adequate nutritional characteristics for cattle feeding. At present, these residues are mainly used for feeding cattle as a complement to alfalfa.

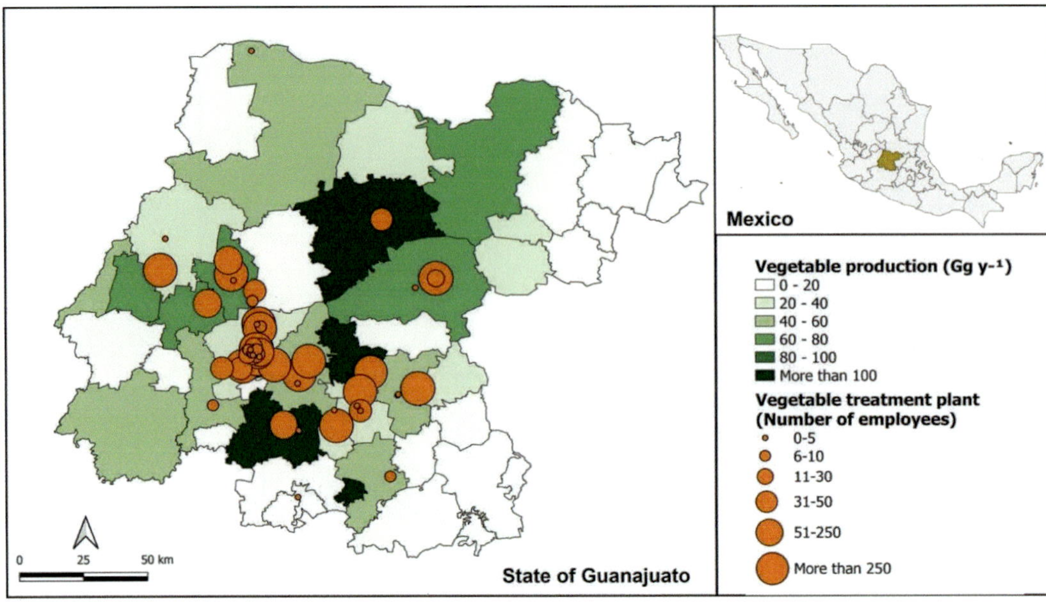

Figure 2. Agroindustrial Corridor of Vegetable Processing Companies in the State of Guanajuato and Its Geographical Relationship With Agricultural Fields.

Another important and concentrated source of FLW is the markets or supply centres. Mexico is the largest supply centre in the world; the Central de Abasto of Mexico City occupies an area of 327 ha and annually markets 10.95 Mt of different foods: fruits, vegetables, legumes, meat products, fish, and seafood, as well as groceries. It is estimated that this distribution centre generates 0.8 Mt/y of waste, consisting mainly of fruits and vegetables. Much of the vegetable waste is recycled for livestock feed; an unquantified part is recovered by scavengers for self-consumption and sale (CEDRESA 2019). Another estimate shows that 643.6 t of waste with MSW characteristics are generated daily, of which 75 t of recyclable products such as plastics, paper, and metal waste are collected; 65 t are used to produce compost, and 503.6 t are disposed of in a sanitary landfill (Morales-Pérez 2011). Surely, a significant amount of FLW that makes up the solid waste generated by the supply centre could be reused and recovered in the production of food or food ingredients.

4. Comprehensive Evaluation and Selection of the Valuation Process

The selection of the best recovery process and product must consider several factors to make the best decision for its implementation. Economic feasibility is a decisive factor of primary importance; however, the serious problem of climate change and its strong relationship with the generation of FLW has led to an opening of possibilities to implement sustainable processes through incentives such as support and government subsidies. In this way, innovation in processes and products must contribute substantially to the use of FLW. On the other hand, new food trends, such as plant-based food, healthy eating, and veganism, open a market for new products that take advantage of FLW.

The evaluation of sustainability in the food industry is a requirement of governments and societies concerned about environmental problems; consumers look for products purchased to have sustainable product labels (Potter and Roos 2021). Life cycle assessment (LCA) is a holistic and standardized methodology that assesses the environmental footprint of products, processes, and services (Rebitzer et al. 2004). There are environmental impact assessment models (ReCiPe—a harmonized life cycle impact assessment method at midpoint and endpoint) which, framed within the LCA methodological framework, assess environmental impacts in three broad impact categories: damage to human health, damage to ecosystems, and depletion of natural resources (called endpoint indicators). These indicators are determined based on information from 18 so-called midpoint indicators based on scientific principles in different environmental areas (Huijbregts et al. 2016). Assessments of the environmental profile of the food industry, with an LCA approach, allow mitigation efforts to be directed in a pertinent way and provide information to clarify decision-making. The economic evaluation is also implemented through indicators such as production cost, investment expense, net present value, and internal rate of return, among others. The ideal selection of recovery alternatives would be to prefer the one that combines low environmental impacts and a convenient economic benefit; however, these attributes are often not achieved.

In the United States in 2016, 52.4 Mt of FLW and 10.1 Mt of unharvested agricultural products were generated. In response to this problem, solutions were proposed in the categories of prevention, recovery, and recycling to reduce the generation of FLW by 20% by 2030. Figure 3 shows the potential amounts of FLW that were intended to be used. In the recycling category, composting and anaerobic digestion solutions stand out. Figure 4 presents environmental and economic indicators for some solutions. FLW processing stands out in obtaining products with added value; it is the option with the best environmental performance and cost-benefit ratio. However, the amounts of FLW that can be exploited in this form of recovery are small. Anaerobic digestion and composting have cost-benefit ratios close to 1; from this perspective, their implementation does not have an

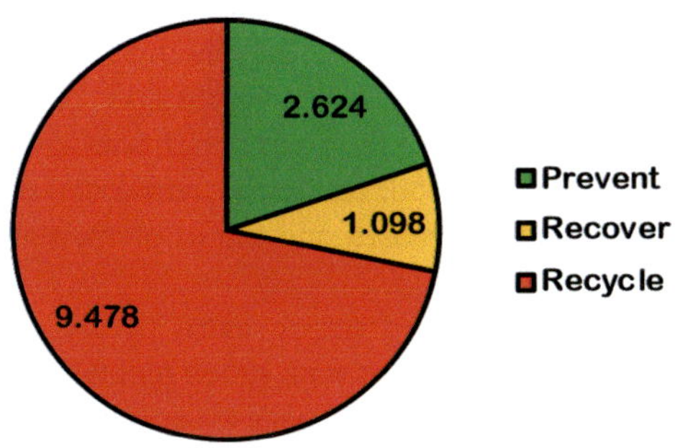

Figure 3. Potential Amounts Avoided From FLW in the US in Mt (ReFED 2016).

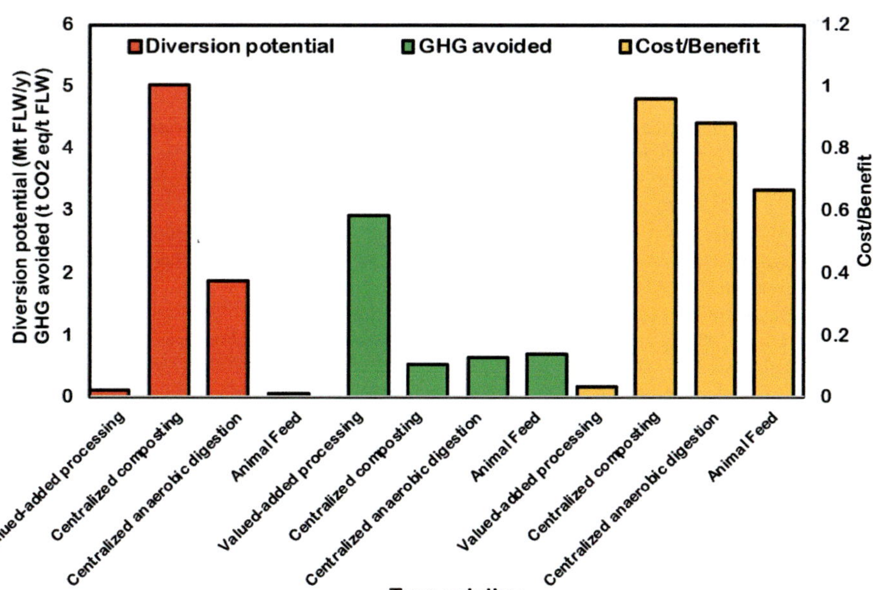

Figure 4. Environmental and Economic Indicators of Some Alternatives for the Valorization of Food and Wastes.

economic appeal; however, they are the recovery options that can be applied to a greater amount of waste (ReFED 2016).

Dou et al. (2018) reported cost/benefit ratios of 3.96 and 2.58 for pig feed (dry and wet feed, respectively) obtained through processing FLW of domestic and commercial origin in South Korea. These relationships are due to the fact that production costs are higher than commercial prices. High production costs are associated with the thermal processing and drying of FLW to ensure its safety and stability. Takata et al. (2012) in Japan obtained a cost/benefit ratio of 0.85 in the production of pig feed using mainly FLW from cereals with low moisture content to reduce processing costs; they also mention that to achieve this cost/benefit ratio, it is necessary to receive economic incentives from local authorities.

5. Recommendations

The final disposal of FLW in landfills and dumps continues to be the most widely used management method. The waste management hierarchy triangle establishes a desirable order of management type options: prevention, reuse, recycling, recovery, and final disposal. However, in the implementation, recycling alternatives, such as composting and anaerobic digestion, are mainly chosen, which process massive amounts of FLW with cost-benefit ratios close to unity; on the other hand, the composition and state of FLW are not relevant to these processes. The valorization of FLW requires novel technological research and development but is pertinent to the regional contexts where FLW is generated. The relevance of an FLW valorization alternative is achieved by establishing and following an adequate strategy that considers the nature and composition of FLW, its potential to originate valuable products, and environmental, economic, and social evaluation. Food security will be closer when the generation of FLW decreases and the implementation of recovery alternatives increases.

The social pressure on food systems is great; a sector of the population demands healthier food with higher nutritional quality and produces with green technologies, and another larger sector of the population demands sufficient, cheap food with basic nutritional quality. To alleviate this pressure, food systems must adopt circular economies, and it is essential to encourage research and technological developments that aim to achieve the valorization of FLW with relevance. Any recovery strategy must start with the recognition that FLW is raw material.

References

Aldana-Espitia, N.C., Botello-Alvarez, J.E., Rivas-García, P., Cerino-Cordova, F.J., Bravo-Sanchez, M.G. Abel-Seabra, J.E. and Estrada-Baltazar, A. 2017. Environmental impact mitigation during the solid waste management in an industrialized city in México: An approach of Life Cycle Assessment. Revista Mexicana de Ingeniería Química. 16(2): 563–580. ISSN electrónico: 2395–8472.

Ali, R., Saravia, F., Hille-Reichel, A., Gescher, J. and Horn, H. 2021. Propionic acid production from food waste in batch reactors: Effect of pH, types of inoculum, and thermal pre-treatment. Bioresource Technology 319: 124166. https://doi.org/10.1016/j.biortech.2020.124166

Avilés-Rios, E.D., Espinosa-García, J.A., Rentería Flores, J.A., Mejía-Guadarrma, C.A., Mariscal-Landín, G. and Cuarón-Ibargüengoytia, J.A. 2009. Non-traditional available ingredients with potential to be used in gestating sow feeding in the Mexican Bajio. Vet. Méx. 40(4): 357–370.

Ballardo, C., Abraham, J., Barrena, R., Artola, A., Gea, T. and Sánchez, A. 2016. Valorization of soy waste through SSF for the production of compost enriched with Bacillus thuringiensis with biopesticide properties. Journal of Environmental Management 169: 126–131. https://doi.org/10.1016/j.jenvman.2015.12.029

Bedoić, R., Špehar, A., Puljko, J., Čuček, L., Ćosić, B., Pukšec, T. and Duić, N. 2020. Opportunities and challenges: Experimental and kinetic analysis of anaerobic co-digestion of food waste and rendering industry streams for biogas production. Renewable and Sustainable Energy Reviews, 130(June). https://doi.org/10.1016/j.rser.2020.109951

Bolaji, I., Nejad, B., Billham, M., Mehta, N., Smyth, B. and Cunningham, E. 2021. Multi-criteria decision analysis of agri-food waste as a feedstock for biopolymer production. Resources, Conservation and Recycling 172(April): 105671. https://doi.org/10.1016/j.resconrec.2021.105671

Brancoli, P., Rousta, K. and Bolton, K. 2017. Life cycle assessment of supermarket food waste. Resources, Conservation and Recycling 118: 39–46. https://doi.org/10.1016/j.resconrec.2016.11.024

Buzby, Jean C., Hodan Farah Wells, Bruce Axtman, and Jana Mickey. Supermarket Loss Estimates for Fresh Fruit, Vegetables, Meat, Poultry, and Seafood and Their Use in the ERS Loss-Adjusted Food Availability Data. EIB-44, U.S. Dept. of Agriculture, Econ. Res. Serv. March 2009

CEC, 2017. Characterization and management of food and waste in North America. Montreal, Canada: Comision for Enviromental Cooperation. 289 pp.

Chen, Y., Wen, J., Deng, Z., Pan, X., Xie, X. and Peng, C. 2020. Effective utilization of food wastes: Bioactivity of grape seed extraction and its application in food industry. Journal of Functional Foods 73(July): 104113. https://doi.org/10.1016/j.jff.2020.104113

Cortés, A., Oliveira, L.F.S., Ferrari, V., Taffarel, S. R., Feijoo, G. and Moreira, M.T. 2020. Environmental assessment of viticulture waste valorisation through composting as a biofertilisation strategy for cereal and fruit crops. Environmental Pollution, 264, 1–8. https://doi.org/10.1016/j.envpol.2020.114794

Dominguez-Candela, I., Garcia-Garcia, D., Perez-Nakai, A., Lerma-Canto, A., Lora, J. and Fombuena, V. 2021. Contribution to a circular economy model: From lignocellulosic wastes from the extraction of vegetable oils to the development of a new composite. Polymers 13(14). https://doi.org/10.3390/polym13142269

Dou, Z., Toth, J.D. and Westendorf, M.L. 2018. Food waste for livestock feeding: Feasibility, safe and sustainability implications. Glob. Food Sec. 17: 154–161

Esparza, I., Jiménez-Moreno, N., Binbela, F., Ancín-Azpilicueta, C. and Gandía, L.M. 2020. Review Fruit and vegetable waste management: Conventional and emerging approaches. Journal of Environmental Management., 265(1),

Etxabide, A., Leceta, I., Cabezudo, S., Guerrero, P. and De La Caba, K. 2016. Sustainable fish gelatin films: From food processing waste to compost. ACS Sustainable Chemistry and Engineering 4(9): 4626–4634. https://doi.org/10.1021/acssuschemeng.6b00750

FAO, 2018. Food Loss and Waste and the Right to Adequate Food: Making the Connection: Rome, p. 48 License: CC BY-NC-SA 3.0 IGO

FAO, 2019. The State of Food and Agriculture 2019. Moving forward on food loss and waste reduction. Rome. Licence: CC BY-NC-SA 3.0 IGO.

FAO, 2021. Food Balances. http://www.fao.org/faostat/en/#data/FBS

Fard, S. H., Toghyani, M. and Tabeidian, S. A. 2014. Effect of oyster mushroom wastes on performance, immune responses and intestinal morphology of broiler chickens. International Journal of Recycling of Organic Waste in Agriculture 3(4): 141–146. https://doi.org/10.1007/s40093-014-0076-9

Fausto-Castro, L., Rivas-García, P., Gomez-Nafte, J. A., Rico-Martínez, R., Rico-Ramírez, V, Gomez-Gonzalez, R., Cuarón-Ibargüengoytia, J.A. and Jose Enrique Botello- Alvarez, J.E. 2020. Selection of food waste with low moisture and high protein content from Mexican restaurants as a supplement to swine feed. Journal of Cleaner Production. Volume 256: 120137.

Fidel Toldrá-Reig, F., Mora, L. and Toldrá, F. 2020. Review Trends in Biodiesel Production from Animal Fat Waste. Appl. Sci. 10(10): 3644. https://doi.org/10.3390/app10103644.

França, A.B., Frota, M.J.M., Ferraz, L.F.C., Silva, M.A., Madeiro, Morenz, D.A., de Faria, B.M. da Silva, C., L., arlos Elysio Moreira da Fonseca, C.E. 2012. Bakery waste in sheep diets: intake, digestibility, nitrogen balance and ruminal parameters. R. Bras. Zootec., v.41, n.1, 147–153

García, A.J., Esteban, M.B., Márquez, M.C. and Ramos, P. 2005. Biodegradable municipal solid waste: Characterization and potential use as animal feedstuffs. Waste Manag. 25, 780–787.

Govindaraju, M., Sathasivam, K.V. and Marimuthu, K. 2021. Waste to Wealth: Value Recovery from Bakery Wastes. Sustainability 13: 2835. https://doi.org/ 10.3390/su13052835

Ho, K.S. and Chu, L.M. 2021. Characterization of food waste from different sources in Hong Kong. Journal of the Air &Waste Management 69(3): 277–288. https://doi.org/10.1080/10962247.2018.1526138

Huijbregts, M. A., Steinmann, Z. J., Elshout, P. M., Stam, G., Verones, F., Vieira, M., and Van Zelm, R., 2017. ReCiPe2016: a harmonised life cycle impact assessment method at midpoint and endpoint level. The International Journal of Life Cycle Assessment 22(2): 138–147.

Ivanchenko, A., Yelatontsev, D. and Savenkov, A. 2021. Anaerobic co-digestion of agro-industrial waste with cheese whey: Impact of centrifuge comminution on biogas release and digestate agrochemical properties. Biomass and Bioenergy, 147: 106010. https://doi.org/10.1016/j.biombioe.2021.106010

Jayesree, N., Hang, P.K., Priyangaa, A., Krishnamurthy, N.P., Ramanan, R.N., Turki, M.S.A., Charis, M.G. and Ooi, C.W. 2021. Valorisation of carrot peel waste by water-induced hydrocolloidal complexation for extraction of carote and pectin. Chemosphere 272: 129919. https://doi.org/10.1016/j.chemosphere.2021.129919

Lin, J., Zuo, J., Gan, L., Li, P., Liu, F., Wang, K., Chen, L. and Gan, H. 2011. Effects of mixture ratio on anaerobic co-digestion with fruit and vegetable waste and food waste of China. Journal of Environmental Sciences 23(8): 1403–1408. https://doi.org/10.1016/S1001-0742(10)60572-4

Long, J.M. and Mohan, A. 2021. Development of meat powder from beef byproduct as value-added food ingredient. Lwt, 146(March), 111460. https://doi.org/10.1016/j.lwt.2021.111460

Margallo, M., Ziegler-Rodriguez, K., Vázquez-Rowe, I., Aldaco, R., Irabien, Á. and Kahhat, R. 2019. Enhancing waste management strategies in Latin America under a holistic environmental assessment perspective: A review for policy support. Science of the Total Environment 689: 1255–1275.

Márquez, M.C. and Ramos, P. 2007. Effect of the inclusion of food waste in pig diets on growth performance, carcass and meat quality. Animal 1(4): 595–599. https://doi.org/10.1017/S1751731107685000

Melikoglu, M. and Turkmen, B. 2019. Food waste to energy: Forecasting Turkey's bioethanol generation potential from wasted crops and cereals till 2030. Sustainable Energy Technologies and Assessments, 36(August): 100553. https://doi.org/10.1016/j.seta.2019.100553

Mo, W.Y., Cheng, Z., Choi, W. M., Lun, C.H.I., Man, Y.B., Wong, J.T.F., Chen, X.W., Lau, S.C.K. and Wong, M.H. 2015. Use of food waste as fish feeds: effects of prebiotic fibers (inulin and mannanoligosaccharide) on growth and non-specific immunity of grass carp (Ctenopharyngodon idella). Environmental Science and Pollution Research 22(22): 17663–17671. https://doi.org/10.1007/s11356-015-4971-z

Mo, W.Y., Man, Y.B. and Wong, M.H. 2018. Use of food waste, fish waste and food processing waste for China's aquaculture industry: Needs and challenge. Science of the Total Environment 613–614: 635–643. https://doi.org/10.1016/j.scitotenv.2017.08.321

Mofijur, M., Fattah, I.M.R., Kumar, P.S., Siddiki, S.Y.A., Rahman, S.M.A., Ahmed, S.F., Ong, H.C., Lam, S.S., Badruddin, I. A., Khan, T. M. Y. and Mahlia, T. M. I., 2021. Bioenergy recovery potential through the treatment of the meat processing industry waste in Australia. Journal of Environmental Chemical Engineering 9(4): 105657. https://doi.org/10.1016/j.jece.2021.105657

Morales, Rosalba. 2011. Planes de manejo de residuos de generadores de alto volumen: El caso de la Central de Abasto del Distrito Federal, México. [Tesis de Maestría en Ciencias en Estudios Ambientales y de la Sustentabilidad]. Instituto Politécnico Nacional.

Mosna, D., Bottani, E., Vignali, G. and Montanari, R. 2021. Environmental benefits of pet food obtained as a result of the valorisation of meat fraction derived from packaged food waste. Waste Management 125: 132–144. https://doi.org/10.1016/j.wasman.2021.02.035

Nzihou, 2010. A. Waste and Biomass Valorization. Waste Biomass Valor. 1: 1–2

Osojnik Črnivec, I. G., Skrt, M., Šeremet, D., Sterniša, M., Farčnik, D., Štrumbelj, E., Poljanšek, A., Cebin, N., Pogačnik, L., Smole Možina, S., Humar, M., Komes, D. and Poklar Ulrih, N., 2021. Waste streams in onion production: Bioactive compounds, quercetin and use of antimicrobial and antioxidative properties. Waste Management 126: 476–486. https://doi.org/10.1016/j.wasman.2021.03.033

Owamah, H.I., Ikpeseni, S.C., Alfa, M.I., Oyebisi, S.O., Gopikumar, S., David Samuel, O. and Ilabor, S.C. 2021. Influence of inoculum/substrate ratio on biogas yield and kinetics from the anaerobic co-digestion of food waste and maize husk. Environmental Nanotechnology, Monitoring & Management, 16(August), 100558. https://doi.org/10.1016/j.enmm.2021.100558

Papargyropoulou, E., Lozano, R., Steinberger, J.K., Wright, N. and Ujang, Z.B. 2014. The food waste hierarchy as a framework for the management of food surplus and food waste. J. Clean. Prod. 76: 106–115.

Polachini, T.C., Sato, A.C.K., Cunha, R.L. and Telis-Romero, J. 2016. Density and rheology of acid suspensions of peanut waste in different conditions: An engineering basis for bioethanol production. Powder Technology, 294, 168–176. https://doi.org/10.1016/j.powtec.2016.02.022

Potter, H.K. and Roos, E. 2021. Multi-criteria evaluation of plant-based foods –use of environmental footprint and LCA data for consumer guidance. Journal of Cleaner Production. 280(1): 124721. https://doi.org/10.1016/j.jclepro.2020.12472

Quintero-Herrera, S., García-León, A.M., Botello-Álvarez, J.E., Estrada-Baltazar, A., Abel-Seabra, J.E., Rivas-García, P., 2021. The use of broccoli agro-industrial waste in dairy cattle diet for environmental mitigation. Cleaner Environmental Systems 2 100035. 1–12. https://doi.org/10.1016/j.cesys.2021.100035. ISSN: 2666–7894.

Rebitzer, G., Ekvall, T., Frischknecht, R., Hunkeler, D., Norris, G., Rydberg, T. and Pennington, D.W. 2004. Life cycle assessment: Part 1: Framework, goal and scope definition, inventory analysis, and applications. Environment International 30(5): 701–720.

ReFED, 2016. A Roadmap to Reduce U.S. Food Waste by 20 Percent. ReFED Steering Committee and Advisory Council, U.S.

Rinne, M., Franco, M., Jalava, T., Järvenpää, E., Kahala, M., Blasco, L., Siljander-Rasi, H. and Kuoppala, K. 2019. Carrot byproduct fermentation quality and aerobic spoilage could be modified with silage additives. Agricultural and Food Science, 28(2), 59–69. https://doi.org/10.23986/afsci.79829

Rudra, S.G., Nishad, J., Jakhar, N. and Kaur, C. 2015 Food industry waste: mine of nutraceuticals. International Journal of Science, Environment and Technology 4(1): 205–29.

Rueda-Avellaneda, J.F., Rivas-García, P., Gomez-Gonzalez, R., Benitez-Bravo, R., Botello-Álvarez, J.E. and Tututi-Avila, S. 2021. Current and prospective situation of municipal solid waste final disposal in Mexico: A spatio-temporal evaluation. Renewable and Sustainable Energy Transition, 1, 100007. https://doi.org/10.1016/j.rset.2021.100007

Saini, A., Panesar, P.S. and Bera, M.B. 2019. Valorization of fruits and vegetables waste through green extraction of bioactive compounds and their nanoemulsions based delivery system. Bioresources and Bioprocessing 6(26): 1–12 https://doi.org/10.1186/s40643-019-0261-9

Salemdeeb, R., zu-Ermgassen, E.K.H.J., Kim M.H., Balmford, A. and Al-Tabbaa, A. 2017. Environmental and health impacts of using food waste as animal feed: a comparative analysis of food waste management options. J. Clean. Prod. 140: 871–880. https://doi.org/10.1016/j.jclepro.2016.05.049

San Martín, D., Ramos, S. and Zufía, J. 2016. Valorisation of food waste to produce new material for animal feed. Food Chemistry 198: 68–74.

SIAP. 2019. Servicio de Información Agroalimentaria y Pesquera. https://www.gob.mx/siap

Siddiqui, Z., Hagare, D., Jayasena, V., Swick, R., Rahman, M.M., Boyle, N. and Ghodrat, M. 2021. Recycling of food waste to produce chicken feed and liquid fertiliser. Waste Management 131(June): 386–393. https://doi.org/10.1016/j.wasman.2021.06.016

Sindhu, R., Manju, A., Mohan, P., Rajesh, R.O., Madhavan, A., Arun, K.B., Hazeena, S.H., Mohandas, A., Rajamani, S.P., Puthiyamadam, A., Binod, P. and Reshmy, R. 2020. Valorization of food and kitchen waste: An integrated strategy adopted for the production of poly-3-hydroxybutyrate, bioethanol, pectinase and 2, 3-butanediol. Bioresource Technology 310(March): 123515. https://doi.org/10.1016/j.biortech.2020.123515

Souza, M.L.R. de, Yoshida, G.M., Campelo, D.A.V., Moura, L.B., Xavier, T.O. and Goes, E.S. and dos R. 2017. Formulation of fish waste meal for human nutrition. Acta Scientiarum. Technology 39(5): 525. https://doi.org/10.4025/actascitechnol.v39i5.29723

Stefanello, C., Vieira, S.L., Xue, P., Ajuwon, K.M. and Adeola, O. 2016. Age-related energy values of bakery meal for broiler chickens determined using the regression method. Poultry Science 95(7): 1582–1590. https://doi.org/10.3382/ps/pew046

Takata, M., Fukushima, K., Kino-Kimata, N., Nagao, N., Niwa, C. and Toda, T. 2012. The effects of recycling loops in food waste management in Japan: Based on the environmental and economic evaluation of food recycling. Sci. Total. Environ. 432,690309–31

Toomer, O.T. 2020. A comprehensive review of the value-added uses of peanut (Arachis hypogaea) skins and byproducts. Critical Reviews in Food Science and Nutrition 60(2): 341–350. https://doi.org/10.1080/10408398.2018.1538101

Tribst, A.A.L., Falcade, L.T.P., Carvalho, N.S., Leite Júnior, B.R. de C. and de Oliveira, M.M. 2020. Manufacture of a fermented dairy product using whey from sheep's milk cheese: An alternative to using the main byproduct of sheep's milk cheese production in small farms. International Dairy Journal, 111. https://doi.org/10.1016/j.idairyj.2020.104833

Weber, C.T., Trierweiler, L.F. and Trierweiler, J.O. 2020. Food waste biorefinery advocating circular economy: Bioethanol and distilled beverage from sweet potato. Journal of Cleaner Production 268: 121788. https://doi.org/10.1016/j.jclepro.2020.121788

Zhang, A.Y., Sun, Z., Leung, C.C.J., Han, W., Lau, K.Y., Li, M. and Lin, C.S.K. 2013. Valorisation of bakery waste for succinic acid production. Green Chem. 15: 690.

20

Food Waste Through Our Body

The Greatest Impact at the End of the Supply Chain

Themistoklis Altintzoglou

1. Introduction

Food wastage takes place throughout the supply chain of each and every food product and category. Measurements of food waste through the supply chain show that there are several bottlenecks that could be targeted to reduce the total loss (Bagherzadeh et al. 2014; Lipinski et al. 2013). Yet the greatest loss within the EU is measured at a household level, accounting for about 76 kg per capita, followed by loss attributed to production, food service, and retail/wholesale at 70 kg, 25 kg and 8 kg per capita, respectively (EC 2011).

A study in the US found that overconsumption of products that are not needed for covering nutritious needs leads to an increased environmental impact because of land use, soil loss, energy expenditure, pollution, body fat, and degenerative diseases (Blair and Sobal 2006). There is a lot of power in the hands of consumers when it comes to the various stages of food waste behaviour (Principato et al. 2021). Consumers' motivation to avoid food waste by managing their shopping and consumption behaviour, combined with supportive governmental regulations and changes in availability at the retail level, would have a stronger influence on the total environmental impact of food waste (Aschemann-Witzel et al. 2015). That synergic effect would support a change in shopping behaviour, which is a critical point of change. Once the food is bought in large amounts, it is brought home where it may be prepared or stored. If it is stored either because of too much food being prepared or simply as a raw food product, the environmental impact in terms of CO_2 emissions is added to the total expenditure of resources attached to the food—either when it is stored to be consumed later, or even worse when it is stored only to be wasted at a later stage (Brown et al. 2014). A question that begs to be asked though is, do consumers care about the environment as a means of service to themselves? Do consumers see themselves as a part of the environment? Or do they realise that environmental impact could lead to natural disasters, such as global warming with consequences which will not make an exception for humans as a species? (Thacker 2010).

Nofima; Muninbakken 9-13, 9291 Tromsø, Norway.
Email: themis.altintzoglou@nofima.no

Food products are targeted towards one destination—i.e., the consumer's shopping bag, then home, and then consumption. Knowing that most food is wasted at home, we could ask the question: What is the ultimate and highest level of waste? Is it a matter of the volume of food? Or should we look at it as a combination of volume and the resources that were used to keep the food product at a safe and pleasant quality? Or should we also add future resources that may be used to take care of the impact of food that is wasted either as a means of household waste management or from the health effects of overeating? Is current research investment across disciplines focusing on issues that support a common goal? How could research and development efforts towards efficient production be harmonised with efforts in food waste reduction and the improvement of public health? Can regulatory efforts increase the synchronicity in goals throughout the supply chain?

This commentary is an effort to connect observations of existing knowledge on food waste into a potent perspective that is aimed at generating constructive debate about resource waste as a more accurate approach to determining the environmental impact of food waste.

2. Commentary

For the sake of simplicity, imagine two hypothetical behaviours, described below as alternatives 1 and 2. These two alternatives are not the only possible behavioural alternatives, yet they represent two possibilities that will clearly exemplify ways in which consumers may behave. They will also assist our discussion and argumentation around the topic of food waste at the consumer level and the impact this may have on consumers, the environment, and society as a whole.

2.1 Behavioural Alternative 1

A person is buying food products in large packages because such products are perceived as cheaper because of the lower price per unit of the food item in the package. The price of a large product is the same as that of a product weighing less. When this product is used in a household, it may lead to larger portions. Larger portions may lead to overconsumption (Wansink et al. 2005).

Finishing up your portion may be a strategy to avoid wasting food, as a resource or a representation of the money spent to buy it. The food that is not used as energy may be stored as overweight, and the energy is not used by the body unless a meal is skipped to balance out the overconsumption or it is spent on exercise. This may lead to strain on the environment in terms of the capacity to feed and regenerate and on the body as overweight. In the meantime, the food that is bought is consumed and the money used to purchase it is essentially spent since the food is eaten. A potential consequence of such behaviour and the resulting overweight may lead to chronic risk for lifestyle diseases. These diseases could lead to increased health costs for the person and society.

Sum result: money spent; food consumed; overweight; increase in health risk.

2.2 Behavioural Alternative 2

A person buys a smaller amount of food for the same price as in Alternative 1, thus at a higher price per unit. The food product follows more or less the same life cycle as in Alternative 1, yet at smaller portion size. All the food is consumed, covering the nutritional requirements of the person without overeating.

Sum result: money spent; food consumed; no change in body weight; no increase in health risk.

The descriptions of these two alternatives exemplify the potential impact of product size on both overeating and avoiding food waste in the household. Behaving like Alternative 1 or 2 is not enforced by any authority. So, what is it that motivates behaviour like these alternatives? Is it that the goal of buying and having more food is attractive in itself? Is it an ulterior goal such as saving money or food for a future plan, such as not having to go to the shop? Is it that friends expect or would appreciate one of the two behavioural directions? Is it a feeling of obligation to behave this way? Or is it a wish to establish or maintain a reputation of being either economically careful or quality-oriented, depending on the behaviour followed (Heider 1958)? Are some people genetically predisposed to overeating and could that be linked to personality traits (Vainik et al. 2019)? Any of these reasons, combinations of them or many more are known justifications for food-related behaviour. And any repeated justification could lead to repeated advocacy fallacy, and automatic rationalisation of the behaviour until it becomes automatic and habitual (Bem 1972).

The two examples of behaviour alternatives exemplify how buying large amounts of food, storing it, and eventually either wasting it directly or through our body is the highest level of food and resource waste. We can argue that unless the portion size is right, the moral of plate clearing in order to avoid food waste does not lead to a reduction of food being wasted but only adds an element of resource waste in terms of potential health risks and overeating behaviour (Sheen et al. 2018). A two-way solution can be suggested as an approach for further investigation and potential implementation: 1) a reduction of product and portion size and 2) an increase in product price per unit.

The reduction of portion size has a known effect on overconsumption, as presented in an overview by Steenhuis and Poelman (2017). They summarise that the mechanisms underlying the portion-size effect on food intake are influenced by the appropriateness mechanism, the 'unit bias' mechanism, the 'previous experience/expectation' mechanism, the 'visual cue' mechanism, and the 'bite size' mechanism. But they add to the explanation of this effect by presenting potential influence by external factors that could lead to the selection and consumption of larger portions of food, such as value for money, mindless eating, levels of awareness, and estimation bias. Actions that can be taken to reduce portion size are also discussed but few studies are done to provide clear measurements that go beyond providing information and reducing choices. The effect of portion size is thus clear and can be studied further to provide recommendations for regulatory, market and behavioural adaptations to support consumers in buying, serving and consuming the right amount of food that can satisfy their needs.

The price of food products in relation to their value and volume is a second approach that could support consumers in making choices that match their needs. Tsalis and colleagues (2021) published a review paper that provides an overview of the impact of price promotions on consumer food waste. They discuss how price promotions result in increased food waste only in part through over-purchase, while some studies show that this link is not present or in fact the reverse. They suggest that even though the price of products at the retail level may facilitate increased purchase volumes, it is the individual characteristics of the consumers that need to be in place in order to result in increased food waste. Consumer price consciousness, attitudes, values, household identities, and household roles are very important factors in predicting food waste reduction. Stepping on the shoulders of Tsalis and colleagues and a relevant discussion by Petrescu-Mag and colleagues (2019), we can conclude that producer and retailer responsibility through product pricing and consumer behaviour at the household level are both critical elements towards the vision of reduced food waste.

In an ideal world, the policy would be in place to facilitate supply chain actors and empower consumers to synchronise goals and provide a common vision of food waste reduction (Roodhuyzen et al. 2017; Schanes et al. 2018). Such a synchronicity in the goals of all stakeholders could ensure profit at production and retail levels through a focus on quality-based value instead of the volume of food sold. Without necessarily imposing a change in household expenditure, but with a better-tuned portion of high-quality food, consumers would simply enjoy meals that satisfy their needs. Such a systemic change would simplify logistics through goal synchronicity, reduced production volumes and shared focus on quality. At a consumer level, such a change would also simplify planning and mental occupation with portions, price and food storage and bring increased pleasure with perishable food products without the shadow of food waste guilt (Roe et al. 2020).

3. Conclusion

Food wasted at the end of the supply chain carries the heaviest resource-use weight, making the impact it has on the environment more significant than food wasted earlier in the supply chain, including expanded production to make use of the whole raw material, freezing storage, etc. Food wasted by consumers' overeating is thus wasted with all the resources used for it and needs to be calculated as a factor that adds weight to the calculation of the total volume of wasted food. A resource-use calculation could also include an estimate of future resource expenditure, including potential health costs from overeating and the related lifestyle diseases that may follow. It is therefore our conclusion that in the total picture related to food waste reduction and the associated behavioural change, cleaning up your plate does not necessarily save food from being wasted. If the portion is not right, it actually adds potential resource use in the future in terms of health. Thus, food waste has a negative impact on the environment, but food wasted through our bodies has an even greater negative impact. What can we do about it? It is the responsibility of society as a whole to enable the production of the right amount of food and empower consumers to make responsible and informed food choices for the benefit of the environment, including humans as an integrated part of the environment.

Acknowledgements

This paper is part of the BlueCC project, funded by BlueBio ERA-NET COFUND, under the European Commission's Horizon 2020 research and innovation programme (Grant Agreement Number 817992). Gratitude goes to the reviewers and several colleagues and friends who constructively argued on this topic to reach a clear and relevant conclusion. Gratitude is also extended to those who critically challenged this issue, which has led to a better-supported justification of this commentary.

References

Aschemann-Witzel, J., De Hooge, I., Amani, P., Bech-Larsen, T. and Oostindjer, M. 2015. Consumer-related food waste: Causes and potential for action. Sustainability 7(6): 6457–6477.

Bagherzadeh, M., Inamura, M. and Jeong, H. 2014. Food waste along the food chain. OECD food, agriculture and fisheries papers (No. 71). OECD Publishing.

Bem, D.J. 1972. Self-perception theory. In: L. Berkowitz (ed.). Advances in experimental social psychology (Vol. 6, pp. 1–62). Academic Press.

Blair, D. and Sobal, J. 2006. Luxus consumption: Wasting food resources through overeating. Agriculture and Human Values, 23(1): 63–74.

Brown, T., Hipps, N.A., Easteal, S., Parry, A. and Evans, J.A. 2014. Reducing domestic food waste by lowering home refrigerator temperatures. International Journal of Refrigeration 40: 246–253.

EC. 2011. Preparatory study on food waste across EU 27 (final report, October 2010). BIO Intelligence Service. http://ec.europa.eu/environment/eussd/pdf/bio_foodwaste_report.pdf

Heider, F. 1958. The psychology of interpersonal relations. Lawrence Erlbeum Associates.

Lipinski, B., Hanson, C., Waite, R., Searchinger, T., Lomax, J. and Kitinoja, L. 2013. Reducing food loss and waste. World Resources Institute.

Petrescu-Mag, R.M., Petrescu, D.C. and Robinson, G.M. 2019. Adopting temperance-oriented behavior? New possibilities for consumers and their food waste. Journal of Agricultural and Environmental Ethics 32(1): 5–26.

Principato, L., Mattia, G., Di Leo, A. and Pratesi, C.A. 2021. The household wasteful behaviour framework: A systematic review of consumer food waste. Industrial Marketing Management 93: 641–649.

Roe, B.E., Qi, D. and Bender, K.E. 2020. Some issues in the ethics of food waste. Physiology & Behavior 219: 112860.

Roodhuyzen, D.M., Luning, P.A., Fogliano, V. and Steenbekkers, L.P.A. 2017. Putting together the puzzle of consumer food waste: Towards an integral perspective. Trends in Food Science & Technology 68: 37–50.

Sheen, F., Hardman, C.A. and Robinson, E. 2018. Plate-clearing tendencies and portion size are independently associated with main meal food intake in women: A laboratory study. Appetite 127: 223–229.

Steenhuis, I. and Poelman, M. 2017. Portion size: Latest developments and interventions. Current Obesity Reports 6(1): 10–17.

Schanes, K., Dobernig, K. and Gözet, B. 2018. Food waste matters-A systematic review of household food waste practices and their policy implications. Journal of Cleaner Production 182: 978–991.

Thacker, E. 2010. Three questions on demonology. *In*: N. Masciandaro (ed.). Hideous gnosis vol. 1 (pp. 179–219).

Tsalis, G., Jensen, B.B., Wakeman, S. and Aschemann-Witzel, J. 2021. Promoting food for the trash bin? A review of the literature on retail price promotions and household-level food waste. Sustainability 13(7): 4018.

Vainik, U., García-García, I. and Dagher, A. 2019. Uncontrolled eating: A unifying heritable trait linked with obesity, overeating, personality and the brain. European Journal of Neuroscience 50(3): 2430–2445.

Wansink, B., Painter, J.E. and North, J. 2005. Bottomless bowls: Why visual cues of portion size may influence intake. Obesity Research 13(1): 93–100.

Index

For Product Safety Concerns and Information please contact our EU representative GPSR@taylorandfrancis.com Taylor & Francis Verlag GmbH, Kaufingerstraße 24, 80331 München, Germany